Y0-CML-868

Theory of Inelastic Structures

THEORY OF INELASTIC STRUCTURES

T. H. Lin

PROFESSOR OF ENGINEERING
UNIVERSITY OF CALIFORNIA
LOS ANGELES

John Wiley and Sons, Inc.
New York · London · Sydney · Toronto

Library of Congress Catalog
Card Number: 68–28837
Printed in the United States of America

To
my
wife,
Susan

Preface

This book presents a unified treatment of thermal, time-independent plastic, and time-dependent creep strains in structures. These strains will be referred to as non-Hookian strains since the stress at a point is directly proportional to the elastic strain and is not proportional to these non-Hookian strains at this point. Analysis of structures with any of these strains will be called inelastic analysis. It is well known from Duhamel's analogy for thermal stress that temperature gradient has the same effect as a body force in causing a strain field in a solid body. In this book, the analogy between thermal strain and body force is generalized to cover creep and plastic strains as well. This analogy reduces the analysis of an inelastic body (a body which has any combinations of non-Hookian strains) to the analysis of an identical elastic body with an additional set of applied forces. This enables the investigator to employ known analytical and numerical methods of elastic analysis to the analysis of the corresponding inelastic bodies.

Ductile structural materials can withstand much additional strain beyond their elastic range. For redundant structures, this strain induces a redistribution of stresses which often results in a considerable additional capacity for carrying loads. To determine the additional load-carrying capacity of such structures, we must take the plastic behavior into consideration. The methods of limit analysis, developed by many distinguished investigators, are applicable mainly to structures of perfectly plastic materials. The elastoplastic analyses given here are applicable to structures of work hardening as well as perfectly plastic materials. The use of structures at elevated temperatures and with large temperature gradients has increased rapidly during the last few decades. To design such inelastic structures, we must analyze the effect of thermal and creep strains on stresses and deformation. Hence, inelastic analysis is expected to become increasingly more important. This book presents a systematic development of a method of elastoplastic and creep analysis of structures to engineers and graduate students whose research and work lie in this important broad field of inelastic structures.

Only quasi-static problems are treated here; that is, dynamics is not discussed. Small displacement theory (except Section 11.8) is assumed. For the thermoelastic problems discussed here, the influence of coupling of temperature with strain field is neglected. Creep and plastic strains considered here are taken to have no dilatation. This is valid for metal structures. For some other materials, such as concrete, creep and plastic strain may have dilatation. For analyzing structures of such materials, we may consider dilatation in the same manner as thermal expansion in structures subject to a given temperature distribution. The treatment of such problems is given in the various chapters. Although most of the examples given are for metal structures, the basic concepts and general methods treated here are applicable to inelastic structures of all materials.

This book is devoted mainly to the utilization of known solutions of elastic bodies to obtain the stress distributions of the corresponding inelastic bodies. In each chapter, the elastic solution is shown first, then the analysis of the corresponding inelastic problem is discussed. A large part of the material in this book is relatively new. It consists of twelve chapters arranged in the following order. In Chapter 1, the conditions of equilibrium, the continuity of displacements, the principle of virtual work, and the elastic stress-strain relations are presented. Chapter 2 presents the basic ideas connecting inelastic analysis and the analogy method. Here the analogy between applied force and inelastic strain is derived in detail and is given in tabular form for easy reference. Chapter 3 gives the uniaxial elastoplastic, as well as creep, stress-strain relations. The former are concerned mainly with metals whereas the latter apply equally well to most metallic and some nonmetallic materials. Chapter 4 presents a generalization of the uniaxial stress-strain relations of the previous chapter to multiaxial stress states. This chapter also includes a thorough discussion of the validity of the most commonly made assumptions in plasticity theories.

Chapters 5 and 6 show the inelastic analysis of structures whose stress state is essentially uniaxial, such as beams, columns, and beam-columns. Chapter 7 treats the inelastic analyses of the axisymmetric structures, thick hollow spheres, and hollow circular cylinders under pressure, as well as rotating circular disks. The inelastic torsion of prismatic bars is covered in Chapter 8, and Chapter 9 treats the inelastic plane strain and plane stress problems. Chapters 10 and 11 provide the theory and analysis of plates under lateral and in-plane loads. The final chapter gives a brief treatment of inelastic shells.

Essentially, the chapters have been arranged for classroom use. It is hoped that the general theories given in Chapters 1 through 4 provide a logical foundation for the analysis of the specific problems given in the succeeding chapters. Chapters 3, 5, and 6, dealing with uniaxial stress, may be read independently of the other chapters. This book is planned for a one-year graduate course in structures or applied mechanics of three hours a week. It may also be used by engineers and researchers for independent study or reference purposes. A good understanding of undergraduate mathematics, including Laplace transforms, is sufficient mathematical background for the material in this book.

It is my pleasure to thank Dr. David Salinas for his careful reading of the manuscript and numerous valuable suggestions; Professor F. R. Shanley for his kind reading of the section on Inelastic Column Buckling in Chapter 6; Professor C. L. Hu for reading Chapter 12; and Professor Karl Pister of the University of California, Berkeley, and other reviewers for valuable suggestions; my students K. S. Chi, Edmond Chow, E. Ho, Marvin Ito, D. Konishi, and others in my classes for proofreading parts of the manuscript; the Educational Development Program Committee of UCLA for sponsoring partly the writing of Chapter 4; my associates for discussions; Estelle Dorsey, Betty Gillis, and their co-workers for typing the manuscript and Claire Ritchot for the typing of rough copies of parts of the manuscript.

Finally, I express my thanks to Albert R. Beckett, editor, and to the editorial and production staffs of the publisher for their contributions to the successful completion of this book.

T. H. LIN

Los Angeles, California
July, 1968

Contents

Theory of Inelastic Structures

chapter

1

State of Stress and Strain

Introduction. This chapter consists of five parts. The first part, Sections 1.1 through 1.4, deals with stress and the general relations of equilibrium of continuous media. The validity of the derived relations is general and is applicable to all continuous media. The second part, Sections 1.5 through 1.7, deals with strain and the deformation of continuous media. The compatibility equations are derived from the uniqueness of displacements. The derived strain and compatibility relations are valid for all continuous media. The third part, Sections 1.8 and 1.9, deals with work and energy, and with the principle of virtual work. The fourth part, Sections 1.10 through 1.12, presents the stress-strain relations of elastic anisotropic and isotropic bodies, the elastic energy function, and the governing conditions for elastic isotropic media. The concluding part, Section 1.13, shows the uniqueness of elastic solutions of stresses and strains.

1.1 Stress

A body deforms when it is subjected to external forces. There are two types of external forces: *Body forces* are forces distributed throughout the volume of the body, such as gravitational force. *Surface forces* are forces distributed over the surface of the body, such as gas pressure acting on the piston of an engine.

In this chapter it will be convenient to denote the usual x, y, z set of rectangular axes by x_1, x_2, and x_3. Referring to Figure 1.1, let Δs denote a surface element within the deformed body B. The portion of the body on one side of Δs exerts a force $\mathbf{f}^* \Delta s$ on the opposite side, where $\mathbf{f}^*$ is the average force per unit area over Δs. Letting Δs approach zero, we have

$$\lim_{\Delta s \to 0} \frac{\mathbf{f}^* \Delta s}{\Delta s} = \mathbf{f}$$

where the vector $\mathbf{f}$ is referred to as the stress vector.

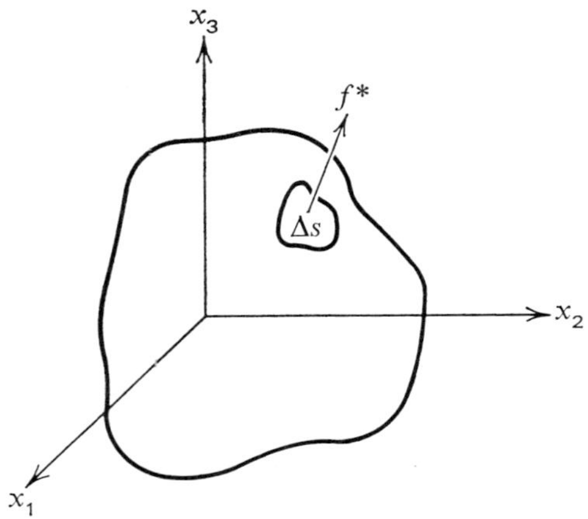

Figure 1.1

If Δs is oriented normal to the x_1 axis, the stress vector on this plane is denoted by $\mathbf{f}_1$. This stress vector can be resolved into three components parallel to the coordinate axes. These components are denoted by τ_{11}, τ_{12}, and τ_{13}. The first subscript denotes the normal of the plane on which the stress vector is acting; the second subscript denotes the direction of the stress component.

Conditions of Equilibrium. Consider a tetrahedon (Figure 1.2) within the body B, with edges dx_1, dx_2, dx_3 parallel to the x_1, x_2, x_3 axes, respectively. Each of the stress vectors on the faces normal to the three coordinate axes is resolved into three components parallel to the axes as shown. Denote the areas OBC, OAC, OAB, and ABC by A_1, A_2, A_3, and A, respectively. Let the normal to ABC be denoted by $\mathbf{n}$. Since the volume of the tetrahedron is equal to one-third the base area times the height, we have

$$\tfrac{1}{3}A_1\, dx_1 = \tfrac{1}{3}A_2\, dx_2 = \tfrac{1}{3}A_3\, dx_3 = \tfrac{1}{3}A\, dh$$

where dh is the length of the normal vector from O to face ABC. Let the direction cosines of $\mathbf{n}$ be l_{n1}, l_{n2}, l_{n3}, where

$$l_{n1} = \frac{dh}{dx_1} = \frac{A_1}{A}$$

$$l_{n2} = \frac{dh}{dx_2} = \frac{A_2}{A}$$

$$l_{n3} = \frac{dh}{dx_3} = \frac{A_3}{A} \tag{1.1.1}$$

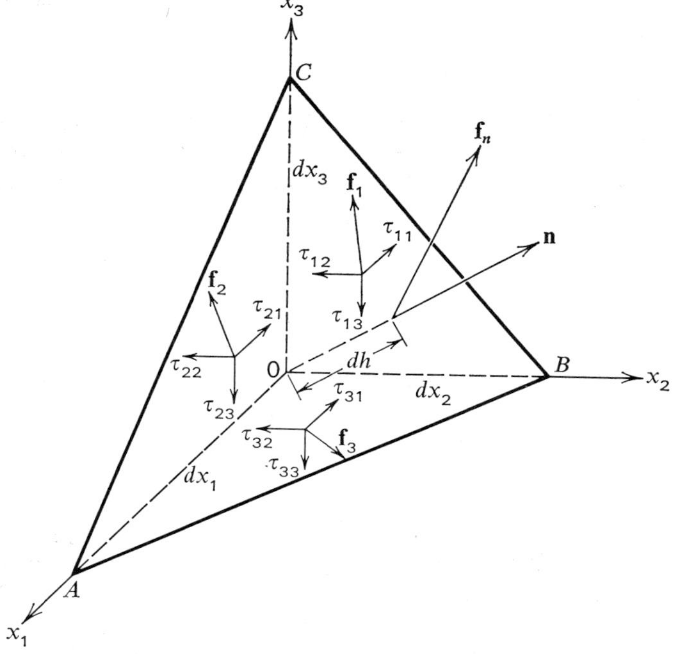

Figure 1.2

Let F_1, F_2, F_3 be the components of the body force per unit volume along the coordinate axes x_1, x_2, x_3, respectively. Similarly, resolve $\mathbf{f}_n$ into components f_{n1}, f_{n2}, f_{n3}. Since the tetrahedron must be in equilibrium, we have

$$\begin{aligned}
f_{n1}\,A - \tau_{11}\,A_1 - \tau_{21}\,A_2 - \tau_{31}\,A_3 + \tfrac{1}{3}F_1\,A\,dh &= 0 \\
f_{n2}\,A - \tau_{12}\,A_1 - \tau_{22}\,A_2 - \tau_{32}\,A_3 + \tfrac{1}{3}F_2\,A\,dh &= 0 \\
f_{n3}\,A - \tau_{13}\,A_1 - \tau_{23}\,A_2 - \tau_{33}\,A_3 + \tfrac{1}{3}F_3\,A\,dh &= 0
\end{aligned} \tag{1.1.2}$$

Dividing the above relations by A and utilizing (1.1.1), we have, as $dh \to 0$,

$$f_{n1} = \sum_{i=1}^{3} \tau_{i1} l_{ni} \qquad f_{n2} = \sum_{i=1}^{3} \tau_{i2} l_{ni} \qquad f_{n3} = \sum_{i=1}^{3} \tau_{i3} l_{ni}$$

These three relations may be written as

$$f_{nj} = \sum_{i=1}^{3} \tau_{ij} l_{ni} \qquad j = 1, 2, 3 \tag{1.1.3}$$

Let the repetition of subscripts in a term, such as i in Equation (1.1.3), denote the summation from 1 to 3. The summation sign may then be deleted, and Equation (1.1.3) may be concisely written as

$$f_{nj} = \tau_{ij} l_{ni} \qquad i, j = 1, 2, 3 \tag{1.1.4}$$

From Equation (1.1.4) it is seen that the stress vector acting on a plane with any orientation can be completely determined once the nine stress components τ_{ij} are specified. The nine stress components are frequently written in matrix form as

$$S = \begin{bmatrix} \tau_{11} & \tau_{12} & \tau_{13} \\ \tau_{21} & \tau_{22} & \tau_{23} \\ \tau_{31} & \tau_{32} & \tau_{33} \end{bmatrix} \tag{1.1.5}$$

On the surface of a body, the f_{nj} in Equation (1.1.4) become the components S_{nj} of the applied surface force $\mathbf{S}_n$ per unit area with normal **n**, along the x_j directions. The *equilibrium condition on the surface* gives

$$S_{nj} = \tau_{ij} l_{ni} \tag{1.1.6}$$

Consider an infinitesimal parallelepiped within the body, with edges dx_1, dx_2, dx_3 parallel to the x_1, x_2, x_3 axes, respectively. The stresses and body forces acting on the volume element are shown in Figure 1.3. Since the body is in equilibrium, the net force along each of the coordinate directions must vanish; hence, along the x_1 direction, we have

$$\left(\tau_{11} + \frac{\partial \tau_{11}}{\partial x_1} dx_1\right) dx_2\, dx_3 - \tau_{11}\, dx_2\, dx_3 + \left(\tau_{21} + \frac{\partial \tau_{21}}{\partial x_2} dx_2\right) dx_1\, dx_3$$
$$- \tau_{21}\, dx_1\, dx_3 + \left(\tau_{31} + \frac{\partial \tau_{31}}{\partial x_3} dx_3\right) dx_1\, dx_2 - \tau_{31}\, dx_1\, dx_2$$
$$+ F_1\, dx_1\, dx_2\, dx_3 = 0$$

which reduces to the equilibrium equation

$$\frac{\partial \tau_{11}}{\partial x_1} + \frac{\partial \tau_{21}}{\partial x_2} + \frac{\partial \tau_{31}}{\partial x_3} + F_1 = 0$$

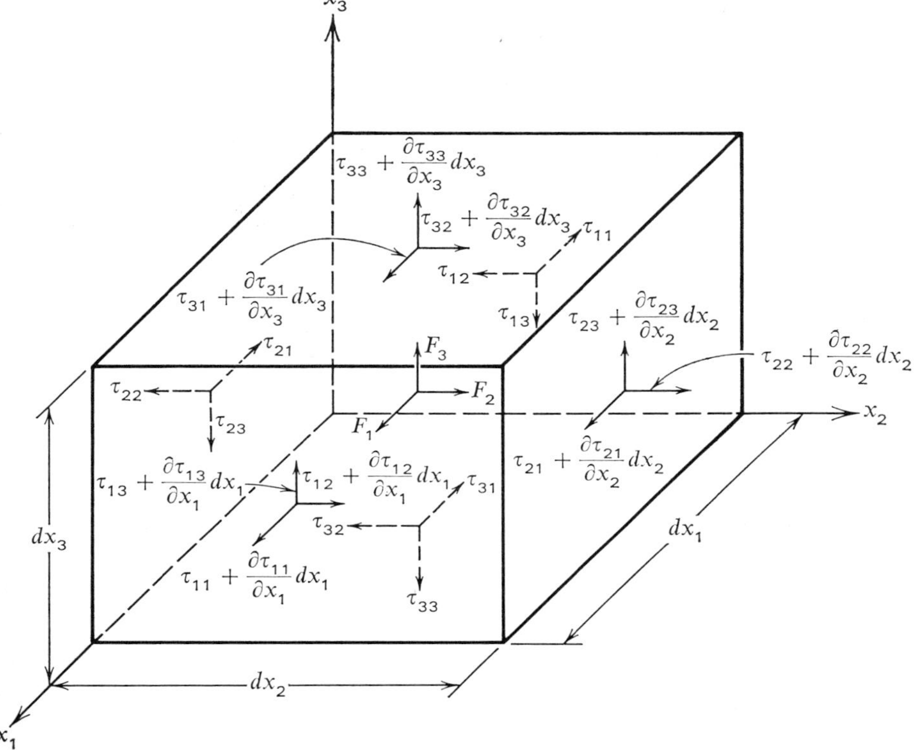

Figure 1.3

A similar calculation for the x_2 and x_3 directions yields the equilibrium equations

$$\frac{\partial \tau_{12}}{\partial x_1} + \frac{\partial \tau_{22}}{\partial x_2} + \frac{\partial \tau_{32}}{\partial x_3} + F_2 = 0$$

$$\frac{\partial \tau_{13}}{\partial x_1} + \frac{\partial \tau_{23}}{\partial x_2} + \frac{\partial \tau_{33}}{\partial x_3} + F_3 = 0$$

These three equilibrium equations can be written as

$$\frac{\partial \tau_{ji}}{\partial x_j} + F_i = 0 \quad \text{or} \quad \tau_{ji,\,j} + F_i = 0 \qquad i, j = 1, 2, 3 \qquad (1.1.7)$$

where the repetition of a subscript denotes summation, and the comma preceding the subscript j denotes differentiation with respect to x_j. *Equations* (1.1.7) *are the equilibrium conditions for a body.*

By considering the moment about an axis parallel to the x_1 axis and passing through the centroid of the parallelepiped, we obtain the following relation:

$$\tau_{23}\, dx_1\, dx_3 \frac{dx_2}{2} + \left(\tau_{23} + \frac{\partial \tau_{23}}{\partial x_2} dx_2\right) dx_1\, dx_3 \frac{dx_2}{2} - \tau_{32}\, dx_1\, dx_2 \frac{dx_3}{2} - \left(\tau_{32} + \frac{\partial \tau_{32}}{\partial x_3} dx_3\right) dx_1\, dx_2 \frac{dx_3}{2} = 0$$

which reduces to

$$\tau_{23} - \tau_{32} + \frac{\partial \tau_{23}}{\partial x_2}\frac{dx_2}{2} - \frac{\partial \tau_{32}}{\partial x_3}\frac{dx_3}{2} = 0$$

As dx_2 and $dx_3 \to 0$, this equation takes the form

$$\tau_{23} = \tau_{32}$$

Similarly, by taking moments about axes parallel to the x_2 and x_3 axes, we obtain

$$\tau_{12} = \tau_{21}$$

$$\tau_{13} = \tau_{31}$$

This may be expressed as

$$\tau_{ij} = \tau_{ji} \tag{1.1.8}$$

The nine components of the stress tensor have now been reduced to six independent components. Thus we have

$$S = \begin{bmatrix} \tau_{11} & \tau_{12} & \tau_{13} \\ \tau_{21} & \tau_{22} & \tau_{23} \\ \tau_{31} & \tau_{32} & \tau_{33} \end{bmatrix} = \begin{bmatrix} \tau_{11} & \tau_{12} & \tau_{13} \\ \tau_{12} & \tau_{22} & \tau_{23} \\ \tau_{13} & \tau_{23} & \tau_{33} \end{bmatrix} \tag{1.1.9}$$

which shows that the matrix of the stress components is symmetric.

The component of the stress vector $\mathbf{f}_n$ on a plane with normal $\mathbf{n}$ along a direction $\mathbf{t}$ is denoted by f_{nt}. It is the sum of the stress components of f_{n1}, f_{n2}, f_{n3}, along $\mathbf{t}$:

$$\begin{aligned} f_{nt} &= f_{n1} \cos(t, 1) + f_{n2} \cos(t, 2) + f_{n3} \cos(t, 3) \\ &= f_{n1}\, l_{t1} + f_{n2}\, l_{t2} + f_{n3}\, l_{t3} \end{aligned} \tag{1.1.10}$$

where l_{ti} is the direction cosine between the x_i axis and $\mathbf{t}$. Substitution of

Equation (1.1.4) for f_{ni} in the above expression yields

$$\begin{aligned} f_{nt} &= (\tau_{i1} l_{ni} l_{t1} + \tau_{i2} l_{ni} l_{t2} + \tau_{i3} l_{ni} l_{t3}) \\ &= l_{ni} l_{tj} \tau_{ij} \end{aligned} \tag{1.1.11}$$

Since $\tau_{ij} = \tau_{ji}$, from Equation (1.1.8), we have

$$f_{nt} = l_{ni} l_{tj} \tau_{ij} = l_{ni} l_{tj} \tau_{ji}$$

Consider now the transformation of the stress components τ_{ij} relative to the x_1, x_2, x_3 axes to another rectangular coordinate system, y_1, y_2, y_3. Denote the stress components in the new system by $\tau_{\alpha\beta}$. By virtue of (1.1.11), we have

$$f_{\alpha\beta} = \tau_{\alpha\beta} = l_{\alpha i} l_{\beta j} \tau_{ij} \tag{1.1.12}$$

The stresses $\tau_{\alpha\beta}$ are referred to as normal stresses when $\alpha = \beta$ and as shear stresses when $\alpha \neq \beta$. The law of transformation given by Equation (1.1.12) is valid even when the coordinate system axes y_1, y_2, y_3 are not perpendicular to each other.

A function f of variables x_i is changed to $\bar{f}$ when the variables x_i are changed to a new set of variables $\bar{x}_i$. If $\bar{f}(\bar{x}_1, \bar{x}_2, \bar{x}_3) = f(x_1, x_2, x_3)$, the functions $f(x_1, x_2, x_3)$ and $\bar{f}(\bar{x}_1, \bar{x}_2, \bar{x}_3)$ are called tensors of zero order, or scalar invariants, or just scalars.

If, corresponding to each point in a three-dimensional space, there is an ordered set of three numbers A^1, A^2, A^3 in variables x_1, x_2, x_3 and $\bar{A}_1{}^1$, $\bar{A}^2$, $\bar{A}^3$ in variables $\bar{x}_1$, $\bar{x}_2$, $\bar{x}_3$, and if A^i is related to $\bar{A}^j$ by the transformation relation

$$\bar{A}^j = A^i \frac{\partial \bar{x}_j}{\partial x_i} \tag{1.1.13}$$

then A^i and $\bar{A}^j$ are components in their respective coordinate system of a contravariant vector.[1]‡

If an ordered set of three numbers A_1, A_2, A_3 in variables x_1, x_2, x_3 and $\bar{A}_1$, $\bar{A}_2$, $\bar{A}_3$ in variables $\bar{x}_1$, $\bar{x}_2$, $\bar{x}_3$ are related by

$$\bar{A}_i = A_j \frac{\partial x_j}{\partial \bar{x}_i} \tag{1.1.14}$$

then $\bar{A}_i$ and A_j are components, in their respective variables, of a covariant vector.[1] Contravariant and covariant vectors are called contravariant and covariant tensors, respectively, of order one.

‡ Superior numerals are used to cite References at the end of the chapter.

If, corresponding to each point in space, there are nine components $A^{mn}(m, n = 1, 2, 3)$ in variables x_1, x_2, x_3 and $\bar{A}^{ij}(i, j = 1, 2, 3)$ in variables $\bar{x}_1, \bar{x}_2, \bar{x}_3$ and if $\bar{A}^{ij}$ and A^{mn} are related by the transformation law

$$\bar{A}^{ij} = A^{mn} \frac{\partial \bar{x}_i}{\partial x_m} \frac{\partial \bar{x}_j}{\partial x_n} \tag{1.1.15}$$

then $\bar{A}^{ij}$ and A^{mn} are called the components of a contravariant tensor[1] of order two. If the nine components $A_{mn}(m, n = 1, 2, 3)$ in variables x_1, x_2, x_3 and the nine components $\bar{A}^{ij}$ in variables $\bar{x}_1, \bar{x}_2, \bar{x}_3$ are related by

$$\bar{A}_{ij} = A_{mn} \frac{\partial x_m}{\partial \bar{x}_i} \frac{\partial x_n}{\partial \bar{x}_j} \tag{1.1.16}$$

then A_{mn} and $\bar{A}_{ij}$ are called the components of a covariant tensor of order two.

If x_1, x_2, x_3 and $\bar{x}_1, \bar{x}_2, \bar{x}_3$ are two sets of rectangular coordinates, then both $\partial \bar{x}_i / \partial x_m$ and $\partial x_m / \partial \bar{x}_i$ give the cosine of the angle between x_m and $\bar{x}_i$, and the transformations given by Equations (1.1.15) and (1.1.16) are the same. Equation (1.1.12) may be written as

$$\tau_{\alpha\beta} = \tau_{ij} \frac{\partial y_\alpha}{\partial x_i} \frac{\partial y_\beta}{\partial x_j} = \tau_{ij} \frac{\partial x_i}{\partial y_\alpha} \frac{\partial x_j}{\partial y_\beta}$$

By comparing this with Equations (1.1.15) and (1.1.16), it is seen that the stress components $\tau_{\alpha\beta}$ obey the law of transformation of a tensor of order two in transforming from one set of coordinates to another. Stress, which is therefore a tensor of order two, is often called a stress tensor.

1.2 Principal Stresses

The normal stress $\tau_{(\alpha)(\alpha)}$ at a point is a function of α, the direction of the normal to the plane at the point, where α in parentheses denotes no summation. In this section we seek to determine the maximum normal stress $\tau_{(\alpha)(\alpha)}$ at a point. From Equation (1.1.12), we have $\tau_{(\alpha)(\alpha)} = l_{(\alpha)i} l_{(\alpha)j} \tau_{ij}$. To facilitate the extremization of $\tau_{(\alpha)(\alpha)}$, the method of Lagrange multipliers is used.†

The direction cosines satisfy $l_{\alpha 1}^2 + l_{\alpha 2}^2 + l_{\alpha 3}^2 = 1$; thus the extremal values of $\tau_{(\alpha)(\alpha)}$ can be found from

$$\frac{\partial}{\partial l_{\alpha k}} [l_{(\alpha)i} l_{(\alpha)j} \tau_{ij} - \lambda(l_{(\alpha)i} l_{(\alpha)i} - 1)] = 0 \tag{1.2.1}$$

† A brief resume of this method is given in the appendix.

where $k = 1$ to 3 and λ is the Lagrange multiplier. Performing the indicated operation, we obtain the following system of equations:

$$\begin{aligned}
(\tau_{11} - \lambda)l_{\alpha 1} + \tau_{12} l_{\alpha 2} + \tau_{13} l_{\alpha 3} &= 0 \\
\tau_{12} l_{\alpha 1} + (\tau_{22} - \lambda)l_{\alpha 2} + \tau_{23} l_{\alpha 3} &= 0 \\
\tau_{13} l_{\alpha 1} + \tau_{23} l_{\alpha 2} + (\tau_{33} - \lambda)l_{\alpha 3} &= 0
\end{aligned} \tag{1.2.2}$$

The condition for the homogeneous system (1.2.2) to have a nontrivial solution is that the determinant of the coefficients vanishes:

$$\begin{vmatrix} (\tau_{11} - \lambda) & \tau_{12} & \tau_{13} \\ \tau_{12} & (\tau_{22} - \lambda) & \tau_{23} \\ \tau_{13} & \tau_{23} & (\tau_{33} - \lambda) \end{vmatrix} = 0 \tag{1.2.3}$$

Denote the three roots of the cubic equation (1.2.3) by $\lambda_{\mathrm{I}}, \lambda_{\mathrm{II}}, \lambda_{\mathrm{III}}$. Let the sets of direction cosines given by Equation (1.2.2) corresponding to λ_{I}, λ_{II}, and λ_{III} for λ in Equation (1.2.2) be denoted by $(l_{\mathrm{I}1}, l_{\mathrm{I}2}, l_{\mathrm{I}3})$, $(l_{\mathrm{II}1}, l_{\mathrm{II}2}, l_{\mathrm{II}3})$ and $(l_{\mathrm{III}1}, l_{\mathrm{III}2}, l_{\mathrm{III}3})$, respectively, and let y_{I}, y_{II}, and y_{III} be the axes defined by these three sets of direction cosines, respectively. Thus, for $\lambda = \lambda_{\mathrm{I}}$, we have

$$\begin{aligned}
\lambda_{\mathrm{I}} l_{\mathrm{I}1} &= l_{\mathrm{I}1}\tau_{11} + l_{\mathrm{I}2}\tau_{12} + l_{\mathrm{I}3}\tau_{13} = f_{\mathrm{I}1} \\
\lambda_{\mathrm{I}} l_{\mathrm{I}2} &= l_{\mathrm{I}1}\tau_{12} + l_{\mathrm{I}2}\tau_{22} + l_{\mathrm{I}3}\tau_{23} = f_{\mathrm{I}2} \\
\lambda_{\mathrm{I}} l_{\mathrm{I}3} &= l_{\mathrm{I}1}\tau_{13} + l_{\mathrm{I}2}\tau_{23} + l_{\mathrm{I}3}\tau_{33} = f_{\mathrm{I}3}
\end{aligned} \tag{1.2.4}$$

where $f_{\mathrm{I}1}, f_{\mathrm{I}2}, f_{\mathrm{I}3}$, are the components of the stress vector on the plane perpendicular to the y_{I} axis along the x_1, x_2, x_3 axes. Since the components of the stress vector are directly proportional to the three direction cosines of the y_{I} axis with the x_i axes, the resultant of the stress vector must be along the y_{I} axis; hence the shear stresses on the plane perpendicular to the y_{I} axis must vanish.

The principal stresses are defined as the extremal values of the normal stresses. The planes on which these stresses occur are called principal planes, and by virtue of the foregoing argument the shear stresses on these planes vanish. Utilizing (1.1.3) and (1.2.4), we have

$$\tau_{\mathrm{II}} = l_{\mathrm{I}1} f_{\mathrm{I}1} + l_{\mathrm{I}2} f_{\mathrm{I}2} + l_{\mathrm{I}3} f_{\mathrm{I}3} \tag{1.2.5}$$

$$\tau_{\mathrm{II}} = \lambda_{\mathrm{I}}[l_{\mathrm{I}1}^2 + l_{\mathrm{I}2}^2 + l_{\mathrm{I}3}^3] \tag{1.2.6}$$

$$\tau_{\mathrm{II}} = \lambda_{\mathrm{I}} \tag{1.2.7}$$

Similarly,

$$\tau_{\mathrm{II\,II}} = \lambda_{\mathrm{II}} \qquad \tau_{\mathrm{III\,III}} = \lambda_{\mathrm{III}} \tag{1.2.8}$$

Results (1.2.7) and (1.2.8) show that the principal stresses are given by the roots of (1.2.3).

When $\lambda = \lambda_{II}$, Equations (1.2.2) take the form

$$\begin{aligned}\lambda_{II} l_{II1} &= l_{II1}\tau_{11} + l_{II2}\tau_{12} + l_{II3}\tau_{13}\\ \lambda_{II} l_{II2} &= l_{II1}\tau_{12} + l_{II2}\tau_{22} + l_{II3}\tau_{23}\\ \lambda_{II} l_{II3} &= l_{II1}\tau_{13} + l_{II2}\tau_{23} + l_{II3}\tau_{33}\end{aligned} \tag{1.2.9}$$

Multiplying the equations of (1.2.9) by l_{I1}, l_{I2}, l_{I3}, respectively we have

$$\begin{aligned}\lambda_{II} l_{I1} l_{II1} &= l_{I1} l_{II1}\tau_{11} + l_{I1} l_{II2}\tau_{12} + l_{I1} l_{II3}\tau_{13}\\ \lambda_{II} l_{I2} l_{II2} &= l_{I2} l_{II1}\tau_{12} + l_{I2} l_{II2}\tau_{22} + l_{I2} l_{II3}\tau_{23}\\ \lambda_{II} l_{I3} l_{II3} &= l_{I3} l_{II1}\tau_{13} + l_{I3} l_{II2}\tau_{23} + l_{I3} l_{II3}\tau_{33}\end{aligned} \tag{1.2.10}$$

Multiplying the equations of (1.2.4) by $l_{II1}, l_{II2}, l_{II3}$, respectively, we have

$$\begin{aligned}\lambda_{I} l_{I1} l_{II1} &= l_{II1} l_{I1}\tau_{11} + l_{II1} l_{I2}\tau_{12} + l_{II1} l_{I3}\tau_{13}\\ \lambda_{I} l_{I2} l_{II2} &= l_{II2} l_{I1}\tau_{12} + l_{II2} l_{I2}\tau_{22} + l_{II2} l_{I3}\tau_{23}\\ \lambda_{i} l_{I3} l_{II3} &= l_{II3} l_{I1}\tau_{13} + l_{II3} l_{I2}\tau_{23} + l_{II3} l_{I3}\tau_{33}\end{aligned} \tag{1.2.11}$$

Adding the three equations of the system (1.2.10) and those of (1.2.11), and subtracting the two sums, we have

$$(\lambda_I - \lambda_{II})(l_{I1} l_{II1} + l_{I2} l_{II2} + l_{I3} l_{II3}) = 0 \tag{1.2.12}$$

If $\lambda_I \neq \lambda_{II}$, Equation (1.2.12) reduces to

$$l_{I1} l_{II1} + l_{I2} l_{II2} + l_{I3} l_{II3} = 0 \tag{1.2.13}$$

which is the condition of orthogonality of the y_I and y_{II} axes. Similarly, if $\lambda_I \neq \lambda_{II} \neq \lambda_{III}$, it can be shown that y_I, y_{II}, y_{III} are mutually perpendicular. The latter result shows that the *principal planes are perpendicular to each other.*

If the rectangular coordinate system (x_1, x_2, x_3) happens to coincide with (y_I, y_{II}, y_{III}), the stress matrix (1.1.5) takes the form

$$S = \begin{bmatrix}\sigma_1 & 0 & 0\\ 0 & \sigma_2 & 0\\ 0 & 0 & \sigma_3\end{bmatrix} \tag{1.2.14}$$

where σ_1, σ_2, and σ_3, are the principal stresses.

1.3 Maximum Shear Stress

The components of the stress vector on a plane with normal **n**, in terms of the principal stresses σ_1, σ_2 and σ_3, are given by

$$f_{n1} = l_{n1}\sigma_1 \qquad f_{n2} = l_{n2}\sigma_2 \qquad f_{n3} = l_{n3}\sigma_3$$

from which it follows that

$$f_{nn} = l_{n1}^2\sigma_1 + l_{n2}^2\sigma_2 + l_{n3}^2\sigma_3 \tag{1.3.1}$$

$$f_n{}^2 = l_{n1}^2\sigma_1{}^2 + l_{n2}^2\sigma_2{}^2 + l_{n3}^2\sigma_3{}^2 \tag{1.3.2}$$

Thus the square of the shearing stress on this plane is

$$f_{ns}^2 = f_n{}^2 - f_{nn}^2 \tag{1.3.3}$$

$$f_{ns}^2 = l_{n1}^2\sigma_1{}^2 + l_{n2}^2\sigma_2{}^2 + l_{n3}^2\sigma_3{}^2 - (l_{n1}^2\sigma_1 + l_{n2}^2\sigma_2 + l_{n3}^2\sigma_3)^2 \tag{1.3.4}$$

Figure 1.4 shows $\mathbf{f}_n$ and its components in various systems. The extremal

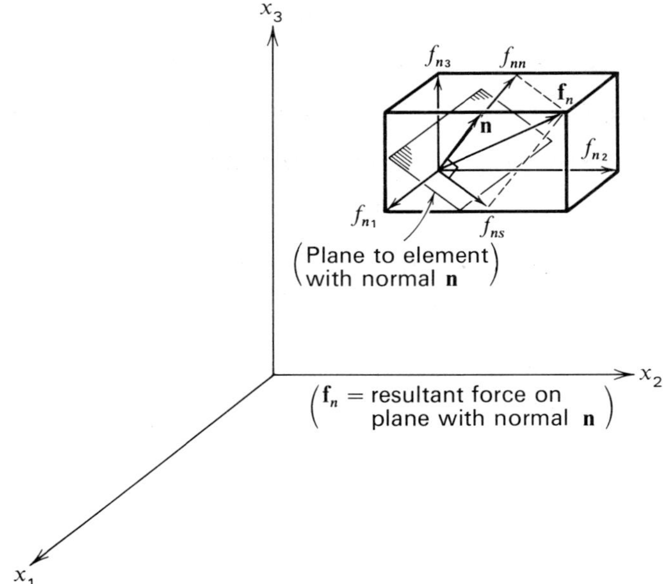

Figure 1.4

values of f_{ns} can be found by applying the method of Lagrange multipliers:

$$\frac{\partial F}{\partial l_{n1}} = 0$$

$$\frac{\partial F}{\partial l_{n2}} = 0$$

$$\frac{\partial F}{\partial l_{n3}} = 0 \tag{1.3.5}$$

where

$$F = f_{ns}^2 - \lambda(l_{n1}^2 + l_{n2}^2 + l_{n3}^2 - 1)$$

Performing the indicated operation, we obtain

$$\begin{aligned} \sigma_1{}^2 l_{n1} - 2l_{n1}\,\sigma_1(l_{n1}^2\sigma_1 + l_{n2}^2\,\sigma_2 + l_{n3}^2\,\sigma_3) - \lambda l_{n1} &= 0 \\ \sigma_2{}^2 l_{n2} - 2l_{n2}\,\sigma_2(l_{n1}^2\sigma_1 + l_{n2}^2\,\sigma_2 + l_{n3}^2\,\sigma_3) - \lambda l_{n2} &= 0 \\ \sigma_3{}^2 l_{n3} - 2l_{n3}\,\sigma_3(l_{n1}^2\sigma_1 + l_{n2}^2\,\sigma_2 + l_{n3}^2\,\sigma_3) - \lambda l_{n3} &= 0 \end{aligned} \tag{1.3.6}$$

Eliminating λ from Equation (1.3.6), we obtain

$$\begin{aligned} l_{n1}l_{n2}(\sigma_1 - \sigma_2)[(\sigma_1 + \sigma_2) - 2(l_{n1}^2\sigma_1 + l_{n2}^2\,\sigma_2 + l_{n3}^2\,\sigma_3)] &= 0 \\ l_{n1}l_{n3}(\sigma_1 - \sigma_3)[(\sigma_1 + \sigma_3) - 2(l_{n1}^2\sigma_1 + l_{n2}^2\,\sigma_2 + l_{n3}^2\,\sigma_3)] &= 0 \\ l_{n2}l_{n3}(\sigma_2 - \sigma_3)[(\sigma_2 + \sigma_3) - 2(l_{n1}^2\sigma_1 + l_{n2}^2\,\sigma_2 + l_{n3}^2\,\sigma_3)] &= 0 \end{aligned} \tag{1.3.7}$$

The first and third of these equations are used to establish the relationship

$$l_{n1}l_{n2}\,l_{n3}(\sigma_1 - \sigma_3) = 0 \tag{1.3.8}$$

which implies that one of the direction cosines must be zero. Supposing that $l_{n1} = 0$, we see that Equation (1.3.7) becomes

$$\sigma_2(1 - 2l_{n2}^2) + \sigma_3(1 - 2l_{n3}^2) = 0$$

which is satisfied if

$$l_{n2} = \pm\frac{\sqrt{2}}{2}$$

and

$$_{n3} = \pm\frac{\sqrt{2}}{2} \tag{1.3.9}$$

Substitution into Equation (1.3.4) yields

$$f_{ns}^2 = \tfrac{1}{2}({\sigma_2}^2 + {\sigma_3}^2) - \left(\frac{\sigma_2}{2} + \frac{\sigma_3}{2}\right)^2$$

or

$$f_{ns} = \pm\tfrac{1}{2}(\sigma_2 - \sigma_3) \tag{1.3.10}$$

In a similar manner, other sets of direction cosines can be obtained. Table 1.1 summarizes these results.

Table 1.1. MAXIMUM SHEAR STRESSES

l_{n1}	l_{n2}	l_{n3}	f_{ns} extremal
0	$\pm\frac{\sqrt{2}}{2}$	$\pm\frac{\sqrt{2}}{2}$	$\frac{1}{2}\lvert\sigma_2 - \sigma_3\rvert$
$\pm\frac{\sqrt{2}}{2}$	0	$\pm\frac{\sqrt{2}}{2}$	$\frac{1}{2}\lvert\sigma_3 - \sigma_1\rvert$
$\pm\frac{\sqrt{2}}{2}$	$\pm\frac{\sqrt{2}}{2}$	0	$\frac{1}{2}\lvert\sigma_1 - \sigma_2\rvert$

1.4 Stress Invariants

The expansion of Equation (1.2.3) gives the cubic equation

$$\lambda^3 - J_1\lambda^2 - J_2\lambda - J_3 = 0 \tag{1.4.1}$$

where

$$\begin{aligned} J_1 &= \tau_{11} + \tau_{22} + \tau_{33} = \tau_{ii} \\ J_2 &= \tau_{12}^2 + \tau_{13}^2 + \tau_{23}^2 - (\tau_{11}\tau_{22} + \tau_{11}\tau_{33} + \tau_{22}\tau_{33}) \\ J_3 &= \tau_{11}\tau_{22}\tau_{33} + 2\tau_{12}\tau_{13}\tau_{23} - (\tau_{11}\tau_{23}^2 + \tau_{22}\tau_{13}^2 + \tau_{33}\tau_{12}^2) \end{aligned} \tag{1.4.2}$$

The three real roots of this equation give the three principal stresses. These principal stresses depend only on the state of stress at the point and clearly do not depend on the orientation of the reference axes; hence the coefficients of the cubic equation are independent of the reference axes. The quantities J_1, J_2, and J_3 are called the first, second, and third stress invariants. If the reference axes coincide with the principal axes, these stress invariants become

$$
\begin{aligned}
J_1 &= \sigma_1 + \sigma_2 + \sigma_3 \\
-J_2 &= \sigma_1\sigma_2 + \sigma_2\sigma_3 + \sigma_3\sigma_1 \\
J_3 &= \sigma_1\sigma_2\sigma_3
\end{aligned}
\tag{1.4.3}
$$

Deviatoric Stress Invariants. The stress matrix given by Equation (1.1.9) may be separated into two parts; one is hydrostatic stress, which causes elastic dilatation, and the other is the deviatoric stress matrix, which causes elastic distortion and plastic strain.

$$
\begin{bmatrix} \tau_{11} & \tau_{12} & \tau_{13} \\ \tau_{21} & \tau_{22} & \tau_{23} \\ \tau_{31} & \tau_{32} & \tau_{33} \end{bmatrix} = \underbrace{\begin{bmatrix} \sigma & 0 & 0 \\ 0 & \sigma & 0 \\ 0 & 0 & \sigma \end{bmatrix}}_{\text{Hydrostatic stress}} + \underbrace{\begin{bmatrix} \tau_{11} - \sigma & \tau_{12} & \tau_{13} \\ \tau_{21} & \tau_{22} - \sigma & \tau_{23} \\ \tau_{31} & \tau_{32} & \tau_{33} - \sigma \end{bmatrix}}_{\text{Deviatoric stress}}
\tag{1.4.4}
$$

where

$$
\sigma = \frac{\tau_{11} + \tau_{22} + \tau_{33}}{3}
$$

The deviatoric stress matrix is written as

$$
\begin{bmatrix} \dfrac{2\tau_{11} - \tau_{22} - \tau_{33}}{3} & \tau_{12} & \tau_{13} \\ \tau_{21} & \dfrac{2\tau_{22} - \tau_{11} - \tau_{33}}{3} & \tau_{23} \\ \tau_{31} & \tau_{32} & \dfrac{2\tau_{33} - \tau_{11} - \tau_{22}}{3} \end{bmatrix} = \begin{bmatrix} S_{11} & S_{12} & S_{1} \\ S_{21} & S_{22} & S_{2} \\ S_{31} & S_{32} & S_{3} \end{bmatrix}
$$

Introducing the Kronecker delta δ_{ij} defined by

$$
\delta_{ij} = \begin{cases} 1 & \text{if} \quad i = j \\ 0 & \text{if} \quad i \neq j \end{cases}
$$

we see that the components of the deviatoric stress matrix are given by

$$
S_{ij} = \tau_{ij} - \delta_{ij}\sigma \tag{1.4.5}
$$

The stress invariants of the deviatoric stresses are denoted by $\bar{J}_1$, $\bar{J}_2$, and $\bar{J}_3$. Substituting S_{ij} for τ_{ij} in the first equation of (1.4.2), we obtain

$$
\bar{J}_1 = S_{11} + S_{22} + S_{33} \equiv 0 \tag{1.4.6}
$$

Squaring both sides, we obtain

$$\bar{J}_1{}^2 = S_{11}^2 + S_{22}^2 + S_{33}^2 + 2(S_{11}S_{22} + S_{22}S_{33} + S_{33}S_{11}) \equiv 0$$

or

$$(S_{11}S_{22} + S_{22}S_{33} + S_{33}S_{11}) = -\frac{S_{11}^2 + S_{22}^2 + S_{33}^2}{2} \tag{1.4.7}$$

Replacing τ_{ij} by S_{ij} in the second equation of (1.4.2) and using Equation (1.4.7), we obtain the second deviatoric stress invariant,

$$\begin{aligned}\bar{J}_2 &= S_{12}^2 + S_{13}^2 + S_{23}^2 - (S_{11}S_{22} + S_{11}S_{33} + S_{22}S_{33}) \\ &= S_{12}^2 + S_{13}^2 + S_{23}^2 + \frac{S_{11}^2 + S_{22}^2 + S_{33}^2}{2} = \frac{1}{2}S_{ij}S_{ij}\end{aligned}$$

In terms of stress components, the second deviatoric stress invariant is

$$\begin{aligned}2\bar{J}_2 &= \tau_{12}^2 + \tau_{13}^2 + \tau_{23}^2 + \tfrac{2}{3}[\tau_{11}^2 + \tau_{22}^2 + \tau_{33}^2 - \tau_{11}\tau_{22} - \tau_{22}\tau_{33} - \tau_{33}\tau_{11}] \\ &= \tau_{12}^2 + \tau_{23}^2 + \tau_{31}^2 + \tfrac{1}{3}[(\tau_{11} - \tau_{22})^2 + (\tau_{22} - \tau_{33})^2 + (\tau_{33} - \tau_{11})^2]\end{aligned} \tag{1.4.8}$$

In terms of principal axes, we have

$$\tau_{11} = \sigma_1 \qquad \tau_{22} = \sigma_2 \qquad \tau_{33} = \sigma_3 \qquad \tau_{12} = \tau_{23} = \tau_{31} = 0$$

which gives

$$\bar{J}_2 = \tfrac{1}{6}[(\sigma_1 - \sigma_2)^2 + (\sigma_2 - \sigma_3)^2 + (\sigma_3 - \sigma_1)^2] \tag{1.4.9}$$

The third deviatoric stress invariant $\bar{J}_3$ can be similarly calculated.

1.5 Strain

Let a body in its undeformed state be referred to a set of rectangular axes x_1, x_2, x_3. Denote the coordinates of a point P in the body by (x_1, x_2, x_3). After deformation the point P is displaced to its new position denoted by (x_1', x_2', x_3'), given by

$$\begin{aligned}x_1' &= x_1 + u_1 \\ x_2' &= x_2 + u_2 \\ x_3' &= x_3 + u_3\end{aligned} \tag{1.5.1}$$

where the u_i are the displacements of P along the x_i axes. The displacements are taken to be continuous functions of the original position, that is,

$$\begin{aligned}u_1 &= u_1(x_1, x_2, x_3) \\ u_2 &= u_2(x_1, x_2, x_3) \\ u_3 &= u_3(x_1, x_2, x_3)\end{aligned} \tag{1.5.2}$$

Consider a neighboring point Q, with coordinates

$$(x_1 + \Delta x_1, x_2 + \Delta x_2, x_3 + \Delta x_3)$$

as shown in Figure 1.5. The line element PQ is denoted by

$$\Delta \mathbf{r} = \mathbf{i}\,\Delta x_1 + \mathbf{j}\,\Delta x_2 + \mathbf{k}\,\Delta x_3$$

where **i**, **j**, **k** are unit vectors along the x_1, x_2, x_3 directions, respectively.

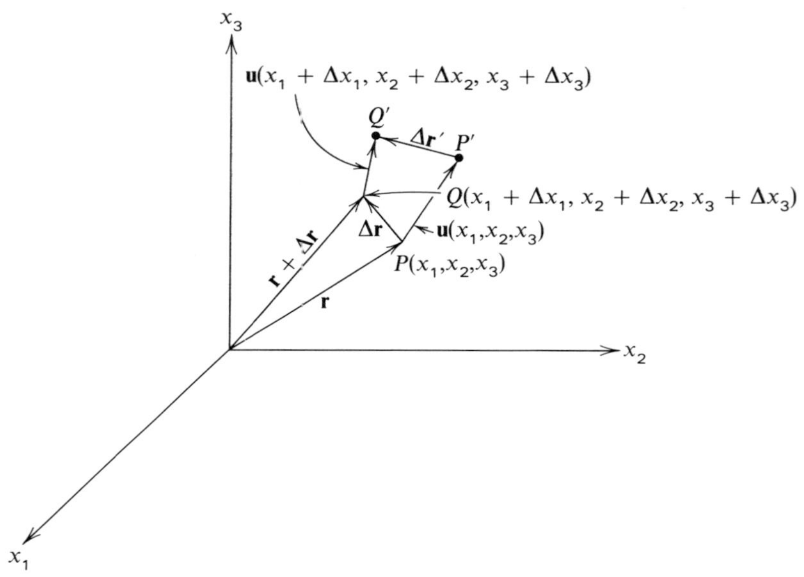

Figure 1.5

The displacements of Q along the three axes are given by

$$u_1 + \Delta u_1 \qquad u_2 + \Delta u_2 \qquad u_3 + \Delta u_3$$

where

$$u_i + \Delta u_i = u_i(x_1 + \Delta x_1, x_2 + \Delta x_2, x_3 + \Delta x_3) \qquad i = 1, 2, 3$$

Expanding the right side of this equation in a Taylor series about P, we obtain

$$u_i + \Delta u_i = u_i(x_1, x_2, x_3) + \left(\frac{\partial u_i}{\partial x_1}\right)_P \Delta x_1 + \left(\frac{\partial u_i}{\partial x_2}\right)_P \Delta x_2 + \left(\frac{\partial u_i}{\partial x_3}\right)_P \Delta x_3 \quad (1.5.3)$$

plus terms of higher order. Letting Q approach P, and neglecting terms of higher order, we see that Equation (1.5.3) becomes

$$du_i = \frac{\partial u_i}{\partial x_j} dx_j \tag{1.5.4}$$

where the repetition of the subscript j denotes summation from 1 to 3.

In the deformed state, the line element takes the form

$$\mathbf{dr}' = \mathbf{i}(dx_1 + du_1) + \mathbf{j}(dx_2 + du_2) + \mathbf{k}(dx_3 + du_3) \tag{1.5.5}$$

Using Equation (1.5.4) and dividing by dr gives

$$\frac{\mathbf{dr}'}{dr} = \mathbf{i}\left(\frac{dx_1}{dr} + \frac{\partial u_1}{\partial x_i}\frac{dx_i}{dr}\right) + \mathbf{j}\left(\frac{dx_2}{dr} + \frac{\partial u_2}{\partial x_i}\frac{dx_i}{dr}\right) + \mathbf{k}\left(\frac{dx_3}{dr} + \frac{\partial u_3}{\partial x_i}\frac{dx_i}{dr}\right) \tag{1.5.6}$$

Hence

$$\left(\frac{dr'}{dr}\right)^2 = \left(\frac{dx_1}{dr} + \frac{\partial u_1}{\partial x_i}\frac{dx_i}{dr}\right)^2 + \left(\frac{dx_2}{dr} + \frac{\partial u_2}{\partial x_i}\frac{dx_i}{dr}\right)^2 + \left(\frac{dx_3}{dr} + \frac{\partial u_3}{\partial x_i}\frac{\partial x_i}{dr}\right)^2 \tag{1.5.7}$$

Let the direction cosines of the $d\mathbf{r}$ line element be

$$\frac{dx_1}{dr} = l_1 \qquad \frac{dx_2}{dr} = l_2 \qquad \frac{dx_3}{dr} = l_3 \tag{1.5.8}$$

Expanding Equation (1.5.7), we obtain

$$\left(\frac{dr'}{dr}\right)^2 = (1 + 2e_{11})l_1{}^2 + (1 + 2e_{22})l_2{}^2 + (1 + 2e_{33})l_3{}^2 + 2\gamma_{12}\, l_1 l_2 + 2\gamma_{13}\, l_1 l_3 + 2\gamma_{23}\, l_2\, l_3 \tag{1.5.9}$$

where

$$\begin{aligned}
e_{11} &= \frac{\partial u_1}{\partial x_1} + \frac{1}{2}\left[\left(\frac{\partial u_1}{\partial x_1}\right)^2 + \left(\frac{\partial u_2}{\partial x_1}\right)^2 + \left(\frac{\partial u_3}{\partial x_1}\right)^2\right] \\
e_{22} &= \frac{\partial u_2}{\partial x_2} + \frac{1}{2}\left[\left(\frac{\partial u_1}{\partial x_2}\right)^2 + \left(\frac{\partial u_2}{\partial x_2}\right)^2 + \left(\frac{\partial u_3}{\partial x_2}\right)^2\right] \\
e_{33} &= \frac{\partial u_3}{\partial x_3} + \frac{1}{2}\left[\left(\frac{\partial u_1}{\partial x_3}\right)^2 + \left(\frac{\partial u_2}{\partial x_3}\right)^2 + \left(\frac{\partial u_3}{\partial x_3}\right)^2\right] \\
\gamma_{12} &= \frac{\partial u_1}{\partial x_2} + \frac{\partial u_2}{\partial x_1} + \frac{\partial u_1}{\partial x_1}\frac{\partial u_1}{\partial x_2} + \frac{\partial u_2}{\partial x_1}\frac{\partial u_2}{\partial x_2} + \frac{\partial u_3}{\partial x_1}\frac{\partial u_3}{\partial x_2} \\
\gamma_{23} &= \frac{\partial u_2}{\partial x_3} + \frac{\partial u_3}{\partial x_2} + \frac{\partial u_1}{\partial x_2}\frac{\partial u_1}{\partial x_3} + \frac{\partial u_2}{\partial x_2}\frac{\partial u_2}{\partial x_3} + \frac{\partial u_3}{\partial x_2}\frac{\partial u_3}{\partial x_3} \\
\gamma_{13} &= \frac{\partial u_1}{\partial x_3} + \frac{\partial u_3}{\partial x_1} + \frac{\partial u_1}{\partial x_1}\frac{\partial u_1}{\partial x_3} + \frac{\partial u_2}{\partial x_1}\frac{\partial u_2}{\partial x_3} + \frac{\partial u_3}{\partial x_1}\frac{\partial u_3}{\partial x_3}
\end{aligned} \tag{1.5.10}$$

Letting $\gamma_{ij} = 2e_{ij}$, we can write

$$e_{ij} = \frac{1}{2}\left(\frac{\partial u_i}{\partial x_j} + \frac{\partial u_j}{\partial x_i}\right) + \frac{1}{2}\left(\frac{\partial u_m}{\partial x_i}\frac{\partial u_m}{\partial x_j}\right) \tag{1.5.11}$$

where it is seen that $e_{ij} = e_{ji}$. Noting that $l_1^2 + l_2^2 + l_3^2 = 1$, we see that Equation (1.5.9) takes the form

$$\frac{dr'}{dr} = [1 + (2e_{11}l_1^2 + 2e_{22}l_2^2 + 2e_{33}l_3^2 + 4e_{12}l_1l_2 + 4e_{13}l_1l_3 + 4e_{23}l_2l_3)]^{1/2} \tag{1.5.12}$$

Expanding this expression by the binomial theorem and retaining only first-order terms, we obtain

$$\frac{dr'}{dr} - 1 \equiv e_{rr} = \tfrac{1}{2}(2e_{11}l_1^2 + 2e_{22}l_2^2 + 2e_{33}l_3^2 + 4e_{12}l_1l_2 + 4e_{13}l_1l_3 + 4e_{23}l_2l_3) \tag{1.5.13}$$

The e_{ij} *are called strain components*. The strain at a point P in any direction PQ is denoted by e_{rr}, and is completely determined by the six strain components e_{11}, e_{22}, e_{33}, e_{12}, e_{13}, e_{23}. If the strains are small, the product

$$\frac{\partial u_m}{\partial x_i} \cdot \frac{\partial u_m}{\partial x_j}$$

is small compared with $\partial u_i/\partial u_j$ and consequently is neglected. Equation (1.5.11) then takes the form

$$e_{ij} = \frac{1}{2}\left(\frac{\partial u_i}{\partial x_j} + \frac{\partial u_j}{\partial x_i}\right) \tag{1.5.14}$$

When $i = j$, for example, $e_{11} = \partial u_1/\partial x_1$, the strain e_{ii} represents the change in length per unit length along the x_i axis. When $i \neq j$, for example, $e_{12} = \frac{1}{2}(\partial u_1/\partial x_2 + \partial u_2/\partial x_1)$, the strain e_{ij} represents the angular departure from the rectangular coordinate system, as shown in Figure 1.6. The rotation about the x_3 axis, $\omega_{12} = -\frac{1}{2}(\partial u_1/\partial x_2 - \partial u_2/\partial x_1)$, is shown in Figure 1.7. Writing this in general form, we see that the rotation components are

$$\omega_{ij} = -\tfrac{1}{2}(u_{i,j} - u_{j,i}) \tag{1.5.15}$$

Consider now a transformation from the x_i coordinate system to the x_α'

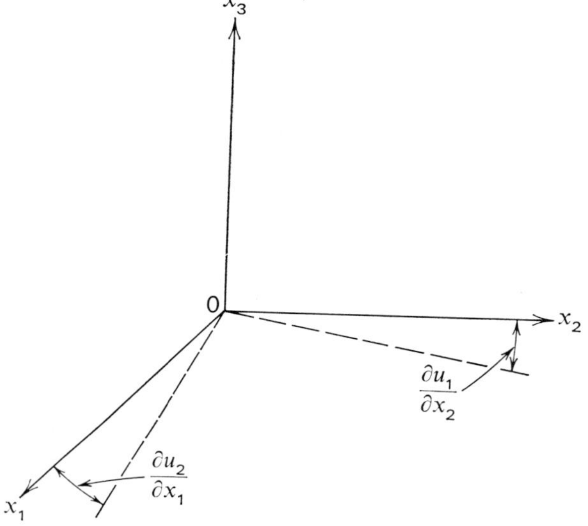

Figure 1.6

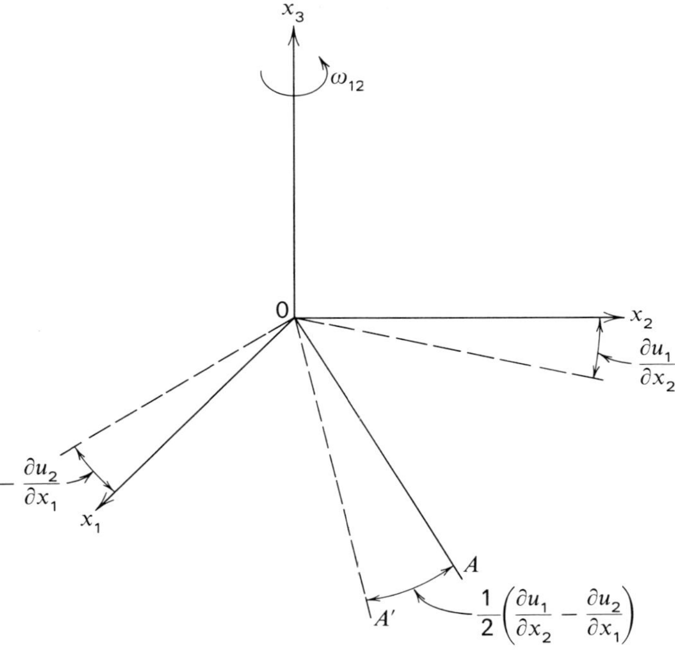

Figure 1.7

coordinate system. We have

$$x_1' = l_{11}x_1 + l_{12}x_2 + l_{13}x_3$$
$$x_2' = l_{21}x_1 + l_{22}x_2 + l_{23}x_3$$
$$x_3' = l_{31}x_1 + l_{32}x_2 + l_{33}x_3$$

or

$$x_\alpha' = l_{\alpha i} x_i \qquad (1.5.16)$$

The displacement vector u_i adheres to the same rules of transformation, hence

$$u_\beta' = l_{\beta 1}u_1 + l_{\beta 2}u_2 + l_{\beta 3}u_3 = l_{\beta j}u_j \qquad (1.5.17)$$

The strains associated with the α, β, γ axes are given by Equations (1.5.14) with i, j, k replaced by α, β, γ. Hence we have

$$\begin{aligned} e'_{\alpha\beta} &= \frac{1}{2}\left(\frac{\partial u_\alpha'}{\partial x_\beta'} + \frac{\partial u_\beta'}{\partial x_\alpha'}\right) \\ &= \frac{1}{2}\left[\frac{\partial(l_{\alpha j}u_j)}{\partial x_\beta'} + \frac{\partial(l_{\beta i}u_i)}{\partial x_\alpha'}\right] \\ &= \frac{1}{2}\left(l_{\alpha j}\frac{\partial u_j}{\partial x_\beta'} + l_{\beta i}\frac{\partial u_i}{\partial x_\alpha'}\right) \end{aligned} \qquad (1.5.18)$$

Noting that

$$\begin{aligned} \frac{\partial u_j}{\partial x_\beta'} &= \frac{\partial u_j}{\partial x_1}\frac{\partial x_1}{\partial x_\beta'} + \frac{\partial u_j}{\partial x_2}\frac{\partial x_2}{\partial x_\beta'} + \frac{\partial u_j}{\partial x_3}\frac{\partial x_3}{\partial x_\beta'} \\ &= \frac{\partial u_j}{\partial x_1}l_{\beta 1} + \frac{\partial u_j}{\partial x_2}l_{\beta 2} + \frac{\partial u_j}{\partial x_3}l_{\beta 3} \\ &= \frac{\partial u_j}{\partial x_i}l_{\beta i} \end{aligned} \qquad (1.5.19)$$

we have

$$\begin{aligned} e'_{\alpha\beta} &= \frac{1}{2}\left(l_{\alpha i}\frac{\partial u_i}{\partial x_j}l_{\beta j} + l_{\beta j}\frac{\partial u_j}{\partial x_i}l_{\alpha i}\right) \\ &= \frac{1}{2}l_{\alpha i}l_{\beta j}\left(\frac{\partial u_i}{\partial x_j} + \frac{\partial u_j}{\partial x_i}\right) \\ &= l_{\alpha i}l_{\beta j}e_{ij} \end{aligned} \qquad (1.5.20)$$

This gives the transformation of strain components from one coordinate system to another. Hence strain, like stress, obeys the law of transformation for tensors of order two, and therefore is a strain tensor.

1.6 Principal Strains and Strain Invariants

Equation (1.5.20) for the transformation of strain components is identical in form with Equation (1.1.12) for the transformation of stress components from one set of axes to another. The planes of principal strain (the extreme values of normal strain) may be found in exactly the same manner as those of principal stresses.

The principal strains, denoted by e_1, e_2, and e_3, are the three roots of the cubic equation

$$\begin{vmatrix} (e_{11}-\varepsilon) & e_{12} & e_{13} \\ e_{12} & (e_{22}-\varepsilon) & e_{23} \\ e_{13} & e_{23} & (e_{33}-\varepsilon) \end{vmatrix} = 0 \tag{1.6.1}$$

Expansion of the determinant yields

$$\varepsilon^3 - I_1\varepsilon^2 - I_2\varepsilon - I_3 = 0 \tag{1.6.2}$$

where

$$\begin{aligned} I_1 &= e_{11} + e_{22} + e_{33} \\ I_2 &= e_{12}^2 + e_{13}^2 + e_{23}^2 - (e_{11}e_{22} + e_{11}e_{33} + e_{22}e_{33}) \\ I_3 &= e_{11}e_{22}e_{33} + 2e_{12}e_{13}e_{23} - (e_{11}e_{23}^2 + e_{22}e_{13}^2 + e_{33}e_{12}^2) \end{aligned} \tag{1.6.3}$$

Since the determination of the principal strains is independent of the choice of coordinate systems, the quantities I_1, I_2, I_3 are independent of reference axes and are called the first, second, and third strain invariants, respectively.

In terms of the principal strains, the strain invariants are

$$\begin{aligned} I_1 &= e_1 + e_2 + e_3 \\ I_2 &= -(e_1e_2 + e_1e_3 + e_2e_3) \\ I_3 &= e_1e_2e_3 \end{aligned} \tag{1.6.4}$$

Deviatoric Strain Invariants. As in the case of the stress matrix, the strain matrix may be separated into two parts. One part is called the volumetric strain, and the other is called the deviatoric strain. The deviatoric strain is associated with change of shape, and is also known as distortional strain

$$\begin{bmatrix} e_{11} & e_{12} & e_{13} \\ e_{21} & e_{22} & e_{23} \\ e_{31} & e_{32} & e_{33} \end{bmatrix} = \underset{\text{Volumetric strain}}{\begin{bmatrix} e & 0 & 0 \\ 0 & e & 0 \\ 0 & 0 & e \end{bmatrix}} + \underset{\text{Deviatoric strain}}{\begin{bmatrix} \varepsilon_{11} & \varepsilon_{12} & \varepsilon_{13} \\ \varepsilon_{21} & \varepsilon_{22} & \varepsilon_{23} \\ \varepsilon_{31} & \varepsilon_{32} & \varepsilon_{33} \end{bmatrix}} \tag{1.6.5}$$

where the volumetric strain component is

$$e = \frac{e_{11} + e_{22} + e_{33}}{3} = \frac{e_{kk}}{3}$$

and the deviatoric strain components are

$$\varepsilon_{ij} = e_{ij} - \delta_{ij} \frac{e_{kk}}{3} \tag{1.6.6}$$

The strain invariants of the deviatoric strain are denoted by $\bar{I}_1$, $\bar{I}_2$, and $\bar{I}_3$. From Equations (1.6.3) and (1.6.6), we have

$$\begin{aligned} \bar{I}_1 &= \varepsilon_{kk} \equiv 0 \\ \bar{I}_2 &= \tfrac{1}{2}\varepsilon_{ij}\,\varepsilon_{ij} \\ \bar{I}_3 &= |\varepsilon_{ij}| \end{aligned} \tag{1.6.7}$$

where $|\varepsilon_{ij}|$ denotes the value of determinant with components ε_{ij}. In terms of principal deviatoric strains, and using $\bar{I}_1 = \varepsilon_{kk} \equiv 0$, we obtain

$$\begin{aligned} \bar{I}_2 = (\varepsilon_1{}^2 + \varepsilon_2{}^2 - \varepsilon_3{}^2) &= -(\varepsilon_1\varepsilon_2 + \varepsilon_2\,\varepsilon_3 + \varepsilon_3\,\varepsilon_1) \\ \bar{I}_3 &= \varepsilon_1\varepsilon_2\,\varepsilon_3 \end{aligned} \tag{1.6.8}$$

1.7 Compatibility Conditions

Strains are expressed as derivatives of displacements. There are six strain components, but there are only three displacements, u_1, u_2, and u_3.

If the strain displacement relations (1.5.14) are such that a given system of strains gives rise to a continuous single-valued system of displacements, then certain relations must exist between the various strains. Let $P_0(x_1{}^\circ, x_2{}^\circ, x_3{}^\circ)$ and $P'(x_1{}', x_2{}', x_3{}')$ be two points in a simply connected region τ. From the definitions of strain and rotation, we have

$$e_{ij} = \tfrac{1}{2}(u_{i,\,j} + u_{j,\,i}) \tag{1.5.14}$$

$$\omega_{ij} = \tfrac{1}{2}(u_{j,\,i} - u_{i,\,j}) \tag{1.5.15}$$

from which we obtain

$$u_{j,i} = e_{ij} + \omega_{ij} \tag{1.7.1}$$

or

$$du_j = (e_{ij} + \omega_{ij})\,dx_i \tag{1.7.2}$$

Integration of Equation (1.7.2) gives

$$\begin{aligned} u_j(x_1', x_2', x_3') &= u_j(x_1^\circ, x_2^\circ, x_3^\circ) + \int_{P^\circ}^{P'} du_j \\ &= u_j(x_1^\circ, x_2^\circ, x_3^\circ) + \int_{P^\circ}^{P'} e_{ij}\,dx_i + \int_{P^\circ}^{P'} \omega_{ij}\,dx_i \end{aligned} \tag{1.7.3}$$

Also,

$$\int_{P^\circ}^{P'} \omega_{ij}\,dx_i = \int_{P^\circ}^{P'} \omega_{ij}\,d(x_i - x_i') \tag{1.7.4}$$

since $dx_i' = 0$ under the integral sign. Integrating by parts, we see that Equation (1.7.4) yields

$$\int_{P^\circ}^{P'} \omega_{ij}\,d(x_i - x_i') = -\omega_{ij}^\circ(x_i^\circ - x_i') + \int_{P^\circ}^{P'} (x_i' - x_i)\omega_{ij,k}\,dx_k \tag{1.7.5}$$

From Equations (1.7.3) through (1.7.5), we have

$$\begin{aligned} u_j(x_1', x_2', x_3') &= u_j(x_1^\circ, x_2^\circ, x_3^\circ) + \omega_{ij}^\circ(x_i' - x_i^\circ) \\ &\quad + \int_{P^\circ}^{P'} [e_{kj} + (x_i' - x_i)\omega_{ij,k}]\,dx_k \end{aligned} \tag{1.7.6}$$

$$\begin{aligned} \omega_{ij,k} &= \tfrac{1}{2}(u_{j,ik} - u_{i,jk}) + \tfrac{1}{2}(u_{k,ij} - u_{k,ij}) \\ &= \tfrac{1}{2}(u_{j,ki} + u_{k,ji}) - \tfrac{1}{2}(u_{k,ij} + u_{i,kj}) \\ &= e_{jk,i} - e_{ki,j} \end{aligned} \tag{1.7.7}$$

$$\begin{aligned} u_j(x_1', x_2', x_3') &= u_j(x_1^\circ, x_2^\circ, x_3^\circ) + (x_i' - x_i^\circ)\omega_{ij}^\circ \\ &\quad + \int_{P^\circ}^{P'} [e_{kj} + (x_i' - x_i)(e_{jk,i} - e_{ki,j})]\,dx_k \end{aligned} \tag{1.7.8}$$

The displacement u_j must be independent of the path of integration, and therefore the integrands in the above integral must be exact differentials. Denoting the bracketed term by

$$U_{jk} = [e_{kj} + (x_i' - x_i)(e_{jk,i} - e_{ki,j})] \tag{1.7.9}$$

we see that the necessary and sufficient condition that the integrands be exact differentials is [6]

$$U_{ji,k} - U_{jk,i} = 0 \tag{1.7.10}$$

Since

$$U_{ji} = [e_{ij} + (x_l' - x_l)(e_{ji,l} - e_{il,j})]$$

we see that

$$\begin{aligned} U_{ji,k} - U_{jk,i} &= e_{ij,k} - \delta_{kl}(e_{ji,l} - e_{il,j}) \\ &\quad + (x_l' - x_l)(e_{ji,lk} - e_{il,jk}) - e_{kj,i} + \delta_{il}(e_{jk,l} - e_{kl,j}) \\ &\quad - (x_l' - x_l)(e_{jk,li} - e_{kl,ji}) \\ &= (e_{ij,k} - e_{kj,i}) - (e_{ji,k} - e_{ik,j}) + (e_{jk,i} - e_{ki,j}) \\ &\quad + (x_l' - x_l)[e_{ji,lk} - e_{il,jk} - e_{jk,li} + e_{kl,ji}] \end{aligned} \tag{1.7.11}$$

Writing $e_{ij} = \frac{1}{2}(u_{i,j} + u_{j,i})$, we see that the sum of the first three terms vanishes. Hence, in order to have unique displacement, the bracketed term must vanish, that is,

$$e_{ij,kl} + e_{kl,ij} - e_{il,jk} - e_{jk,il} = 0 \tag{1.7.12}$$

Equation (1.7.12) *is called the compatibility condition in terms of strain.* Since Equation (1.7.12) contains the third derivative of displacement $u_i(x_1, x_2, x_3)$, the displacement u_i, together with the first, second, and third derivatives, must be continuous. Expanding Equation (1.7.12), we obtain the following six equations of compatibility:

$$\frac{\partial^2 e_{11}}{\partial x_2^{\,2}} + \frac{\partial^2 e_{22}}{\partial x_1^{\,2}} = 2\,\frac{\partial^2 e_{12}}{\partial x_1\,\partial x_2} \tag{1.7.13}$$

$$\frac{\partial^2 e_{22}}{\partial x_3^{\,2}} + \frac{\partial^2 e_{33}}{\partial x_2^{\,2}} = 2\,\frac{\partial^2 e_{23}}{\partial x_2\,\partial x_3} \tag{1.7.14}$$

$$\frac{\partial^2 e_{33}}{\partial x_1^{\,2}} + \frac{\partial^2 e_{11}}{\partial x_3^{\,2}} = 2\,\frac{\partial^2 e_{13}}{\partial x_1\,\partial x_3} \tag{1.7.15}$$

$$\frac{\partial^2 e_{11}}{\partial x_2\,\partial x_3} = \frac{\partial^3 u_1}{\partial x_1\,\partial x_2\,\partial x_3} = \frac{\partial}{\partial x_1}\left(-\frac{\partial e_{23}}{\partial x_1} + \frac{\partial e_{13}}{\partial x_2} + \frac{\partial e_{12}}{\partial x_3}\right) \tag{1.7.16}$$

$$\frac{\partial^2 e_{22}}{\partial x_1\,\partial x_3} = \frac{\partial}{\partial x_2}\left(\frac{\partial e_{23}}{\partial x_1} - \frac{\partial e_{13}}{\partial x_2} + \frac{\partial e_{12}}{\partial x_3}\right) \tag{1.7.17}$$

$$\frac{\partial^2 e_{33}}{\partial x_1\,\partial x_2} = \frac{\partial}{\partial x_3}\left(\frac{\partial e_{23}}{\partial x_1} + \frac{\partial e_{13}}{\partial x_2} - \frac{\partial e_{12}}{\partial x_3}\right) \tag{1.7.18}$$

These conditions have to be satisfied for the displacement functions to be continuous.

1.8 Work and Energy

The increment of work done by a force **F**, when acting through a displacement $d\mathbf{r}$, is

$$dw = \mathbf{F} \cdot d\mathbf{r}$$

Consider now a rectangular parallelepiped, within the body, with edges Δx_1, Δx_2, Δx_3, subjected to uniform stresses τ_{11}, τ_{22}, τ_{33}, τ_{12}, τ_{13}, and τ_{23}. The rate of displacement per unit time at the center of this volume element has three components $\dot{u}_1$, $\dot{u}_2$, $\dot{u}_3$ along the x_1, x_2, x_3 axes, respectively. For an infinitesimal element, we have, by Equation (1.5.4),

$$\begin{aligned}
\dot{u}_1 + d\dot{u}_1 &= \dot{u}_1 + \frac{\partial \dot{u}_1}{\partial x_1} dx_1 + \frac{\partial \dot{u}_1}{\partial x_2} dx_2 + \frac{\partial \dot{u}_1}{\partial x_3} dx_3 \\
\dot{u}_2 + d\dot{u}_2 &= \dot{u}_2 + \frac{\partial \dot{u}_2}{\partial x_1} dx_1 + \frac{\partial \dot{u}_2}{\partial x_2} dx_2 + \frac{\partial \dot{u}_2}{\partial x_3} dx_3 \\
\dot{u}_3 + d\dot{u}_3 &= \dot{u}_3 + \frac{\partial \dot{u}_3}{\partial x_1} dx_1 + \frac{\partial \dot{u}_3}{\partial x_2} dx_2 + \frac{\partial \dot{u}_3}{\partial x_3} dx_3
\end{aligned} \tag{1.8.1}$$

where the dot denotes differentiation with respect to time. At the center of the top face,

$$\begin{aligned}
\dot{u}_1 + d\dot{u}_1 &= \dot{u}_1 + \frac{\partial \dot{u}_1}{\partial x_3} \frac{dx_3}{2} \\
\dot{u}_2 + d\dot{u}_2 &= \dot{u}_2 + \frac{\partial \dot{u}_2}{\partial x_3} \frac{dx_3}{2} \\
\dot{u}_3 + d\dot{u}_3 &= \dot{u}_3 + \frac{\partial \dot{u}_3}{\partial x_3} \frac{dx_3}{2}
\end{aligned}$$

and at the center of the bottom face,

$$\begin{aligned}
\dot{u}_1 + d\dot{u}_1 &= \dot{u}_1 - \frac{\partial \dot{u}_1}{\partial x_3} \frac{dx_3}{2} \\
\dot{u}_2 + d\dot{u}_2 &= \dot{u}_2 - \frac{\partial \dot{u}_2}{\partial x_3} \frac{dx_3}{2} \\
\dot{u}_3 + d\dot{u}_3 &= \dot{u}_3 - \frac{\partial \dot{u}_3}{\partial x_3} \frac{dx_3}{2}
\end{aligned}$$

The forces acting on the top face along the x_1, x_2, and x_3 coordinate axes are

$$\tau_{31}\,dx_1\,dx_2 \qquad \tau_{32}\,dx_1\,dx_2 \qquad \text{and} \qquad \tau_{33}\,dx_{31}\,dx_2$$

respectively. Similarly those on the bottom faces along the x_1, x_2, and x_3 coordinate axes are

$$-\tau_{31}\,dx_1\,dx_2 \qquad -\tau_{32}\,dx_1\,dx_2 \qquad \text{and} \qquad -\tau_{33}\,dx_1\,dx_2$$

respectively. The work done per unit time due to forces and displacements on the top and bottom faces is

$$\left[\tau_{31}\left(\dot{u}_1 + \frac{\partial \dot{u}_1}{\partial x_3}\frac{dx_3}{2}\right) + \tau_{32}\left(\dot{u}_2 + \frac{\partial \dot{u}_2}{\partial x_3}\frac{dx_3}{2}\right) + \tau_{33}\left(\dot{u}_3 + \frac{\partial \dot{u}_3}{\partial x_3}\frac{dx_3}{2}\right)\right.$$

$$\left. - \tau_{31}\left(\dot{u}_1 - \frac{\partial \dot{u}_1}{\partial x_3}\frac{dx_3}{2}\right) - \tau_{32}\left(\dot{u}_2 - \frac{\partial \dot{u}_2}{\partial x_3}\frac{dx_3}{2}\right) - \tau_{33}\left(\dot{u}_3 - \frac{\partial \dot{u}_3}{\partial x_3}\frac{dx_3}{2}\right)\right] dx_1\,dx_2$$

$$= \left(\tau_{31}\frac{\partial \dot{u}_1}{\partial x_3} + \tau_{32}\frac{\partial \dot{u}_2}{\partial x_3} + \tau_{33}\frac{\partial \dot{u}_3}{\partial x_3}\right) dx_1\,dx_2\,dx_3$$

$$= \left(\tau_{31}\frac{\partial \dot{u}_1}{\partial x_3} + \tau_{32}\frac{\partial \dot{u}_2}{\partial x_3} + \tau_{33}\frac{\partial \dot{u}_3}{\partial x_3}\right) dv \tag{1.8.2}$$

where $dv = dx_1\,dx_2\,dx_3$ is the element of volume. Similarly the work done per unit time on the right and left faces is,

$$\left(\tau_{21}\frac{\partial \dot{u}_1}{\partial x_2} + \tau_{22}\frac{\partial \dot{u}_2}{\partial x_2} + \tau_{23}\frac{\partial \dot{u}_3}{\partial x_2}\right) dv \tag{1.8.3}$$

and the work done per unit time on the front and rear faces is

$$\left(\tau_{11}\frac{\partial \dot{u}_1}{\partial x_1} + \tau_{12}\frac{\partial \dot{u}_2}{\partial x_1} + \tau_{13}\frac{\partial \dot{u}_3}{\partial x_1}\right) dv \tag{1.8.4}$$

Noting that $\tau_{ij} = \tau_{ji}$, we find the work done per unit time per unit volume by adding the expressions (1.8.2) through (1.8.4), giving

$$\dot{W} = \tau_{11}\frac{\partial \dot{u}_1}{\partial x_1} + \tau_{22}\frac{\partial \dot{u}_2}{\partial x_2} + \tau_{33}\frac{\partial \dot{u}_3}{\partial x_3} + \tau_{12}\left(\frac{\partial \dot{u}_2}{\partial x_1} + \frac{\partial \dot{u}_1}{\partial x_2}\right)$$

$$+ \tau_{23}\left(\frac{\partial \dot{u}_2}{\partial x_3} + \frac{\partial \dot{u}_3}{\partial x_2}\right) + \tau_{31}\left(\frac{\partial \dot{u}_3}{\partial x_1} + \frac{\partial \dot{u}_1}{\partial x_3}\right) \tag{1.8.5}$$

Using Equation (1.5.14), we obtain for the element of work

$$\dot{W} = \tau_{11}\dot{e}_{11} + \tau_{22}\dot{e}_{22} + \tau_{33}\dot{e}_{33} + 2\tau_{12}\dot{e}_{12} + 2\tau_{13}\dot{e}_{13} + 2\tau_{23}\dot{e}_{23}$$

$$= \tau_{ij}\dot{e}_{ij} \tag{1.8.6}$$

or

$$dW = \tau_{ij}\, de_{ij} \tag{1.8.7}$$

For a coordinate system transformation given by Equations (1.5.16), the increment of work per unit volume due to the change of coordinates is invariant:

$$dW = \tau_{ij}\, de_{ij} = \tau_{\alpha\beta}\, de_{\alpha\beta} \tag{1.8.8}$$

From Equation (1.1.12), we have

$$\tau_{\alpha\beta} = l_{\alpha i}\, l_{\beta j}\, \tau_{ij}$$

and hence

$$\tau_{\alpha\beta}\, de_{\alpha\beta} = l_{\alpha i}\, l_{\beta j}\, \tau_{ij}\, de_{\alpha\beta}$$

Since $l_{\alpha i}$ is the cosine of the angle between the α and i axes,

$$l_{\alpha i} = l_{i\alpha}$$

and

$$_{\beta j} = l_{j\beta}$$

Then, from Equation (1.8.8), we have

$$(de_{ij} - l_{i\alpha}\, l_{j\beta}\, de_{\alpha\beta})\tau_{ij} = 0$$

for all values of τ_{ij}. This last equation requires that

$$de_{ij} = l_{i\alpha}\, l_{j\beta}\, de_{\alpha\beta}$$

Interchanging i, j with α, β, we obtain

$$de_{\alpha\beta} = l_{\alpha i}\, l_{\beta j}\, de_{ij} \tag{1.8.9}$$

This gives the same result given by Equation (1.5.20).

1.9 Principle of Virtual Work

When a particle is in a state of equilibrium, the sum of all forces acting on the particle equals zero:

$$\sum \mathbf{F} = 0 \tag{1.9.1}$$

If any arbitrary displacement $\delta\mathbf{u}$ is imposed on a particle in equilibrium, the work done is zero; that is,

$$\left(\sum \mathbf{F}\right) \cdot \delta\mathbf{u} = 0 \tag{1.9.2}$$

In three-dimensional space **F** has three components F_1, F_2, and F_3. Similarly $\delta\mathbf{u}$ has three components δu_1, δu_2, and δu_3. Equation (1.9.2) may then be written

$$(\textstyle\sum F_1)\,\delta u_1 = 0 \qquad (\sum F_2)\,\delta u_2 = 0 \qquad (\sum F_3)\,\delta u_3 = 0 \tag{1.9.3}$$

Since $\delta\mathbf{u}$ may be any imaginary displacement, it is called a *virtual displacement*. The product of a force vector with a virtual displacement vector is called *virtual work*. Equation (1.9.2) states that *the virtual work done by a body in equilibrium going through a virtual displacement is zero.*

In this chapter we consider only bodies with continuous deformations, i.e., bodies in which no cracks are produced by a deformation. Hence the displacement functions $\mathbf{u} = \mathbf{u}(x_1, x_2, x_3)$ are continuous functions of (x_1, x_2, x_3). Although virtual displacements $\delta\mathbf{u}$ may be arbitrary—that is, they need not be the result of the forces on the body—it is required that they be continuous and unique functions of (x_1, x_2, x_3) and also that they satisfy the displacement boundary conditions imposed on the body. Expressing strain as derivatives of displacement, we have

$$e_{ij} = \tfrac{1}{2}(u_{i,\,j} + u_{j,\,i}) \qquad \text{and} \qquad \delta e_{ij} = \tfrac{1}{2}(\delta u_{i,\,j} + \delta u_{j,\,i}) \tag{1.9.4}$$

For a body in equilibrium, we have the equilibrium equations

$$\tau_{ij,\,j} + F_i = 0 \qquad \text{in the interior of the body} \tag{1.1.7}$$

and

$$S_i = \tau_{ij}\,\nu_j \qquad \text{on the boundary of the body} \tag{1.1.6}$$

Multiplying the first of the above equations by an arbitrary virtual displacement δu_i and integrating over the volume of the body yields

$$\int (\tau_{ij,\,j} + F_i)\,\delta u_i\,dv = 0 \tag{1.9.5}$$

Since $\tau_{ij,j}\,\delta u_i = (\tau_{ij}\,\delta u_i)_{,\,j} - \tau_{ij}\,\delta u_{i,\,j}$, Equation (1.9.5) becomes

$$\int F_i\,\delta u_i\,dv + \int [(\tau_{ij}\,\delta u_i)_{,\,j} - \tau_{ij}\,\delta u_{i,\,j}]\,dv = 0 \tag{1.9.6}$$

Since

$$\tau_{ij}\,u_{i,\,j} = \tau_{ij}\,\tfrac{1}{2}u_{i,\,j} + \tau_{ji}\,\tfrac{1}{2}u_{j,\,i} = \tau_{ij}\,\tfrac{1}{2}(u_{i,\,j} + u_{j,\,i}) = \tau_{ij}\,e_{ij}$$

we have

$$\tau_{ij}\,\delta u_{i,\,j} = \tau_{ij}\,\delta e_{ij} \tag{1.9.7}$$

By the divergence theorem, and the boundary equilibrium equation

$$\int (\tau_{ij}\,\delta u_i)_{,j}\,dv = \int \tau_{ij}\,\delta u_i\,v_j\,d\Gamma = \int S_i\,\delta u_i\,d\Gamma$$

where Γ is the boundary surface.

Substituting these into Equation (1.9.6) and noting Equation (1.9.7), we obtain

$$\int F_i\,\delta u_i\,dv + \int S_i\,\delta u_i\,d\Gamma - \int \tau_{ij}\,\delta e_{ij}\,dv = 0 \tag{1.9.8}$$

Since $\delta\mathbf{u}$ need not be related to τ_{ij}, δu_i and δe_{ij} *are not required to be related to* τ_{ij}. *This is known as the principle of virtual work.*

1.10 Elastic Stress-Strain Relations

When a molten metal cools, the atoms in the liquid metal continue to lose their kinetic energy, the specific volume decreases, and the viscosity increases. Finally, upon solidification, the kinetic energy and average interatomic distance are so small that the attraction of neighboring atoms becomes predominant. In the solid state, the atoms are arranged in an orderly three-dimensional pattern, in which rows of atoms run in different directions and are regularly spaced, forming a crystal. The locations of the atoms are described by means of a space lattice, which is an infinite array of points in space. Actually each atom oscillates randomly about a fixed point. At atmospheric temperature, the amplitude of oscillation is about 5 to 10% of the average interatomic spacing.[2] This amplitude increases with temperature. The geometrical pattern of atoms in the space lattice of the crystal causes physical properties to be different in various directions in that space. Hence *single crystals are intrinsically anisotropic* and have anisotropic elastic constants.

The tensile stress-strain curve of most metals actually varies with its temperature and rate of loading. Nevertheless, for a number of metals commonly used in structures, the stress-strain curve at room temperature does not vary appreciably with rate of loading (except at very high rates of loading). Within the proportional limit, the stress-strain relation is linear, and hence any strain component e_{ij} referring to a set of rectangular coordinates x_1, x_2, and x_3 may be expressed as a linear function of stress components τ_{ij}:

$$e_{ij} = S_{ijmn}\tau_{mn} \tag{1.10.1}$$

where the S_{ijmn} are known as the elastic compliances with a dimension reciprocal to the elastic modulus, and the repetition of subscripts denotes summation from 1 to 3.

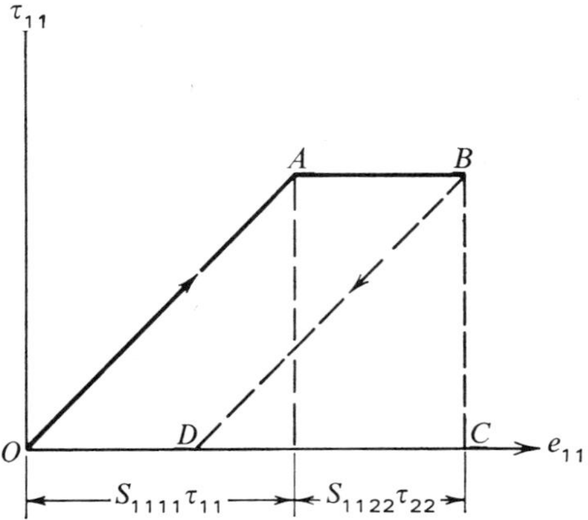

Figure 1.8

We can also express stress in terms of strain:

$$\tau_{ij} = C_{ijkl}\, e_{kl} \tag{1.10.2}$$

where the C_{ijkl} are known as elastic constants. Since the stress and strain components are symmetrical, $e_{ij} = e_{ji}$ and $\tau_{mn} = \tau_{nm}$, and therefore

$$S_{ijmn} = S_{jimn} = S_{ijnm} = S_{jinm}$$

$$C_{ijmn} = C_{jimn} = C_{ijnm} = C_{jinm} \tag{1.10.3}$$

There are six stress components and six strain components; hence there are 36 elastic compliances S_{ijmn}.

Consider a crystal subject to a uniaxial stress τ_{11}, that is, one in which all stress components except τ_{11} vanish:

$$e_{ij} = S_{ij11}\tau_{11}$$

$$e_{11} = S_{1111}\tau_{11}$$

$$e_{22} = S_{2211}\tau_{11}$$

If now τ_{22} is applied, e_{11} is increased by $S_{1122}\,\tau_{22}$ and e_{22} by $S_{2222}\,\tau_{22}$. The work done per unit volume will be the areas under the τ_{11} vs. e_{11} and τ_{22} vs. e_{22} curves. Referring to Figures 1.8 and 1.9, we see that the work is given by the sum of the areas *OABCO* and *D′B′C′D′*. If τ_{11} and then τ_{22} are removed, the negative work done is *CBDC* and *C′B′A′O′C′*. With loads removed, deformation is completely recovered and no strain energy is stored, and therefore

$$OABCO + D'B'C'D' = CBDC + C'B'A'O'C'$$

$$OABDO = O'A'B'D'O'$$

$$\tau_{11} S_{1122} \tau_{22} = \tau_{22} S_{2211} \tau_{11}$$

$$S_{1122} = S_{2211}$$

By the same reasoning, we can show that

$$S_{ijmn} = S_{mnij} \tag{1.10.4}$$

Substituting Equation (1.10.3) into Equation (1.10.2) we obtain

$$\tau_{ij} = C_{ijkl} e_{kl} = C_{ijkl} S_{klmn} \tau_{mn}$$

$$(C_{ijkl} S_{klmn} - \delta_{im} \delta_{jn}) \tau_{mn} = 0$$

$$C_{ijkl} S_{klmn} = \delta_{im} \delta_{jn} \tag{1.10.5}$$

From Equation (1.10.4), we have

$$S_{mnkl} C_{ijkl} = \delta_{im} \delta_{jn}$$

Multiplying the above equation by C_{pqmn} on both sides and summing over m and n, we obtain

$$C_{pqmn} S_{mnkl} C_{ijkl} = \delta_{im} \delta_{jn} C_{pqmn}$$

$$\delta_{pk} \delta_{ql} C_{ijkl} = C_{pqij}$$

$$C_{ijpq} = C_{pqij} \tag{1.10.6}$$

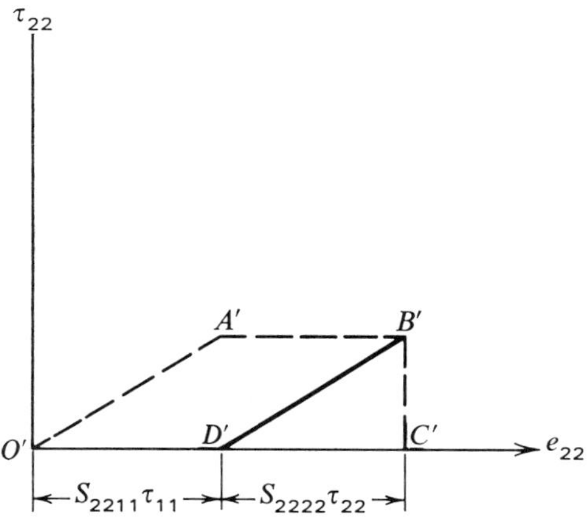

Figure 1.9

Thus the 36 constants of the S_{ijkl} and of the C_{ijkl} are reduced to 21 each. If there are planes of symmetry in the medium, the number of elastic constants is further reduced.[3] Since a piece of metal generally consists of numerous crystals at random orientation, *the elastic constants of the bulk of the metal may be considered isotropic*. For an isotropic material, there are only two elastic constants, and the elastic stress-strain relations may be represented as

$$\tau_{ij} = \lambda\, \delta_{ij}\theta + 2\mu e_{ij} \tag{1.10.7}$$

where δ_{ij} is the Kronecker delta, $\theta = e_{11} + e_{22} + e_{33}$, and λ and μ are called Lame's constants.

From Equation (1.4.5) we may write the stress component τ_{ij} as the sum of the hydrostatic stress σ and the deviatoric stress components S_{ij}:

$$\tau_{ij} = \sigma + S_{ij} = \delta_{ij}\frac{\tau_{kk}}{3} + S_{ij}$$

Similarly the strain component e_{ij} may be written as the sum of the volumetric strain e and the deviatoric strain components ε_{ij}

$$e_{ij} = e + \varepsilon_{ij} = \delta_{ij}\frac{\theta}{3} + \varepsilon_{ij}$$

where

$$\theta = 3e = e_{ii}$$

Equation (1.10.7) becomes

$$S_{ij} + \delta_{ij}\frac{\tau_{kk}}{3} = \delta_{ij}\lambda\theta + 2\mu\left(\varepsilon_{ij} + \delta_{ij}\frac{\theta}{3}\right) \tag{1.10.8}$$

Let $i = j$, and sum over j from 1 to 3, noting that $S_{jj} = 0$, $\varepsilon_{jj} = 0$, and $\delta_{jj} = 3$.

$$\tau_{kk} = 3\lambda\theta + 2\mu\theta = (3\lambda + 2\mu)\theta \tag{1.10.9}$$

$$\theta = \frac{\tau_{kk}}{3\lambda + 2\mu} = \frac{3}{3\lambda + 2\mu}\frac{\tau_{kk}}{3} = \frac{\tau_{kk}}{3K} \tag{1.10.10}$$

Since $\tau_{kk}/3$ is the hydrostatic stress and θ the dilatation, $(3\lambda + 2\mu)/3$ is called the bulk modulus and is denoted by K.

Substituting Equation (1.10.10) into Equation (1.10.7) and Equation (1.10.8), we obtain the following two equations:

$$e_{ij} = \frac{\tau_{ij}}{2\mu} - \delta_{ij}\frac{\lambda}{2\mu(3\lambda + 2\mu)}\tau_{kk} \tag{1.10.11}$$

$$S_{ij} = 2\mu\varepsilon_{ij} \tag{1.10.12}$$

Equation (1.10.10) gives the relation between volumetric strain and hydrostatic stress, while Equation (1.10.12) gives the relation between deviatoric strain and deviatoric stress.

Now consider strain due to τ_{11} alone. From Equation (1.10.11), we have

$$e_{11} = \frac{\tau_{11}}{2\mu} - \frac{\lambda}{2\mu(3\lambda + 2\mu)}\tau_{11} = \frac{\lambda + \mu}{\mu(3\lambda + 2\mu)}\tau_{11}$$

$$e_{22} = -\frac{\lambda}{2\mu(3\lambda + 2\mu)}\tau_{11}$$

It is known that $Ee_{11} = \tau_{11}$ and $e_{22} = -\nu(\tau_{11}/E)$ for tensile stress τ_{11}, where E is the Young's modulus and ν is Poisson's ratio, hence

$$E = \frac{\mu(3\lambda + 2\mu)}{\lambda + \mu} \tag{1.10.13}$$

and

$$\nu = \frac{\lambda}{2(\lambda + \mu)} \tag{1.10.14}$$

which yields

$$\mu = \frac{E}{2(1 + \nu)} \tag{1.10.15}$$

Here μ is the shear modulus and is commonly denoted by G. Other relations between λ, E, G, and ν are given in a number of textbooks.[4, 5] Writing the elastic isotropic stress-strain relationship in a different form, we obtain

$$e_{11} = \frac{\tau_{11} - \nu(\tau_{22} + \tau_{33})}{E}$$

$$e_{22} = \frac{\tau_{22} - \nu(\tau_{11} + \tau_{33})}{E}$$

$$e_{33} = \frac{\tau_{33} - \nu(\tau_{11} + \tau_{22})}{E}$$

$$e_{12} = \frac{\tau_{12}}{2G} \qquad e_{23} = \frac{\tau_{23}}{2G} \qquad e_{31} = \frac{\tau_{31}}{2G} \tag{1.10.16}$$

Equations (1.10.16), relating stress and elastic strain and using Young's modulus E and Poisson's ratio ν, may be written concisely as

$$e_{ij} = \frac{1 + \nu}{E}\tau_{ij} - \frac{\nu}{E}\delta_{ij}\tau_{kk} \tag{1.10.17}$$

1.11 Elastic Energy Function

Consider an elastic medium subject to a stress state defined by the six stress components τ_{ij}. The corresponding strain components are given by

$$e_{ij} = S_{ijkl}\tau_{kl}$$

Let U be the elastic strain energy of the medium per unit volume. Now let the τ_{kl} be proportionally decreased to zero. From the above equation, the e_{ij} are also proportionally decreased to zero, so during this unloading process e_{ij} is always proportional to τ_{ij}. The work of the medium per unit volume, from Equation (1.8.7), is

$$W = \int \tau_{ij}\, de_{ij} = \tfrac{1}{2}\tau_{ij}e_{ij} \tag{1.11.1}$$

After the stress τ_{ij} is completely removed, $e_{ij} = 0$ and the strain energy of this elastic medium is completely removed. This energy is transformed into the external work

$$U = W = \tfrac{1}{2}\tau_{ij}e_{ij}$$

The stress state given by the τ_{ij} before unloading may be obtained through any sequence of loading, i.e., any order of application of the different τ_{ij}. *Hence the elastic energy for a given set of τ_{ij} depends only on the state of the stress τ_{ij} and is independent of how the stress state is reached,*

$$U = \tfrac{1}{2}\tau_{ij}e_{ij} = \tfrac{1}{2}C_{ijkl}e_{kl}e_{ij} \tag{1.11.2}$$

Noting Equation (1.10.6), we obtain

$$\begin{aligned} \frac{\partial U}{\partial e_{mn}} &= \tfrac{1}{2}C_{ijkl}(e_{kl}\delta_{im}\delta_{nj} + \delta_{km}\delta_{ln}e_{ij}) \\ &= \tfrac{1}{2}(C_{mnkl}e_{kl} + C_{ijmn}e_{ij}) \\ &= C_{mnkl}e_{kl} = \tau_{mn} \end{aligned} \tag{1.11.3}$$

Similarly we can show that

$$\frac{\partial U}{\partial \tau_{mn}} = e_{mn} \tag{1.11.4}$$

1.12 Governing Equations for an Elastic Isotropic Solid

Recall that for a body to be in equilibrium throughout, we must have, from Equation (1.1.7), $\tau_{ij,j} + F_i = 0$ in the interior of the body and, from Equation (1.1.6), $S_{nj} = \tau_{ij}l_{ni}$ on the boundary.

For the displacement field in the body to be unique, i.e., for continuity of displacement, the strain $e_{ij} = \frac{1}{2}(u_{i,j} + u_{j,i})$ must satisfy the compatibility conditions

$$e_{ij,kl} + e_{kl,ij} - e_{il,jk} - e_{jk,il} = 0 \tag{1.7.12}$$

where the u_i, together with their first, second, and third derivatives, are continuous.

If the body is elastic and isotropic, the stress-strain relationship is given by

$$\tau_{ij} = \lambda\, \delta_{ij}\theta + 2\mu e_{ij} \tag{1.10.7}$$

If the internal body force F_i and the displacement u_i on the surface are given, it is desirable to express the equilibrium condition in the interior of the body in terms of displacement. We obtain, from Equations (1.5.14),

$$\tau_{ij} = \lambda\, \delta_{ij} u_{k,k} + \mu(u_{i,j} + u_{j,i}) \tag{1.12.1}$$

Substituting this into Equation (1.1.7), we obtain

$$\delta_{ij} \lambda u_{k,kj} + \mu(u_{i,jj} + u_{j,ij}) + F_i = 0 \tag{1.12.2}$$

or

$$\mu\nabla^2 u_i + (\lambda + \mu)\theta_{,i} + F_i = 0 \tag{1.12.3}$$

where $\nabla^2 u_i$ denotes $u_{i,jj}$ and $\theta = u_{k,k}$.

The reason for the compatibility condition is to ensure that the strain field e_{ij} obtained gives a unique value to u_i. Since the equilibrium condition is expressed in terms of displacement in Equation (1.12.3), the compatibility condition is no longer required. The displacement $u_i(x_1, x_2, x_3)$ is solved from the differential equation (1.12.3) to satisfy the given displacement u_i at the boundary. After $u_i(x_1, x_2, x_3)$ is known throughout the body, the strain e_{ij} is obtained from the strain-displacement relations, and the stress distribution is obtained from Equation (1.12.1).

If S_{nj}, instead of u_i, is given on the boundary of the body, the differential equation of equilibrium in terms of stress is commonly used.

The six independent compatibility equations are shown in Equations (1.7.13) through (1.7.18). Substituting Equation (1.10.16) into (1.7.13), we obtain

$$(1 + \nu)\left(\frac{\partial^2 \tau_{11}}{\partial x_2{}^2} + \frac{\partial^2 \tau_{22}}{\partial x_1{}^2}\right) - \nu\left(\frac{\partial^2}{\partial x_2{}^2} + \frac{\partial^2}{\partial x_1{}^2}\right)\tau_{mm} = 2(1 + \nu)\frac{\partial^2 \tau_{12}}{\partial x_1\, \partial x_2} \tag{1.12.4}$$

From the equilibrium conditions $\tau_{ij,j} + F_i = 0$, we have

$$\tau_{12,21} = -\tau_{11,11} - \tau_{13,31} - F_{1,1} \tag{1.12.5}$$

$$\tau_{21,12} = -\tau_{22,22} - \tau_{23,32} - F_{2,2} \tag{1.12.6}$$

$$2\tau_{12,21} = -\tau_{11,11} - \tau_{22,22} - (\tau_{13,1} + \tau_{23,2})_{,3} - F_{1,1} - F_{2,2} \tag{1.12.7}$$

From the equilibrium condition $\tau_{31,1} + \tau_{32,2} + \tau_{33,3} + F_3 = 0$, we have

$$\tau_{13,1} + \tau_{23,2} = -\tau_{33,3} - F_3$$

Substituting this into Equation (1.12.7) gives

$$2\tau_{12,21} = -\tau_{11,11} - \tau_{22,22} + \tau_{33,33} - F_{1,1} - F_{2,2} + F_{3,3} \tag{1.12.8}$$

From Equations (1.12.8) and (1.12.4), we obtain

$$\begin{aligned}(1 + \nu)(\tau_{11,22} + \tau_{22,11}) - \nu(\tau_{mm,22} + \tau_{mm,11}) \\ = (1 + \nu)[-\tau_{11,11} - \tau_{22,22} + \tau_{33,33} - F_{1,1} - F_{2,2} + F_{3,3}]\end{aligned}$$

$$\begin{aligned}(1 + \nu)(\nabla^2\tau_{mm} - \nabla^2\tau_{33} - \tau_{mm,33}) - \nu(\nabla^2\tau_{mm} - \tau_{mm,33}) \\ = (1 + \nu)(F_{3,3} - F_{1,1} - F_{2,2})\end{aligned} \tag{1.12.9}$$

where

$$\nabla^2 = \left(\frac{\partial^2}{\partial x_1{}^2} + \frac{\partial^2}{\partial x_2{}^2} + \frac{\partial^2}{\partial x_3{}^2}\right)$$

Similarly, two other equations may be written by permuting the subscripts:

$$\begin{aligned}(1 + \nu)(\nabla^2\tau_{mm} - \nabla^2\tau_{11} - \tau_{mm,11}) - \nu(\nabla^2\tau_{mm} - \tau_{mm,11}) \\ = (1 + \nu)(F_{1,1} - F_{2,2} - F_{3,3})\end{aligned} \tag{1.12.10}$$

$$\begin{aligned}(1 + \nu)(\nabla^2\tau_{mm} - \nabla^2\tau_{22} - \tau_{mm,22}) - \nu(\nabla^2\tau_{mm} - \tau_{mm,22}) \\ = (1 + \nu)(F_{2,2} - F_{3,3} - F_{1,1})\end{aligned} \tag{1.12.11}$$

Adding Equations (1.12.9), (1.12.10), and (1.12.11) together, we obtain

$$(1 - \nu)\nabla^2\tau_{mm} = -(1 + \nu)(F_{1,1} + F_{2,2} + F_{3,3}) \tag{1.12.12}$$

Substituting this into Equation (1.12.10), we find that

$$\nabla^2\tau_{11} + \frac{1}{1 + \nu}\tau_{mm,11} = -\frac{\nu}{1 - \nu}(F_{1,1} + F_{2,2} + F_{3,3}) - 2F_{1,1} \tag{1.12.13}$$

Two other, similar equations are obtained from Equations (1.12.9) and (1.12.11). These three equations may be written as

$$\nabla^2 \tau_{(i)(i)} + \frac{1}{1+\nu} \tau_{mm,\,(i)(i)} = -\frac{\nu}{1-\nu} F_{m,\,m} - 2F_{(i),\,(i)} \qquad (1.12.14)$$

where the parentheses on the subscript denote no summation.

In the same way, Equations (1.7.16) through (1.7.18) are expressed in terms of stress. They are of the following type:

$$\nabla^2 \tau_{ij} + \frac{1}{1+\nu} \tau_{mm,\,ij} = -(F_{i,\,j} + F_{j,\,i}) \qquad i \neq j \qquad (1.12.15)$$

Equations (1.12.14) and (1.12.15) may be written as a single equation:

$$\nabla^2 \tau_{ij} + \frac{1}{1+\nu} \tau_{mm,\,ij} = -(F_{i,\,j} + F_{j,\,i}) - \delta_{ij} \frac{\nu}{1-\nu} F_{m,\,m} \qquad (1.12.16)$$

These six equations are the equations of compatibility in terms of stress, and are called the Beltrami-Michell equations. Stresses in an isotropic elastic body must satisfy the condition-of-equilibrium equations (1.1.7) and the compatibility equations (1.12.16).

1.13 Uniqueness of Solution of Elastic Bodies

In this section we shall show that the solution of stress and strain for an elastic body is unique. The proof is based on the assumptions that the displacements u_i are single-valued functions and that there are no initial stresses. In addition, infinitesimal-strain-and-displacement theory is assumed to hold. The proof is not valid for buckling solutions.

Since the actual displacements u_i satisfy the continuity and uniqueness conditions, we can replace δu_i by u_i and δe_{ij} by e_{ij} in Equation (1.9.8) and obtain

$$\int_v F_i u_i \, dv + \int_\Gamma S_i u_i \, d\Gamma - \int_v \tau_{ij} e_{ij} \, dv = 0 \qquad (1.13.1)$$

Let Γ_s be that part of the boundary surface on which the surface force S_i is specified and Γ_u be the remaining part of the surface on which the displacement u_i is prescribed. Then

$$\int_\Gamma S_i u_i \, d\Gamma = \int_{\Gamma_s} S_i u_i \, d\Gamma + \int_{\Gamma_u} S_i u_i \, d\Gamma \qquad (1.13.2)$$

Assume that there are two sets of stress and strain $\tau_{ij}^{(1)}$, $u_i^{(1)}$, $e_{ij}^{(1)}$ and $\tau_{ij}^{(2)}$, $u_i^{(2)}$, $e_{ij}^{(2)}$ which satisfy the conditions of equilibrium, continuity, and the elastic stress-strain relations of the material. Then $(\tau_{ij}^{(2)} - \tau_{ij}^{(1)})$,

$(u_i^{(2)} - u_i^{(1)})$, and $(e_{ij}^{(2)} - e_{ij}^{(1)})$ must also satisfy these conditions with body force $(F_i^{(2)} - F_i^{(1)})$ in the body, $(S_i^{(2)} - S_i^{(1)})$ on Γ_s and $(u_i^{(2)} - u_i^{(1)})$ on Γ_u. From Equation (1.13.1), we have

$$\begin{aligned}\int_v (F_i^{(2)} - F_i^{(1)})(u_i^{(2)} - u_i^{(1)})\,dv + \int_{\Gamma_s} (S_i^{(2)} - S_i^{(1)})(u_i^{(2)} - u_i^{(1)})\,d\Gamma \\ + \int_{\Gamma_u} (S_i^{(2)} - S_i^{(1)})(u_i^{(2)} - u_i^{(1)})\,d\Gamma \\ = \int_v (\tau_{ij}^{(2)} - \tau_{ij}^{(1)})(e_{ij}^{(2)} - e_{ij}^{(1)})\,dv \qquad (1.13.3)\end{aligned}$$

Both $F_i^{(2)}$ and $F_i^{(1)}$ are equal to the given body force, so $F_i^{(2)} = F_i^{(1)}$, and the first integral vanishes. On Γ_s, we have $S_i^{(2)} = S_i^{(1)}$ and on Γ_u, we have $u_i^{(2)} = u_i^{(1)}$, so the second and third integrals vanish. Therefore Equation (1.13.3) becomes

$$\int_v (\tau_{ij}^{(2)} - \tau_{ij}^{(1)})(e_{ij}^{(2)} - e_{ij}^{(1)})\,dv = 0 \qquad (1.13.4)$$

For isotropic elastic materials,

$$\tau_{ij} = \lambda\,\delta_{ij}\theta + 2\mu e_{ij} \qquad (1.13.5)$$

Writing $e_{ij} = \delta_{ij}(\theta/3) + \varepsilon_{ij}$ where ε_{ij} is the deviatoric strain component, we see that Equation (1.13.5) becomes

$$\tau_{ij} = \lambda\,\delta_{ij}\theta + 2\mu\left(\varepsilon_{ij} + \delta_{ij}\frac{\theta}{3}\right) = \frac{3\lambda + 2\mu}{3}\,\delta_{ij}\theta + 2\mu\varepsilon_{ij} \qquad (1.13.6)$$

Multiplying Equation (1.13.6) through by e_{ij} yields

$$\begin{aligned}\tau_{ij}e_{ij} &= \left(\frac{3\lambda + 2\mu}{3}\,\delta_{ij}\theta + 2\mu\varepsilon_{ij}\right)\left(\frac{\delta_{ij}}{3}\,\theta + \varepsilon_{ij}\right) \\ &= \frac{3\lambda + 2\mu}{9}\,\delta_{ij}\delta_{ij}\theta^2 + \frac{3\lambda + 2\mu}{3}\,\delta_{ij}\theta\varepsilon_{ij} + \frac{2\mu}{3}\,\delta_{ij}\theta\varepsilon_{ij} + 2\mu\varepsilon_{ij}\varepsilon_{ij} \qquad (1.13.7)\end{aligned}$$

Since $\varepsilon_{ii} = 0$, the second and the third terms vanish. Also, noting that $\delta_{ij}\delta_{ij} = \delta_{ii} = 3$, we see that Equation (1.13.7) reduces to

$$\tau_{ij}e_{ij} = \frac{3\lambda + 2\mu}{3}\,\theta^2 + 2\mu\varepsilon_{ij}\varepsilon_{ij} \qquad (1.13.8)$$

The integrand of Equation (1.13.4) then becomes

$$(\tau_{ij}^{(2)} - \tau_{ij}^{(1)})(e_{ij}^{(2)} - e_{ij}^{(1)}) = \frac{3\lambda + 2\mu}{3}\,(\theta^{(2)} - \theta^{(1)})^2 + 2\mu(\varepsilon_{ij}^{(2)} - \varepsilon_{ij}^{(1)})(\varepsilon_{ij}^{(2)} - \varepsilon_{ij}^{(1)})$$
$$(1.13.9)$$

Since the right-hand side of Equation (1.13.9) is the sum of squares, it is always equal to or greater than zero. In order to satisfy Equation (1.13.4), the above expression must vanish throughout the body. Hence

$$\theta^{(2)} = \theta^{(1)} \qquad \varepsilon_{ij}^{(2)} = \varepsilon_{ij}^{(1)}, \qquad e_{ij}^{(2)} = e_{ij}^{(1)} \qquad \text{and} \qquad \tau_{ij}^{(2)} = \tau_{ij}^{(1)}$$

This proves the uniqueness of elastic solution of stress and strain.

REFERENCES

1. Sokolnikoff, I. S., *Tensor Analyses*, Wiley, New York, pp. 60–63, 1951.
2. Smith, M. C., *Principles of Physical Metallurgy*, Harper & Row, New York, pp. 61–62, 1956.
3. Kittel, C., *Introduction to Solid State Physics*, Wiley, New York, Chap. 4, p. 85, 1956.
4. Durelli, A. J., E. A. Phillips, and C. N. Tsao, *Introduction to the Theoretical and Experimental Analysis of Stress and Strain*, McGraw-Hill, New York, pp. 82 and 90, 1958.
5. Sokolnikoff, I. S., *Mathematical Theory of Elasticity*, McGraw-Hill, New York, pp. 71, 336–337, 1956.
6. Sokolnikoff, I. S., and R. M. Redheffer, *Mathematics of Physics and Modern Engineering*, McGraw-Hill, New York, pp. 378–382, 1958.

Appendix

The Method of Lagrange Multipliers. Consider the problem of finding the extremal values of $F(x, y, z)$, subject to a constraint $\phi(x, y, z) = 0$. If $\phi(x, y, z) = 0$ can be solved for one of the variables, say z, the problem is reduced to extremization of $F[x, y, z(x, y)]$, where the independent variables are x, y. Nevertheless, it is frequently quite laborious, or even impossible, to solve for one of the variables; for this reason the following method is used.

The differential of $F(x, y, z)$ is

$$\begin{aligned} dF &= \frac{\partial F}{\partial x} dx + \frac{\partial F}{\partial y} dy + \frac{\partial F}{\partial z} dz \\ &= \frac{\partial F}{\partial x} dx + \frac{\partial F}{\partial y} dy + \frac{\partial F}{\partial z} \left(\frac{\partial z}{\partial x} dx + \frac{\partial z}{\partial y} dy \right) \\ &= \left(\frac{\partial F}{\partial x} + \frac{\partial F}{\partial z} \frac{\partial z}{\partial x} \right) dx + \left(\frac{\partial F}{\partial y} + \frac{\partial F}{\partial z} \frac{\partial z}{\partial y} \right) dy \end{aligned} \tag{1A.1}$$

At the extremals, dF must vanish for all ratios of dy and dx, so

$$\frac{\partial F}{\partial x} + \frac{\partial F}{\partial z} \frac{\partial z}{\partial x} = 0$$

$$\frac{\partial F}{\partial y} + \frac{\partial F}{\partial z} \frac{\partial z}{\partial y} = 0$$

Differentiating $\phi(x, y, z) = 0$ yields

$$\frac{\partial \phi}{\partial x} + \frac{\partial \phi}{\partial z} \frac{\partial z}{\partial x} = 0$$

$$\frac{\partial \phi}{\partial y} + \frac{\partial \phi}{\partial z} \frac{\partial z}{\partial y} = 0 \tag{1A.2}$$

Solving (1A.2) for $\partial z/\partial x$, $\partial z/\partial y$, and substituting into (A1.1), we obtain

$$\frac{\partial F}{\partial x} - \frac{\partial F/\partial z}{\partial \phi/\partial z}\frac{\partial \phi}{\partial x} = 0$$

$$\frac{\partial F}{\partial y} - \frac{\partial F/\partial z}{\partial \phi/\partial z}\frac{\partial \phi}{\partial y} = 0 \tag{1A.3}$$

Defining

$$\lambda = \frac{\partial F/\partial z}{\partial \phi/\partial z}$$

and substituting into (1A.3), we obtain

$$\frac{\partial F}{\partial x} - \lambda \frac{\partial \phi}{\partial x} = 0$$

$$\frac{\partial F}{\partial y} - \lambda \frac{\partial \phi}{\partial y} = 0$$

$$\frac{\partial F}{\partial z} - \lambda \frac{\partial \phi}{\partial z} = 0 \tag{1A.4}$$

The three equations of (1A.4) with the equation of constraint $\phi(x, y, z) = 0$ give four equations to solve the four unknowns x, y, z, and λ. Thus the problem of finding the extremals of $F(x, y, z)$, subject to $\phi(x, y, z) = 0$, reduces to finding the extremals of an auxiliary function $u(x, y, z) = F(x, y, z) - \lambda\phi(x, y, z)$.

chapter

2

Analogy between Inelastic Strain and Applied Forces

Introduction. Engineers are generally more familiar with methods of analyzing stresses in elastic bodies subject to surface and body forces than with the corresponding analysis for inelastic bodies. It is the aim of this chapter to show the analogy between inelastic strain and applied forces, so that bodies with inelastic strain may be analyzed by the methods used for elastic bodies. This analogy is well known for thermal stresses (Duhamel's analogy). It is here generalized to include creep and plastic strains.

2.1 Inelastic Strain

When a rod is stretched within the elastic limit, the elastic strain e' is related to stress σ by Hooke's Law $\sigma = Ee'$, where E is Young's modulus. If the rod is stretched beyond the elastic limit, then in addition to elastic strain e', time-independent plastic strain e_p—hereafter simply referred to as plastic strain—occurs. If the temperature of a metal rod is raised by T, the length of the rod, l, is increased by $\alpha T l$, where α is the thermal coefficient of expansion. The resulting strain $e_T = \alpha T$ is called the thermal strain. Should the rod be stressed at an elevated-temperature, then time-dependent creep strain e_c, hereafter simply referred to as creep strain, occurs. This creep strain increases with time. The analysis of a body with only thermal and elastic strain is generally called thermoelastic analysis. However, *since plastic strain, thermal strain, and creep strain can be treated mathematically in the same manner* (as shown in the sections which follow) *in causing stresses in a body*, these *three strains together are hereafter called inelastic strain* and denoted by double prime

$$e'' = e_p + e_T + e_c \tag{2.1.1}$$

The sum of elastic and inelastic strains gives the total strain e,

$$e = e' + e'' \tag{2.1.2}$$

The above definition of tensile elastic and inelastic strains can be generalized to multiaxial strains as

$$e''_{ij} = e_{ij_p} + e_{ij_T} + e_{ij_c} \tag{2.1.3}$$

and

$$e_{ij} = e'_{ij} + e''_{ij} \tag{2.1.4}$$

2.2 Analogy between Body Force and Inelastic Strain Gradient

Consider strain to consist of two parts; one is the elastic part e'_{ij}, which for an isotropic body is related to stress by Hooke's law,

$$\tau_{ij} = \lambda\, \delta_{ij}\theta' + 2\mu e'_{ij} \tag{2.2.1}$$

where θ' is elastic dilatation. The other part is the inelastic part e''_{ij}, which consists of the thermal, creep, and plastic strains. Thus

$$e_{ij} = e'_{ij} + e''_{ij} \tag{2.1.4}$$

Taking the elastic strain as the difference between total and inelastic strain, we have

$$\tau_{ij} = \lambda\, \delta_{ij}(\theta - \theta'') + 2\mu(e_{ij} - e''_{ij}) \tag{2.2.2}$$

where μ is the shear modulus, $\lambda = 2\mu v/(1 - 2v)$, and v is Poisson's ratio.

The condition of equilibrium is

$$\tau_{ij,\,j} + F_i = 0 \tag{2.2.3}$$

in the interior of a body v, and

$$S_i = \tau_{ij} n_j \tag{2.2.4}$$

on the boundary surface Γ, where S_i is the surface traction per unit area along the x_i direction and n_j is the cosine of the angle between the surface normal $\mathbf{n}$ and the x_j axis. Substituting Equation (2.2.2) into (2.2.3) and (2.2.4), we obtain

$$\lambda\,\delta_{ij}(\theta-\theta''),_{j}+2\mu(e_{ij}-e''_{ij}),_{j}+F_i=0$$

$$S_i=\lambda\,\delta_{ij}(\theta-\theta'')n_j+2\mu(e_{ij}-e''_{ij})n_j$$

Rearranging terms, we obtain

$$\lambda\,\delta_{ij}\theta,_{j}+2\mu e_{ij,j}+(-\lambda\,\delta_{ij}\theta''_{,j}-2\mu e''_{ij,j})+F_i=0 \qquad (2.2.5)$$

$$S_i+(\lambda\,\delta_{ij}n_j\theta''+2\mu n_j e''_{ij})=\lambda\,\delta_{ij}n_j\theta+2\mu n_j e_{ij} \qquad (2.2.6)$$

Substituting $e_{ij}=\frac{1}{2}(u_{i,j}+u_{j,i})$ into Equations (2.2.5) and (2.2.6) yields

$$\lambda\,\delta_{ij}u_{k,kj}+\mu(u_{i,jj}+u_{j,ij})+(-\lambda\,\delta_{ij}\theta''_{,j}-2\mu e''_{ij,j})+F_i=0$$

or

$$(\lambda+\mu)u_{k,ki}+\mu\nabla^2u_i+(-\lambda\theta''_{,i}-2\mu e''_{ij,j})+F_i=0 \qquad (2.2.7)$$

and

$$S_i+(\lambda n_i\theta''+2\mu n_j e''_{ij})=\lambda n_i u_{k,k}+\mu n_j(u_{i,j}+u_{j,i}) \qquad (2.2.8)$$

From Equations (2.2.5) through (2.2.8), it is clearly seen that $(-\lambda\delta_{ij}\theta''_{,j}-2\mu e''_{ij,j})$ is equivalent[1–4] to F_i, and that $(\lambda n_i\theta''+2\mu n_j e''_{ij})$ is equivalent to S_i in causing strain e_{ij} or displacement u_i. Hence the term $(-\lambda\delta_{ij}\theta''_{,j}-2\mu e''_{ij,j})$ is called the *equivalent body force* and $(\lambda n_i\theta''+2\mu n_j e''_{ij})$ the *equivalent surface force.*

Reisner[5,6] in 1931 obtained the same equations from a physical approach in his study of initial stresses in a body. He imagined a body subdivided into small elements and assumed that each element has only plastic deformation represented by e''_{ij}. If the e''_{ij} of the body satisfy the compatibility conditions, the elements fit each other well and no initial stresses will be produced. If e''_{ij} does not satisfy the compatibility condition, the elements do not fit each other. Then in order to make them fit each other, surface forces are applied to each element to restore them to the volume and shape they had before the plastic strain occurred. Assuming that the elements remain perfectly elastic during the application of the restoring surface forces, the plastic strain may be eliminated by applying a stress $\tau_{ij}=-\lambda\delta_{ij}\theta''-2\mu e''_{ij}$, which corresponds to a surface force

$$\bar{S}_i=\tau_{ij}n_j=-\lambda n_i\theta''-2\mu e''_{ij}n_j \qquad (2.2.9)$$

Under these surface forces the elements now fit each other well. Imagine[7] now that these elements are welded together. Since the plastic strains θ'' and e''_{ij} vary from one element to another, $\bar{S}_i$ also varies from element to element, and there will be an unbalanced force at the interface of adjacent elements. Consider the four adjacent differential volume elements A, B, C, and D as shown in Figure 2.1, where

$$e''_{ij_B} = e''_{ij_A} + \frac{\partial e''_{ij}}{\partial x_1} dx_1 \qquad e''_{ij_C} = e''_{ij_A} + \frac{\partial e''_{ij}}{\partial x_2} dx_2 \qquad e''_{ij_D} = e''_{ij_A} + \frac{\partial e''_{ij}}{\partial x_3} dx_3$$

The surface forces along the x_1 direction on the interfaces between A and B, A and C, and A and D are shown in Figure 2.1. Each face is common to two adjacent cubes. The unbalanced force on each face is considered to be equally shared by the two adjacent cubes. Each cube is bounded by

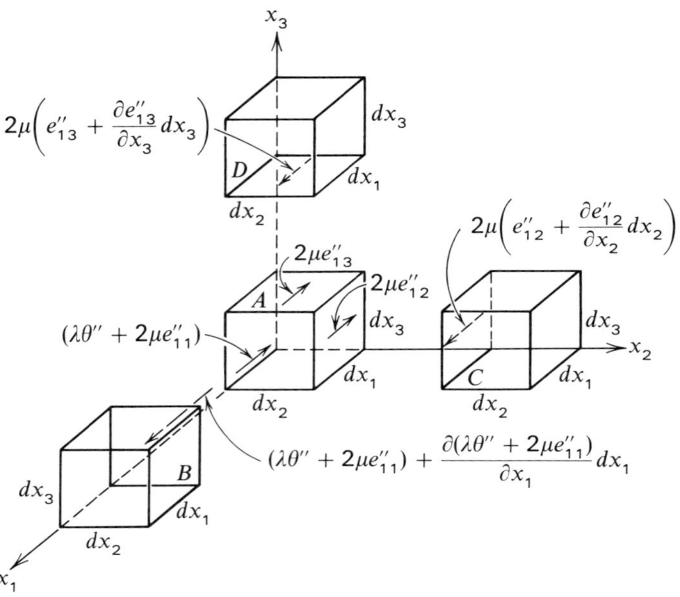

Figure 2.1. Elements under restoring surface forces along the x_1 *direction.*

six faces, hence the unbalanced forces on three faces are considered to be associated with each cube. These three faces are taken to be the right, front, and top faces of the cube (Figure 2.1). The resultant of these interface forces along the x_1 direction of the cubical element A is

$$\left[\frac{\partial(\lambda\theta'' + 2\mu e''_{11})}{\partial x_1} dx_1\right] dx_2\, dx_3 + \left[\frac{\partial(2\mu e''_{12})}{\partial x_2} dx_2\right] dx_1\, dx_3 + \left[\frac{\partial(2\mu e''_{13})}{\partial x_3} dx_3\right] dx_1\, dx_2$$

and hence is equivalent to a body force per unit volume of

$$\lambda\theta''_{,1} + 2\mu(e''_{11,1} + e''_{12,2} + e''_{13,3}) = \lambda\theta''_{,1} + 2\mu e''_{1j,j}$$

These are the forces required to keep the body strain-free, that is, with $e_{ij} = 0$. Since actually there are no such forces, these body forces must be relaxed by applying an equal and opposite body force,

$$\underline{F}_1 = -\lambda\theta''_{,1} - 2\mu e''_{1j,j}$$

Similarly the equivalent body forces along the x_2 and x_3 axes due to inelastic strain e''_{ij} are

$$\underline{F}_2 = -\lambda\theta''_{,2} - 2\mu e''_{2j,j} \qquad \underline{F}_3 = -\lambda\theta''_{,3} - 2\mu e''_{3j,j}$$

These three expressions may be concisely written as

$$\underline{F}_i = -[\lambda\theta''_{,i} + 2\mu e''_{ij,j}] \tag{2.2.10}$$

which is identical to the bracketed term in Equations (2.2.5) and (2.2.7). This is the equivalent body force caused by the inelastic strain. Similarly the equivalent surface force caused by inelastic strain, from Equation (2.2.9), is

$$\underline{S}_i = \lambda n_i\theta'' + 2\mu n_j e''_{ij} \tag{2.2.11}$$

Using Equations (2.2.10) and (2.2.11) enables us to calculate the strain distribution of a body with known inelastic strain distribution e''_{ij} *by calculating the strain distribution of an identical body with no inelastic strain but with the equivalent body and surface forces.*

An alternative approach to the solution of inelastic problems is to consider the displacement component, u_i, as being composed of two parts: u_i^p, a particular solution of Equation (2.2.7), and a complementary part u_i^c,

$$u_i = u_i^p + u_i^c \tag{2.2.12}$$

Equations (2.2.7) and (2.2.8) may then be written as

$$(\lambda + \mu)u^p_{k,ki} + \mu\,\nabla u^p_i - \lambda\theta''_{,i} - 2\mu e''_{ij,j} + F_i = 0 \qquad \text{in } v \tag{2.2.13}$$

$$(\lambda + \mu)u^c_{k,ki} + \mu\,\nabla u^c_i = 0 \qquad \text{in } v \tag{2.2.14}$$

$$S_i + \lambda n_i\theta'' + 2\mu n_j e''_{ij} = \lambda n_i(u^p_{k,k} + u^c_{k,k}) + \mu n_j(u^p_{i,j} + u^c_{i,j} + u^p_{j,i} + u^c_{j,i}) \qquad \text{on } \Gamma \tag{2.2.15}$$

The solution of strain in an elastic body reduces to a particular solution of u_i^p satisfying Equation (2.2.13) and a complementary solution u_i^c

satisfying Equations (2.2.14) and (2.2.15). The resulting displacement is the sum of the particular and complementary solutions. This is an extension of Goodier's method[8] of reducing a thermoelastic problem to one at constant temperature with no body force. It is seen that Goodier's concept may be applied just as well to cases with creep or plastic strains, or both.

2.3 Methods of Analogy Analysis

Duhamel's Analogy.[9–12] In an isotropic body with thermal strain only, we have

$$e_{ij_T} = \delta_{ij}\alpha T, \qquad \theta'' = 3\alpha T$$

Equations (2.2.10) and (2.2.11) for equivalent body and surface forces reduce to

$$\underline{F}_i = -3\lambda\alpha T,_i - 2\mu\alpha T,_i = -(3\lambda + 2\mu)\alpha T,_i \tag{2.3.1}$$

$$\underline{S}_i = (3\lambda\alpha T + 2\mu\alpha T)n_i = (3\lambda + 2\mu)\alpha T n_i \tag{2.3.2}$$

The strain distribution e_{ij} in a body subject to a given set of body and surface forces F_i and S_i with thermal strain e_{ij_T} is the same as the strain distribution in an identical body with no thermal strain but with the additional set of equivalent body and surface forces $\underline{F}_i$ and $\underline{S}_i$ as given by Equations (2.3.1) and (2.3.2). The thermal stresses caused by an arbitrary temperature distribution $T(x_1, x_2, x_3)$ can be calculated from an identical body at uniform temperature as follows:

(1) Apply a body force $F_i + \underline{F}_i$ and surface force $S_i + \underline{S}_i$ to the body. Consider the body to be purely elastic and solve for the stresses. These stresses are denoted by τ^{I}_{ij} and are related to the total strain by Hooke's law as

$$\tau^{\mathrm{I}}_{ij} = \lambda\,\delta_{ij}\theta + 2\mu e_{ij} \tag{2.3.3}$$

(2) From Equation (2.2.3), the actual stress in the body is given by

$$\tau_{ij} = \tau^{\mathrm{I}}_{ij} - \delta_{ij}(3\lambda + 2\mu)\alpha T \tag{2.3.4}$$

This is the well-known Duhamel's analogy or Duhamel-Neumman law.[13,14]

The equivalent body and surface forces are fictitious and should not be confused with actual body and surface forces which might be present. The stresses caused by the actual body and surface forces may be determined separately and then superposed on the thermal stress (for structures with no buckling). For convenience, Duhamel's analogy is also shown in

Table 2.1 with suffixes I and II, denoting the actual and fictitious (analogous) bodies, respectively.[13]

Table 2.1. DUHAMEL'S ANALOGY FOR THERMAL STRESSES[14]

	Body I (Actual)	Body II (Fictitious)
Temperature	$T^{\mathrm{I}} = T \neq 0$	$T^{\mathrm{II}} = 0$
Body force	F_i^{I}	$F_i^{\mathrm{II}} = F_i^{\mathrm{I}} - (3\lambda + 2\mu)\alpha T,_i$
Surface force	S_i^{I}	$S_i^{\mathrm{II}} = S_i^{\mathrm{I}} + (3\lambda + 2\mu)\alpha T n_i$
Stresses	$\tau_{ij}^{\mathrm{I}} = \tau_{ij}^{\mathrm{II}} - \delta_{ij}(3\lambda + 2\mu)\alpha T$	$\tau_{ij}^{\mathrm{II}} = \tau_{ij}^{\mathrm{I}} + \dfrac{\alpha E T}{1 - 2v}\delta_{ij}$
Strains	$e_{ij}^{\mathrm{I}} = e_{ij}^{\mathrm{II}}$	$e_{ij}^{\mathrm{II}} = e_{ij}^{\mathrm{I}}$

Uniqueness of Thermoelastic Solutions. From Table 2.1, it is seen that the strain field of a body with a given temperature distribution is the same as the strain field of an identical elastic body with zero temperature distribution but with an additional set of equivalent applied forces. Since the strain distribution of an elastic body subject to applied forces is unique (except in the buckling problems, as shown in Section 1.13), the strain field and hence the stress field of a body with a given temperature distribution are also unique. *This establishes the uniqueness of stress and strain solutions of thermoelastic bodies.*

Analogy for Creep and Plastic Strains. Now consider a body with creep or plastic strain, or both, but with no thermal strain. Since there is no dilatation associated with creep and plastic strain of metals, that is, $\theta'' = 0$, the equivalent body and surface forces for metal structures reduce to

$$\underline{F}_i = -2\mu e''_{ij,j} \tag{2.3.5}$$

$$\underline{S}_i = 2\mu e''_{ij} n_j \tag{2.3.6}$$

The strain distribution e_{ij} in a body with known creep or plastic strain, or both, subject to a set of given body and boundary forces F_i and S_i is the same as the strain distribution in an identical body with no creep or time-independent plastic strain, subject to, in addition to the applied forces F_i and S_i, the set of equivalent body and surface forces $\underline{F}_i$ and $\underline{S}_i$, given by Equations (2.3.5) and (2.3.6). The stress distribution in a body with known creep or plastic strains, or both, may be found by a procedure similar to the one for finding thermal stresses, as follows:

(1) Apply the equivalent body force $\underline{F}_i$, given by Equation (2.3.5), and the equivalent surface force $\underline{S}_i$, given by Equation (2.3.6), to the body. The elastic stress solution under this fictitious loading, denoted by τ_{ij}^{I}, may be obtained.

(2) From Equation (2.2.3), the stresses caused by the presence of creep or plastic strains, or both, are

$$\tau_{ij} = \tau_{ij}^{\mathrm{I}} - 2\mu e_{ij}''$$

This analogy is also shown in Table 2.2, with suffixes I and II now denoting the actual and fictitious (analogous) bodies.

It is seen that, from the general analogy between body force and inelastic strain gradient, the analogy between body force and temperature gradient can easily be derived. *Hence Duhamel's analogy is a special case of the general analogy between body force and inelastic strain gradient.*

Table 2.2. ANALOGY BETWEEN BODY FORCE AND CREEP OR PLASTIC STRAIN GRADIENT, OR BOTH

	Body I (Actual)	Body II (Fictitious)
Creep or plastic strains	$e_{ij}''^{\mathrm{I}} = e_{ij}'' \neq 0$	$e_{ij}''^{\mathrm{II}} = 0$
	$\theta''^{\mathrm{I}} = 0$	$\theta''^{\mathrm{II}} = 0$
Body force	F_i^{I}	$F_i^{\mathrm{II}} = F_i^{\mathrm{I}} - 2\mu e_{ij,j}''$
Surface force	S_i^{I}	$S_i^{\mathrm{II}} = S_i^{\mathrm{I}} + 2\mu e_{ij}'' n_j$
Stresses	$\tau_{ij}^{\mathrm{I}} = \tau_{ij}^{\mathrm{II}} - 2\mu e_{ij}''$	$\tau_{ij}^{\mathrm{II}} = \tau_{ij}^{\mathrm{I}} + 2\mu e_{ij}''$
Strain	$e_{ij}^{\mathrm{I}} = e_{ij}^{\mathrm{II}}$	$e_{ij}^{\mathrm{II}} = e_{ij}^{\mathrm{I}}$

By this analogy, the solutions of inelastic problems are reduced to those of elastic problems. Although elastic solutions are not available for many three-dimensional problems, there are many elastic solutions that can be used to find the solutions of the corresponding inelastic problems.[2,3,15–17]

For the case with known thermal strain and with known creep or plastic strain, or both, a similar table can be constructed by combining Tables 2.1 and 2.2. Alternatively, separate analyses may be obtained for thermal strain, and for creep or plastic strain, or both, and for the actual applied loads and then superposed to obtain the stresses caused by the actual loads, the thermal strain, and the creep or plastic strains, or both.

There are problems for which the elastic solutions are either not available or are much too unwieldy to apply. In such cases approximate methods, such as the methods of finite differences and finite elements,[18–24] have often been used. These numerical methods give the approximate stress and strain fields of elastic bodies under arbitrary loads. They can also be extended to find the approximate stress and strain fields of the corresponding inelastic bodies by applying the analogy procedure of converting the inelastic strains into a set of equivalent loads.

2.4 Displacement of Bodies Caused by a Known Distribution of Inelastic Strain[25]

In the previous section, the equivalence of inelastic strain to external force has been shown. By considering this equivalent external force applied to an elastic body of identical shape and supporting conditions, the displacement of the body with inelastic strain can be calculated by the method of elastic analysis. In this section, a more direct method of finding the displacement caused by a given distribution of inelastic strain is presented.

Let u_i and $u_i{}^*$ be two sets of displacement functions which are single-valued and continuous throughout the body. Let F_i, $F_i{}^*$, S_i, $S_i{}^*$ be the corresponding body and surface forces. Let v denote the volume occupied by the body and Γ the boundary surface. The conditions of static equilibrium are

$$\tau_{ij,j} + F_i = 0 \qquad \text{in } v$$

$$S_i = \tau_{ij} n_j \qquad \text{on } \Gamma$$

Multiplying the above equations through by $u_i{}^*$, integrating over v and Γ, respectively, and adding, yields

$$\int_v F_i u_i{}^* \, dv + \int_\Gamma S_i u_i{}^* \, d\Gamma = -\int_v \tau_{ij,j} u_i{}^* \, dv + \int_\Gamma \tau_{ij} n_j u_i{}^* \, d\Gamma \qquad (2.4.1)$$

By the divergence theorem, we have

$$\int_\Gamma \tau_{ij} u_i{}^* n_j \, d\Gamma = \int_v \tau_{ij,j} u_i{}^* \, dv + \int_v \tau_{ij} u_{i,j}^* \, dv$$

$$= \int_v \tau_{ij,j} u_i{}^* \, dv + \int_v \tau_{ij} e_{ij}^* \, dv$$

where use was made of the relation

$$\tau_{ij} u_{i,j}^* = \tfrac{1}{2}(\tau_{ij} u_{i,j}^* + \tau_{ji} u_{j,i}^*)$$

$$= \tfrac{1}{2}\tau_{ij}(u_{i,j}^* + u_{j,i}^*) = \tau_{ij} e_{ij}^*$$

Substituting this into Equation (2.4.1), we obtain

$$\int_v F_i u_i{}^* \, dv + \int_\Gamma S_i u_i{}^* \, d\Gamma = \int_v \tau_{ij} e_{ij}^* \, dv \qquad (2.4.2)$$

Recalling Equations (1.9.2) and (1.9.6) for *elastic bodies,* we see that

$$\int_v \tau_{ij} e_{ij}^* \, dv = \int_v C_{ijkl} e_{kl} e_{ij}^* \, dv$$

$$= \int C_{klij} e_{kl} e_{ij}^* \, dv$$

$$= \int_v \tau_{kl}^* e_{kl} \, dv \qquad (2.4.3)$$

From Equations (2.4.2) and (2.4.3), we obtain

$$\int_v F_i u_i^* \, dv + \int_\Gamma S_i u_i^* \, d\Gamma = \int_v F_i^* u_i \, dv + \int_\Gamma S_i^* u_i \, d\Gamma \qquad (2.4.4)$$

Now *consider a body with inelastic strain* e''_{ij}, which may be a combination of thermal, creep, and plastic strain. The elastic strain e'_{ij} is the difference between the total and the inelastic strain; that is,

$$e'_{ij} = e_{ij} - e''_{ij} \qquad (2.1.4)$$

From Equation (1.9.2), we have

$$\tau_{ij} = C_{ijkl}(e_{kl} - e''_{kl})$$

Substituting this in the equilibrium conditions $\tau_{ij,j} + F_i = 0$ and $S_i = \tau_{ij} n_j$, we obtain

$$C_{ijkl}(e_{kl,j} - e''_{kl,j}) + F_i = 0 \qquad \text{in } v$$

$$S_i = C_{ijkl} e''_{kl} n_j \qquad \text{on } \Gamma$$

Hence the equivalent body and surface forces are

$$\underline{F}_i = -C_{ijkl} e''_{kl,j} \qquad (2.4.5)$$

$$\underline{S}_i = C_{ijkl} e''_{kl} n_j \qquad (2.4.6)$$

Replacing F_i and S_i in Equation (2.4.4) by $F_i + \underline{F}_i$ and $S_i + \underline{S}_i$, respectively, we obtain

$$\int_v (F_i - C_{ijkl} e''_{kl,j}) u_i^* \, dv + \int_\Gamma (S_i + C_{ijkl} e''_{kl} n_j) u_i^* \, d\Gamma$$

$$= \int_v F_i^* u_i \, dv + \int_\Gamma S_i^* u_i \, d\Gamma$$

To find the displacement u_i'' due to inelastic strain alone, we let $F_i = S_i = 0$. Applying the divergence theorem to the second term on the left, we find that

$$-\int_v C_{ijkl} e''_{kl,j} u_i^* \, dv + \int_v [C_{ijkl} e''_{kl} u_i^*]_{,j} \, dv = \int C_{ijkl} e''_{kl} e^*_{ij} \, dv$$
$$= \int F_i^* u_i'' \, dv + \int_\Gamma S_i^* u_i'' \, d\Gamma \tag{2.4.7}$$

From Equation (1.9.6), we have $C_{ijkl} = C_{klij}$, and therefore

$$\int C_{klij} e''_{kl} e^*_{ij} \, dv = \int_v \tau^*_{kl} e''_{kl} \, dv$$
$$= \int_v F_i^* u_i'' \, dv + \int_\Gamma S_i^* u_i'' \, d\Gamma \tag{2.4.8}$$

Now let $\tau^*_{kl}(x', x)$ be the stress at $x'(x_1', x_2', x_3')$ caused by a concentrated unit force acting at point $x(x_1, x_2, x_3)$ along the direction of the x_i axis. The right-hand side of Equation (2.4.8) then gives the displacement at x along the x_i direction caused by $e''_{kl}(x_1', x_2', x_3')$ throughout the body; that is,

$$u_i''(x) = \int_{v'} \tau^*_{kl}(x', x) e''_{kl}(x') \, dv' \tag{2.4.9}$$

where $dv' = dx_1' \, dx_2' \, dx_3'$.

For the case of a body with thermal strain only and with an isotropic coefficient of expansion, we have

$$e''_{kl} = \delta_{kl} \alpha T$$

Then Equation (2.4.9) reduces to

$$u_i''(x) = \int_{v'} \tau^*_{kl}(x', x) \, \delta_{kl} \alpha T(x') \, dv' = \int_{v'} \tau^*_{kk}(x', x) \alpha T(x') \, dv' \tag{2.4.10}$$

This is Maysel's[28] expression for displacement in a body caused by a steady temperature field in an isotropic body.

For the case with no thermal strain but with creep or plastic strain, or both, there is no volume change in the inelastic strain;

$$e''_{ii} = 0 \qquad e''_{ij} = \varepsilon''_{ij}$$

where ε_{ij} is the deviatoric strain component. Writing τ^*_{kl} as $S^*_{kl} + \delta_{kl}(\tau^*_{mm}/3)$, where S^*_{ij} is the deviatoric stress component, we see that Equation (2.4.9) reduces to

$$u_i''(x) = \int_{v'} \left(S^*_{kl} + \delta_{kl} \frac{\tau^*_{mm}}{3} \right) \varepsilon''_{kl} \, dv'$$
$$u_i''(x) = \int_{v'} S^*_{kl} \varepsilon''_{kl} \, dv' \tag{2.4.11}$$

Equations (2.4.9) to (2.4.11) give the displacement at a point x in the body along the x_i direction caused by a steady temperature field $T(x')$, creep or plastic strain, or both, $\varepsilon''_{ij}(x')$. The τ^*_{kl}, S^*_{kl} components are the stress components and deviated stress components, respectively, at point x' in an elastic body, having the same shape and the same kinematic boundary conditions,† caused by a unit force along the x_i direction applied at x. For a number of problems the elastic solutions are available, so τ^*_{kl} and S^*_{kl} are readily found.

REFERENCES

1. Lin, T. H., "On the Associated Flow Rule of Plasticity Based on Crystal Slip," *J. Franklin Institute*, **270**, 291–300, 1960.
2. Lin, T. H., S. Uchiyama, and D. Martin, "Stress Field in Metals at Initial Stage of Plastic Deformation," *J. Mech. Phys. Solids*, **9**, 200–209, 1961.
3. Lin, T. H., and T. K. Tung, "The Stress Field Produced by Localized Plastic Slip at a Free Surface," *J. Appl. Mech.*, 523–532, September 1962.
4. Lin, T. H., "Analogy Between the Inelastic Strain Gradient and Body Force in Cubic Crystals and Isotropic Media," *Proc. Nat. Acad. Sci.*, March 1966.
5. Reissner, H., "Eigenspannungen und Eigenspannungsquellen," *Z. Angew. Math. Mech.*, **11**, 1–8, 1931.
6. Timoshenko, S., *Theory of Elasticity*, McGraw-Hill, New York, pp. 213–215, 1934.
7. Eshelby, J. D., "The Determination of the Elastic Field of an Ellipsoidal Inclusion and Related Problems," *Proc. Roy. Soc. A*, **241**, 396, 1957.
8. Goodier, J. N., "Integration of Thermoelectric Equations," *Phil. Mag.*, **23**, 1017, 1937.
9. Duhamel, J. M. C., "Mémoire sur le calcul des actions moleculaires developpées par les changements de temperature dans les corps solides," *Mem. Inst.*, France, **5**, 440–498, 1838.
10. Duhamel, J. M. C., "Seconde mémoire sur les phénomènes thermomécaniques," *J. Ecole Polytech.* (Paris), **15**, 1–57, 1837.
11. Duhamel, J. M. C., "Mémoire sur le mouvement des differents points d'une barre cylindrique dont la temperature varie," *J. Ecole Polytech.* (Paris), **21**, 1–33, 1856.
12. Neumann, F., *Abhandl. Akad. Wiss*, Part 2, p. 1, Berlin, 1841.
13. Rydewski, J. R., *Introduction to Structural Problems in Nuclear Reactor Engineering*, Macmillan, New York, pp. 211–257, 1962.
14. Morgan, A. J. A., "A Proof of Duhamel's Analogy for Thermal Stresses," *J. Aerospace Sci.*, **25**, 466–467, July 1958.

† On a part of the boundary of the body, the displacement u_i is specified. "Same kinematic boundary conditions" means that u_i on this part of boundary is the same as the actual body.

15. Lin, T. H., and S. Uchiyama, "Stress Field Caused by a Slid Crystal at a Free Surface," *J. Mech. Phys. Solids*, **11**, 327–343, 1963.
16. Lin, T. H., "Bending of a Plate with Nonlinear Strain Hardening Creep," *Proceedings of the International Union of Theoretical and Applied Mech., Colloquium of Creep of Structures*, Springer-Verlag, Berlin, Germany, 1962.
17. Lin, T. H., and J. K. Ganoung, "Bending of Rectangular Plates with Nonlinear Strain-Hardening Creep," *Intern. J. Mech. Sci.*, **6**, 337–348, 1964.
18. Zienkiewicz, O. C., *The Finite Element Method in Structural and Continuum Mechanics*, McGraw Hill, New York, 1967.
19. Turner, M. J., R. W. Clough, H. C. Martin, and L. J. Topp, "Stiffness and Deflection Analysis of Complex Structures," *J. Aerospace Sci.*, **23**, pp. 805–823, 1956.
20. Zienkiewicz, O. C., and E. I. Holister, *Stress Analysis*, Wiley, New York, 1965.
21. Argyris, J. H., "Triangular elements with Linearly Varying Strain for the Matrix Displacement Method," *J. Roy. Aero. Soc.* Tech. Note, **69**, 711–13, October 1965.
22. Argyris, J. H., "Matrix Analysis of Three Dimensional Elastic Media—Small and Large Displacements," *J. Am. Inst. Aeronautics and Astronautics*, **3**, 45–51, 1965.
23. Rubinstein, M. F., *Matrix Computer Analysis of Structures*, Prentice-Hall, Englewood Cliffs, N.J., 1966.
24. Greenbaum, G. A., Creep Analysis of Axisymmetric Bodies, Ph.D. Dissertation, University of California, Los Angeles, 1966.
25. Lin, T. H., "Reciprocal Theorem of Displacement of Inelastic Bodies," *J. Composite Materials*, April 1967.
26. Nowacki, W., *Thermoelasticity*, Pergamon, London, pp. 32–38, 1962.

chapter

3

Inelastic Uniaxial Stress-Strain Relations

Introduction. In the previous chapter a general analogy between inelastic strain and applied force was presented. This analogy reduces the solution of stress and strain distributions in a body with known distribution of inelastic strains to the solution of an identical elastic body with no inelastic strain but with an additional set of body and surface forces. Among the three types of inelastic strain previously discussed, thermal strain is generally obtained from a given temperature distribution, while creep strain depends on the history of stress and temperature, and plastic strain depends on the history of stress. Plastic strain-stress relations and creep-strain/stress/time relations vary from one metal to another. The general characteristics of these relations for commonly used engineering materials under uniaxial loading is presented. This is followed by a discussion of linear viscoelastic models. These models are particularly applicable for the representation of the creep behavior of high polymers. Solutions of stress and strain distributions in a body have to satisfy the stress-strain relations of the material in addition to the conditions of equilibrium and compatibility discussed in Chapter 1.

3.1 Elastic-Plastic Uniaxial Stress-Strain Relations

When an aluminum or steel bar of initial length l_0 and constant cross section A_0 is subjected to a tensile load P in a tensile testing machine, the length of the bar increases to l and its cross-sectional area is reduced to A.

The extension $\Delta l = l - l_0$ divided by the original length l_0 is known as the axial strain e. The load P divided by A_0 is called the nominal stress $\sigma = P/A_0$. The curve $\sigma = f(e)$ is called the nominal or conventional stress-strain curve of the metal. If a bar is tested at a temperature far below its melting point, its stress-strain curve does not vary appreciably with the rate of loading. When the strain e is increased monotonically, it may be assumed that the stress σ is a unique function of strain e. A typical stress-strain curve of an aluminum alloy is shown in Figure 3.1 and that of a mild steel is shown in Figure 3.2 overleaf.

The stress σ at first increases proportionally with strain. The initial portion of this stress-strain curve is considered a straight line. The slope of this line is known as Young's modulus of elasticity. At some point A in Figure 3.1 or Figure 3.2, the curve starts to deviate from this straight line. If the bar is loaded up to or below A and then unloaded, it will return to its original length l_0, and hence the strain e is completely recovered. *Strain which is recoverable upon unloading is known as elastic strain.* If the bar is loaded beyond A, say to B, and then unloaded, the stress-strain curve will follow B to C with the slope of BC the same as that of AO. If the bar is completely unloaded, a permanent plastic strain e_p of the amount OC remains. *Time-independent strain which is not recoverable upon unloading is known as plastic strain.* The point A is known as the elastic, or proportional, limit of the material. At A' in Figure 3.2, a sudden decrease in slope to zero occurs in the stress-strain curve. The point A' is

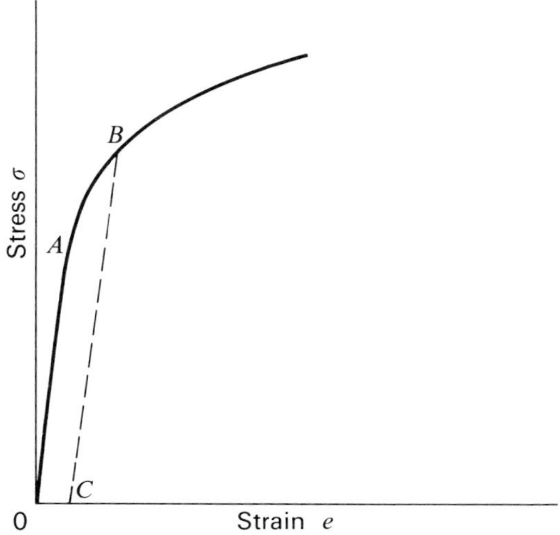

Figure 3.1. Stress-strain curve of an aluminum alloy.

called the yield stress of the material. In those stress-strain curves, like Figure 3.1, in which there is no such distinct point, the yield stress is arbitrarily defined as the stress giving a permanent strain e_p of 0.002. When the load is increased beyond the yield load, the conventional stress generally increases with strain to a maximum value and then decreases. This maximum conventional stress is called the ultimate tensile strength of the material.

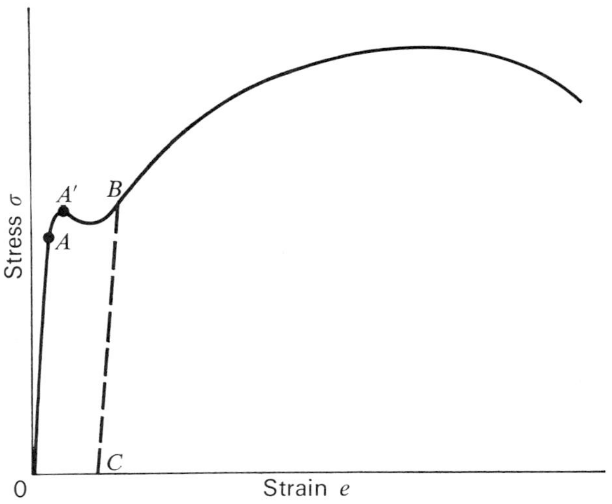

Figure 3.2. Stress-strain curve of a mild steel.

If the bar is stretched beyond the elastic limit and the loading then reversed, the straight-line portion of the stress-strain curve during unloading or reversed loading (CD of Figure 3.3) has the same slope as Young's modulus. Beyond a certain point D, the curve bends and crosses the zero-stress line at F. Under reversed loading, 0.002 permanent compressive strain occurs at G. The magnitude of the tensile stress at B which produces a 0.002 permanent tensile strain is larger than the magnitude of the compressive stress at G which produces a 0.002 permanent compressive strain. This phenomenon observed in tension followed by compression also occurs for compression followed by tension. It is called the Bauschinger effect.

During elongation of a bar the original cross-sectional area A_0 reduces to A, and thus the true stress is equal to P/A instead of P/A_0, where A is the instantaneous cross-sectional area. Similarly, by defining the increment

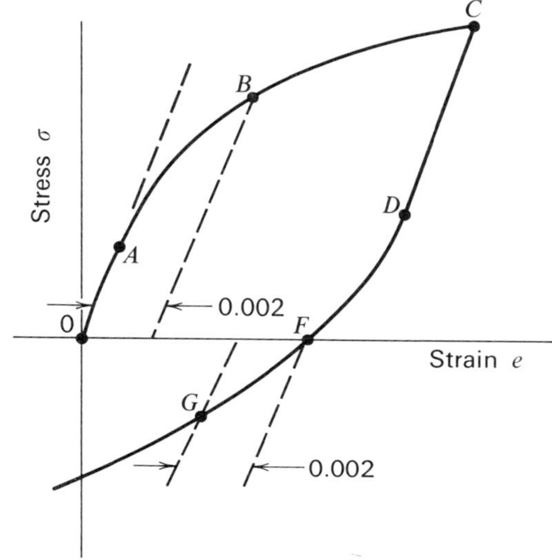

Figure 3.3. Stress-strain curve under reversed loading.

of true strain as the incremental elongation divided by the instantaneous length,

$$de = \frac{dl}{l}$$

we obtain, upon integration,

$$e = \int_{l_0}^{l} \frac{dl}{l} = \ln \frac{l}{l_0}$$

Denoting the conventional strain by $\bar{e}$, where

$$\bar{e} = \frac{l - l_0}{l_0} = \frac{l}{l_0} - 1$$

or (3.1.1)

$$1 + \bar{e} = \frac{l}{l_0}$$

we obtain the relation between true strain and conventional strain:

$$e = \ln (1 + \bar{e}) \tag{3.1.2}$$

When the strain is small, these two strains are essentially the same. At a conventional strain of 0.01, the difference between the two is only 0.00005. For small deformation, conventional strain is sufficiently accurate, and is generally used. For large deformation, true strain should be used.

3.2 Uniaxial Stress-Strain Time Relationships at Elevated Temperatures

At room temperature the stress-strain curves of *most commonly used metal alloys* are essentially time-independent, i.e., independent of the rate or duration of loading. Generally, however, when the temperature of the environment exceeds half the melting-point temperature of the alloy, departure from this idealization becomes noticeable, and the strain increases under constant load, or stress. *Strain which increases with time at constant load is known as creep strain.* A typical curve of creep strain vs. time under constant load for most metals used in engineering structures is shown in Figure 3.4.

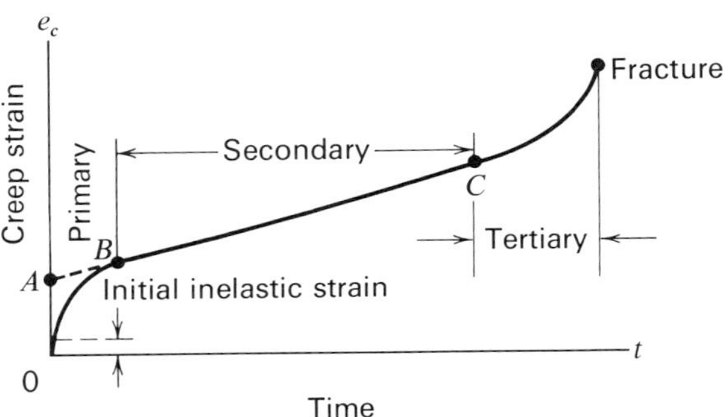

Figure 3.4. A typical creep curve under constant load.

The first portion of the creep curve is known as primary or transient creep. A large portion of this creep strain is recoverable after unloading. The secondary stage, associated with the straight-line portion of the curve, is sometimes called the minimum-creep-rate region or the steady creep. The last stage, associated with an increase in the rate of creep strain, is called tertiary creep. During tertiary creep the strain rate increases until fracture occurs. Because the cross-sectional area decreases as elongation proceeds, a constant tensile load causes increasing true stress. Andrade[1,2] first pointed out that the upward curvature in the tertiary stage is partly due to the reduction of cross section at large strains. Creep tests have been made by Andrade on lead wire for both constant load and

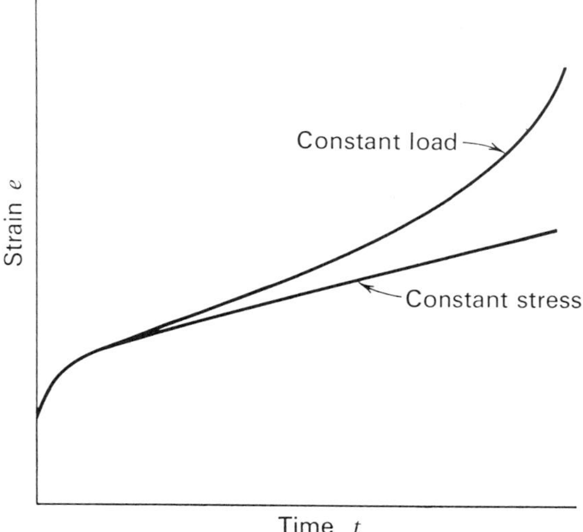

Figure 3.5. Creep tests of lead wire (Andrade, 1910).

constant stress. The resulting curves are shown in Figure 3.5. It is seen that creep under constant stress seems to show no tertiary stage.

If, after a specimen is loaded under either constant load or constant stress, it is unloaded at a certain time instant A in Figure 3.6, time-independent elastic strain AB is instantly recovered, followed by a time-dependent recovery of a portion of the creep strain. This is called the recovery effect.

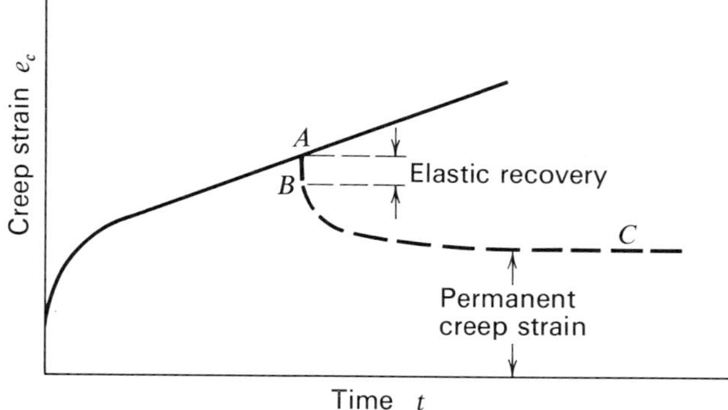

Figure 3.6. Creep recovery of metals.

3.3 Representation and Extrapolation of Empirical Creep Data

Empirical Creep Data Representation. Many types of functions have been proposed to represent the empirical uniaxial creep-strain vs. time relationships of metals under constant stress and constant temperature. Some of these are shown below:

$$e_c = a[\log(1 + bt)]^{2/3} \quad \text{by Mott and Nabarro[3]} \tag{3.3.1}$$

$$e_c = a \log t + bt + c \quad \text{by Weaver[4]} \tag{3.3.2}$$

$$e_c = a + bt - c \exp(-kt) \quad \text{by McVetty[5]} \tag{3.3.3}$$

$$e_c = at + b[(1 - \exp(-ct)] \quad \text{by Soderberg[6]} \tag{3.3.4}$$

$$e_c = a + bt^n \quad \text{by Swift and Tyndall[7]} \tag{3.3.5}$$

$$e_c = a(1 + bt^{1/3}) \exp(ct) \quad \text{by Andrade[8]} \tag{3.3.6}$$

$$e_c = at^m + bt^n + ct^p \quad \text{by Graham[9]} \tag{3.3.7}$$

$$e_c = a \log t + bt^n + ct \quad \text{by Wyatt[10]} \tag{3.3.8}$$

where a, b, c, k are constants, e_c is the creep strain, and t is time. For a detailed discussion of these relationships, see *Processes of Creep and Fatigue in Metals*, by A. J. Kennedy.[11]

As the temperature is raised, the steady creep rate increases with temperature. This temperature dependence has been expressed as

$$\dot{e}_c = a \exp(-Q/RT) \quad \text{by Mott[12]} \tag{3.3.9}$$

$$\dot{e}_c = a[t \exp(-Q/RT]^n \quad \text{by Dorn[13]} \tag{3.3.10}$$

where Q, R, a, and n are constants of the material, and T is the absolute temperature.

A number of researches have been made on the relation between stress and steady creep. The commonly used empirical formulas relating the steady creep rate to stress with the same notations as before, are listed as follows:

$$\dot{e}_c = a[\exp(b\sigma) - 1] \quad \text{by Soderberg[6]} \tag{3.3.11}$$

$$\dot{e}_c = a\sigma^n \quad \text{by Norton,[14] Bailey[15]} \tag{3.3.12}$$

$$\dot{e}_c = a \exp(b + c\sigma) \quad \text{by Dushman[16]} \tag{3.3.13}$$

$$\dot{e}_c = a \sinh b\sigma \quad \text{by Nadai[17]} \tag{3.3.14}$$

$$\dot{e}_c = a \exp(b\sigma) \quad \text{by Ludwik[18]} \tag{3.3.15}$$

Equation (3.3.12) gives a straight line in the log σ vs. log $\dot{e}_c$ plot, while Equations (3.3.13) and (3.3.15) give a straight line in σ vs. log $\dot{e}_c$ plots.[19] Of these two types, Equation (3.3.12) has been found to better represent certain austenitic heat-resistant alloy steels (Cr 26% Ni 12%; Cr 26% Ni 20%; Cr 12% Ni 60%; Cr 21% Ni 9%). For wrought 18% Cr 8% Ni steel tested at 1500° F, Equation (3.3.14) has been shown to fit the empirical data better[20] than Equation (3.3.12). Equations (3.3.13), (3.3.14), and (3.3.15) have the advantage of being conservative for the low stress levels, since the hyperbolic sine relation predicts a higher creep rate at low stress than Equation (3.3.12).

In constant-stress tests of short duration, primary creep predominates. One commonly used relation[21] to represent this primary creep is

$$e_c = D\sigma^n t^k \tag{3.3.16}$$

where t is the time, and D, n, and k are material constants.

To include the recoverable creep strain in an empirical expression, the total strain has been written as the sum of the elastic, transient, and steady creep components, and the transient component is assumed to be recoverable,

$$\dot{e} = \dot{e}_e + \dot{e}_t + \dot{e}_s \tag{3.3.17}$$

These components have been expressed in the following forms by McVetty[22] and Davis.[23]

$$\dot{e}_e = \frac{\dot{\sigma}}{E} \qquad \dot{e}_t = a(K\sigma^b - e_t) \qquad \dot{e}_s = B\sigma^n \tag{3.3.18}$$

where K, a, b, B, and n are material constants. The transient creep e_t is assumed to have a maximum value of $K\sigma^b$, and $\dot{e}_s$ is the steady creep rate. Letting $b = n$ in Equations (3.3.18), Pao and Marin[24] obtained, after integration,

$$e = \frac{\sigma}{E} + K\sigma^n[1 - \exp(-at)] + B\sigma^n t \tag{3.3.19}$$

If a specimen is under tension stress σ for time t', at which time unloading reduces the stress to σ', there will be an immediate elastic recovery of $(\sigma - \sigma')/E$, and the new creep rate will be $a(K\sigma'^n - e_t') + B\sigma'^n$, where e_t' is the transient creep strain reached at the time of unloading. If this transient creep is less than $K\sigma'^n$, the transient creep will continue to increase. On the other hand, if the transient creep is larger than $K\sigma'^n$, the first term is negative and decreases. If σ' is zero, the total strain rate is negative until the transient creep strain is recovered.

Odqvist[25] has proposed that when the primary creep is only a small part of the total creep, the curve *OBC* in Figure 3.4 may be replaced by the straight line *ABC*. It is assumed here that the transient phase of the creep takes place instantaneously after loading. This assumption is satisfactory if most of the creep strain is due to steady creep. The strain rate may then be written in the form

$$\begin{aligned} \dot{e} &= \frac{\dot{\sigma}}{E} + K_1 \sigma^m \dot{\sigma} + K_2 \sigma^n \qquad \text{if } \sigma\, d\sigma > 0 \\ \dot{e} &= \frac{\dot{\sigma}}{E} + K_2 \sigma^n \qquad \text{if } \sigma\, d\sigma < 0 \end{aligned} \tag{3.3.20}$$

where E is Young's modulus, and K_1, K_2, m, and n are material constants.

Extrapolation of Creep Data to Low Stress Levels. In general, at low stress levels, the creep rate is very small, and therefore creep tests would require long durations of time. These tests are rather scarce. To predict creep rate at low stress, the data at higher stresses have to be extrapolated. The steady creep rate vs. stress relation is generally represented by a power law $\dot{e}_c = A\sigma^n$, where A and n are constants, or a hyperbolic sine law $\dot{e}_c = a \sinh b\sigma$, where a and b are constants. Each of these relations has two constants, which are determined from two values of stresses σ_1 and σ_2 corresponding to two creep rates $\dot{e}_{c_1}$ and $\dot{e}_{c_2}$. Sketches of the power and hyperbolic sine relations passing through $\dot{e}_{c_1}$ and $\dot{e}_{c_2}$ are shown by Figure

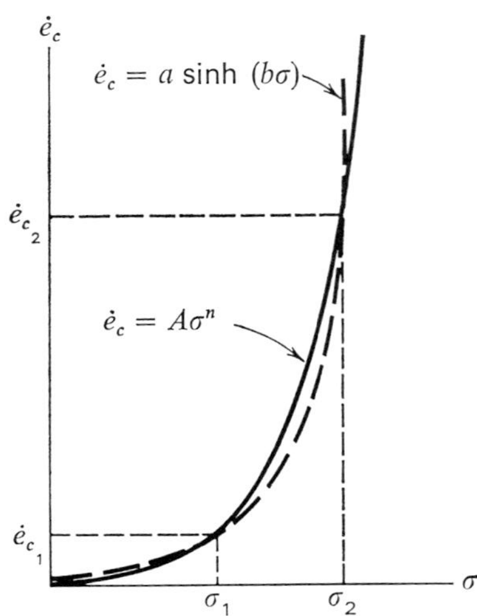

Figure 3.7. Comparison of two creep relations.

3.7. It is seen that for $\sigma < \sigma_1$ and $\sigma > \sigma_2$ the hyperbolic sine law is more conservative, while between σ_1 and σ_2 the power law is more conservative. McVetty[21] has shown that the power law may predict unconservative creep rates at low stress levels. Hence, for extrapolating the creep rate at low stress, the hyperbolic sine law is generally preferable. When $\sigma > \sigma_1$, the power law usually better represents the creep properties of metals.

3.4 The Mechanical Equation of State[26]

Most structures are subject to stresses varying with time. Creep tests have generally been done under constant load or constant stress. In applying these test data to structures in which stress varies with time, certain assumptions have to be made. One commonly used assumption for creep analysis of structures is the existence of a mechanical equation of state between creep rate, stress, temperature, and the current creep strain. *This mechanical equation of state is not derived from the physics of the metals and therefore may not be valid for some metals.* Hence, for the particular metal used, the validity of this assumption should be verified by tests. Under constant temperature, this equation may be written as[18,26]

$$\dot{e}_c = f(e_c, \sigma) \tag{3.4.1}$$

where f denotes a function. Here the stress is assumed to depend only on the current creep strain and its rate, and not on the strain rate during earlier stages of deformation. Creep and relaxation tests have been conducted by Johnson[27] on 0.5 % Mo steel at 575° C. In creep tests the load or stress is kept constant, while in relaxation tests the length of the specimen is maintained constant and the stress decreases with time. The results of Johnson's tests are shown in Figure 3.8 overleaf and listed in Table 3.1.

Table 3.1

$\text{Log}_{10} e$	$\text{Log}_{10} \dot{e}$ (in. hr^{-1})	Creep Stress (tons/sq in.)	Relaxation Stress (tons/sq in.)
−3.775	−5.25	5.00	3.85
−3.575	−5.40	5.00	5.95
			4.90 average

The constant-stress (5 tons per sq in.) curve of $\log_{10} \dot{e}_c$ vs. $\log_{10} e_c$ intercepts the two relaxation test curves at $\log_{10} e_c$ of about −3.775 and −3.575 corresponding to $\log_{10} \sigma$ of 0.586 and 0.777, giving stresses of

3.85 and 5.98 tsi (tons per sq in.), respectively. It is seen that the average of these two stresses (4.90 tsi) in the relaxation tests is equal, approximately, to the stress (5 tsi) in creep tests for the same strain and strain rate. For a number of metals within certain temperature and stress ranges, the strain rate history has no significant effect on the current strain rate. For these metals the simple relation, Equation (3.4.1), may be used. This relation

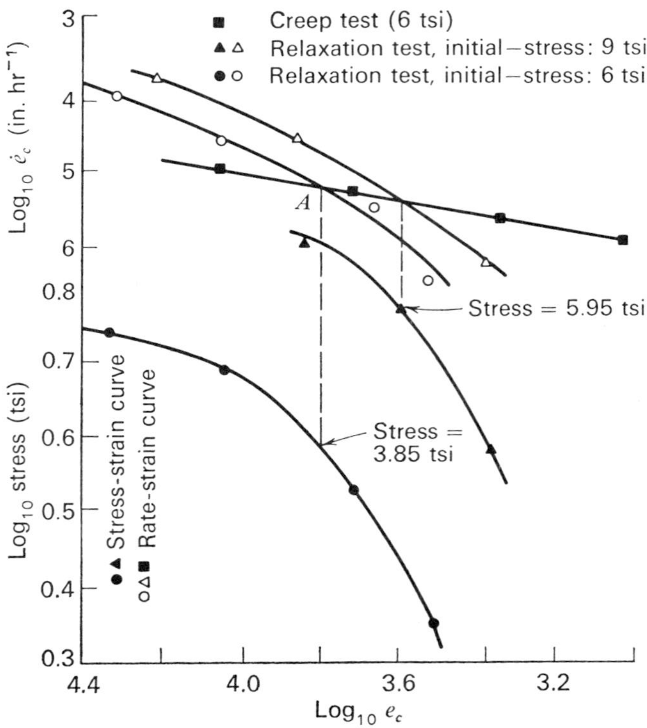

Figure 3.8. Creep and relaxation behavior of a 0.5% MO steel at 575°C (after Johnson, 1954).

agrees approximately with tests of many metals.[28] With this relation, strain vs. time may be found for cases of varying stress, such as the relaxation test, where the total strain remains constant but the stress decreases.

To calculate the creep strain vs. time curve of a specimen under varying tensile stress, the smooth stress-time curve is often approximated by horizontal and vertical segments, as shown in Figure 3.9. This was

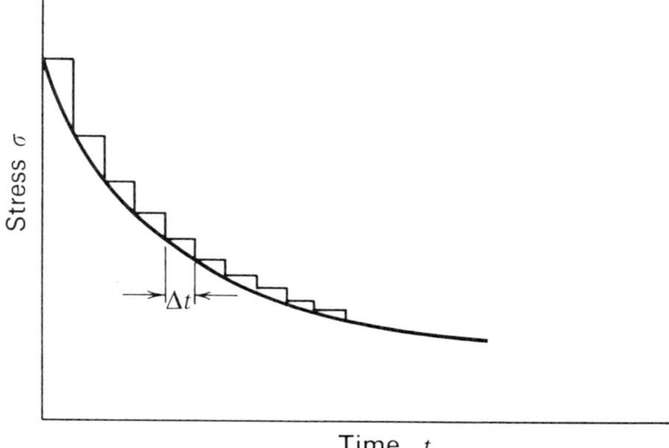

Figure 3.9. Approximation of the smooth stress-time curve by steps.

first proposed by Shanley.[29] Within each Δt, we assume σ is constant so that the empirical constant-stress curve may be used. In the limit, as Δt approaches zero, the step stress-time curve approaches the smooth stress-time curve. To calculate the creep strain vs. time curve of the relaxation test, $(\sigma/E) + e_c = \text{constant}$, the following procedure is used. At $t = 0$ we have $e_c = 0$ and $\sigma = Ee$. This gives the initial stress σ_1. For the first time interval, the creep strain follows the σ_1 constant-stress curve shown in Figure 3.10. At the end of this time interval, e_c would correspond to A, and then the stress would be reduced by Ee_c, giving a stress σ_2, but e_c

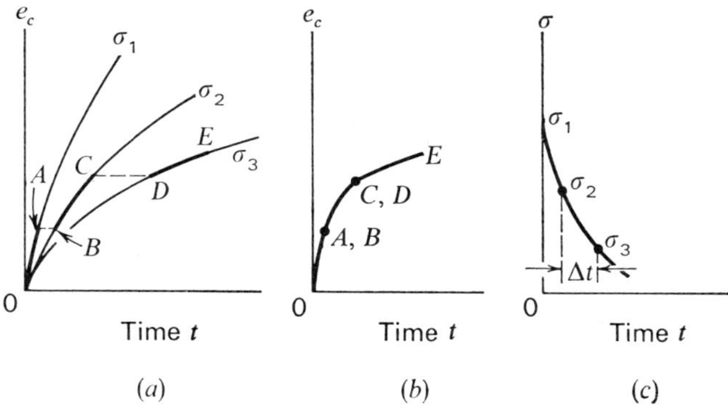

Figure 3.10. (a) Calculation of strain-time curve for relaxation test from constant stress tests; (b) creep-strain vs. time; (c) stress vs. time.

would still correspond to A. For the next time interval, the strain increment vs. time would follow BC on the σ_2 curve. Continuing this process, the strain-time curve for the relaxation test is computed. With the creep strain vs. time known, the stress-time curve, as shown in Figure 3.10(c), is readily obtained.

3.5 Representation of Creep Characteristics by Linear Mechanical Models

The extension of an elastic material subject to tension is instantaneous and directly proportional to the stress. This elastic characteristic may be represented by a spring, as shown in Figure 3.11, where the spring constant

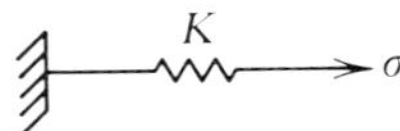

Figure 3.11. Mechanical model for an elastic solid.

K corresponds to Young's Modulus E. In a viscous fluid, the rate of strain $\dot{e}$ is directly proportional to the stress, $\sigma = C\dot{e}$. This viscous characteristic may be represented by a dashpot with damping constant C, as shown in Figure 3.12. Strain in an elastic solid remains constant with time, while

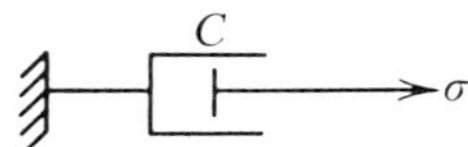

Figure 3.12. Mechanical model for a perfect viscous fluid.

strain in a viscous fluid varies linearly with time. In both cases, however, the strain is directly proportional to stress. Different combinations of springs and dashpots in series or parallel, or both, have linear response to stress, as described below.[30–33]

If a stress σ_0 is applied at $t = 0$, the strain as a function of time will be

$$e(t) = \sigma_0 F(t)$$

If a higher stress $\sigma_0 + \sigma_1$ is applied at $t = 0$, the stress will be proportionally larger at all instants of time,

$$e(t) = (\sigma_0 + \sigma_1)F(t)$$

However, if σ_0 is applied at $t = 0$ and an additional stress σ_1 is added at $t = t_1$, the total strain will be

$$e(t) = \sigma_0 F(t) + \sigma_1 F(t - t_1)$$

where F is known as the creep memory function. *This behavior is known as viscoelasticity.* If the stress is increased by $(d\sigma/d\tau)\,d\tau$ at time τ, the increment of the present strain corresponding to this is

$$de(t) = F(t-\tau)\,\frac{d\sigma(\tau)}{d\tau}\,d\tau$$

For a continuous strain variation beginning at zero time,

$$e(t) = \int_0^t F(t-\tau)\,\frac{d\sigma(\tau)}{d\tau}\,d\tau \tag{3.5.1}$$

Similarly the stress at present may be considered to be related to past strain in the same way. Suppose at some time τ the strain is increased by $[de(\tau)/d\tau]\,d\tau$. Then the increment of present stress corresponding to this is

$$d\sigma(\tau) = f(t-\tau)\,\frac{de(\tau)}{d\tau}\,d\tau$$

For a continuous stress variation beginning at zero time,

$$\sigma(t) = \int_0^t f(t-\tau)\,\frac{de(\tau)}{d\tau}\,d\tau \tag{3.5.2}$$

where $f(t)$ is known as the relaxation memory function. Equation (3.5.1) implies that the effects on the present strain of increments of stress at past instants of time are additive. This demonstrates that the principle of superposition[31,32] is valid for viscoelastic materials.

The relation between creep-strain rate and stress of metals is generally nonlinear, as shown by Equations (3.3.11) through (3.3.15). However, for most metals there is a range of stress and temperature within which this relation is approximately linear, and therefore in these ranges, the creep behavior may be approximately represented by these linear mechanical models. *For polymers,*[30] *creep-strain rate vs. stress is much more linear, and hence polymers are much better represented by these models.*

Different arrangements of springs and dashpots have been proposed; some of the common ones are shown below.

Maxwell Solid. Figure 3.13 shows a spring and a dashpot in series. This

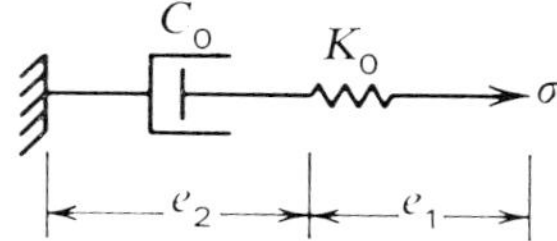

Figure 3.13. Maxwell solid.

model is known as the Maxwell solid, and its relations are easily seen to be

$$e = e_1 + e_2$$

$$\sigma = K_0 e_1 = C_0 \dot{e}_2$$

where the dot denotes differentiation with respect to time t. From the above relations, we obtain

$$\dot{\sigma} = K_0 \dot{e}_1$$

$$\dot{e} = \dot{e}_1 + \dot{e}_2$$

$$\dot{e} = \frac{\dot{\sigma}}{K_0} + \frac{\sigma}{C_0}$$

$$\dot{\sigma} = -\frac{K_0}{C_0}\sigma + K_0 \dot{e} \tag{3.5.3}$$

If a constant strain e_0 is maintained, $\dot{e} = 0$, and Equation (3.5.3) becomes

$$\dot{\sigma} = -\frac{K_0}{C_0}\sigma$$

which yields, upon integration,

$$\ln \sigma = -\frac{K_0}{C_0} t + C$$

where C is a constant of integration. Then

$$\sigma = \exp\left(\frac{K_0}{C_0} t + C\right) = K_0 e_0 \exp\left(-\frac{K_0}{C_0} t\right)$$

Under constant strain e_0, the initial stress $K_0 e_0$ is gradually relaxed. This is known as a relaxation test. The rate of relaxation is governed by the ratio C_0/K_0, which has the dimension of time and is known as the *relaxation time*. Taking the Laplace transform[34–36] of Equation (3.5.3) with respect to time, i.e., multiplying the above equations by $\exp(-st)\,dt$ and integrating with respect to t from zero to infinity, where s is the transform variable, we obtain

$$s\bar{\sigma} - \sigma(0) + \frac{K_0}{C_0}\bar{\sigma} = K_0[s\bar{e} - e(0)] \tag{3.5.4}$$

where the bar above a variable denotes the Laplace transform of it and $\sigma(0)$ and $e(0)$ denote the initial stress and strain, respectively. With $\sigma(0) = K_0 e(0)$, we have

$$\bar{e} = \frac{1}{K_0}\left(1 + \frac{K_0}{C_0 s}\right)\bar{\sigma} \tag{3.5.5}$$

The Laplace transform of $F(t - t_0)u(t - t_0)$, where $u(t - t_0)$ is the unit step function defined by

$$u(t - t_0) = 0 \quad \text{for} \quad t < t_0$$
$$u(t - t_0) = 1 \quad \text{for} \quad t > t_0$$

is given by[34]

$$L[F(t - t_0)u(t - t_0)] = \int_{t_0}^{\infty} \exp(-st)F(t - t_0)\,dt$$

Let $t - t_0 = t'$ and $t = t' + t_0$; then

$$\int_{t_0}^{\infty} \exp(-st)F(t - t_0)\,dt = \int_0^{\infty} \exp[-s(t_0 + t')]F(t')\,dt' = \exp(-st_0)\bar{F}(s)$$

$$L[F(t - t_0)u(t - t_0)] = \exp(-st_0)\bar{F}(s) \tag{3.5.6}$$

For the case corresponding to constant stress for $0 < t < t_0$, and $\sigma = 0$ for $t \geq t_0$, we have

$$\sigma(t) = \sigma_0 \quad \bar{\sigma} = \frac{\sigma_0}{s} \qquad 0 \leq t \leq t_0$$
$$= 0 \qquad\qquad t > t_0$$

Using the unit step function, we can write $\sigma(t)$ as

$$\sigma(t) = \sigma_0 - \sigma_0\, u(t - t_0)$$

whose Laplace transform is given by

$$\bar{\sigma} = \frac{\sigma_0}{s} - \frac{\sigma_0}{s}\exp(-st_0)$$

Then Equation (3.5.5) becomes

$$\bar{e} = \frac{1}{K_0}\left(1 + \frac{K_0}{C_0 s}\right)\frac{\sigma_0}{s}[1 - \exp(-st_0)] \tag{3.5.7}$$

The inverse of Equation (3.5.7) is

$$e = \frac{\sigma_0}{K_0} + \frac{\sigma_0}{C_0}t - \left[\frac{\sigma_0}{K_0} + \frac{\sigma_0}{C_0}(t - t_0)\right]u(t - t_0) \tag{3.5.8}$$

The strain-time curve of a Maxwell solid, subject to constant stress σ_0 for $0 \leq t \leq t_0$, is shown in Figure 3.14 overleaf.

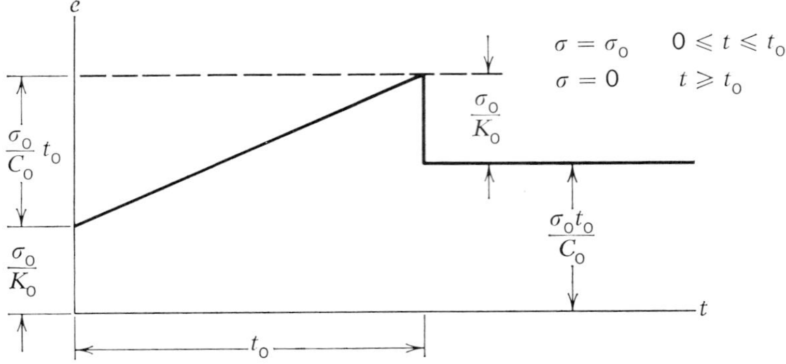

Figure 3.14. Strain-time curve of a Maxwell solid.

Kelvin Solid. The model which consists of a spring in parallel with a dashpot, as shown in Figure 3.15, is known as the Kelvin solid. The governing equations are given by

$$\sigma = \sigma_1 + \sigma_2$$

$$\sigma_1 = K_1 e$$

$$\sigma_2 = C_1 \dot{e}$$

Thus

$$\sigma = K_1 e + C_1 \dot{e}$$

Rewriting the last expression as

$$\dot{e} + \frac{K_1}{C_1} e = \frac{\sigma}{C_1}$$

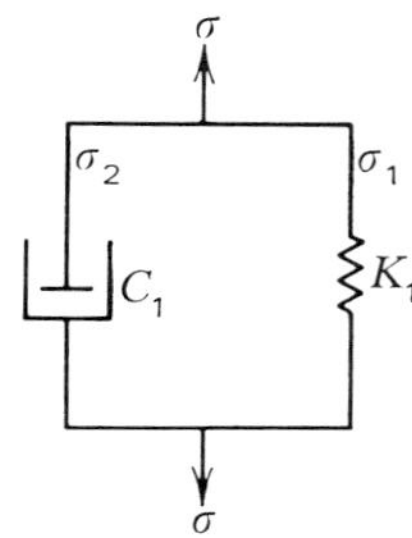

Figure 3.15. Kelvin solid.

and taking its Laplace transform, we obtain

$$s\bar{e} - e(0) + \frac{K_1}{C_1}\bar{e} = \frac{\bar{\sigma}}{C_1}$$
$$\bar{e} = \frac{\bar{\sigma}}{C_1}\left(\frac{1}{s + (K_1/C_1)}\right) + \left(\frac{e(0)}{s + (K_1/C_1)}\right) \tag{3.5.9}$$

For a creep test under constant stress $\sigma(t) = \sigma_0$, with initial strain $e(0) = 0$, we have

$$e = \frac{1}{K_1}\sigma_0\{1 - \exp\left[-(K_1/C_1)t\right]\}$$

For $\sigma(t) = \sigma_0 - \sigma_0 u(t - t_0)$ as before, then,

$$\bar{\sigma} = \frac{\sigma_0}{s}\left[1 - \exp(-st_0)\right]$$

With $e_0 = 0$, Equation (3.5.9) becomes

$$\bar{e} = \frac{\sigma_0}{C_1 s}\left(\frac{1}{s + (K_1/C_1)}\right)\left[1 - \exp(-st_0)\right]$$
$$= \frac{\sigma_0}{K_1}\left[\frac{1}{s} - \frac{1}{s + (K_1/C_1)}\right]\left[1 - \exp(-st_0)\right] \tag{3.5.10}$$

The inverse of this gives

$$e(t) = \frac{\sigma_0}{K_1}\left[1 - \exp(-K_1 t/C_1)\right] - \frac{\sigma_0}{K_1}\{1 - \exp\left[(K_1/C_1)(t - t_0)\right]\}u(t - t_0) \tag{3.5.11}$$

The ratio C_1/K_1 has the dimension of time and is called the *retardation time*. It is the time required to produce $(1 - 1/\exp)$ of the full elastic deformation under the applied constant stress σ_0 as shown by Equation (3.5.8). The strain-time curve of Equation (3.5.9) is shown in Figure 3.16.

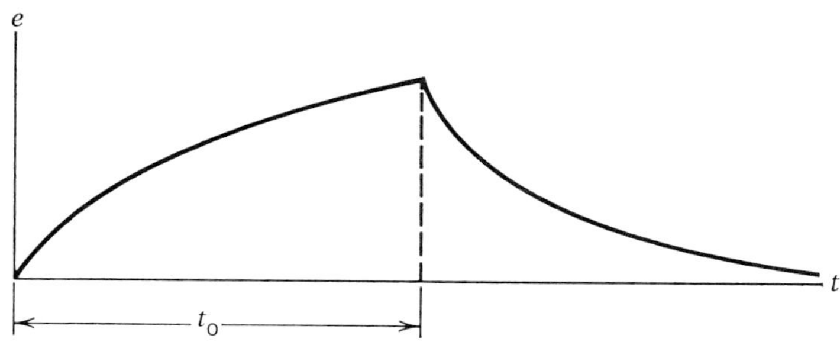

Figure 3.16. Strain-time curve of a Kelvin solid:

$\sigma = \sigma_0 \qquad 0 < t < t_0$

$\sigma = 0 \qquad t > t_0.$

It is seen that under constant stress the Kelvin solid gives a decreasing rate of strain corresponding to the transient creep, and the Maxwell solid gives a constant strain rate corresponding to steady creep. It is therefore expected that a combination of Maxwell and Kelvin models would yield both the transient and the steady-creep behavior exhibited by many materials.

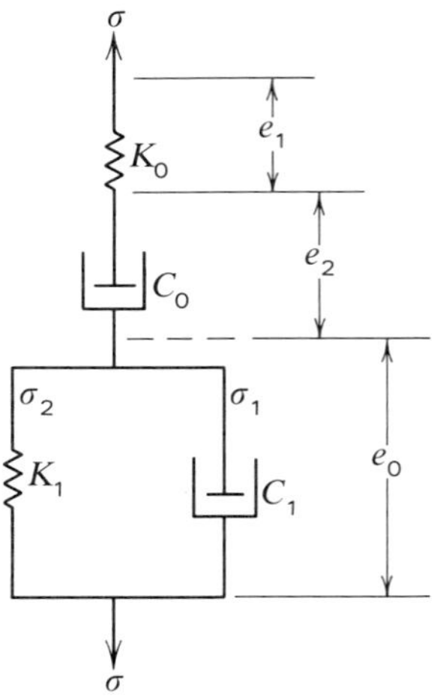

Figure 3.17. Maxwell-Kelvin solid.

Maxwell-Kelvin Solid. Figure 3.17 shows a Maxwell-Kelvin solid whose governing strain equation is given by

$$e = e_0 + e_1 + e_2$$

From Equation (3.5.4) with e replaced by $e_1 + e_2$, we have

$$s\bar{\sigma} - \sigma(0) + \frac{K_0}{C_0}\bar{\sigma} = K_0[s(\bar{e}_1 + \bar{e}_2) - e_1(0) - e_2(0)]$$

Since the $\bar{e}$ in Equation (3.5.9) corresponds to e_0 in Figure 3.17, we have

$$\bar{e}_0 = \frac{\bar{\sigma}}{C_1}\left(\frac{1}{s + (K_1/C_1)}\right) + \frac{e_0(0)}{(s + (K_1/C_1))}$$

With $e_0(0) = 0$, $e_1(0) = 0$, $e_2(0) = 0$, and $\sigma(0) = 0$, we have

$$\begin{aligned}\bar{e} &= \bar{e}_1 + \bar{e}_2 + \bar{e}_0 \\ &= \left[\frac{(s + (K_0/C_0))}{K_0\, s} + \frac{1}{C_1 s + K_1}\right] \cdot \bar{\sigma}\end{aligned}$$

or

$$\bar{e} = \left[\frac{(K_0 + C_0 s)(K_1 + C_1 s) + K_0 C_0 s}{C_0 K_0 s(K_1 + C_1 s)}\right] \cdot \bar{\sigma} \tag{3.5.12}$$

$$\bar{\sigma} = \frac{C_0 K_0 (K_1 + C_1 s)s}{(K_0 + C_0 s)(K_1 + C_1 s) + K_0 C_0 s} \cdot \bar{e} = \phi(s)\bar{e} \tag{3.5.13}$$

For the case $\sigma = \sigma_0$ for $0 < t < t_0$, and $\sigma = 0$ for $t \geq t_0$, that is, when $\sigma = \sigma_0 - \sigma_0 \cdot u(t - t_0)$, we have

$$\bar{\sigma} = \frac{\sigma_0}{s}[1 - \exp(-st_0)]$$

$$\bar{e} = \left[\frac{(K_0 + C_0 s)\sigma_0}{C_0 K_0 s^2} + \frac{\sigma_0}{s(K_1 + C_1 s)}\right][1 - \exp(-st_0)]$$

$$= \left[\frac{\sigma_0}{K_0 s} + \frac{\sigma_0}{C_0 s^2} + \frac{\sigma_0}{K_1 s} - \frac{\sigma_0}{K_1(s + K_1/C_1)}\right][1 - \exp(-st_0)] \tag{3.5.14}$$

The inverse of Equation (3.5.14) is

$$e = \frac{\sigma_0}{K_0} + \frac{\sigma_0}{C_0} t + \frac{\sigma_0}{K_1} - \frac{\sigma_0}{K_1} \exp[-(K_1/C_1)t]$$

$$- \left[\frac{\sigma_0}{K_0} + \frac{\sigma_0}{C_0}(t - t_0) + \frac{\sigma_0}{K_1} - \frac{\sigma_0}{K_1} \exp[-(K_1/C_1)(t - t_0)]\right] u(t - t_0) \tag{3.5.15}$$

At $t = t_0$, we have

$$e(t_0^-) = \frac{\sigma_0}{K_0} + \frac{\sigma_0}{C_0} t_0 + \frac{\sigma_0}{K_1}\{1 - \exp[-(K_1/C_1)t_0]\} \tag{3.5.16}$$

and at $t = t_0^+$, $\sigma = 0$, we have

$$e(t_0^+) = \frac{\sigma_0}{C_0} t_0 + \frac{\sigma_0}{K_1}\{1 - \exp[-(K_1/C_1)t_0]\} = e_2 + e_0 \tag{3.5.17}$$

From Equations (3.5.16) and (3.5.17), the change in strain at $t = t_0$ is

$$\Delta e\Big|_{t = t_0} = e(t_0^+) - e(t_0^-)$$

$$= -\frac{\sigma_0}{K_0} = -e_1$$

The slope of the strain vs. time curve at $t = 0$ is given by

$$\dot{e}(0) = \frac{\sigma_0}{C_0} + \frac{\sigma_0}{C_1} = \sigma_0\left(\frac{1}{C_0} + \frac{1}{C_1}\right) = \dot{e}_0(0) + \dot{e}_2(0) \qquad 0 < t < t_0 \quad (3.5.18)$$

For the case $\sigma = 0$ when $t > t_0$, we have

$$e = \frac{\sigma_0}{C_0} t_0 + \frac{\sigma_0}{K_1} \{\exp[-(K_1/C_1)(t - t_0)] - \exp[-(K_1/C_1)t]\} \quad (3.5.19)$$

At $t = \infty$, we have

$$e = \frac{\sigma_0}{C_0} t_0$$

The strain vs. time curve for $\sigma = \sigma_0$ in $0 < t < t_0$ and $\sigma = 0$ for $t > t_0$ is shown in Figure 3.18. From Equation (3.5.18), we have

$$\dot{e}(0) = \cot \alpha = \sigma_0\left(\frac{1}{C_1} + \frac{1}{C_0}\right)$$

The creep curves of most engineering metals are similar to Figure 3.18. From a particular material creep curve, the corresponding constants of the Maxwell-Kelvin solid may be determined. The K_0 is determined by the instantaneous strain and corresponds to E. The steady creep rate corresponds to $\tan\theta = \sigma_0/C_0$. From this, C_0 is determined. From the initial

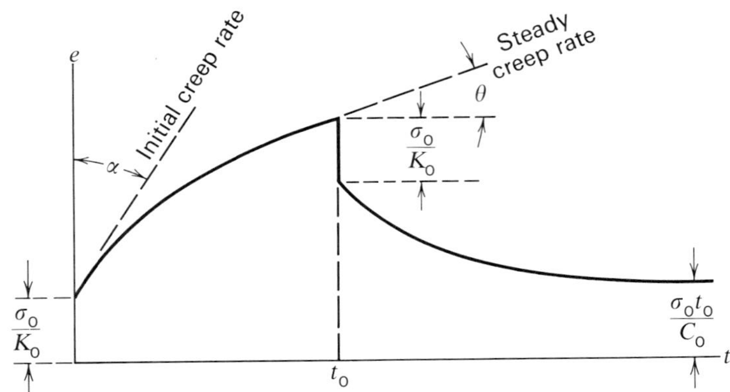

Figure 3.18. Strain-time curve of Maxwell-Kelvin solid:

$\sigma = \sigma_0 \qquad 0 < t < t_0$

$\sigma = 0 \qquad t > t_0.$

strain rate $\dot{e}(0)$, the value of C_1 is found. From e at $t = t_0$, we may determine K_1 from Equation (3.5.17). In this manner, the creep characteristic of the material may be approximated by a Maxwell-Kelvin model.

The transforms of stress and strain are related in simple algebraic forms as shown in Equations (3.5.12) and (3.5.13). With a known stress transform, the strain may be found by taking the inverse transform of Equation (3.5.12). In many cases the right-hand sides of (3.5.12) and (3.5.13) may be treated as products of two transforms, and the theorem of convolution may be applied. Rewriting Equation (3.5.13), we obtain

$$\bar{\sigma} = \frac{K_0 s\left(s + \dfrac{K_1}{C_1}\right)\bar{e}}{\left(s + \dfrac{K_0}{C_0}\right)\left(s + \dfrac{K_1}{C_1}\right) + \dfrac{K_0}{C_1}s} = \frac{K_0 s\left(s + \dfrac{K_1}{C_1}\right)\bar{e}}{s^2 + \left(\dfrac{K_0}{C_0} + \dfrac{K_0 + K_1}{C_1}\right)s + \dfrac{K_0 K_1}{C_0 C_1}} \tag{3.5.20}$$

The ratio K/C has the dimension of the reciprocal of time. Denoting $K_0/C_0 + (K_0 + K_1/C_1)$ by $1/T_0 + 1/T_2$; C_1/K_1 by T_1; K_0 by E; and $K_1 K_0/C_1 C_0$ by $1/T_0 T_2$, we see that Equation (3.5.20) becomes

$$\bar{\sigma} = \frac{E\left(s + \dfrac{1}{T_1}\right)s}{\left(s + \dfrac{1}{T_0}\right)\left(s + \dfrac{1}{T_2}\right)}\bar{e} \tag{3.5.21}$$

The uniaxial stress-strain-time relation of general linear viscoelastic materials may be represented by

$$P\sigma = Qe \tag{3.5.22}$$

where P and Q are linear operators of the form $\sum_0^M a_n D^n$ and D is the time derivative $\partial/\partial t$. The coefficients a_n are constants. We consider the cases in which the bodies are initially undisturbed. With zero values for initial stress, strain, and their derivatives, a function of the operator D^n becomes, after we take its Laplace transform, the same function of the Laplace transform parameter s^n. Taking Laplace transforms of the above equation, we obtain[37,39]

$$P(s)\bar{\sigma} = Q(s)\bar{e} \tag{3.5.23}$$

Equation (3.5.23) may be written as

$$\bar{\sigma} = \frac{Q(s)}{sP(s)}s\bar{e} = \bar{\psi}(s)s\bar{e}$$

where $\bar{\psi}(s) = Q(s)/sP(s)$. Taking the inverse transform of both sides and using the convolution theorem,[34] we obtain

$$\sigma = \int_0^t \psi(t-\tau)\,\frac{de(\tau)}{d\tau}\,d\tau$$

This equation is the same as Equation (3.5.2) with $\psi(t)$ corresponding to $f(t)$, hence linear viscoelastic behavior may be represented by either Equation (3.5.2) or Equation (3.5.23). It is seen that Equations (3.5.8) and (3.5.21) are special cases of Equation (3.5.23). The use of many Maxwell or Kelvin units or both to represent the creep characteristics of materials has been proposed.[39] A detail discussion of the responses of the different mechanical models when subjected to harmonic stresses is given by Kennedy.[11] For a more comprehensive discussion of viscoelastic solids, see References 37 through 41.

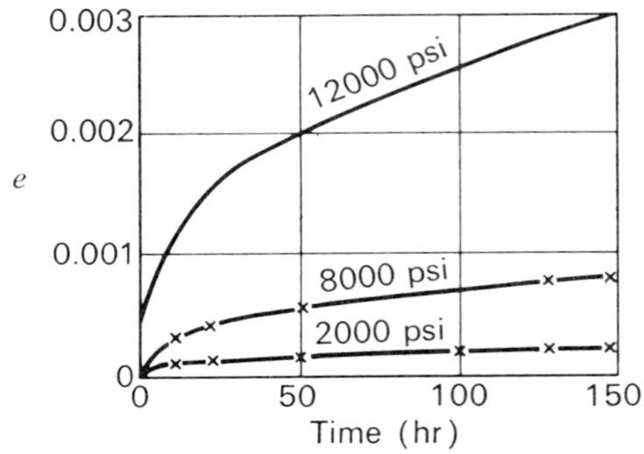

Figure 3.19. Creep curves of Nimonic 75 at 650°C.

The creep curves of Nimonic 75 at 650° C are shown in Figure 3.19. The creep-strain curve at a stress of 2000 psi is about one-quarter of that at a stress of 8000 psi, and hence the creep strain rate vs. stress is fairly linear up to a stress of 8000 psi. Within this stress, the creep characteristic may be represented by a linear mechanical model. Calculations have been made, from a Maxwell-Kelvin model, for creep strain vs. time curves at these two stresses. The calculated values are shown as crosses in Figure 3.19. It is seen that these two test curves agree with the calculated values. However, at a stress of 12,000 psi the increase in creep strain is no longer directly proportional to the stress, and the linear mechanical model is no longer adequate. It has been proposed[39,41] that a nonlinear dissipative element which gives a strain rate proportional to a hyperbolic function of stress in parallel with a Maxwell model be used to represent the nonlinearity of the creep behavior of actual metals. However, the complexity of the mathematical expressions obtained from this nonlinear model makes its practical application to structural problems rather limited.[39]

REFERENCES

1. Andrade, E. N. da C., "The Viscous Flow in Metals and Allied Phenomena," *Proc. Roy. Soc. A*, **84**, 1, 1910.
2. Andrade, E. N. da C., "The Flow in Metals under Large Constant Stresses," *Proc. Roy. Soc. A*, **90**, 329, 1914.
3. Mott, N. F., and F. R. N. Nabarro, "Report of a Conference on the Strength of Solids," The Physical Society, London, 1949.
4. Weaver, S. H., "The Creep Curve and Stability of Steels at Constant Stress and Temperature," *Trans. ASME*, **58**, 745, 1936.
5. McVetty, P. G., "Factors Affecting Choice of Working Stresses for High-Temperature Service," *Trans. ASME*, **55**, 99, 1933.
6. Soderberg, C. R., "The Interpretation of Creep Tests for Machine Design," *Trans. ASME*, **58**, 733, 1936.
7. Swift, S. H., and E. P. T. Tyndal, "Elasticity and Creep of Pb Single Crystals," *Phys. Rev.*, **61**, 359, 1942.
8. Andrade, E. N. da C., "On the Viscous Flow in Metals, and Allied Phenomena," *Proc. Roy. Soc. A.*, **84**, 1, 1910.
9. Graham, A., "The Phenomenological Method in Rheology," *Research*, **6**, 92, 1953.
   Graham, A., and K. F. A. Walles, "Relationships between Long- and Short-Time Creep and Tensile Properties of a Commercial Alloy," *J. Iron Steel Inst.*, **179**, 105, 1955.
10. Wyatt, O. H., "Transient Creep in Pure Metals," *Proc. Phys. Soc. London*, **B66**, 459, 1953.
11. Kennedy, A. J., *Processes of Creep and Fatigue in Metals*, Oliver and Boyd, Edinburgh, 1962.
12. Mott, N. F., "A Theory of Work-Hardening of Metals II: Flow Without Slip-Lines, Recovery and Creep," *Phil. Mag.*, **44**, 742, 1953.
13. Dorn, J. E., "Some Fundamental Experiments on High Temperature Creep," *J. Mech. Phys. Solids*, **3**, 85, 1955.
14. Norton, F. H., *The Creep of Steel at High Temperatures*," McGraw-Hill, New York, 1929.
15. Bailey, R. W., "Creep of Steel under Simple and Compound Stresses," *Engineering*, **129**, 265, 1930; "The Utilization of Creep Test Data in Engineering Design," *Proc. Inst. Mech. Engrs.*, **131**, 1935.
16. Dushman, S., L. W. Dunbar, and H. Huthsteiner, "Creep of Metals," *J. Appl. Phys.*, **15**, 108, 1944.

17. Nadai, A., *The Influence of Time Upon Creep, The Hyperbolic Sine Law*, S. Timoshenko Anniversary Volume, Macmillan, New York, 155, 1938.
18. Ludwik, P., *Elemente Der Technologischen Mechanik*, Springer, Berlin, 1909.
19. Machlin, E. S., and A. S. Noswick, "Stress Rupture of Heat Resistant Alloys as a Rate Process," *Trans. AIME*, **172**, 386–407, 1947.
20. McVetty, P. G., "Creep of Metals at Elevated Temperature—The Hyperbolic-Sine Relation Between Stress and Creep Rate," *Trans. ASME*, **65**, 761, 1943.
21. Hoff, N. J., "A Survey of the Theories of Creep Buckling," *Proc. Third U.S. Nat. Cong. Appl. Mech.*, 29–49, 1958.
22. McVetty, P. G., "Working Stresses for High Temperature Service," *Mech. Eng.*, **56**, 149, 1934.
23. Davis, E. A., "Creep and Relaxation of Oxygen Free Copper," *Trans. ASME*, **65**, A101, 1943.
24. Pao, Y. H., and J. Marin, "An Analytical Theory of Creep Deformation of Materials," *J. Appl. Mech.*, **20**, 245, 1953.
25. Odqvist, F. K. G., "Influence of Primary Creep on Stresses in Structural Parts," *Trans. Royal Inst. of Tech.*, No. 66, Stockholm, Sweden, 1953.
26. Hollomon, J. H., "The Mechanical Equation of State," *Metals Technology*, T. P. 2034, 1946.
27. Johnson, A. E., "Creep and Relaxation of Metals at High Temperatures," *Engineering*, **168**, 237, 1949.
28. Johnson, A. E., and J. Henderson, "Complex-Stress Creep Relaxation, and Fracture of Metallic Alloys," Dept. of Scientific and Industrial Research, National Engineering Laboratory, Edinburgh, p. 41, 1962.
29. Shanley, F. R., *Weight-Strength Analysis of Aircraft Structures*, McGraw-Hill, New York, 1952.
30. Alfrey, T., *Mechanical Behavior of High Polymers*, Interscience, New York, 1948.
31. Ferry, J. D., *Viscoelastic Properties of Polymers*, Wiley, New York, pp. 15–19, 1961.
32. Finnie, I., and W. R. Heller, *Creep of Engineering Materials*, McGraw-Hill, New York, pp. 33–38, 1959.
33. Bland, D. R., *The Theory of Linear Viscoelasticity*, Pergamon, London, pp. 1–12, 1960.
34. Churchill, R. V., *Modern Operational Mathematics in Engineering*, McGraw-Hill, New York, pp. 1–37, 1944.
35. Carslaw, H. S., and J. C. Jaeger, *Operational Methods in Applied Mathematics*, Dover, New York, p. 80, 1947.
36. Thomson, W. T., *Laplace Transformation*, Prentice-Hall, Englewood Cliffs, N.J., p. 3, 1962.
37. Brull, M. A., *A Structural Theory Incorporating the Effect of Time Dependent Elasticity*, Ph.D. Dissertation, University of Michigan, 1952; *Proc. First Midwestern Conference on Solid Mechanics*, 1953.
38. Lee, E. H., "Stress Analysis in Visco-Elastic Bodies," *Quart. Appl. Math.*, **13**, 183–190, 1955.
39. Freudenthal, A. M., *The Inelastic Behavior of Engineering Materials and Structures*, Wiley, New York, pp. 220–247, 1950.
40. Hunter, S. C., "Viscoelastic Waves," in I. N. Snedden and R. Hill (eds.), *Progress in Solid Mechanics*, Vol. 1, North Holland, Amsterdam, pp. 4–11, 1964.
41. Williams, M. "Structural Analysis of Viscoelastic Materials," *AIAA J.*, **2**, pp. 785–808, 1964.

chapter

4

Inelastic Multiaxial Stress-Strain Relations

Introduction. In the analyses of beams, columns, and frames of structures, the stress state is generally considered uniaxial. Therefore the inelastic stress-strain relationships given in Chapter 3 are generally sufficient for such simple structures. For more complex structures, such as plates and shells, that are subject to general loading, biaxial or multiaxial stress states result, and hence multiaxial inelastic stress-strain relations are necessary. Since there are six stress and six strain components, the inelastic stress-strain relations for multiaxial stress states are much more complicated than the corresponding relations for the uniaxial stress-state. This chapter gives the multiaxial plastic strain-stress relations for metals, and generalizes the uniaxial creep strain-stress-time relations of metals at elevated temperatures to the multiaxial stress state at elevated temperatures.

The mechanical characteristics of polycrystals are determined from those of the individual crystals. A number of assumptions have been made by the various investigators in formulating the stress-strain relationships of polycrystalline metals. The limitations and justifications of these assumptions are essential in the analysis of some inelastic structures. To provide a better understanding of these justifications and limitations, the inelastic stress-strain relationships of single crystals are first presented. From the single-crystal presentation, certain conclusions are obtained for the polycrystal stress-strain relations. From these conclusions, the justifications and limitations of the commonly used inelastic stress-strain relations are shown and the discrepancies between the theoretical and experimental results of inelastic stress-strain relations are discussed. Finally, the uniaxial viscoelastic relations of Chapter 3 are generalized for multiaxial stress states.

4.1 Plastic Stress-Strain Relations of Single Crystals

Since the early 1920's single crystals of sizes adequate for strain measurement have been successfully grown. It has been found that single crystals, under high stress, slide along certain crystallographic directions on particular crystal planes.[1–5] The sliding takes place in the form of movement of lamellae of the crystal over one another. The movement is concentrated on a set of parallel planes, like the displacement of cards in a pack. Denoting the normal to the set of parallel sliding planes by α and the direction of slip by β, we see that the slip produces plastic strain $e''_{\alpha\beta}$, as shown in Figure 4.1. The planes on which slip occurs are usually those of densest

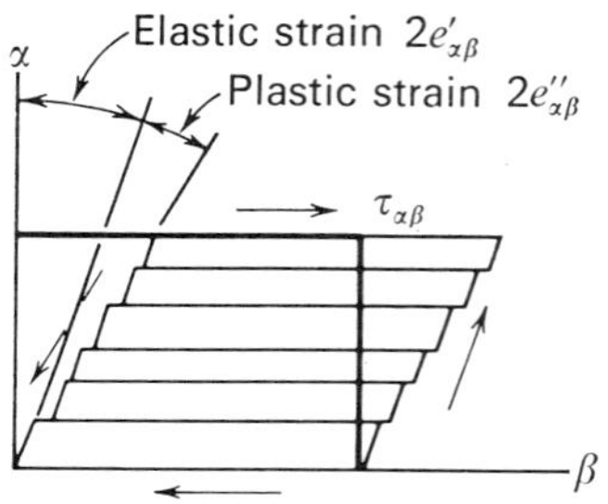

Figure 4.1. Plastic strain caused by slip.

atomic packing, i.e., of highest atomic density, and the directions of slip correspond to the directions of densest atomic packing in the slip plane. *A slip direction and its corresponding slip plane together form a slip system.* Some crystals, such as aluminum, have four sets of parallel planes of densest packing, and on each set of these planes there are three directions of highest atomic density. These crystals have four sets of slip planes; on each plane there are three slip directions, and hence such crystals have twelve slip systems.[5] Other metals have other arrangements of atoms and other slip systems.

Consider[6] a circular cylinder of a single crystal loaded in axial tension P. The normals to the sliding plane and the sliding direction make angles ϕ and λ, respectively, with the axis of the cylinder as shown in Figure 4.2. The force P has a component in the direction of slip $P \cos \lambda$ which is distributed over an area $A' = A/\cos \phi$, where A is the cross-sectional area of the cylinder. The shear stress on the sliding plane along the slip direction is called the *resolved shear stress* and in this case is equal to

$$\frac{P}{A'} \cos \lambda = \frac{P}{A} \cos \phi \cos \lambda$$

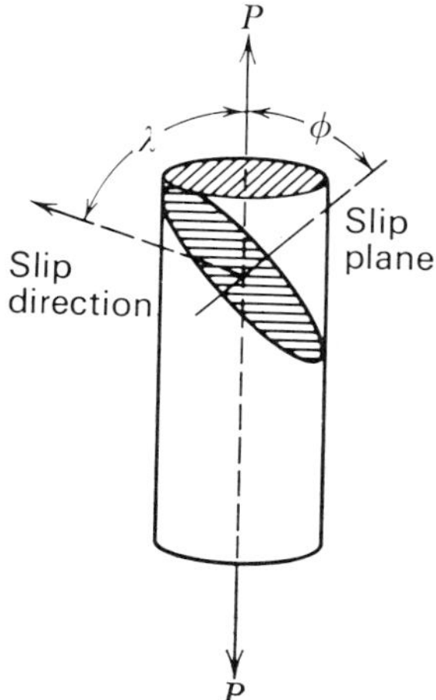

Figure 4.2. Coordinates for calculating resolved shear stress.

Single-crystal tests have shown that *slip takes place when the resolved shear stress reaches a critical value (known as the critical shear stress), and that the slip is not affected by the magnitude of the normal stress on the slip plane*. This is known as Schmid's law of critical shear stress.[5,6] Since slip is the main mechanism of plastic deformation of many metals, *there is no appreciable change of volume associated with plastic deformation of metals*.

Initial Yield Surface of Single Crystals. Before the critical shear stress is reached, the deformation in the crystal is purely elastic. After sliding, generally the critical shear stress in the sliding system increases with the amount of slip. This is known as strain-hardening. The property by which the critical shear stress in the nonslid systems increases is known as latent hardening. Strain-hardening and latent hardening characteristics vary from one metal to another.

Consider a close-packed hexagonal crystal, such as a magnesium crystal. The basic element which makes up its lattice structure is shown in Figure 4.3 overleaf, where the heavy dots denote the atoms of the basic element. This crystal has one main set of slip planes (parallel to plane $ABCDEF$), and on these planes there are three slip directions spaced 60° apart. These give the three *slip systems* of the crystal. Referring to Figure 4.3, we see that x_1, x_2, and x_3 are a set of rectangular coordinates, and that β_1, β_2, and β_3 denote the three slip directions. The normal to the set of parallel sliding planes, denoted by α, is along the x_3 axis and has the direction cosine (0, 0, 1). These planes are here referred to as α planes. The slip directions have direction cosines

$$\beta_1(1, 0, 0) \qquad \beta_2\left(\frac{1}{2}, \frac{\sqrt{3}}{2}, 0\right) \qquad \beta_3\left(-\frac{1}{2}, \frac{\sqrt{3}}{2}, 0\right)$$

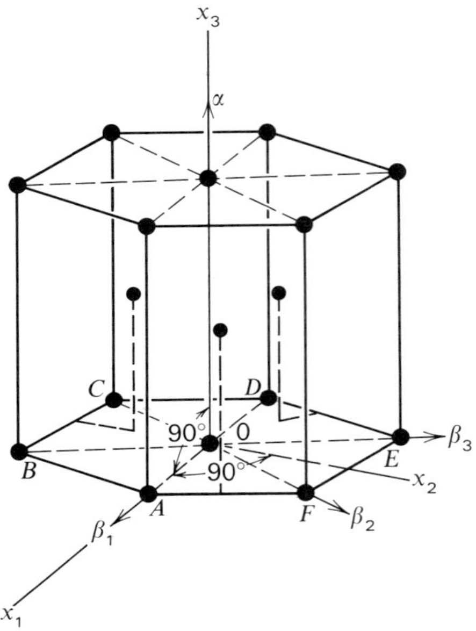

Figure 4.3. Basic element of close-packed hexagonal crystal (circles denote atoms).

For the crystal to slide in the β_1 direction on the α plane, the resolved shear stress in this slip system must equal the critical shear stress,

$$\tau_{\alpha\beta_1} = l_{\alpha i}\, l_{\beta_1 j}\, \tau_{ij} = \pm c_{\alpha\beta_1}. \tag{4.1.1}$$

where $l_{\alpha i}$ is the cosine between the α and i axes, $l_{\beta_1 j}$ is the cosine between the β_1 and j axes, $c_{\alpha\beta_1}$ is the critical shear stress of the $\alpha\beta_1$ slip system, and the repetition of subscripts i and j denotes summation from 1 to 3. Writing Equation (4.1.1) in its expanded form, we see that

$$\begin{aligned}\tau_{\alpha\beta_1} = {} & l_{\alpha 1}(l_{\beta_1 1}\tau_{11} + l_{\beta_1 2}\tau_{12} + l_{\beta_1 3}\tau_{13}) \\ & + l_{\alpha 2}(l_{\beta_1 1}\tau_{21} + l_{\beta_1 2}\tau_{22} + l_{\beta_1 3}\tau_{23}) \\ & + l_{\alpha 3}(l_{\beta_1 1}\tau_{31} + l_{\beta_1 2}\tau_{32} + l_{\beta_1 3}\tau_{33}) = \pm c_{\alpha\beta_1}\end{aligned}$$

where the $\pm$ sign indicates that sliding can take place in either the positive or negative direction. Referring to Figure 4.3, we see that $l_{\alpha 1} = l_{\alpha 2} = l_{\beta_1 2} = l_{\beta_1 3} = 0$ and $l_{\alpha 3} = l_{\beta_1 1} = 1$, and we obtain

$$\tau_{\alpha\beta_1} = \tau_{31} = \pm c_{\alpha\beta_1} \tag{4.1.2}$$

Similarly, for sliding to start along the β_2 direction, we must have

$$\tau_{\alpha\beta_2} = \frac{1}{2}\tau_{31} + \frac{\sqrt{3}}{2}\tau_{32} = \pm c_{\alpha\beta_2} \tag{4.1.3}$$

and for sliding to start along the β_3 direction, we must have

$$\tau_{\alpha\beta_3} = -\frac{1}{2}\tau_{31} + \frac{\sqrt{3}}{2}\tau_{32} = \pm c_{\alpha\beta_3} \tag{4.1.4}$$

As seen from Equations (4.1.2) through (4.1.4), the initiation of slip depends only on τ_{31} and τ_{32} and is independent of all other stress components. Using τ_{31} and τ_{32} as a pair of horizontal and vertical rectangular axes, as shown in Figure 4.4, we see that Equations (4.1.2) through

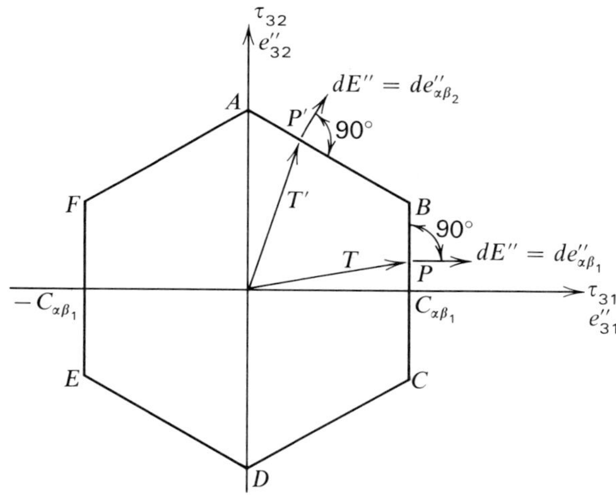

Figure 4.4. Initial yield locus of a closed-packed hexagonal crystal.

(4.1.4) give three pairs of straight lines. Within the bounded area of these lines there is no slip and hence no plastic strain. *A yield surface is defined as the surface in stress space, with stress components as coordinates, within which the stress vector may change without any plastic-strain increment, and exterior to which an incremental plastic strain is produced.* The intersection of a yield surface with a plane in the stress space gives a curve on the plane which is called a yield locus. Hence the boundary of this hexagonal area is the initial yield locus of this crystal. With $c_{\alpha\beta_1} = c_{\alpha\beta_2} = c_{\alpha\beta_3}$, the initial yield locus becomes a regular hexagon.

Coincidence of Yield Surface and Plastic Potential of Single Crystals. Now let us superpose the plastic strain coordinates e''_{31} and e''_{32} on the stress coordinates τ_{31} and τ_{32}, respectively, as shown in Figure 4.4. When the stress vector T reaches the boundary BC at P, we have $\tau_{\alpha\beta_1} = c_{\alpha\beta_1}$, and the $\alpha\beta_1$ slip system may begin to slide, giving incremental plastic strain $de''_{\alpha\beta_1} = de''_{31}$. It is seen that the incremental plastic strain de''_{31} is normal to BC of the yield locus. If the stress vector T' reaches the boundary of the yield locus at P', then $\tau_{\alpha\beta_2} = c_{\alpha\beta_2}$, and the $\alpha\beta_2$ slip system may begin to slide, giving the plastic-strain increment $de''_{\alpha\beta_2}$. The plastic-strain increment is

$$de''_{ij} = (l_{i\alpha} l_{j\beta_2} + l_{i\beta_2} l_{j\alpha})\, de''_{\alpha\beta_2} \qquad i, j = 1, 2, 3$$

where there is no summation on α. With $l_{\alpha 1} = l_{\alpha 2} = l_{\beta_2 3} = 0$, we obtain

$$de''_{11} = de''_{22} = de''_{33} = de''_{12} = 0$$

$$de''_{32} = \frac{\sqrt{3}}{2}\, de''_{\alpha\beta_2}$$

$$de''_{31} = \frac{1}{2}\, de''_{\alpha\beta_2}$$

The resultant incremental plastic-strain vector $dE''(de''_{31}, de''_{32})$ is normal to AB of the yield locus. Similarly, it can be shown that $de''_{\alpha\beta_3}$ is normal to FA, and hence that the incremental plastic-strain vector is always normal to the yield locus. It is well known that the direction of the gradient of a newtonian potential function coincides with that of the attractive force.[7] The yield locus is said to coincide with a *plastic potential* function, since its gradient gives the direction of the plastic-strain increment.

When slip takes place in slip system $\alpha\beta_1$ (see Figure 4.5), strain-hardening increases the critical shear stress in the active slip system $\alpha\beta_1$. This moves BC outward to $B'C'$, and FE to $F'E'$. Latent hardening increases the critical shear stress in the unslid systems $\alpha\beta_2$ and $\alpha\beta_3$. This moves AB, CD, DE, and FA outward to $A'B'$, $C'D'$, $D'E'$, and $F'A'$. If active hardening is equal to latent hardening, the yield polygon remains regular in shape but increases in size. If active hardening is less than or greater than latent hardening, the yield polygon changes in both size and shape as shown in Figures 4.5(b) and (c). Bauschinger effect (Section 3.1) has also been observed in single-crystal tests.[8] If this effect is taken into account, both BC and FE (Figure 4.5) will move toward the right.

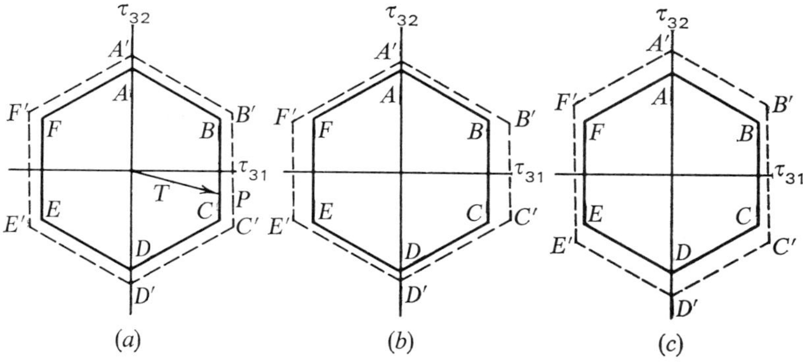

Figure 4.5. Change of yield locus with loading. $\alpha\beta_1$ is the active slip system. (a) active = latent hardening; (b) active > latent hardening; (c) active < latent hardening.

When the stress vector T is at a point P on the yield locus (Figure 4.5), the application of an incremental stress vector ΔT with an outward normal component to BC, is called *loading* and that with an inward normal component to BC is called *unloading*. If the normal component of ΔT to BC is zero, the application of this ΔT is here called *neutral loading*. Incremental plastic strain can occur only for loading and neutral loading, but nor for unloading.

When the stress vector is at a corner point on the polygon, say at B (Figure 4.5), both slip systems $\alpha\beta_1$ and $\alpha\beta_2$ have reached the critical shear stress. Hence both systems may slide, and the direction of the resultant incremental plastic strain lies between the normals to AB and BC. Let a_{11} and a_{22} be the rates of strain-hardening in slip systems $\alpha\beta_1$ and $\alpha\beta_2$, respectively; let a_{12} be the latent hardening in $\alpha\beta_1$ due to slip in $\alpha\beta_2$, and let a_{21} be the latent hardening in $\alpha\beta_2$ due to slip in $\alpha\beta_1$. Then

$$\begin{aligned} d\tau_{\alpha\beta_1} &= a_{11} de''_{\alpha\beta_1} + a_{12}\, de''_{\alpha\beta_2} \\ d\tau_{\alpha\beta_2} &= a_{21} de''_{\alpha\beta} \ + a_{22}\, de''_{\alpha\beta_2} \end{aligned} \tag{4.1.5}$$

This gives

$$de''_{\alpha\beta_1} = \frac{\begin{vmatrix} d\tau_{\alpha\beta_1} & a_{12} \\ d\tau_{\alpha\beta_2} & a_{22} \end{vmatrix}}{\begin{vmatrix} a_{11} & a_{12} \\ a_{21} & a_{22} \end{vmatrix}} \qquad de''_{\alpha\beta_2} = \frac{\begin{vmatrix} a_{11} & d\tau_{\alpha\beta_1} \\ a_{21} & d\tau_{\alpha\beta_2} \end{vmatrix}}{\begin{vmatrix} a_{11} & a_{12} \\ a_{21} & a_{22} \end{vmatrix}} \tag{4.1.6}$$

The incremental plastic-strain components are uniquely determined by the incremental stress $d\tau_{\alpha\beta_1}$ and $d\tau_{\alpha\beta_2}$ if the denominator does not vanish. If the rates of active and latent hardening are equal, that is, if $a_{11} = a_{21}$ and $a_{22} = a_{12}$, the denominators of Equations (4.1.6) vanish, and therefore $de''_{\alpha\beta_1}$ and $de''_{\alpha\beta_2}$ are not uniquely determined.

When the stress vector T is on the regular line (not corner) of the yield polygon, the incremental plastic strain vector dE'' is normal to the yield polygon, *so the direction of the incremental plastic strain vector depends on the stress vector T but is independent of the incremental stresses. When the stress vector T is at a corner point of the yield locus, the direction of the incremental plastic strain vector dE'' depends on the direction of incremental stress $d\tau_{ij}$* (except for the case with equal rates of latent and active hardening in both slip systems).

These results may be extended to crystals with more than one slip plane. When a slip system in such a crystal with a set of parallel sliding planes of normal α and a sliding direction β slides, the plastic strain produced by this slip is $e''_{\alpha\beta}$. With $l_{\alpha i}$ and $l_{\beta j}$ denoting the cosines of the angles between the α and i axes and between the β and j axes, respectively, and noting that $de''_{\alpha\beta} = de''_{\beta\alpha}$, we have

$$de''_{ij} = (l_{i(\alpha)} l_{j(\beta)} + l_{i(\beta)} l_{j(\alpha)})\, de''_{(\alpha)(\beta)} \tag{4.1.7}$$

where the parentheses on the subscript denote no summation. When a given stress is referred to the x_i axes, the resolved shear stress in the $\alpha\beta$ slip system is

$$\tau_{\alpha\beta} = l_{\alpha i}\, l_{\beta j}\, \tau_{ij}$$

During sliding, $\tau_{\alpha\beta}$ must be equal to its critical shear stress, $\pm c_{\alpha\beta}$. Hence the condition for slip to occur in this slip system is given by

$$f(\tau_{ij}) = l_{\alpha i}\, l_{\beta j}\, \tau_{ij} - (\pm c_{\alpha\beta}) = 0$$

or equivalently,

$$f(\tau_{ij}) = \tfrac{1}{2}(l_{\alpha i}\, l_{\beta j} + l_{\alpha j}\, l_{\beta i})\tau_{ij} - (\pm c_{\alpha\beta}) = 0 \tag{4.1.8}$$

Consider a scalar function $F(x_1, x_2, x_3)$ in a three-dimensional space; $F(x_1, x_2, x_3) = 0$ defines a surface in the space. On this surface

$$dF = \frac{\partial F}{\partial x_1} dx_1 + \frac{\partial F}{\partial x_2} dx_2 + \frac{\partial F}{\partial x_3} dx_3 = 0 \tag{4.1.9}$$

If unit vectors along the x_1, x_2, and x_3 axes are denoted by **i**, **j**, and **k**, respectively, the gradient of the function may be written as

$$\nabla F = \mathbf{i}\,\frac{\partial F}{\partial x_1} + \mathbf{j}\,\frac{\partial F}{\partial x_2} + \mathbf{k}\,\frac{\partial F}{\partial x_3} \tag{4.1.10}$$

The position vector of a point P on the surface is denoted by $\mathbf{r} = \mathbf{i}x_1 + \mathbf{j}x_2 + \mathbf{k}x_3$. Then

$$d\mathbf{r} = \mathbf{i}\,dx_1 + \mathbf{j}\,dx_2 + \mathbf{k}\,dx_3 \tag{4.1.11}$$

Equation (4.1.9) may be written as

$$dF = \nabla F \cdot d\mathbf{r} = 0 \tag{4.1.12}$$

Hence ∇F is normal to $d\mathbf{r}$. Since $d\mathbf{r}$ is any incremental position vector on the surface at P, we see that ∇F is normal to the tangent plane of the surface $F = 0$ at point P.

The above discussion may be generalized for an n-dimensional space as follows. For a scalar function of n variables $\phi(x_1, x_2, \ldots, x_n)$ in n-dimensional space, $\phi(x_1, x_2, \ldots, x_n) = 0$ may be considered as a surface in that n-dimensional space. On this surface, at a point P, we have

$$d\phi = \frac{\partial \phi}{\partial x_1}\,dx_1 + \frac{\partial \phi}{\partial x_2}\,dx_2 + \cdots + \frac{\partial \phi}{\partial x_n}\,dx_n = 0 \tag{4.1.13}$$

This may be written as

$$\nabla\phi \cdot d\mathbf{r} = 0 \tag{4.1.14}$$

where $d\mathbf{r}$ is the differential position vector in the n-dimensional space, and therefore the gradient of ϕ—that is, $\nabla\phi$—is perpendicular to the tangent plane of the surface $\phi(x_1, x_2, \ldots, x_n) = 0$.

In Equation (4.1.8) there are nine stress components τ_{ij} (we are formally treating τ_{ij} and τ_{ji} as distinct components). Considering these stress components as coordinates, we have a nine-dimensional space; that is, $n = 9$ in Equation (4.1.13). Equation (4.1.8) defines a surface in this space. Since this equation is linear in the τ_{ij}, the surface consists of pairs of nine-dimensional parallel planes. The pairs of planes give the conditions of sliding and are called *yield planes*. These yield planes are planes in a nine-dimensional space with stresses τ_{ij} as coordinates. The reader should not confuse these yield planes with slip planes in a crystal. Each component of the normal to the yield plane is proportional to the corresponding component of the gradient of the function $f(\tau_{ij})$. The component of the gradient along the τ_{ij} axis is

$$\frac{\partial f}{\partial \tau_{ij}} = \frac{(l_{\alpha i} l_{\beta j} + l_{\alpha j} l_{\beta i})}{2} \tag{4.1.15}$$

Superposing the plastic strain coordinates e''_{ij} on the corresponding stress coordinates τ_{ij} in the nine-dimensional stress space results in a corresponding nine-dimensional strain space. *It is seen from Equations* (4.1.7) *and* (4.1.15) *that the vector dE'' with de''_{ij} as components is normal to the yield* *planes given by Equation* (4.1.8). If there are m slip systems in a crystal, there are m pairs of yield planes in the stress space. These yield planes generally form a polyhedron called a yield polyhedron. The bounding surface of this polyhedron is called the yield surface of the crystal. When the stress vector is on a plane of the yield polyhedron, the incremental plastic-strain vector is normal to this plane. When the stress vector is on an edge of the polyhedron, the incremental plastic-strain vector lies between the normals to the two adjacent yield planes. When the stress vector is at a vertex of the polyhedron, the incremental plastic-strain vector lies within the region bounded by the normals[9] to the yield planes intersecting at the vertex.

This normality has been shown by Bishop and Hill[9] in a more elegant manner as follows. Let c_k be the critical shear stress in the kth slip system, T a stress vector in the stress space with τ_{ij} as coordinates which will produce a given incremental plastic-strain dE'' with components de''_{ij}, and T^* any other stress vector with τ^*_{ij} as coordinates within the yield surface of the crystal. Since T produces the incremental plastic strain dE'', its τ_{ij} components are on the yield surface. Let τ_k and $\tau_k{}^*$ be the resolved shear stresses of the kth active slip system corresponding to τ_{ij} and τ^*_{ij}, respectively. The τ_k in the active slip systems are equal to the critical shear stress values. Since τ^*_{ij} is within the yield surface, $\tau_k{}^*$ is less than the critical shear stress. Hence

$$(\tau_k - \tau_k{}^*)\, d\gamma_k{}'' \geq 0$$

where $d\gamma_k{}''$ is the incremental plastic strain due to slip in the kth slip system. The incremental work due to the plastic strain increment $d\gamma_k{}''$ is

$$dW - dW^* = (T - T^*) \cdot dE'' = (\tau_{ij} - \tau^*_{ij})\, de''_{ij} = \sum_{k=1}^{m} (\tau_k - \tau_k{}^*)\, d\gamma_k{}'' \geq 0$$

where m denotes the number of active slip systems. Then

$$(T - T^*) \cdot dE'' \geq 0 \tag{4.1.16}$$

This shows that among all stress states lying within the yield surface, the actual stress state giving dE'' is the one which must lie on the yield surface and gives maximum work[9] for the given dE''. Consider $(T - T^*)$ as a stress vector in the stress space with T on the yield surface and T^* within the yield surface, and consider dE'' as an incremental plastic-strain vector. $(T - T^*) \cdot dE''$ is the scalar product of these two vectors. If θ is the angle between $(T - T^*)$ and dE'', the scalar product becomes $|T - T^*| \times |dE''| \cos\theta$. The intersection of the plane containing these two vectors with the yield surface gives a yield locus on the plane as shown in Figure 4.6. Consider two stress vectors $T_1{}^*$ and $T_2{}^*$ in this plane which are very close to T, one on the right and one on the left of T. $(T - T_1{}^*)$ is represented by BA and $(T - T_2{}^*)$ by CA. If the incremental plastic strain vector is represented by dE'' as shown, the condition given by Equation (4.1.16) requires that

$$\theta_1 \leq 90^\circ \qquad \text{and} \qquad \theta_2 \leq 90^\circ \tag{4.1.17}$$

If this portion of the yield locus is a portion of a smooth curve, BAC approaches the tangent line as C and B approach A. In the limit as C and B approach A, we have

$$\theta_1 + \theta_2 = 180^\circ \tag{4.1.18}$$

In order to satisfy both Equations (4.1.17) and (4.1.18), we must have

$$\theta_1 = \theta_2 = 90^\circ$$

Hence the plastic-strain increment vector dE'' must be normal to the yield surface at a smooth point and must lie between adjacent normals at an edge or a corner.[10]

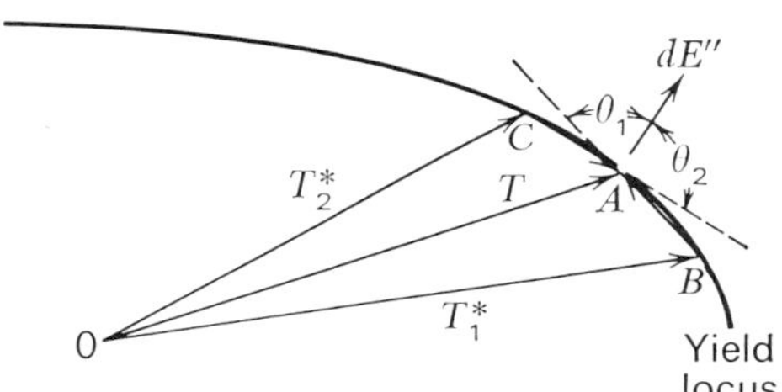

Figure 4.6. A plane in stress space.

When $\tau_{\alpha\beta} = c_{\alpha\beta}$, the stress vector is on the yield surface. When loading takes place, $d\tau_{\alpha\beta} > 0$, causing incremental plastic strain $de''_{\alpha\beta}$ to occur. When $d\tau_{\alpha\beta} < 0$, unloading takes place and the resulting incremental strain is purely elastic. For a crystal with no strain-hardening, $c_{\alpha\beta}$ remains constant and $de''_{\alpha\beta}$ occurs at $d\tau_{\alpha\beta} = 0$. With strain- and latent-hardening rates known, the stress-strain relation of the crystal is uniquely determined.[11]

This section shows that (1) *plastic deformation in single crystals causes no significant change in volume*, and (2) *the incremental plastic-strain vector is normal to the initial and subsequent loading surfaces*. These characteristics will be shown to remain valid for polycrystals. Single crystals are rarely used as engineering materials. As stated before, the presentation of the theory of plasticity for single crystals just given is mainly for the derivation of plastic behavior of polycrystals in the following section.

4.2 Initial Yield Surface of Polycrystalline Solids

Actual annealed metals consist of many randomly oriented crystals. Experiments on such polycrystalline metals show that its material properties are independent of any special planes or directions. The initial plastic deformation characteristic is isotropic, and hence the initial yield surface must be independent of the orientation of the reference axes. By choosing the axes of the principal stresses as the reference axes, the initial yield surface may be expressed in terms of the principal stresses and represented by a surface in a stress space with σ_1, σ_2, and σ_3 as coordinate axes.[12–15] From Equation (1.4.3), we see that stress invariants J_1, J_2, and J_3 depend only on the principal stresses σ_1, σ_2, and σ_3, and therefore the initial yield condition may also be expressed in terms of J_1, J_2, and J_3.

The grain boundaries in a polycrystalline metal have been estimated to be only a few interatomic distances thick;[8] hence in the calculation of stress-strain relations of a polycrystalline solid, the boundaries may be considered as surfaces of zero thickness, across which the crystal orientation changes from one to another.[16] The elastic constants of single crystals are generally anisotropic,[8] and this anisotropy varies from one metal to another. Neglecting this anisotropy in order to simplify the following discussion, we have an aggregate of *elastically* isotropic crystals. Hydrostatic pressure applied to this elastically isotropic aggregate causes uniform hydrostatic stress and zero shear stress in all the individual crystals, and hence no crystal slides. Tests on polycrystalline metals also reveal that hydrostatic pressure has no significant effect on plastic deformation.[17] *Hence hydrostatic pressure is considered to have no effect on plastic deformation of the polycrystals.* Let OC represent the line through the origin which makes equal angles with the principal stress axes σ_1, σ_2, and σ_3 as shown in Figure 4.7. Then all points on OC represent pure hydrostatic stress states, since, on OC, we have $\sigma_1 = \sigma_2 = \sigma_3$ and $S_1 = (2\sigma_1 - \sigma_2 - \sigma_3)/3 = 0$, and similarly $S_2 = S_3 = 0$. Any stress vector from O to any point P on the yield surface is composed of two components, one parallel to OC and the other perpendicular to OC. Since initial yielding is independent of OC, the yield surface must be a cylinder with its generator axis parallel to OC. Hence the initial yield condition may be expressed in terms of the deviatoric stress invariants $\bar{J}_2$ and $\bar{J}_3$ ($\bar{J}_1 = 0$; see Section 1.4); that is, $f(\bar{J}_2, \bar{J}_3) = 0$. Consider the intersection of this cylinder by a plane perpendicular to OC. This plane is called the *π plane*, and the intersection of the yield surface on this plane is called the yield locus.[12–15] The projections of the three axes σ_1, σ_2, and σ_3 on the π plane are spaced 120° apart and are $\sqrt{2/3}\,\sigma_1$, $\sqrt{2/3}\,\sigma_2$, and $\sqrt{2/3}\,\sigma_3$, respectively, as shown in Figure 4.8.

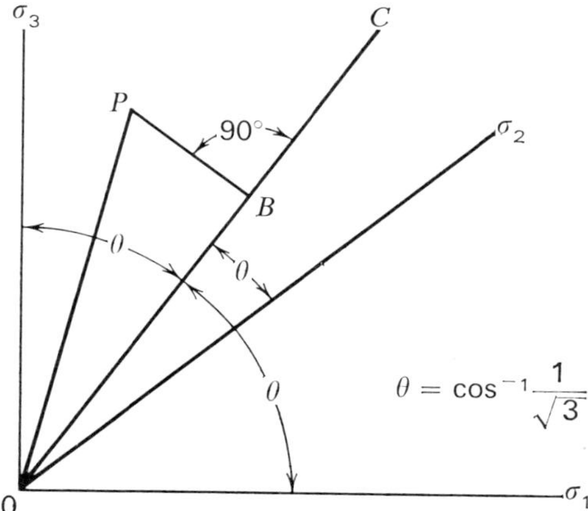

Figure 4.7

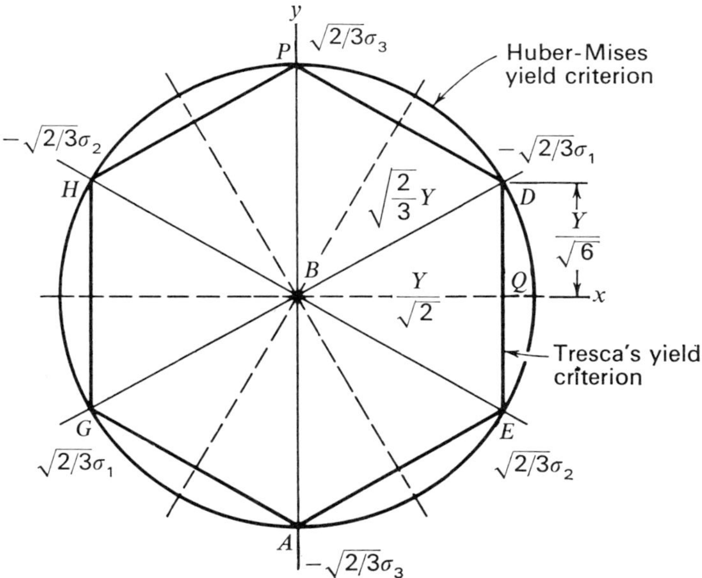

Figure 4.8. π plane.

The initial yield locus on the π plane must be symmetrical with respect to the σ_1, σ_2, and σ_3 axes and therefore must intersect the projection of these three axes at equal distance from the center B in Figure 4.8; that is $BG = BE = BP$. The material is assumed to have the same property in compression as in tension, and hence reversing the signs of the stresses at a point on the yield surface must give a corresponding point on the surface. Therefore the yield locus must pass through the three points D, H, and A with $BD = BG$, $BH = BE$, and $BA = BP$. Hence the initial yield surface intersects the three axes on the π plane at six points equidistant from the center B and is symmetrical to those dotted lines (Figure 4.8) passing through the center B and bisecting the adjacent two solid lines.[12] The initial yield locus on the π plane has a six-fold symmetry. Two simple curves, among others, which satisfy this symmetry are the regular hexagon and the circle shown in Figure 4.8. These correspond to the two yield criteria of great historical and practical interest. The former was advanced by H. Tresca[18] in 1864 and the latter by Huber[19] (1904) and Von Mises.[20] Superposing cartesian coordinates x and y on the π plane of Figure 4.8 and projecting the three axes $\sqrt{2/3}\,\sigma_1$, $\sqrt{2/3}\,\sigma_2$, and $\sqrt{2/3}\,\sigma_3$ on the x and y axes, respectively, we obtain the relations

$$x = \sqrt{\frac{2}{3}}(\sigma_2 - \sigma_1)\frac{\sqrt{3}}{2} = \frac{\sigma_2 - \sigma_1}{\sqrt{2}} \tag{4.2.1}$$

$$y = \sqrt{\frac{2}{3}}[\sigma_3 - (\sigma_1 + \sigma_2)\tfrac{1}{2}] = \frac{1}{\sqrt{6}}(2\sigma_3 - \sigma_2 - \sigma_1) \tag{4.2.2}$$

Let Y be the initial tensile yield stress. If $\sigma_1 = Y$, $\sigma_2 = \sigma_3 = 0$, then $BG = +\sqrt{2/3}\,\sigma_1 = \sqrt{2/3}\,Y$.

Consider a segment of the hexagon, say DE, in which

$$x = \frac{\sigma_2 - \sigma_1}{\sqrt{2}} = \frac{Y}{\sqrt{2}} \qquad |y| < \frac{Y}{\sqrt{6}}$$

Since hydrostatic pressure is taken to have no effect on plastic deformation, the yield locus remains the same as the π plane moves along OC (Figure 4.7). If we take the π plane to pass through the origin O, then $\sigma_1 + \sigma_2 + \sigma_3 = 0$, and Equation (4.2.2) becomes

$$|y| = \frac{1}{\sqrt{6}}\,3\,|\sigma_3| < \frac{Y}{\sqrt{6}} \qquad \text{or} \qquad |\sigma_3| < \frac{1}{3}\,Y \tag{4.2.3}$$

Tresca's Yield Criterion. On DE, we see that σ_3 has extreme values at D and E of $+\frac{1}{3}Y$, $-\frac{1}{3}Y$, respectively. At D, we have $y = (1/\sqrt{6})\,Y$,

$x = (1/\sqrt{2})Y$. From Equations (4.2.1) and (4.2.2), we have

$$\sigma_2 - \sigma_1 - Y = 0$$
$$\sigma_2 + \sigma_1 + \tfrac{1}{3}Y = 0$$

and these give

$$\sigma_1 = -\tfrac{2}{3}Y$$
$$\sigma_2 = \tfrac{1}{3}Y$$
$$\sigma_3 = \tfrac{1}{3}Y$$

Hence on QD, we have $\sigma_2 \geqq \sigma_3 > \sigma_1$. The maximum shear stress is $(\sigma_2 - \sigma_1)/2 = Y/2$. Similarly QE and other segments of the hexagon also correspond to maximum shear stresses. This hexagon, shown in Figure 4.8, may be expressed mathematically as[21]

$$[(\sigma_2 - \sigma_1)^2 - Y^2][(\sigma_1 - \sigma_3)^2 - Y^2][(\sigma_3 - \sigma_2)^2 - Y^2] = 0 \qquad (4.2.4)$$

and is known as Tresca's yield surface.

Consider again an aggregate of crystals which have the same isotropic elastic constants. When the aggregate is uniformly loaded within the initial yield surface, the strain is purely elastic, hence both the stress and strain in all crystals are homogeneous and are the same as the aggregate. As the load on the aggregate is increased, the resolved shear stress of one slip system of some crystal in the aggregate reaches the critical shear stress, and this crystal starts to slide. At the start of slide the stress must be a point on the initial yield surface of the aggregate. *Since the stress in each crystal is the same as that of the aggregate, the yield polyhedrons of all crystals in the aggregate are the same as if these crystals were loaded separately.* The initial yield surface of the aggregate is defined as the surface in stress space at which plastic strain (no matter how small) starts. Slip in any slip system in any crystal will give some plastic strain to the aggregate. Hence the initial yield surface of the aggregate is bounded by the yield planes of all the individual crystals as if they were loaded separately.

The orientations of crystals in the aggregate are taken to be randomly distributed. As a limit idealization the slip systems of the individual crystals of a fine-grained aggregate cover all possible directions on all planes. The initial critical shear stress of each crystal is taken to be the same. When the maximum shear stress of the aggregate, occurring on a plane with normal α along a direction β, reaches the critical shear stress of the crystals, the resolved shear stress of the crystal with a slip system on the α plane along the β direction also reaches this critical shear stress and starts to slide. Hence this maximum shear stress must lie on the initial yield surface. *Therefore the theoretical initial yield surface of an aggregate of crystals with homogeneous isotropic elastic constants and homogeneous initial critical shear stress coincides with Tresca's yield surface of maximum shear stress.*

Incremental Plastic Strain of the Aggregate at Initial Yielding. Consider an aggregate of crystals with one crystal sliding. Let e''_{ij} denote the plastic strain in the sliding region v' caused by sliding. To find the plastic strain of the aggregate due to this e''_{ij}, we imagine that the aggregate is subject to a traction of $S_i{}^* = \tau^*_{ij} n_j$ over its boundary, where τ^*_{ij} is constant over the boundary surface Γ and n_j is the cosine of the angle between the normal to Γ and the x_j axis. The crystals are assumed to have *isotropic and homogeneous elastic constants*. Then, corresponding to $S_i{}^*$, the stress in the body (assuming that it is purely elastic), is τ^*_{ij} constant throughout the body. Then, from Equation (2.4.8), we have

$$\tau^*_{ij} \int_{v'} e''_{ij}\, dv' = \int_{v'} \tau^*_{ij} e''_{ij}\, dv' = \int_\Gamma S_i{}^* U_i{}''\, d\Gamma$$

$$= \int_\Gamma \tau^*_{ij} n_j U_i{}''\, d\Gamma = \tau^*_{ij} \int_\Gamma U_i{}'' n_j\, d\Gamma \tag{4.2.5}$$

where $U_i{}''$ is the displacement of the aggregate along the i axis caused by e''_{ij} in v', and τ^*_{ij} is taken out of the integral sign since it is constant. By the divergence theorem,

$$\tau^*_{ij} \int_\Gamma U_i{}'' n_j\, d\Gamma = \tau^*_{ij} \int_v U''_{i,j}\, dv = \tau^*_{ij} \int_v E''_{ij}\, dv = \tau^*_{ij} \bar{E}''_{ij} v$$

where $E''_{ij} = \frac{1}{2}(U''_{i,j} + U''_{j,i})$ and $\bar{E}''_{ij}$, the average plastic strain in the aggregate, is called the plastic strain of the aggregate. Let the volume v be the unit volume; then

$$\tau^*_{ij} \int_{v'} e''_{ij}\, dv' = \tau^*_{ij} \bar{E}''_{ij} \tag{4.2.6}$$

This demonstrates that for an aggregate of crystals with homogeneous isotropic elastic constants, the plastic strain of the aggregate is the average of the plastic strain of all the crystals in the aggregate. The incremental plastic strain vector dE'' of the aggregate due to de''_{ij} of a crystal has the same direction in the nine-dimensional strain space (counting e_{ij} and e_{ji} as two separate components) as de''_{ij}.

Since the yield plane of a slid crystal is tangent to the yield surface of the aggregate, and the incremental plastic strain vector caused by slip in the crystal is normal to its yield plane, the initial incremental plastic-strain vector dE'' of the aggregate is normal to the initial yield surface of the aggregate. *Hence the initial yield surface of the aggregate coincides with a plastic potential function.* If the principal plastic strains e_1'', e_2'', and e_3'' are superposed on the σ_1, σ_2, and σ_3 axes of the π plane (Figure 4.8), then for stress on DE, the incremental plastic strain vector dE'' will be perpendicular to DE, that is, $de_2'' = -de_1''$ and $de_3'' = 0$. When the stress is at point D of the hexagon, the normal to the hexagon is not unique and the incremental plastic strain vector dE'' will lie between the normals to PD and to DE. The direction of dE'' in this case is not uniquely determined from the yield locus.

Huber-Mises Yield Criterion. The other commonly used simple yield locus is the Huber-Mises criterion of yielding. It gives an initial yield locus in the form of a circle on the π plane. This circle corresponds to the constant value of BP in Figure 4.7. The following relations are readily seen:

$$
\begin{aligned}
OP &= (\sigma_1{}^2 + \sigma_2{}^2 + \sigma_3{}^2)^{1/2} \\
OB &= \frac{1}{\sqrt{3}}(\sigma_1 + \sigma_2 + \sigma_3) \\
BP^2 &= OP^2 - OB^2 \\
&= \sigma_1{}^2 + \sigma_2{}^2 + \sigma_3{}^2 - \frac{(\sigma_1 + \sigma_2 + \sigma_3)^2}{3} \\
&= \tfrac{2}{3}(\sigma_1{}^2 + \sigma_2{}^2 + \sigma_3{}^2 - \sigma_1\sigma_2 - \sigma_2\sigma_3 - \sigma_3\sigma_1) \\
&= \tfrac{1}{3}[(\sigma_1 - \sigma_2)^2 + (\sigma_2 - \sigma_3)^2 + (\sigma_3 - \sigma_1)^2] = \tfrac{2}{3}Y^2 \qquad (4.2.7)
\end{aligned}
$$

since, from Figure 4.8, we have $BP = \sqrt{2/3}\,Y$. Comparing this with Equation (1.4.9), we see that yielding starts when

$$\bar{J}_2 = \tfrac{1}{3}Y^2 \qquad (4.2.8)$$

For uniaxial stress, $\sigma_2 = \sigma_3 = 0$, and Equation (4.2.7) reduces to $\sigma_1 = \pm Y$. This was proposed by Huber (1904) and Von Mises, and is known as the Huber-Mises criterion of yielding. This criterion has been widely used in structural analysis. A number of physical interpretations of this criterion are given in the following discussions.

Physical Interpretations of Huber-Mises Criterion of Yielding. Writing $\bar{J}_2$ in terms of maximum shear stresses, from Equation (1.4.9) and Table 1.1, we obtain

$$
\begin{aligned}
\bar{J}_2 &= \tfrac{1}{6}[(\sigma_1 - \sigma_2)^2 + (\sigma_2 - \sigma_3)^2 + (\sigma_3 - \sigma_1)^2] \\
&= \tfrac{1}{6}[(2\tau_1)^2 + (2\tau_2)^2 + (2\tau_3)^2] \qquad (4.2.9)
\end{aligned}
$$

where τ_1, τ_2, and τ_3 are the three extreme values of shear stresses. The elastic shear or distortional energy is

$$E_d = \frac{1}{2}\left(\tau_1 \frac{\tau_1}{G} + \tau_2 \frac{\tau_2}{G} + \tau_3 \frac{\tau_3}{G}\right) = \frac{1}{2G}(\tau_1{}^2 + \tau_2{}^2 + \tau_3{}^2)$$

Comparing the above equation with Equation (4.2.9) shows that

$$\bar{J}_2 \propto E_d \tag{4.2.10}$$

From Equations (4.2.9) and (4.2.10), $\bar{J}_2$ represents the elastic-distortion energy. Hence the Huber-Mises criterion of yielding, Equation (4.2.7), means that *yielding begins when the elastic-distortion energy reaches a critical value*. This was first shown by Hencky.

Referring to axes of principal stresses, we see that the stress vector on a plane with unit normal $\mathbf{v}$ is

$$\overset{v}{\mathbf{T}} = \mathbf{v}_1\sigma_1 + \mathbf{v}_2\sigma_2 + \mathbf{v}_3\sigma_3$$

where $\mathbf{v}_1$, $\mathbf{v}_2$, and $\mathbf{v}_3$ are the component vectors of $\mathbf{v}$ along the x_1, x_2, and x_3 axes.

$$(\overset{v}{T})^2 = \overset{v}{\mathbf{T}} \cdot \overset{v}{\mathbf{T}} = (v_1\sigma_1)^2 + (v_2\sigma_2)^2 + (v_3\sigma_3)^2$$

The normal stress on the plane with normal $\mathbf{v}$ is

$$\tau_{vv} = \mathbf{v} \cdot \overset{v}{\mathbf{T}} = v_1{}^2\sigma_1 + v_2{}^2\sigma_2 + v_3{}^2\sigma_3$$

The shear stress on this plane is

$$\begin{aligned}(\tau_v)^2 = (\overset{v}{T})^2 - (\tau_{vv})^2 &= \sigma_1{}^2 v_1{}^2 + \sigma_2{}^2 v_2{}^2 + \sigma_3{}^2 v_3{}^2 \\ &\quad - (\sigma_1 v_1{}^2 + \sigma_2 v_2{}^2 + \sigma_3 v_3{}^2)\end{aligned} \tag{4.2.11}$$

If $\mathbf{v}$ is the vector making equal angles with the three axes, that is, if

$$v_1 = v_2 = v_3 = \frac{1}{\sqrt{3}}$$

then

$$\begin{aligned}(\tau_v)^2 &= \tfrac{1}{3}(\sigma_1{}^2 + \sigma_2{}^2 + \sigma_3{}^2) - \tfrac{1}{9}(\sigma_1 + \sigma_2 + \sigma_3)^2 \\ &= \tfrac{1}{9}[(\sigma_1 - \sigma_2)^2 + (\sigma_2 - \sigma_3)^2 + (\sigma_3 - \sigma_1)^2]\end{aligned} \tag{4.2.12}$$

Comparison of Equations (4.2.12) and (4.2.9) shows that $\bar{J}_2$ corresponds to the shear stress on the octahedral plane, where the octahedral plane is defined as that plane whose normal is at equal angles to the principal axes.

The Huber-Mises criterion means that *yielding occurs when the shear stress on the octahedral plane reaches the critical value*. This was first shown by A. Nadai.[22]

Another intepretation of the Huber-Mises yield criterion is as follows. Consider the shear stresses on all planes. The normals to all possible planes may be represented by the radial lines to a unit sphere. In spherical coordinates (Figure 4.9) the components of the unit normal **v** are

$$\begin{aligned} \nu_1 &= \sin\theta \sin\phi \\ \nu_2 &= \sin\theta \cos\phi \\ \nu_3 &= \cos\theta \end{aligned} \tag{4.2.13}$$

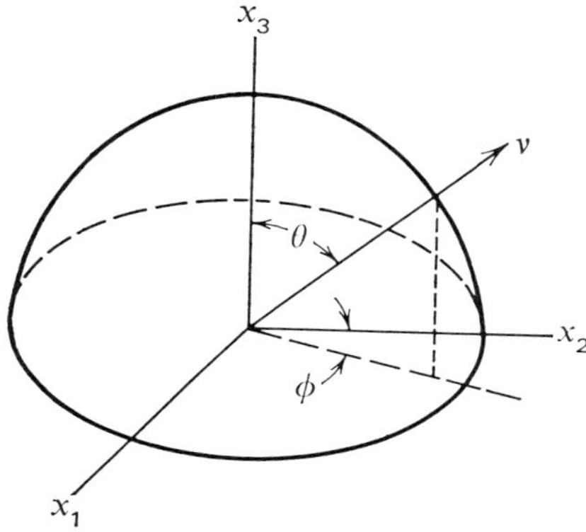

Figure 4.9. Spherical coordinates.

The mean square of the shear stresses over all planes over the sphere is

$$\overline{(\tau_\nu{}^2)} = \frac{1}{\Omega} \int (\tau_\nu)^2 \, d\Omega \tag{4.2.14}$$

where Ω is the surface area of the unit sphere,

$$d\Omega = \sin\theta \, d\theta \, d\phi \tag{4.2.15}$$

Substituting Equation (4.2.13) into Equation (4.2.11) gives

$$\begin{aligned} \tau_\nu{}^2 = {} & \sigma_1{}^2 \sin^2\theta \sin^2\phi + \sigma_2{}^2 \sin^2\theta \cos^2\phi + \sigma_3{}^2 \cos^2\theta \\ & - (\sigma_1 \sin^2\theta \sin^2\phi + \sigma_2 \sin^2\theta \cos^2\phi + \sigma_3 \cos^2\theta)^2 \end{aligned} \tag{4.2.16}$$

Substituting Equations (4.2.15) and (4.2.16) into Equation (4.2.14) and integrating over the sphere yields

$$\overline{(\tau_v^2)} = \frac{1}{4\pi}\int_0^{2\pi} d\phi \int_0^{\pi} \tau_v^2 \sin\theta \, d\theta$$

$$= \frac{1}{15}[(\sigma_1 - \sigma_2)^2 + (\sigma_2 - \sigma_3)^2 + (\sigma_3 - \sigma_1)^2] \qquad (4.2.17)$$

Hence $\bar{J}_2$ is directly proportional to $\overline{(\tau_v^2)}$. Therefore the Huber-Mises criterion of yielding also means that *yielding starts when the mean square of the shear stresses on all planes reaches the critical value*. This was pointed out by Novozhilov[23] in 1952.

Experimental Data on Initial Yielding. In the case of plane stress, we have $\sigma_3 = 0$, and the initial yield criteria may be represented with σ_1 and σ_2 as coordinates. Mises' yield criterion, Equation (4.2.7), reduces to $\sigma_1^2 + \sigma_2^2 - \sigma_1\sigma_2 = Y^2$. Tresca's criterion, Equation (4.2.4), reduces to

$$|\sigma_1 - \sigma_2| = Y$$

$$|\sigma_2| = Y \qquad (4.2.18)$$

$$|\sigma_1| = Y$$

These yield criterion are shown in Figure 4.10.

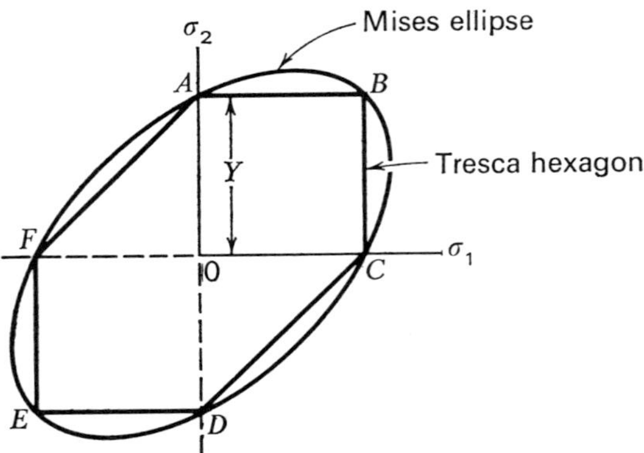

Figure 4.10. Representation of yield criteria for case of plane stress.

The yield criterion has been studied experimentally by loading a thin-walled tube subject to combined tension and torsion. With the tensile stress denoted by τ_{11} and the shearing stress due to torsion denoted by

τ_{12}, the principal stresses become

$$\sigma_1 = \tfrac{1}{2}\tau_{11} + (\tfrac{1}{4}\tau_{11}^2 + \tau_{12}^2)^{1/2}$$
$$\sigma_2 = \tfrac{1}{2}\tau_{11} - (\tfrac{1}{4}\tau_{11}^2 + \tau_{12}^2)^{1/2}$$
$$\sigma_3 = 0$$

Tresca's yield criterion gives

$$\sigma_1 - \sigma_2 = 2(\tfrac{1}{4}\tau_{11}^2 + \tau_{12}^2)^{1/2}$$
$$= (\tau_{11}^2 + 4\tau_{12}^2)^{1/2} = Y \qquad (4.2.19)$$

Mises' yield criterion gives

$$(\sigma_1 - \sigma_2)^2 + (\sigma_2 - \sigma_3)^2 + (\sigma_3 - \sigma_1)^2 = 2(\tau_{11}^2 + 3\tau_{12}^2) = 2Y^2 \qquad (4.2.20)$$

Results of the famous experiments of Taylor and Quinney[24] with copper, steel, and aluminum are shown below with the two theoretical curves (Figure 4.11). *These data, along with most other test results, are in*

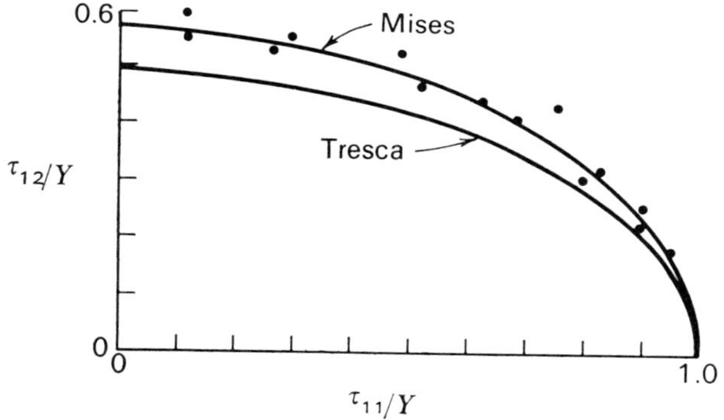

Figure 4.11. Yield curves under combined stresses. τ_{12} = torsional shear stress; τ_{11} = tensile stress; Y = yield stress under pure tension.

better agreement with Mises' than with Tresca's criterion. The deviation of test data from theoretical results may be understood from the following discussion.

The theoretical initial yield surface denotes the start of infinitesimal plastic strain. Experimentally, however, plastic strain is not observed until it grows to a measurable value. An experimental yield surface can only approach the theoretical one when the measuring instrument can detect infinitesimal plastic strain, which, of course, has never been achieved. *It has been shown*[25,26] *analytically that although the start of plastic strain* *in polycrystalline solids depends more on the maximum shear stress than on $\bar{J}_2$, the stress causing the measured amount of plastic strain depends more on the mean square of shear stresses on all planes than on the maximum shear stress.*[25,26] *This explains why the experimental results on initial yield surface agree better with Mises' than with Tresca's yield surface.*

4.3 Subsequent Loading Surfaces of Polycrystals

When a metal is subjected to combined loading beyond the initial yield surface, the new yield surface, also known as the *loading surface*, changes in both size and shape in a complicated manner. Consider again an aggregate of crystals with *isotropic* and homogeneous elastic constants. When the aggregate is initially loaded within the elastic range, all crystals are subject to the same stress. Since different crystals have different orientations, the resolved shear stress varies from one crystal to another. The crystal with the highest resolved shear stress, called the most favorably oriented crystal, slides first. After slip occurs in this crystal locally, the stress in the aggregate is no longer homogeneous. As the loading is further increased, more crystals slide. The slip planes and directions vary from one crystal to another. The slip distribution and the stress distribution are heterogeneous.

Imagine that the aggregate is unloaded. The slip remains and causes nonhomogeneous residual stress,[27] τ_{ij_r}. Now *imagine* that the aggregate is reloaded. Assume that no additional plastic strain occurred during the *imaginary* process of unloading and reloading. Since there is no additional plastic strain during reloading, the stress caused by reloading, τ_{ij_A}, is uniform throughout the aggregate. Then

$$\tau_{ij} = \tau_{ij_r} + \tau_{ij_A} \tag{4.3.1}$$

Now consider, in the aggregate, one crystal with a slip plane denoted by its normal α, and on this plane a slip direction β. The resolved shear stress in this slip system is

$$\tau_{\alpha\beta} = l_{\alpha i}\, l_{\beta j}(\tau_{ij_A} + \tau_{ij_r}) \tag{4.3.2}$$

Denote the initial critical shear stress in this slip system by $c_{\alpha\beta_0}$ and the increase of critical shear stress due to strain-hardening by $\Delta c_{\alpha\beta}$. The condition for this slip system to start, continue, or reverse sliding is

$$l_{\alpha i}\, l_{\beta j}(\tau_{ij_A} + \tau_{ij_r}) = (c_{\alpha\beta_0} + \Delta c_{\alpha\beta}) \qquad \text{or} \qquad -(c_{\alpha\beta_0} + \Delta c_{\alpha\beta})$$

This is written as

$$|l_{\alpha i}\, l_{\beta j}(\tau_{ij_A} + \tau_{ij_r})| = c_{\alpha\beta_0} + \Delta c_{\alpha\beta} = c_{\alpha\beta} \tag{4.3.3}$$

For those crystals which have not slid, $\Delta c_{\alpha\beta} = 0$, and we have

$$|l_{\alpha i} l_{\beta j}(\tau_{ij_A} + \tau_{ij_r})| = c_{\alpha\beta_0} \tag{4.3.4}$$

Before slip occurs in the aggregate, $\tau_{ij_r} = 0$ and $\Delta c_{\alpha\beta} = 0$, and Equation (4.3.3) reduces to

$$|l_{\alpha i} l_{\beta j} \tau_{ij_A}| = c_{\alpha\beta_0}$$

This gives the yield planes for the *initial* yielding of the aggregate. After slip occurs in the aggregate, the yield planes in crystals will be given by Equation (4.3.3). In Equation (4.3.3), we see that $\Delta c_{\alpha\beta}$ has the effect of increasing the distance between the pair of parallel yield planes, while τ_{ij_r} has the effect of translating the pair of yield planes in the stress space. For aluminum crystals, there are twelve slip systems. Each system gives a pair of parallel yield planes. Twelve slip systems give a polyhedron of twelve pairs of parallel yield planes. The initial yield surface of an aluminum aggregate is bounded by the initial yield polyhedrons of the individual crystals. After slip occurs, the yield polyhedrons of the slid crystals are expanded and translated, and those of the unslid crystals are translated. The new loading surface of the aggregate is bounded by the faces of these new yield polyhedrons. One pair of parallel faces of the yield polyhedron is written as

$$F(\tau_{ij_A}) = |l_{\alpha i} l_{\beta j}(\tau_{ij_A} + \tau_{ij_r})| - c_{\alpha\beta_0} - \Delta c_{\alpha\beta} = 0 \tag{4.3.5}$$

The elastic domain in this stress space bounded by n pairs of yield planes from different crystals may be written as

$$F_p(\tau_{ij_A}) < 0 \qquad p = 1, 2, \ldots, n \tag{4.3.6}$$

where n is the number of slip systems of all crystals in the aggregate. Hence, as loading proceeds beyond the initial yield surface, *the initial yield surface neither expands uniformly* [*a process commonly known as isotropic hardening*[12,28–30] (e.g., under loading the Huber-Mises circle of initial yielding expands uniformly into another concentric circle on the π plane)] *nor translates as a whole* [*a process commonly known as kinematic hardening*[31,32] (e.g., the Huber-Mises circle of initial yielding translates as a rigid body)]. In other words, *the yield planes bounding the loading surface move differently*.

For isotropic hardening, the size of the yield surface increases, but the shape remains the same during loading. For kinematic hardening, both size and shape of the yield surface remain unchanged during loading, but the origin undergoes a translation. Actually, as shown in the previous discussions, both the shape and size of yield surfaces change as plastic strain develops.

As loading increases, more crystals have reached critical shear stress ($|\tau_{\alpha\beta}| = c_{\alpha\beta}$) and have slid. The yield planes of the active slip systems of these crystals pass through the loading point in stress space. The incremental stress giving no further slip in the $\alpha\beta$ slip system requires $d|\tau_{\alpha\beta_A}| \leqq 0$. Hence $d|\tau_{\alpha\beta_A}| = 0$, for each of these sliding systems, in each sliding crystal, gives one bounding plane to the yield surface. All these bounding planes pass through the loading point in stress space. Since the yield surface is bounded by a number of yield planes of different orientations passing through the same point, there is clearly a vertex at the loading point.[25,26,34] Hence, *on the basis of the slip characteristics of single crystals, corners exist in the subsequent yield surface of polycrystals at the loading point.*

The detail method of calculating the theoretical initial yield locus and subsequent theoretical yield loci of a polycrystalline aggregate subjected to combined tensile and shear stresses, τ_{11} and τ_{12}, has been shown.[26] The calculated loci of an aggregate of face-centered-cubic crystals are shown in Figure 4.12. It is clearly seen from this figure that a

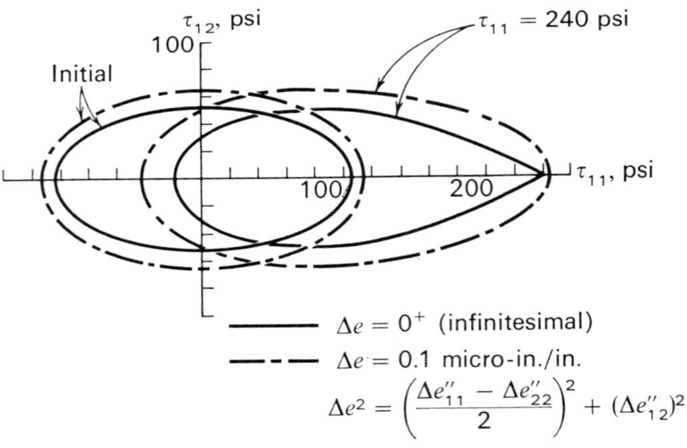

Figure 4.12. Yield surfaces: initial and tensile loading at $\tau_{11} = 240$ psi.

vertex exists at the loading point. The dotted lines denote the loci corresponding to a finite plastic-strain increment of small magnitude. The theoretical loci are for infinitesimal plastic-strain increment. Since an infinitesimal plastic-strain increment is not measurable by actual strain gauges, the experimental loci actually correspond to the dotted lines. Note that the vertex at the loading point does not appear in the loci with finite incremental plastic strain. This explains why so many experimental data[35-38] show the existence of large curvature near the loading point, but do not conclusively demonstrate the existence of the vertex that theory predicts.

The above discussion is limited to aggregates of crystals with isotropic elastic constants. Similar results may be obtained for aggregates of crystals with anisotropic elastic constants. Consider now such an aggregate subject to a uniform stress state τ_{ij_A} applied on the boundary of this aggregate. Within the elastic limit, the stress at any point x of the aggregate is proportional to the applied stress τ_{ij_A}. Let $a_{ijkl}(x)$ be the stress τ_{ij} at the point x in the aggregate due to such a unit applied uniform stress τ_{kl_A}, that is,

$$\tau_{ij}(x) = a_{ijkl}(x)\tau_{kl_A} \tag{4.3.7}$$

The resolved shear stress in a slip system of sliding plane with normal α and sliding direction β at point x is then

$$\tau_{\alpha\beta}(x) = l_{\alpha i}\, l_{\beta j}\tau_{ij}(x) = l_{\alpha i}\, l_{\beta j}\, a_{ijkl}(x)\tau_{kl_A} \tag{4.3.8}$$

For this slip system to start sliding, the resolved shear stress must reach the initial critical shear stress in this slip system $c_{\alpha\beta_0}$. (This $c_{\alpha\beta_0}$ is generally the same in all slip systems in a crystal.) This gives

$$|l_{\alpha i}\, l_{\beta j}\, a_{ijkl}(x)\tau_{kl_A}| = c_{\alpha\beta_0} \tag{4.3.9}$$

which corresponds to a pair of parallel planes in stress-space with τ_{ij_A} as coordinates. Since $a_{ijkl}(x)$ changes from point to point, the orientation and the distance between the pair of parallel planes given by Equation (4.3.9) change from point to point. Let n be the number of slip systems in each crystal ($n = 12$ for face-centered cubic crystals). There are n pairs of these parallel planes at each point. The region containing the origin in stress-space bounded by all these planes is the elastic region, and its boundary surface is the initial yield surface. The parallel planes tangent to the yield surface are known as yield planes.

After the aggregate is stressed beyond the elastic limit, some crystals slide. *Imagine* that the applied stress to the aggregate is removed. The slip remains and causes residual stress τ_{ij_r}. Now *imagine* that the original uniform stress field τ_{ij_A} on the aggregate is reapplied. There is no additional slip during the imaginary process of unloading and reloading, and the resolved shear stress may be written as

$$\tau_{\alpha\beta}(x) = l_{\alpha i}\, l_{\beta j}[\tau_{ij_r}(x) + a_{ijkl}(x)\tau_{kl_A}]$$

For any point x in the $\alpha\beta$ slip system to slide, the absolute value of $\tau_{\alpha\beta}(x)$ must equal the critical shear stress,

$$|l_{\alpha i}\, l_{\beta j}[\tau_{ij_r}(x) + a_{ijkl}(x)\tau_{kl_A}]| = c_{\alpha\beta}(x) \tag{4.3.10}$$

For aggregates of crystals with anisotropic elastic constants, Equation (4.3.3) is replaced by Equation (4.3.10). The elastic domain in stress space is again bounded by numerous pairs of yield planes from the yield conditions of the various points in different crystals, and a *vertex* exists at the loading point of the loading surface.

4.4 Coincidence of Loading Surface with Plastic Potential[9]

Consider an aggregate of crystals, with either isotropic or anisotropic elastic constants, loaded beyond the initial yield surface. The initial yield surface changes into a new loading surface. Let T be a stress vector in stress space that produces a given incremental plastic-strain vector dE'' and let T^* be any other stress vector within the loading surface of the aggregate. Let τ_k and $\tau_k{}^*$ be the resolved shear stresses in the kth active slip system corresponding to τ_{ij} and τ_{ij}^*, respectively. The τ_k in the active slip systems are equal to the critical values, while the $\tau_k{}^*$ are less than the critical values. Hence $(\tau_k - \tau_k{}^*)\, d\gamma_k{}'' \geq 0$ in all crystals, where $d\gamma_k{}''$ is the incremental plastic strain due to slip in the kth slip system. The difference of the incremental work due to the plastic strain increment is

$$dW - dW^* = (T - T^*)\, dE''v = \int (\tau_{ij} - \tau_{ij}^*)\, de_{ij}''\, dv$$

$$= \int \sum_{k=1}^{m} (\tau_k - \tau_k{}^*)\, d\gamma_k{}''\, dv \geq 0 \qquad (4.4.1)$$

Hence $(T - T^*)dE'' \geq 0$. By the same reasoning as in Section 4.1, the plastic-strain increment vector dE'' must be normal to the loading surface at a smooth point and lie between the adjacent normals at a corner. In this derivation, no assumption is made on the isotropy of the elastic constants, so the loading surface coincides with plastic potential for aggregates of crystals with either isotropic or anisotropic elastic constants. This proof was first given by Bishop and Hill[9] in 1951.

An incremental stress-strain relationship incorporating the normality of incremental plastic strain to loading surface property has been suggested by Hill.[12] Since the yield surface coincides with a plastic potential function,

$$de_{ij}'' = h \frac{\partial F}{\partial \tau_{ij}}\, dF \qquad (4.4.2)$$

at a point on the smooth part of the loading surface, where $F(\tau_{ij})$ is the loading surface and h is a scalar depending on the loading history. At a corner point of the loading surface, the surface is represented by its bounding planes,

$$F_p(\tau_{ij}) = 0 \qquad p = 1, 2, \ldots, n \tag{4.4.3}$$

$$de''_{ij} = \sum_{p=1}^{n} h_p \frac{\partial F_p}{\partial \tau_{ij}} dF_p \tag{4.4.4}$$

In structural analysis much simpler forms of $F(\tau_{ij})$ have been used.

In the above we have shown, from the physical slip characteristics of single crystals, the following plastic deformation characteristics of polycrystalline metals: (1) *Plastic deformation of a polycrystal causes no appreciable change in volume.* (2) *The incremental plastic-strain vector is normal to the yielding (loading) surface in stress space*; i.e., the loading surface coincides with the plastic potential. (3) *A vertex exists at the loading point of the loading surface.* The first and second conclusions are implied in nearly all the theories of plasticity used in structural analysis, but the third is seldom considered. The physical significance of the existence of this vertex may be seen from accompanying figures. Suppose the polycrystal has been loaded in tension beyond the elastic limit. We have shown that the loading surface coincides with plastic potential. Therefore, if the yield locus has no vertex [the dotted line locus of Figure 4.12; also Figure 4.13(a)], an incremental stress $\Delta\tau$ will cause purely plastic tensile strain and no incremental plastic shear strain. On the other hand, if the yield locus follows the solid-line locus of Figure 4.12 [also Figure 4.13(b)],

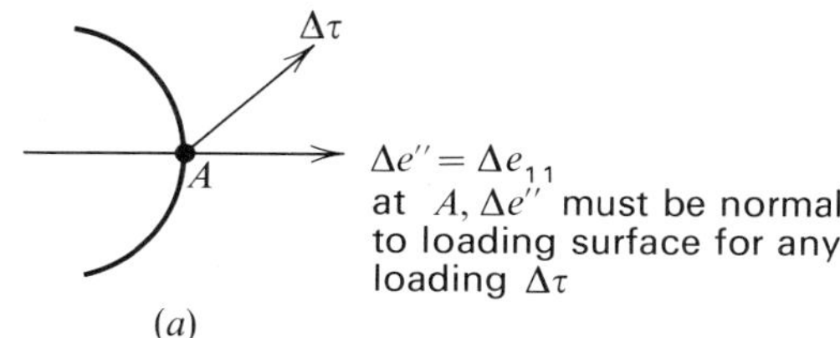

(*a*)

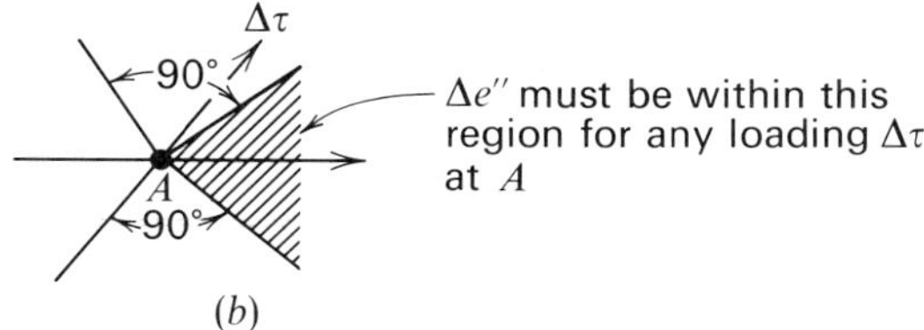

(*b*)

Figure 4.13

which has a vertex at the loading point, an incremental stress $\Delta\tau$ will cause an incremental plastic shear strain $\Delta e''_{12}$ as well as incremental plastic tensile strain $\Delta e''_{11}$. This difference may have an important effect on the buckling strength of rectangular plates.[39]

4.5 Plastic Stress-Strain Relations

As shown in Figure 4.12, after plastic strain occurs, the yield surface of a polycrystalline metal changes both in size and in shape.[25,26,29] The calculation of the subsequent yield surfaces and the derivation of the stress-strain relation of an aggregate from the single-crystal characteristics are very complicated,[26] and therefore the resulting expression of the stress-strain relation would necessarily be complicated and lengthy[25,26] and not at all convenient for structural analysis. Hence some simple idealized stress-strain relations for polycrystalline metals have been commonly used in stress analysis and are given as follows.

Strain-Hardening Materials. In 1870 Saint Venant[40] proposed that the principal axes of strain increment coincided with the axes of principal stresses. Levy (1871) and subsequently Mises[20] (1913) introduced the general relation between strain increment and the deviatoric stresses. Prandtl[82] (1924) and Reuss[41] (1930) assumed that the plastic-strain increment at any instant is proportional to the deviatoric stress, giving

$$\frac{de''_{11}}{S_{11}} = \frac{de''_{22}}{S_{22}} = \frac{de''_{33}}{S_{33}} = \frac{de''_{12}}{S_{12}} = \frac{de''_{23}}{S_{23}} = \frac{de''_{31}}{S_{31}} = d\kappa \tag{4.5.1}$$

where $d\kappa$ is an instantaneous positive constant of proportionality which may vary during the loading process and S_{ij} are the deviatoric stress components. This may be written in short form as

$$de''_{ij} = S_{ij}\, d\kappa \tag{4.5.2}$$

This satisfies the condition of zero plastic dilatation. Referring to the principal stress axes, we see that Equations (4.5.1) reduce to

$$\frac{de_1'' - de_2''}{\sigma_1 - \sigma_2} = \frac{de_2'' - de_3''}{\sigma_2 - \sigma_3} = \frac{de_3'' - de_1''}{\sigma_3 - \sigma_1} = d\kappa$$

The ratios of the incremental plastic-strain components depend on the ratios of the deviatoric-stress components but not on the ratios of the incremental-stress components. Loading and unloading here denotes whether $\bar{J}_2$ is increasing or decreasing. Incremental plastic strain occurs only in loading or neutral loading. In these cases, the total strain increment is the sum of the elastic and plastic parts, hence

$$de_{ij} = de'_{ij} + de''_{ij}$$
$$= \frac{dS_{ij}}{2G} + \frac{(1-2\nu)}{3E}\delta_{ij}\,d\tau_{kk} + S_{ij}\,d\kappa \tag{4.5.3}$$

where $d\kappa = 0$ for unloading. This loading and unloading condition is often stated in a different manner.

Let us define effective stress σ^* and strain e^* as

$$\sigma^* = \frac{1}{\sqrt{2}}[(\tau_{11}-\tau_{22})^2 + (\tau_{22}-\tau_{33})^2 + (\tau_{33}-\tau_{11})^2 + 6(\tau_{12}^2 + \tau_{23}^2 + \tau_{31}^2)]^{1/2}$$
$$= \frac{1}{\sqrt{2}}[(S_{11}-S_{22})^2 + (S_{22}-S_{33})^2 + (S_{33}-S_{11})^2 + 6(S_{12}^2 + S_{23}^2 + S_{31}^2)]^{1/2} \tag{4.5.4}$$

$$e^* = \frac{\sqrt{2}}{3}[(e_{11}-e_{22})^2 + (e_{22}-e_{33})^2 + (e_{33}-e_{11})^2 + 6(e_{12}^2 + e_{23}^2 + e_{31}^2)]^{1/2}$$
$$= \frac{\sqrt{2}}{3}[(\varepsilon_{11}-\varepsilon_{22})^2 + (\varepsilon_{22}-\varepsilon_{33})^2 + (\varepsilon_{33}-\varepsilon_{11})^2 + 6(\varepsilon_{12}^2 + \varepsilon_{23}^2 + \varepsilon_{31}^2)]^{1/2} \tag{4.5.5}$$

where ε_{ij} are the deviatoric-strain components. Loading and unloading then depend on whether σ^* increases or decreases. In terms of principal stresses and plastic strains, these become

$$\sigma^* = \sqrt{\tfrac{1}{2}[(\sigma_1-\sigma_2)^2 + (\sigma_2-\sigma_3)^2 + (\sigma_3-\sigma_1)^2]} \tag{4.5.6}$$

$$e''^* = \sqrt{\tfrac{2}{9}[(e_1''-e_2'')^2 + (e_2''-e_3'')^2 + (e_3''-e_1'')^2]} \tag{4.5.7}$$

For the uniaxial case, σ^* reduces to σ_1 and e''^* to e_1''. Equation (4.5.1) gives the proportionality of the incremental plastic-strain components to the corresponding deviatoric-stress components, but does not give the constant of proportionality. From uniaxial tests, we obtain the function $e_1'' = f(\sigma_1)$, which for the multiaxial stress state is generalized to

$$e''^* = f(\sigma^*) \tag{4.5.8}$$

Then $de''^* = f'(\sigma^*)\,d\sigma^*$ when σ^* increases and vanishes when σ^* decreases, where $f'(\sigma^*)$ is the derivative of f with respect to its argument. From uniaxial tests, we have $\sigma^* = \tau_{11} = \frac{3}{2}S_{11}$, $e''^* = e''_{11}$, this gives $de''_{11}/S_{11} = 3de^{*''}/2\sigma^*$. Hence the ratios in (4.5.1) are

$$d\kappa = \frac{3de''^*}{2\sigma^*} = \frac{3(de''^*/d\sigma^*)\,d\sigma^*}{2\sigma^*} = \frac{3(de''_{11}/d\tau_{11})\,d\tau_{11}}{2\tau_{11}} \tag{4.5.9}$$

Then, by Equation (4.5.2), the de''_{ij} are uniquely determined. For example, suppose the tension test data for monotonic loading are represented by

$$e_{11} = \frac{\tau_{11}}{E} + K\left(\frac{\tau_{11}}{E}\right)^n \tag{4.5.10}$$

and

$$e''_{11} = K\left(\frac{\tau_{11}}{E}\right)^n \tag{4.5.11}$$

Replacing e''_{11} and τ_{11} in Equation (4.5.11) by e''^* and σ^* gives

$$e''^* = K\left(\frac{\sigma^*}{E}\right)^n$$

and hence

$$de''^* = Kn\left(\frac{\sigma^*}{E}\right)^{n-1}\frac{d\sigma^*}{E}$$

From this and Equation (4.5.9) we obtain

$$d\kappa = \frac{3de''^*}{2\sigma^*} = \frac{3Kn}{2E^n}(\sigma^*)^{n-2}\,d\sigma^* \tag{4.5.12}$$

Substituting this into Equation (4.5.2) yields $de''_{ij} = S_{ij}(3Kn/2E^n)(\sigma^*)^{n-2}\,d\sigma^*$; hence $de''_{ij} \propto S_{ij} = (\partial \bar{J}_2/\partial S_{ij})$. This is equivalent to assuming the subsequent yield surfaces to be represented by $\bar{J}_2 = F(e''^*)$, where F is a function. The incremental plastic-strain vector is parallel to the gradient of the loading surface; that is, $de''_{ij} \propto (\partial \bar{J}_2/\partial S_{ij})$:

$$de''_{ij} \propto \frac{\partial \bar{J}_2}{\partial S_{ij}} = S_{ij}$$

Thus this implies the coincidence of the yield surface with plastic potential as theoretically derived in Section 4.4. Therefore the yield surfaces are of the same shape as the Huber-Mises initial yield surface, and the incremental plastic strain de''_{ij} is always normal to the yield surface. Although the Mises-Reuss incremental plastic stress-strain relation is simple in form, *it does not account for the Bauschinger effect* or predict the existence of corners on the yield surface, which were shown to exist in the previous section. This is not adequate for cases involving a reversal of loading beyond the elastic range.

Initial and subsequent yield surfaces of Tresca's type (maximum shear stress) have also been considered as the plastic potential to determine the incremental plastic strain. In this case, the effective stress and effective plastic strain are defined by

$$\sigma^* = 2\tau_{\max} \tag{4.5.13}$$

$$de''^* = d\gamma''_{\max} \tag{4.5.14}$$

where $\gamma''_{\max}$ is the maximum plastic shear strain corresponding to $\tau_{\max}$, the maximum shear stress. For biaxial stress with $\sigma_3 = 0$, the yield surface and the normality of the incremental plastic-strain increment de''_{ij} to the yield surface are shown in Figure 4.14. Experiments have shown that

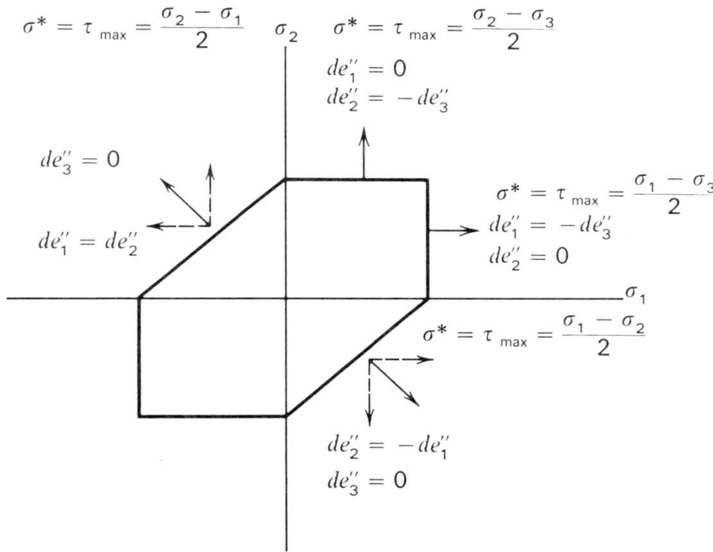

Figure 4.14. Yield surface and incremental plastic strain for a biaxial stress state, using the Tresca criterion.

Huber-Mises yield surfaces as plastic potential give incremental plastic strain closer to the actual results than Tresca's yield surfaces.[12] However, Tresca's yielding condition is simpler to apply in a number of problems in structural analyses.

For the special cases in which the deviatoric-stress components are proportionally increased, S_{ij} is proportional to σ^*. This is known as *radial loading*. Denoting this ratio by α_{ij}, we have

$$S_{ij} = \alpha_{ij}\sigma^*$$

From Equation (4.5.2), we have $de''_{ij} = \alpha_{ij}\sigma^*\, d\kappa$. From Equations (4.5.8) and (4.5.9), we have

$$d\kappa = \frac{3(de''^*/d\sigma^*)\, d\sigma^*}{2\sigma^*} = \frac{3}{2\sigma^*}\frac{1}{E_t}\, d\sigma^* = \frac{3}{2\sigma^*} f'(\sigma^*)\, d\sigma^*$$

where E_t is the tangent modulus of the σ^* vs. e''^* curve. Then

$$de''_{ij} = \alpha_{ij}\tfrac{3}{2}f'(\sigma^*)\, d\sigma^*$$

$$e''_{ij} = \alpha_{ij}\tfrac{3}{2}f(\sigma^*) = \frac{3S_{ij}}{2\sigma^*}f(\sigma^*) = F(\sigma^*)S_{ij} \tag{4.5.15}$$

where $F(\sigma^*) = (3/2\sigma^*)f(\sigma^*)$. This relation, however, has been applied to the cases in which S_{ij} are not proportionally increased, giving the deviatoric strain as

$$\varepsilon_{ij} = \varepsilon'_{ij} + \varepsilon''_{ij} = \frac{S_{ij}}{2G} + F(\sigma^*)S_{ij} \tag{4.5.16}$$

for the loading process ($d\sigma^* > 0$), where the single prime denotes the elastic part of strain. Since the volumetric strain is purely elastic, it is given by

$$e_{ii} = \frac{1 - 2v}{E}\tau_{ii} \tag{4.5.17}$$

These relations [Equations (4.5.16) and (4.5.17)] form the basis of the deformation theory of plasticity, which was originally proposed by Hencky[42] in 1924. From the discussions in previous sections, it is clear that plastic strain depends greatly on the path of loading. Deformation theory of plasticity predicts, for the process of loading, that is, for $d\sigma^* \geq 0$, a plastic strain independent of the path of loading. Hence this theory generally does not accurately describe the plastic behavior of metal. Other serious objections have been raised by Handleman, Prager, and Lin.[43]

It is generally agreed that the Huber-Reuss incremental theory of plasticity as given by Equations (4.5.1) and (4.5.8) is a much better representation of the plastic behavior of metals than the Hencky plastic deformation theory as given by Equation (4.5.16). However, Equation (4.5.16) is much simpler to apply in actual analyses and hence is still much used in practice. The discrepancy between the deformation theory [Equation

(4.5.16); it is also known as the total theory of plasticity] and the incremental theory [Equations (4.5.1) and (4.5.8)] reduces and vanishes in the limit, as the loading approaches radial, i.e., the ratios of the deviatoric stress components are constant during loading. Sometimes the stress components at different points at different stages of loadings are analyzed by the deformation theory, and then the ratios of the components of the deviatoric stress components are calculated at different stages of loadings. The variations of these ratios at a given point serve to indicate the amount of error introduced through the use of deformation theory instead of the incremental theory. Further comparative discussion between deformation theories and incremental theories is given by Budiansky[44].

Perfectly Plastic Materials.[21] Structural steel[45] (low carbon steel) has the characteristic that its tensile strain increases from 0.001 to 0.014 when under a constant stress equal to its elastic limit, as shown in Figure 4.15.

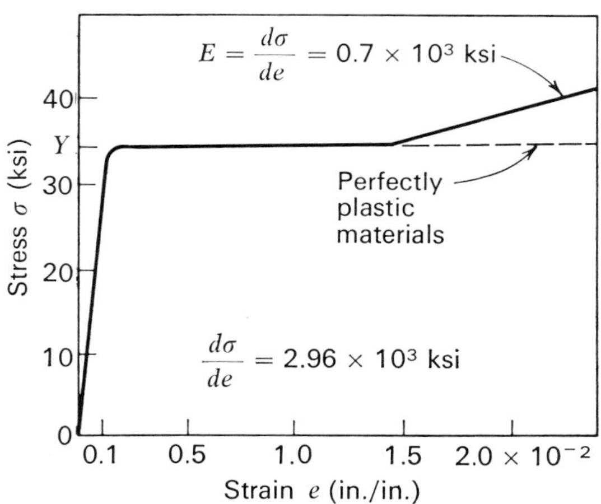

Figure 4.15. Stress-strain curve of A7 structural steel idealized[45].

The stress thereafter increases slowly with the strain. This slow increase of stress with strain is often neglected to simplify the structural analysis. The resulting simplified stress-strain relation is shown by the dashed line in Figure 4.15 and is called *perfect plasticity*. The stress-strain relation of this idealized perfectly plastic material under loading, unloading, and reversed loading is represented by Figure 4.16 overleaf. Before the tensile stress reaches the yielding stress Y, the stress is proportional to strain. Once the stress reaches the yielding stress, the strain increases without limit under constant stress Y as shown by ABC. Along ABC, there is no unique relation

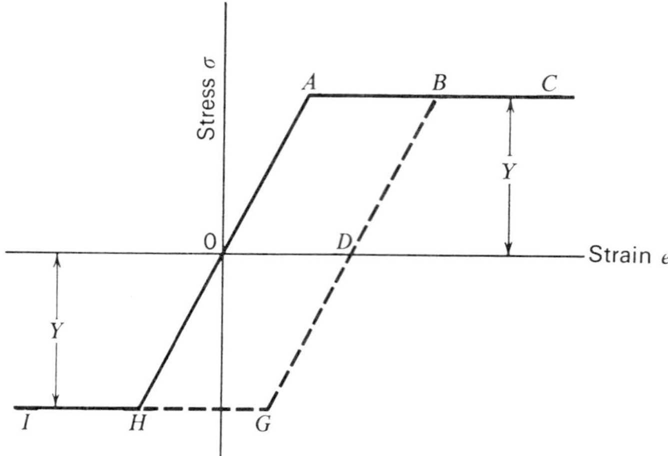

Figure 4.16. Stress-strain curve of a perfectly plastic material.

between stress and strain. When the specimen is unloaded at B, the unloading curve is represented by BD, which has the same slope as OA, and the resultant permanent strain is OD. If the material is then loaded in the reversed direction, the reversed loading is represented by $DGHI$, where the yield stress in compression is taken to be the same as in tension.

When a perfectly plastic material is under combined loading, only elastic strain occurs if σ^* is less than the tensile yield stress Y. Once σ^* reaches Y, incremental plastic strain occurs. The incremental plastic strain at any instant is assumed to be proportional to the instantaneous deviatoric stress as given by Equation (4.5.2),

$$de''_{ij} = S_{ij}\, d\kappa \tag{4.5.2}$$

It is again assumed that $\bar{J}_2 = \text{constant}$ is the yield surface and that it coincides with plastic potential as shown in Section 4.4. From Equations (1.9.11) and (1.4.5), we have

$$\begin{aligned} e'_{ij} &= \frac{\tau_{ij}}{2\mu} - \delta_{ij} \frac{\lambda}{2\mu(3\lambda + 2\mu)} \tau_{kk} \\ &= \frac{S_{ij}}{2\mu} + \delta_{ij} \frac{\tau_{kk}}{3(3\lambda + 2\mu)} \end{aligned} \tag{4.5.18}$$

$$de'_{ij} = \frac{dS_{ij}}{2\mu} + \delta_{ij} \frac{d\tau_{kk}}{3(3\lambda + 2\mu)} \tag{4.5.19}$$

Hence the total incremental strain is

$$de_{ij} = de'_{ij} + de''_{ij} = \frac{dS_{ij}}{2\mu} + \delta_{ij} \frac{d\tau_{kk}}{3(3\lambda + 2\mu)} + S_{ij}\, d\kappa \qquad (4.5.20)$$

Here $d\kappa$ is a positive parameter and is not determined by S_{ij} or dS_{ij} for perfectly plastic materials.

For the cases in which the plastic strain is much larger than the elastic, elastic strain is neglected. These are generally known as *rigid-plastic materials*.[45] In this case

$$de_{ij} = S_{ij}\, d\kappa \qquad (4.5.21)$$

Many studies[25–77] on the inelastic multiaxial stress-strain relations are listed at the end of this chapter for those readers who are particularly interested.

4.6 Uniqueness of Solution of Elastic-Plastic Bodies[12,56]

Consider a state of stress τ_{ij} and strain e_{ij} in equilibrium with a given body force F_i in the body and surface force S_i on part of the boundary Γ_s, and satisfying a given displacement u_i on the remaining part of the boundary Γ_u. Assume that in a part or whole of the body τ_{ij} is beyond the elastic limit. Now suppose that an incremental set of body force dF_i, surface force dS_i on Γ_s, and displacement du_i on Γ_u is applied, and let the corresponding increments of stress and strain be denoted by $d\tau_{ij}$ and de_{ij}. To prove the *uniqueness of the set* $(d\tau_{ij}, de_{ij})$, we assume that there are two sets of incremental stress and strain ($d\tau_{ij}^{(1)}, de_{ij}^{(1)}$ and $d\tau_{ij}^{(2)}, de_{ij}^{(2)}$) satisfying the condition of equilibrium with dF_i in the body, dS_i on Γ_s, and du_i on Γ_u. Then $(d\tau_{ij}^{(2)} - d\tau_{ij}^{(1)})$ must satisfy the condition of equilibrium throughout the body with $dF_i = 0$ and $dS_i = 0$ on Γ_s. Also, $(de_{ij}^{(2)} - de_{ij}^{(1)})$ must satisfy the condition of continuity and the condition $du_i = 0$ on Γ_u.

Replacing F_i by dF_i, S_i by dS_i, and τ_{ij} by $d\tau_{ij}$ in Equation (1.13.1), we obtain

$$\int_v dF_i\, du_i\, dv + \int_{\Gamma_s} dS_i\, du_i\, d\Gamma + \int_{\Gamma_u} dS_i\, du_i\, d\Gamma - \int_v d\tau_{ij}\, de_{ij}\, dv = 0$$

With du_i replaced by $(du_i^{(2)} - du_i^{(1)})$, $d\tau_{ij}$ by $(d\tau_i^{(2)} - d\tau_i^{(1)})$, and de_{ij} by $(de_{ij}^{(2)} - de_{ij}^{(1)})$, we have

$$\int_v (d\tau_{ij}^{(2)} - d\tau_{ij}^{(1)})(de_{ij}^{(2)} - de_{ij}^{(1)})\, dv = 0 \qquad (4.6.1)$$

since $dF_i = 0$, we have $dS_i = 0$ on Γ_s and $du_i = 0$ on Γ_u.

Writing the incremental strain as the sum of the elastic and plastic parts, we see that Equation (4.6.1) becomes

$$\int_v (d\tau_{ij}^{(2)} - d\tau_{ij}^{(1)})[(de_{ij}'^{(2)} - de_{ij}'^{(1)}) + (de_{ij}''^{(2)} - de_{ij}''^{(1)})]\, dv \qquad (4.6.2)$$

where the single prime and double prime denote the elastic and plastic parts of strain, respectively.

From Equation (1.13.9), we have

$$(d\tau_{ij}^{(2)} - d\tau_{ij}^{(1)})(de_{ij}'^{(2)} - de_{ij}'^{(1)})$$
$$= \frac{3\lambda + 2\mu}{3}(d\theta^{(2)} - d\theta^{(1)})^2 + 2\mu(de_{ij}'^{(2)} - de_{ij}'^{(1)})(de_{ij}'^{(2)} - de_{ij}'^{(1)}) \geq 0 \qquad (4.6.3)$$

In calculating the incremental plastic strain, the loading surface is considered to coincide with the plastic potential (Sections 4.3 and 4.4). When the stress vector is on a smooth part of the loading surface, $f(\tau_{ij}) = 0$ in the stress space, the incremental plastic strain may be written for work-hardening materials as

$$de_{ij}'' = h\frac{\partial f}{\partial \tau_{ij}}\, df \qquad \text{when } f = 0 \text{ and } df > 0$$
$$= 0 \qquad \text{when } f < 0 \text{ or } f = 0 \text{ and } df \leq 0$$

where h is a scalar function of stress, plastic strain, and strain history. Letting

$$I = (d\tau_{ij}^{(2)} - d\tau_{ij}^{(1)})(de_{ij}''^{(2)} - de_{ij}''^{(1)})$$

we consider the following cases:

(i) When $f < 0$, then $I = 0$.
(ii) When $f = 0$, $df^{(1)} > 0$, and $df^{(2)} < 0$, then

$$I = (d\tau_{ij}^{(2)} - d\tau_{ij}^{(1)})\left(-h\frac{\partial f}{\partial \tau_{ij}}\, df^{(1)}\right)$$
$$= h\, d\tau_{ij}^{(1)}\frac{\partial f}{\partial \tau_{ij}}\, df^{(1)} - h\, d\tau_{ij}^{(2)}\frac{\partial f}{\partial \tau_{ij}}\, df^{(1)}$$
$$= h(df^{(1)}\, df^{(1)} - df^{(2)}\, df^{(1)}) \geq 0$$

since

$$\frac{\partial f}{\partial \tau_{ij}}\, d\tau_{ij}^{(1)} = df^{(1)} \qquad \text{and} \qquad \frac{\partial f}{\partial \tau_{ij}}\, d\tau_{ij}^{(2)} = df^{(2)}$$

(iii) When $f = 0$, $df^{(2)} > 0$, and $df^{(1)} < 0$, then, as in (ii), $I \geq 0$.
(iv) When $f = 0$, $df^{(1)} > 0$, and $df^{(2)} > 0$, then

$$I = (d\tau_{ij}^{(2)} - d\tau_{ij}^{(1)})h \frac{\partial f}{\partial \tau_{ij}} (df^{(2)} - df^{(1)})$$

$$= h(df^{(2)} - df^{(1)})(df^{(2)} - df^{(1)}) \geq 0 \tag{4.6.4}$$

Hence I is always positive when the stress vector is at a smooth part of the loading surface.

When the stress vector is at a vertex or singular point of the yield surface, the yield surface at that point is represented by a family of yield planes,

$$f_p(\tau_{ij}) = 0 \qquad p = 1, 2, \ldots, n$$

and we replace Equation (4.4.4) by

$$de''_{ij} = \sum_{p=1}^{n} c_p h_p \frac{\partial f_p}{\partial \tau_{ij}} df_p \tag{4.6.5}$$

where

$$c_p = \begin{cases} 0 & \text{if } f_p < 0 \text{ or } df_p \leq 0 \\ 1 & \text{if } f_p = 0 \text{ and } df_p > 0 \end{cases}$$

and h_p is *assumed* to be a positive function of stress, plastic strain, and strain history, *but not of* $d\tau_{ij}$. Let

$$I = (d\tau_{ij}^{(2)} - d\tau_{ij}^{(1)}) \sum_{p=1}^{n} c_p h_p \frac{\partial f_p}{\partial \tau_{ij}} (df_p^{(2)} - df_p^{(1)})$$

and

$$I_p = (d\tau_{ij}^{(2)} - d\tau_{ij}^{(1)}) c_p h_p \frac{\partial f_p}{\partial \tau_{ij}} (df_p^{(2)} - df_p^{(1)})$$

and again consider the following four cases for each p:

(i) When $f_p < 0$, then $c_p = 0$, and hence $I = 0$.
(ii) When $f_p = 0$, $df_p^{(1)} > 0$, and $df_p^{(2)} < 0$, then

$$I_p = (d\tau_{ij}^{(2)} - d\tau_{ij}^{(1)}) h_p \frac{\partial f_p}{\partial \tau_{ij}} (-df_p^{(1)})$$

$$= h_p(df_p^{(1)}\, df_p^{(1)} - df_p^{(1)}\, df_p^{(2)}) \geq 0$$

(iii) When $f_p = 0$, $df_p^{(2)} > 0$, and $df_p^{(1)} < 0$, then, as in (ii), $I_p \geq 0$.
(iv) When $f_p = 0$, $df_p^{(1)} > 0$, and $df_p^{(2)} > 0$, then

$$I_p = h_p(df_p^{(2)} - df_p^{(1)})(df_p^{(2)} - df_p^{(1)}) \geq 0 \tag{4.6.6}$$

Since for all cases $I_p \geq 0$, we have

$$I = \sum_{p=1}^{h} I_p \geq 0$$

and therefore

$$(d\tau_{ij}^{(2)} - d\tau_{ij}^{(1)})(de_{ij}^{(2)} - de_{ij}^{(1)}) \geq 0 \tag{4.6.7}$$

whether the stress vector is on a smooth part or at a singular point of the loading surface.

From this and Equation (4.6.1), we have the following result:

$$(d\tau_{ij}^{(2)} - d\tau_{ij}^{(1)})(de_{ij}^{(2)} - de_{ij}^{(1)}) = 0 \tag{4.6.8}$$

Equation (4.6.8) establishes uniqueness on the set (τ_{ij}, e_{ij}) to the extent that one or more of the following conditions are satisfied:

(i) $d\tau_{ij}^{(2)} = d\tau_{ij}^{(1)}$
(ii) $de_{ij}^{(2)} = de_{ij}^{(1)}$
(iii) $(d\tau_{ij}^{(2)} - d\tau_{ij}^{(1)})$ is orthogonal to $(de_{ij}^{(2)} - de_{ij}^{(1)})$

For a perfectly plastic material, $df = 0$, and Equation (4.6.5) is replaced by

$$de''_{ij} = \sum_{p=1}^{n} \lambda_p \frac{\partial f_p}{\partial \tau_{ij}} \tag{4.6.9}$$

where

$$\begin{aligned} \lambda_p &= 0 \qquad \text{if } f_p < 0 \text{ or } df_p < 0 \\ \lambda_p &\geq 0 \qquad \text{if } f_p = 0 \text{ and } df_p = 0 \end{aligned} \tag{4.6.10}$$

Letting

$$I_p = (d\tau_{ij}^{(2)} - d\tau_{ij}^{(1)})\left(\lambda_p^{(2)} \frac{\partial f_p}{\partial \tau_{ij}} - \lambda_p^{(1)} \frac{\partial f_p}{\partial \tau_{ij}}\right) \tag{4.6.11}$$

we can show as in the previous cases that $I_p \geq 0$, and uniqueness to the extent given by Equation (4.6.8) is established also for perfectly plastic materials.

4.7 Theorems of Limit Analysis[13,21,45,78–81]

For a structure subject to boundary force only, $F_i = 0$, and Equation (1.9.8) reduces to

$$\int_{\Gamma_s} S_i \, \delta u_i \, d\Gamma = \int_v \tau_{ij} \, \delta e_{ij} \, dv \tag{4.7.1}$$

Replacing the virtual displacement δu_i and the virtual strain δe_{ij} by the actual incremental displacement du_i and the actual incremental strains de_{ij}, we have

$$\int_{\Gamma_s} S_i \, du_i \, d\Gamma = \int_v \tau_{ij} \, de_{ij} \, dv \tag{4.7.2}$$

Assume the structure to be rigid, and perfectly plastic, so that $de_{ij} = de''_{ij}$; that is, the total strain is purely plastic. As the S_i are increased to certain values, the stresses in parts of the structure reach the elastic limit, but deformation does not take place because of the support of the surrounding rigid material. When the S_i reach the collapse load, the structure will undergo a plastic deformation of arbitrary magnitude. In the following discussion, however, we assume that the deformations are sufficiently small that changes of geometry may be ignored. It is further assumed that the deformations take place quasi-statically and thus inertia forces may be neglected.

Let τ_{ij}^E be a continuous stress field which satisfies the equilibrium and boundary conditions for the surface traction denoted by $\lambda_s S_i$ and that $\lambda_s S_i$ is of such magnitude that τ_{ij}^E does not exceed the yield limit at any point in the structure. Any such stress field is known as a *statically admissible stress field.* Equation (4.7.2) becomes

$$\int_v \tau_{ij}^E \, de_{ij} \, dv = \lambda_s \int_\Gamma S_i \, du_i \, d\Gamma \tag{4.7.3}$$

Let the boundary Γ of the body be composed of three parts: Γ_s, on which the applied force is specified; one part on which the displacement is *zero*; and one part on which the boundary force vanishes. Then from Equations (4.7.2) and (4.7.3), noting that $de_{ij} = de''_{ij}$, we obtain

$$\int_v (\tau_{ij}^E - \tau_{ij}) \, de''_{ij} \, dv = (\lambda_s - 1) \int_{\Gamma_s} S_i \, du_i \, d\Gamma$$

Assuming the coincidence of the plastic potential with the yielding surface, we have, from Section 4.5,

$$(\tau_{ij} - \tau_{ij}^E) \, de''_{ij} \geq 0$$

Hence

$$(\lambda_s - 1) \int_{\Gamma_s} S_i \, du_i \, d\Gamma = \int_v (\tau_{ij}^E - \tau_{ij}) \, de''_{ij} \, dv \leq 0 \tag{4.7.4}$$

Since

$$\int_{\Gamma_s} S_i \, du_i \, d\Gamma \geq 0$$

the quantity $(\lambda_s - 1)$ must be less than or equal to zero, and hence $\lambda_s \leq 1$. The load $\lambda_s S_i$ corresponding to any statically admissible stress field is *less* than the limiting loads the structure can carry. This is the *first theorem of limit analysis.*

Let de_{ij}^c denote an incremental strain field which satisfies the condition of continuity of displacement and du_i^c be the displacement corresponding to de_{ij}^c. From Equation (4.7.2), we have

$$\int_{\Gamma_s} S_i \, du_i^c \, d\Gamma = \int_v \tau_{ij} \, de_{ij}^c \, dv \tag{4.7.5}$$

Let τ_{ij}^c denote the stress corresponding to de_{ij}^c. We write

$$\int_v \tau_{ij}^c \, de_{ij}^c \, dv = \lambda_k \int_{\Gamma_s} S_i \, du_i^c \, d\Gamma \tag{4.7.6}$$

where λ_k is defined as the ratio of the above two integrals. From Equations (4.7.5) and (4.7.6), we have

$$(\lambda_k - 1) \int_{\Gamma_s} S_i \, du_i^c \, d\Gamma = \int_v (\tau_{ij}^c - \tau_{ij}) \, de_{ij}^c \, dv \tag{4.7.7}$$

Again assume the coincidence of the plastic potential with the yielding surface, and consider that τ_{ij}^c is such as to cause yielding and that τ_{ij} is within or on the yield surface. Then, from Section 4.5, recall that

$$(\tau_{ij}^c - \tau_{ij}) \, de_{ij}'' \geq 0$$

and therefore Equation (4.7.7) becomes

$$(\lambda_k - 1) \int_{\Gamma_s} S_i \, du_i^c \, d\Gamma \geq 0$$

Since the integral is positive, $(\lambda_k - 1)$ must be greater than or equal to zero, hence $\lambda_k \geqq 1$.

The parameter λ_k is known as a *kinematically admissible multiplier.* The load corresponding to a kinematic admissible deformation for a rigid-plastic structure gives an upper bound to the actual load that the structure can carry. This is known as the *second theorem of limit analysis.* Much[13,21,45,78–81] has been written by distinguished writers on the application of limit analysis to structures. Detail applications of limit analysis are given in many of the works listed at the end of this chapter.

4.8 Multiaxial Stress-Strain-Time Relations of Creep

As was shown in Section 3.4, for uniaxial stress the creep rate $\dot{e}_c$ under constant temperature may be approximately expressed as a function of the current stress and the current amount of creep strain,

$$\dot{e}_c = f(e_c, \sigma) \tag{3.4.1}$$

Under multiaxial stress, this equation of state has been generalized[67] by replacing σ with σ^*, as given in Equation (3.6.4), and e_c by

$$e_c^* = \frac{\sqrt{2}}{3}[(e_{11_c} - e_{22_c})^2 + (e_{22_c} - e_{33_c})^2 + (e_{33_c} - e_{11_c})^2 + 6(e_{12_c}^2 + e_{23_c}^2 + e_{31_c}^2)]^{1/2} \tag{4.8.1}$$

or in terms of principal strains as

$$e_c^* = \sqrt{\tfrac{2}{9}\{(e_{1_c} - e_{2_c})^2 + (e_{2_c} - e_{3_c})^2 + (e_{3_c} - e_{1_c})^2\}} \tag{4.8.2}$$

The generalized creep relation becomes

$$\dot{e}_c^* = f(e_c^*, \sigma^*) \tag{4.8.3}$$

$\dot{e}_c^*$ can be obtained by taking the time derivative of Equation (4.8.1) or (4.8.2); however, the following simplified expression for $\dot{e}_c^*$ has also been used:

$$\dot{e}_c^* = \frac{\sqrt{2}}{3}[(\dot{e}_{11_c} - \dot{e}_{23_c})^2 + (\dot{e}_{22_c} - \dot{e}_{33_c})^2 + (\dot{e}_{33_c} - \dot{e}_{11_c})^2 + 6(\dot{e}_{12_c}^2 + \dot{e}_{23_c}^2 + \dot{e}_{31_c}^2)]^{1/2} \tag{4.8.4}$$

or, in terms of principal strain rates,

$$\dot{e}_c^* = \frac{\sqrt{2}}{3}[(\dot{e}_{1_c} - \dot{e}_{2_c})^2 + (\dot{e}_{2_c} - \dot{e}_{3_c})^2 + (\dot{e}_{3_c} - \dot{e}_{1_c})^2]^{1/2} \tag{4.8.5}$$

For uniaxial tests, $\sigma^* = \tau_{11} = \frac{3}{2}S_{11}$. Since there is no change in volume associated with creep strain, we have

$$\left.\begin{aligned} \dot{e}_{22_c} &= \dot{e}_{33_c} = -\tfrac{1}{2}\dot{e}_{11_c} \\ \dot{e}_c &= \dot{e}_{11_c} \\ \frac{\dot{e}_{11_c}}{S_{11}} &= \frac{3\dot{e}_c^*}{2\sigma^*} \end{aligned}\right\} \tag{4.8.6}$$

The principle axes of strain rate are assumed to coincide with those of principal stresses as given in Equation (4.5.1). Then

$$\frac{\dot{e}_{11_c}}{S_{11}} = \frac{\dot{e}_{22_c}}{S_{22}} = \frac{\dot{e}_{33_c}}{S_{33}} = \frac{\dot{e}_{12_c}}{S_{12}} = \frac{\dot{e}_{23_c}}{S_{23}} = \frac{\dot{e}_{31_c}}{S_{31}} = \frac{3\dot{e}_c^*}{2\sigma^*} \tag{4.8.7}$$

or, in terms of principal stresses,

$$\frac{\dot{e}_{1_c} - \dot{e}_{2_c}}{\sigma_1 - \sigma_2} = \frac{\dot{e}_{2_c} - \dot{e}_{3_c}}{\sigma_2 - \sigma_3} = \frac{\dot{e}_{3_c} - \dot{e}_{1_c}}{\sigma_3 - \sigma_1} \tag{4.8.8}$$

This generalization of the uniaxial stress-strain-rate relation to the multiaxial stress state is exactly the same as in the plastic stress-strain relation. For the case of steady creep, Equation (4.8.3) reduces to

$$\dot{e}_c^{\,*} = F(\sigma^*) \tag{4.8.9}$$

These relations [Equations (4.8.3) and (4.8.7)] seem to represent—approximately—the properties of a number of metals. For the comparison of these relations with experimental results, the reader is referred to the works by Johnson[67,68] and Kennedy.[69]

The time-dependent strain (creep strain) of metals has no change in volume, so the volumetric strain is purely elastic, hence

$$e_{ii_c} = 0$$

and

$$e_{ii} = \frac{1 - 2\nu}{E}\,\tau_{ii} \tag{4.8.10}$$

The strain is composed of volumetric strain e_{ii} and the deviatoric strain ε_{ij}. Equations (4.8.7) and (1.10.17) give the complete stress-strain-time relations of materials with nonlinear creep.

For isotropic *viscoelastic materials*,[70–77] the deviatoric strain component ε_{ij} is related to the deviatoric stress S_{ij} as in Equation (3.8.21),

$$\overline{S_{ij}} = 2G[1 - \bar{\phi}(s)]\bar{\varepsilon}_{ij} \tag{4.8.11}$$

where the bar above a quantity denotes the Laplace transform of the quantity. Taking the inverse transform, we obtain

$$S_{ij} = 2G\left\{\varepsilon_{ij} - \int_0^t \phi(t - \tau)\varepsilon_{ij}(\tau)\,d\tau\right\} \tag{4.8.12}$$

Also,

$$\tau_{ij} = S_{ij} + \delta_{ij}\frac{\tau_{kk}}{3}$$

$$\overline{\tau_{ij}} = \overline{S_{ij}} + \delta_{ij}\frac{\overline{\tau_{kk}}}{3} \tag{4.8.13}$$

For viscoelastic materials with no time-dependent volumetric strain,

$$e_{kk} = \frac{1 - 2v}{E} \tau_{kk}$$

$$\bar{\tau}_{ij} = 2G[1 - \bar{\phi}(s)]\bar{\varepsilon}_{ij} + \delta_{ij} \frac{E}{3(1 - 2v)} \overline{e_{kk}}$$

$$= 2G[1 - \bar{\phi}(s)]\overline{e_{ij}} + \left\{\frac{E}{3(1 - 2v)} - \frac{2G}{3} [1 - \bar{\phi}(s)]\right\} \delta_{ij} \overline{e_{kk}}$$

By taking the inverse transform, we obtain, upon noting that $E = 2G(1 + v)$,

$$\tau_{ij}(t) = 2Ge_{ij}(t) + \frac{E}{3(1 - 2v)} \delta_{ij} e_{kk}(t) - \frac{2G}{3} \delta_{ij} e_{kk}(t)$$

$$- 2G \int_0^t \phi(t - \tau)\left[e_{ij}(\tau) - \delta_{ij} \frac{e_{kk}(\tau)}{3}\right] d\tau$$

$$= 2Ge_{ij}(t) + \frac{2Gv}{1 - 2v} \delta_{ij} e_{kk}(t)$$

$$- 2G \int_0^t \phi(t - \tau)\left[e_{ij}(\tau) - \delta_{ij} \frac{e_{kk}(\tau)}{3}\right] d\tau \qquad (4.8.14)$$

The above two equations give the multiaxial stress-strain-time relation of viscoelastic materials. If the volumetric strain is also time-dependent and has a linear relation with the hydrostatic stress, as the deviatoric strain with the deviatoric stress, then this relation may be similarly represented by Equation (4.8.11),

$$\bar{\tau}_{kk} = \frac{E}{1 - 2v} [1 - \bar{\psi}(s)]\bar{e}_{kk}$$

where $[1 - \bar{\psi}(s)]$ is a function similar to $\phi(s)$ in Equation (3.6.21):

$$\bar{\tau}_{ij} = \bar{S}_{ij} + \delta_{ij} \frac{\bar{\tau}_{kk}}{3}$$

$$= 2G[1 - \phi(s)]\bar{\varepsilon}_{ij} + \delta_{ij} \frac{E}{3(1 - 2v)} [1 - \bar{\psi}(s)]\bar{e}_{kk}$$

The stress-strain-time relation may be similarly obtained by taking the inverse transform.

Uniqueness of Creep Solution. Creep strain is time-dependent for both viscoelastic materials and materials with other creep characteristics. Let t

be the time from the instant of load application. Since creep strain is zero at $t = 0$, the initial stress distribution is an elastic one and is therefore unique. If one considers the stress field to remain constant during an infinitesimal time increment, the incremental creep strain at different points in the body for this time increment is uniquely determined by the creep characteristics of the material. This incremental creep strain may be considered as an additional set of incremental applied forces on an identical elastic body as shown in Section 2.3. The strain field (and hence the stress field) caused by this set of equivalent applied forces on this identical elastic body is unique, and therefore the stress and strain field for the actual body is also unique. This reasoning is valid for each subsequent time increment, and hence the incremental stress and strain solutions for all time increments are unique. This establishes the uniqueness of creep solutions for stresses and strains.

For further discussion of viscoelastic materials, see References 70 through 77 and the many other excellent works listed in these references.

REFERENCES

1. Taylor, G. I., and C. F. Elam, "The Distortion of an Aluminum Crystal During a Tensile Test," *Proc. Roy. Soc. A.*, **102**, 643–667, 1923.
2. Taylor, G. I., and C. F. Elam, "The Plastic Extension and Fracture of Aluminum Crystals," *Proc. Roy. Soc. A.*, **108**, 28–51, 1925.
3. Taylor, G. I., and C. F. Elam, "The Distortion of Iron Crystals," *Proc. Roy. Soc. A.*, **112**, 337–361, 1926.
4. Taylor, G. I., "The Deformation of Crystals of β-Brass," *Proc. Roy. Soc. A.*, **114**, 121–125, 1928.
5. Taylor, G. I., "Plastic Strain in Metals," *J. Inst. Metals*, **62**, 306–324, 1938.
6. Elam, C. F., *Distortion of Metal Crystals*, Clarendon Press, Oxford, England, 1935.
7. MacMillan, W. D., *The Theory of the Potential*, Dover, New York, pp. 24–25, 1930.
8. Barrett, C. S., *Structure of Metals*, McGraw-Hill, New York, pp. 354, 360, 533, 1950.
9. Bishop, J. F. W., and R. Hill, "A Theory of Plastic Distortion of Polycrystalline Aggregate Under Combined Stresses," *Phil. Mag.*, **42**, 414, 1298, 1951.
10. Drucker, D. C., "A More Fundamental Approach to Plastic Stress-Strain Relations," *Proc. 1st U.S. Nat. Congr. Appl. Mech.*, 487–491, 1952.
11. Lin, T. H., "Analysis of Elastic and Plastic Strain in a Face-Centered Cubic Crystal," *J. Mech. Phys. Solids*, **5**, 143–149, 1957.

12. Hill, R., *The Mathematical Theory of Plasticity*, Clarendon Press, Oxford, England, pp. 14–49, 1950.
13. Olszak, W., Z. Mroz, and P. Perzyna, *Recent Trends in the Development of the Theory of Plasticity*, Macmillan, New York, pp. 13–49, 84–152, 1963.
14. Johnson, W., and P. B. Mellow, *Plasticity for Mechanical Engineers*, Van Nostrand, Princeton, N.J., pp. 41–55, 1961.
15. Hoffman, O., and G. Sachs, *Introduction to the Theory of Plasticity for Engineers*, McGraw-Hill, New York, 1953.
16. Taylor, G. I., "Strains in Crystalline Aggregates," *Proc. Colloq. on Deformation and Flow of Solids*, Springer-Verlag, Berlin, pp. 3–12, 1956.
17. Crossland, B., "The Effect of Fluid Pressure on the Shear Properties of Metals," *Proc. Inst. Mech. Eng.*, **169**, 935–944, 1954.
18. Tresca, H., *Comptes Rendus*, Acad. Sci., Paris, **59**, 754, 1864.
19. Huber, M. T., "Czasopismo Techniczne," **22**, 81, Lemberg, Poland, 1904.
20. Mises, R. Von, "Gottinger Nachrichten," *Math. Phys.*, **Kl 582**, 1913.
21. Prager, W., and P. G. Hodge, Jr., *Theory of Perfectly Plastic Solids*, Wiley, New York, 1951.
22. Nadai, A., *Theory of Flow and Fracture of Solids*, Vol. 1, McGraw-Hill, New York, 1950.
23. Novozhilov, V. V., "The Physical Meaning of Stress Invariants of the Theory of Plasticity" (in Russian), *Prikl. Mat. i. Mekh.*, translated by Irving Patten, **16**, 617, 1952.
24. Taylor, G. I., and H. Quinney, "The Plastic Distortion of Metals," *Phil. Trans. Roy. Soc.*, **230**, 323, 1921.
25. Lin, T. H., and M. Ito, "Theoretical Plastic Distortion of a Polycrystalline Aggregate under Combined and Reversed Stresses," *J. Mech. Phys. Solids*, **13**, 103–115, 1965.
26. Lin, T. H., and M. Ito, "Theoretical Plastic Stress-Strain Relationship of a Polycrystal and Comparisons with Von Mises and the Tresca Plasticity Theories," *Intern. J. Eng. Sci.*, **4**, 543–561, 1966.
27. Colonnetti, G., "Elastic Equilibrium in the Presence of Permanent Set," *Quart. Appl. Math.*, **7**, 353–362, 1950.
28. Drucker, D. C., *Stress-Strain Relations in the Plastic Range, A Survey of Theory and Experiment*, Rept. A11-S1, Grad. Div. of Appl. Math., Brown University, December 1950.
29. Naghdi, P. M., "Stress-Strain Relations in Plasticity and Thermoplasticity," *Plasticity* (Proc. Second Symposium on Naval Structural Mechanics, edited by E. H. Lee and P. S. Symonds), Pergamon Press, London, pp. 121–169, 1960.
30. Havner, K. S., *Fundamental Aspects of the Theory of Stress Hardening Solids*, Douglas Aircraft Co. Rept. SM45897, TM161, February 1964.
31. Prager, W., "The Theory of Plasticity: A Survey of Recent Achievements," *Proc. Inst. Mech. Engr.*, London, 1955.
32. Ziegler, H., "A Modification of Prager's Hardening Rule," *Quart. Appl. Math.*, **17**, 55–65, 1959.
33. Sanders, J. L., Jr., "Plastic Stress-Strain Relations Based on Linear Loading Functions," *Proc. 2nd U.S. Nat. Congr. Appl. Mech.* (Ann Arbor, Mich., 1954), pp. 455–460, ASME, New York, 1951.
34. Lin, T. H., "On the Associated Flow Rule of Plasticity Based on Crystal Slip," *J. Franklin Inst.*, **270**, 291–300, 1960.

35. Naghdi, P. M., F. Essenburg, and W. Koff, "An Experimental Study of Initial and Subsequent Yield Surfaces in Plasticity," *J. Appl. Mech.*, **25**, 201–209, 1958.
36. Phillips, A., "Pointed Vertices in the Yield Surface," ONR Tech. Report No. 7, Contract Nonr-609(12), Yale University, September 1959.
37. Phillips, A., and G. Gray, "Experimental Investigation of Corners in the Yield Surface," ONR Tech. Report No. 5, Yale University, December 1958.
38. Phillips, A., and L. Kaechele, "Combined Stress Tests in Plasticity," *J. Appl. Mech.*, **23**, 43–48, 1958.
39. Handelman, E. H., and W. Prager, "Plastic Buckling of a Rectangular Plate Under Edge Thrusts," NACA Tech. Report 946, 1946.
40. de Saint-Venant, B., *Comptes Rendus*, Acad. Sci., Paris, **70**, 473, 1870.
41. Reuss, A., *Z. Angew. Math. Mech.*, **10**, 266, 1930.
42. Hencky, H., "Zur Theorie Plastischer Deformationen und des hier durch im Material hervorgerufenen Dachspanrungen," *Z. Angew. Math. Mech.*, **4**, 323, 1924.
43. Handleman, G. H., C. C. Lin, and W. Prager, "On the Mechanical Behavior of Metals in Strain-Hardening Range," *Quart. Appl. Math.*, **4**, 397–407, 1947.
44. Budiansky, B., "A Reassessment of Deformation Theory of Plasticity," *J. Appl. Mech.*, **26**, 259–264, 1959.
45. Beedle, L. S., *Plastic Design of Steel Frames*, Wiley, New York, p. 4, 1958.
46. Hodge, P. G., Jr., "Piecewise Linear Plasticity," *Proc. 9th Intern. Congr. Appl. Mech.* (Brussels, 1956), **8**, 65–72, 1951.
47. Hodge, P. G., Jr., "Theory of Piecewise Linear Isotropic Plasticity," *Proc. Colloquium on Deformation and Flow of Solids*, Madrid, Spain, pp. 147–169, 1955.
48. Feigen, M., "Inelastic Behavior Under Combined Tension and Torsion," *Proc. 2nd U.S. Nat. Congr. Appl. Mech.*, ASME, New York, 469–476, 1954.
49. Kliushnikov, V. D., "On Plasticity Laws for Work-Hardening Materials," *PMM J. Appl. Math. Mech.*, **22**, 129–160, 1958.
50. Koiter, W. T., "Stress-Strain Relations, Uniqueness and Variational Theorems for Elastic-Plastic Materials with a Singular Yield Surface," *Quart. Appl. Math.*, **11**, 350–354, 1953.
51. Shield, R. T., and H. Ziegler, "On Prager's Hardening Rule," *Z. Angew. Math. Phys.*, **9a**, 260–276, 1958.
52. Kliushnikov, V. D., "On Laws of Plasticity for a Particular Class of Loading Paths" (in Russian), *Prikl. Mat. Mekh.*, **22**, 533–544, 1957.
53. Budiansky, B., N. F. Dow, R. W. Peters, and R. P. Shepherd, "Experimental Studies of Polyaxial Stress-Strain Laws and Plasticity," *Proc. 1st U.S. Nat. Congr. Appl. Mech.* (Chicago, 1951), pp. 503–512, New York, 1952.
54. Batdorf, S. B., and B. Budiansky, "A Mathematical Theory of Plasticity Based on Concept of Slip," NACA TN 1871, 1949.
55. Lin, T. H., "A Proposed Theory of Plasticity Based on Slip," *Proc. 2nd U.S. Nat. Congr. Appl. Mech.*, ASME, New York, pp. 461–463, 1954.
56. Lin, T. H., "On Stress-Strain Relations Based on Slip," *Proc. 3rd U.S. Nat. Congr. Appl. Mech.*, pp. 581–587, 1958.
57. Kliushnikov, V. D., "On Plasticity Laws for Work-Hardening Materials," PMM (Translation of *Prikl. Mat. i. Mekh.*), **22**, 129–160, 1958.
58. Drucker, D. C., "Plasticity," *Proc. 1st Symp. Naval Structural Mechanics*, Pergamon Press, London, pp. 407–455, 1960.
59. Koiter, W. T., "General Theorems for Elastic-Plastic Solids," *Progress in Solid Mechanics*, Vol. 1, I. N. Sneddon and R. Hill, (eds.), Chap. 4, North-Holland, 1960.

60. Sokolvskii, V. V., *Theorie der Plastizitat* (translated from the Russian 1945 edition), Berlin, 1955.
61. Bland, D. R., "The Associated Flow Rule of Plasticity," *J. Mech. Phys. Solids*, **6**, 71–78, 1957.
62. Kliushnikov, V. D., "On a Possible Manner of Establishing the Plasticity Relations," PMM (Translation of *Prikl. Mat. i. Mekh.*), **23**, 405–418, 1959.
63. Kliushnikov, V. D., "New Concepts in Plasticity and Deformation Theory," PMM (Translation of *Prikl. Mat. i. Mekh.*), **23**, pp. 1030–42, 1959.
64. Lensky, V. S., "Analysis of Plastic Behavior of Metals Under Complex Loading, Plasticity," *Proc. 2nd Symp. on Naval Structural Mechanics*, Pergamon, London, pp. 259–278, 1960.
65. Ilyushin, A. A., *Prikl. Mat. Mekh.*, **18**, 641–666, 1954.
66. Ilyushin, A. A., *Plastichnosti*, Moscow, 1948.
67. Johnson, A. E., "Creep Under Complex Stress Systems at Elevated Temperatures," *Proc. Inst. Mech. Engr.*, **164**, 432, 1951.
68. Johnson, A. E., J. Henderson, and B. Khan, "Complex Stress-Creep Relaxation and Fracture of Metallic Alloys," H. M. Stationery Office, Edinburgh, 1962.
69. Kennedy, C. R., W. O. Harms, and D. A. Douglas, "Multiaxial Creep Studies on Inconel at 1500° F," *J. Basic Eng.*, **81**, 599, 1959.
70. Brull, M. A., *A Structural Theory Incorporating the Effect of Time-Dependent Elasticity*, Ph.D. Dissertation, University of Michigan, 1952.
71. Freudenthal, A. M., *The Inelastic Behavior of Engineering Materials and Structures*, Wiley, New York, pp. 214–248, 1950.
72. Bland, D. R., *The Theory of Linear Viscoelasticity*, Pergamon, London, 1960.
73. Alfrey, T., *Mechanical Behavior of High Polymers*, Interscience, New York, 1948.
74. Fung, Y. C., *Foundations of Solid Mechanics*, Prentice-Hall, Englewood Cliffs, N.J., pp. 20–30, 412–433, 1965.
75. Williams, M. L., "Structural Analysis of Viscoelastic Materials," *AIAA Journal*, **2**, 785, 1964.
76. Lee, E. H., "Some Recent Developments in Linear Viscoelastic Stress Analysis," *Proc. 11th. Intern. Congr. Appl. Mech.*, edited by H. Gortler, Springer-Verlag, Berlin, Germany, p. 396, 1964.
77. Hilton, H. H., "A Summary of Linear Viscoelastic Stress Analysis," Tech. Rept. AAE 65–2, 1965.
78. Goodier, J. N., and P. G. Hodge, Jr., *Elasticity and Plasticity*, Wiley, New York, pp. 51–120, 1958.
79. Hodge, P. G., Plastic Analysis of Structures, McGraw Hill, New York, 1959.
80. Baker, J. F., M. R. Horne, and J. Heyman: *Plastic Behavior and Design*, Vol. 2: The Steel Skeleton, Cambridge University Press, New York, 1956.
81. Drucker, D. C., W. Prager, and H. J. Greenberg, "Extended Limit Design Theorems for Continuous Media," *Quart. Appl. Math.* **9**, pp. 381–389, January 1952.
82. Prandtl, L., "Spannungsverteilung in plastischen Körpern," *Proc. 1st Intern. Congr. Appl. Mech.*, Delft, The Netherlands, pp. 43–54, 1924.

chapter

5

Inelastic Beams

Introduction. In this and the next chapter, inelastic structures with uniaxial stress are considered. In this chapter methods for determining deflection curves of determinate beams and for analyzing indeterminate beams with inelastic strain are given. The analogy between inelastic strains and externally applied moments is derived and tabulated. The steps for the analysis of an elastoplastic indeterminate beam under reversed loading are shown. The method may be extended to elastoplastic analysis of rigid frames.

5.1 Curvature-Moment Relation

Classical beam theory is based on the Bernoulli-Euler assumption that plane sections before bending remain plane after bending. This assumption has good experimental corroboration[1,2,3,4] and is used in the present analysis.

Let x, y, and z be a set of rectangular coordinate axes with the x axis along the span of the beam and passing through the centroids of the sections. Consider two parallel planes perpendicular to the x axis and spaced at a distance Δx apart. Let u denote the displacement along the x axis and Δu denote the relative displacement of one plane to the other. Since these two planes remain planes after loading, Δu and hence $\Delta u/x\Delta$ are linear functions of y and z. With Δx approaching zero, $\Delta u/\Delta x$ gives the tensile axial strain e. Hence the axial strain e is a linear function of y and z, and may be expressed as

$$e = a_1 y + a_2 z + c$$

where a_1, a_2, and c are constants.

Let the strain e at the origin be denoted by e_0; then

$$e = a_1 y + a_2 z + e_0 \tag{5.1.1}$$

Consider the strain in the xz plane. From the similarity of OAB and OCD in Figure 5.1, we have

$$\frac{AB}{OA} = \frac{CD}{OA + AC}$$

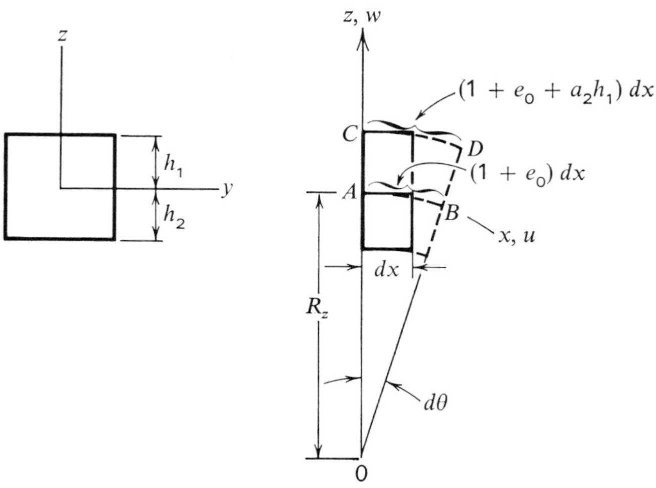

Figure 5.1. Curvature and bending strains.

Noting that $AB = dx + e_0\, dx$, $CD = AB + a_2 h_1\, dx$, $OA = R_z$, and $OC = (R_z + h_1)$, we obtain

$$\frac{dx(1 + e_0)}{R_z} = \frac{dx(1 + a_2 h_1 + e_0)}{R_z + h_1}$$

$$R_z(1 + e_0 + a_2 h_1) = (R_z + h_1)(1 + e_0)$$

$$R_z a_2 h_1 = h_1(1 + e_0)$$

$$a_2 = \frac{1 + e_0}{R_z} \cong \frac{1}{R_z}$$

Let w denote the upward vertical deflection of the beam and K_z the curvature. The curvature is given by the well-known equation,

$$a_2 = \frac{1}{R_z} = K_z = \frac{-(\partial^2 w/\partial x^2)}{[1 + (\partial w/\partial x)^2]^{3/2}}$$

For small deflections $\partial w/\partial x$ [and hence $(\partial w/\partial x)^2$] is small compared to unity, and hence

$$a_2 = K_z \cong -\frac{\partial^2 w}{\partial x^2} \tag{5.1.2}$$

Similarly

$$a_1 = K_y \cong -\frac{\partial^2 v}{\partial x^2} \tag{5.1.3}$$

where K_y is the curvature of the beam in the xy plane and v is the displacement along the y axis. Substituting Equations (5.1.2) and (5.1.3) into Equation (5.1.1) yields

$$e = K_y y + K_z z + e_0 \tag{5.1.4}$$

Substituting Equations (5.1.3) and (5.1.4) into the uniaxial elastic stress-strain relation, $\sigma = Ee' = E(e - e'')$, gives

$$\sigma = E(K_y y + K_z z + e_0 - e'') \tag{5.1.5}$$

where e is the total axial strain and e'' is the inelastic axial strain. The axial force on an element of area dA is

$$dF = \sigma\, dA$$

Upon integration of dF over the area, we obtain

$$F = \int_A \sigma\, dA = \int_A E(K_y y + K_z z + e_0 - e'')\, dA \tag{5.1.6}$$

Since the y and z axes are chosen to pass through the centroid of each section, we have

$$\int_A y\, dA = 0 \qquad \int_A z\, dA = 0 \tag{5.1.7}$$

Substitution of Equations (5.1.7) into Equation (5.1.6) yields

$$F = Ee_0 A - E\int_A e''\, dA \tag{5.1.8}$$

Upon solving for e_0, we have

$$e_0 = \frac{F + F_I}{EA} \tag{5.1.9}$$

where

$$F_I = E\int_A e''\, dA \tag{5.1.10}$$

is referred to as *inelastic force*.

Referring to Figure 5.2, we see that the moments about the y and z axes are denoted by M_y and M_z. The sign convention is such that compressive stress in the first quadrant gives positive M_y and M_z. The moment about the y axis contributed by an element of area dA is then

$$dM_y = -z\sigma\, dA = -zE(K_y y + K_z z + e_0 - e'')\, dA \qquad (5.1.11)$$

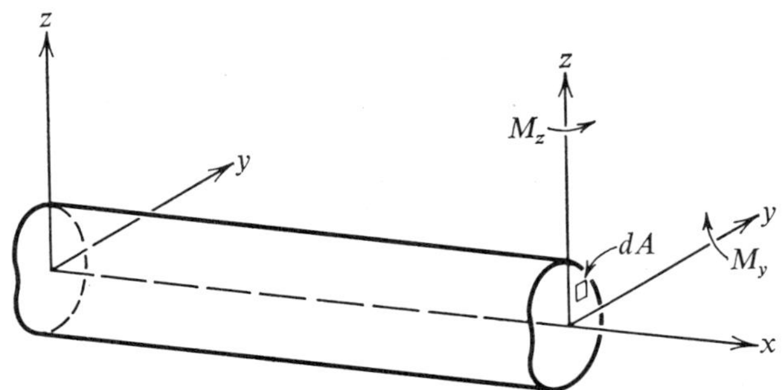

Figure 5.2. Sign convention of bending moments.

Upon integration, and using Equation (5.1.7), we obtain

$$M_y = -E\left(K_y I_{yz} + K_z I_{yy} - \int_A ze''\, dA\right) \qquad (5.1.12)$$

where I_{yy} denotes the moment of inertia about the y axis and I_{yz} denotes the product of inertia, that is,

$$I_{yy} = \int_A z^2\, dA \qquad I_{yz} = \int_A yz\, dA$$

Similarly for the moment about the z axis, M_z, we have

$$M_z = -E\left(K_y I_{zz} + K_z I_{yz} - \int_A ye''\, dA\right) \qquad (5.1.13)$$

where I_{zz}, the moment of inertia about the z axis, is

$$I_{zz} = \int_A y^2\, dA$$

Let

$$-E\int_A ze''\, dA = M_{y_I}$$

$$-E\int_A ye''\, dA = M_{z_I} \qquad (5.1.14)$$

where M_{y_I} and M_{z_I} are referred to as *inelastic moments.* Using these expressions, we may write Equations (5.1.12) and (5.1.13) as

$$M_y + M_{y_I} = -E(K_y I_{yz} + K_z I_{yy})$$
$$M_z + M_{z_I} = -E(K_y I_{zz} + K_z I_{yz}) \tag{5.1.15}$$

Solution of these two simultaneous equations yields

$$K_y = \frac{(M_y + M_{y_I}) I_{yz} - (M_z + M_{z_I}) I_{yy}}{E(I_{yy} I_{zz} - I_{yz}^2)} \cong -\frac{\partial^2 v}{\partial x^2} \tag{5.1.16}$$

$$K_z = \frac{(M_z + M_{z_I}) I_{yz} - (M_y + M_{y_I}) I_{zz}}{E(I_{yy} I_{zz} - I_{yz}^2)} \cong -\frac{\partial^2 w}{\partial x^2} \tag{5.1.17}$$

Substituting Equations (5.1.9), (5.1.16), and (5.1.17) into Equation (5.1.5) yields

$$\sigma = \frac{(M_y + M_{y_I}) I_{yz} - (M_z + M_{z_I}) I_{yy}}{(I_{yy} I_{zz} - I_{yz}^2)} \cdot y + \frac{(M_z + M_{z_I}) I_{yz} - (M_y + M_{y_I}) I_{zz}}{(I_{yy} I_{zz} - I_{yz}^2)} \cdot z + \frac{F + F_I}{A} - Ee'' \tag{5.1.18}$$

If the section has one plane of symmetry, or the y and z axes are principal axes, the product of inertia I_{yz} vanishes, and Equations (5.1.16) through (5.1.18) reduce to

$$K_y = -\frac{(M_z + M_{z_I})}{EI_{zz}} \cong -\frac{\partial^2 v}{\partial x^2} \tag{5.1.19}$$

$$K_z = -\frac{(M_y + M_{y_I})}{EI_{yy}} \cong -\frac{\partial^2 w}{\partial x^2} \tag{5.1.20}$$

$$\sigma = -\frac{(M_z + M_{z_I})}{I_{zz}} \cdot y - \frac{(M_y + M_{y_I})}{I_{yy}} \cdot z + \frac{F + F_I}{A} - Ee'' \tag{5.1.21}$$

Equations (5.1.18) and (5.1.21) give the stress distribution of beams with a given arbitrary variation of inelastic strain.

5.2 Determinate Beams Subject to Lateral Load and Thermal Strain

Any part of a structure under static load must satisfy the conditions of equilibrium. If a beam is supported so that the external reactions and internal moments are completely determinable by the conditions of equilibrium, the beam is said to be *statically determinate.* If, on the other hand, the beam has a greater number of unknown reactions or internal moments, or both, than the number of conditions of equilibrium, the beam is said to be *statically indeterminate*. For determinate beams, the moment at any section of the beam depends only on the external loads and is independent of the inelastic moments M_{z_I}, M_{y_I} in the beam. On the other hand, the moment in an indeterminate beam depends on both the external load and the distribution of inelastic moment throughout the beam. Hence the analysis of inelastic indeterminate beams is more difficult than the analysis of inelastic determinate beams.

Determinate Beams with Thermal Strain. Consider a determinate beam subject to a given load and temperature distribution. We assume that the temperature and load distribution are such that plastic and creep strains do not occur. The moments M at different sections are obtained from the conditions of equilibrium. From the thermal strain e_T, the inelastic forces and moments, F_I, M_{y_I}, and M_{z_I} at different sections are obtained from Equations (5.1.10) and (5.1.14). Inserting these inelastic loads and moments into Equations (5.1.16), (5.1.17), and (5.1.18), we obtain the curvatures and stresses at different sections. The w and v deflections may be obtained by integration of Equations (5.1.16) and (5.1.17), or (5.1.19) and (5.1.20), with the constants of integration determined from the end conditions of the beam. This is illustrated in the following example.

Example. In a beam of span l and of uniform rectangular section with simply supported ends, as shown in Figure 5.3, the temperature distribution in degrees Fahrenheit is given by $T = cxz$, where c is a constant. The external uniform load is $p = 16$ lb per in. Young's modulus is denoted by E. The moment of inertia of the section is $I = (1/12)bh^3$. The coefficient of expansion is denoted by α per degree Fahrenheit. Find the deflection curve of the beam.

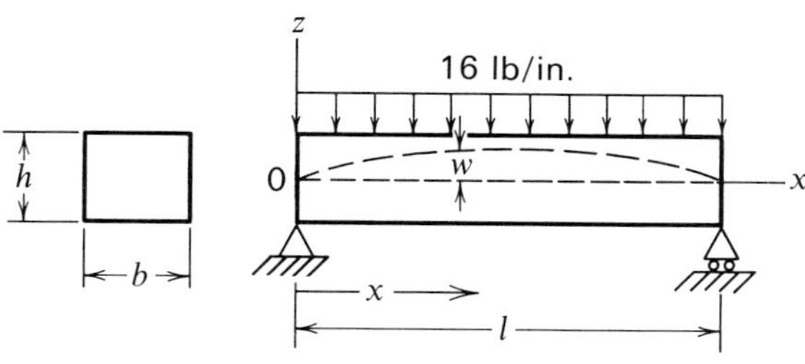

Figure 5.3

Solution. The inelastic strain is given as

$$e'' = e_T = \alpha T = \alpha cxz$$

From Equation (5.1.10), we have

$$F_I = E\int_A e''\,dA = E\int_{-h/2}^{h/2} \alpha cxzb\,dz = E\alpha cxb\int_{-h/2}^{h/2} z\,dz = 0 \quad (5.2.1)$$

From Equation (5.1.14), we have

$$\begin{aligned} M_{y_I} &= -E\int_A ze''\,dA = -E\int_{-h/2}^{h/2} z\alpha cxzb\,dz \\ &= -E\alpha cxb\int_{-h/2}^{h/2} z^2\,dz = -E\alpha cx\frac{bh^3}{12} \\ &= -EI\alpha cx \end{aligned} \quad (5.2.2)$$

From equilibrium conditions, the moment M_y at any section is

$$M_y = \left(\frac{plx}{2} - \frac{px^2}{2}\right) \quad (5.2.3)$$

Substituting M_y and M_{y_I} into Equation (5.1.20) gives

$$\left.\begin{aligned} \frac{\partial^2 w}{\partial x^2} &= \frac{M_y + M_{y_I}}{EI} = \frac{pl}{2EI}x - \frac{px^2}{2EI} - \alpha cx \\ \frac{\partial w}{\partial x} &= \frac{plx^2}{4EI} - \frac{px^3}{6EI} - \frac{\alpha cx^2}{2} + c_1 \\ w &= \frac{plx^3}{12EI} - \frac{px^4}{24EI} - \frac{\alpha cx^3}{6} + c_1x + c_2 \end{aligned}\right\} \quad (5.2.4)$$

From the end conditions,

$$w = 0 \qquad \text{at } x = 0 \quad (5.2.5)$$

and

$$w = 0 \qquad \text{at } x = l$$

we have $c_2 = 0$, and

$$0 = \frac{pl^4}{12EI} - \frac{pl^4}{24EI} - \frac{\alpha cl^3}{6} + c_1 l$$

$$c_1 = \frac{\alpha cl^2}{6} - \frac{pl^3}{24EI}$$

hence

$$w = \frac{1}{EI}\left[\frac{p}{24}(2lx^3 - x^4 - l^3x) + \frac{EI\alpha c}{6}(l^2x - x^3)\right] \quad (5.2.6)$$

This gives the lateral deflection of the beam.

5.3 Determinate Beams Stressed in the Plastic Range

When a beam section is stressed beyond the elastic limit, plastic strain occurs. The plastic strain at a point on the section depends on the stress at the point, which in turn depends on the plastic strain distribution on the section. Consider a beam section with two axes of symmetry, such as an I beam, with load applied in a plane of symmetry. The beam material is assumed to have identical tensile and compressive stress-strain curves. Then under bending, the variation of compressive stress above the centroidal axis is the same as the variation of tensile stress below this axis. The neutral axis coincides with the centroidal axis. Hence $e_0 = 0$, and the strain distribution reduces to $e = K_z z$.

Let h and b denote the depth and width of a beam of rectangular section, and ε the strain of the top extreme fiber. From the Bernoulli-Euler assumption that plane sections before bending remain plane during bending, we have

$$\frac{e}{\varepsilon} = \frac{z}{h/2}$$

$$z = \frac{he}{2\varepsilon} \tag{5.3.1}$$

$$dz = \frac{h}{2\varepsilon}\, de \tag{5.3.2}$$

$$M = -2\int_0^{h/2} \sigma z\, dA$$

where $dA = b\, dz$. By changing the variable of integration from z to e, we obtain

$$M = -2\int_0^{\varepsilon} \sigma \frac{he}{2\varepsilon}\, b\, \frac{h}{2\varepsilon}\, de \tag{5.3.3}$$

$$M = \frac{-bh^2}{2}\left(\frac{1}{\varepsilon^2}\int_0^{\varepsilon} \sigma e\, de\right) = \frac{-Ah}{2}\, m \tag{5.3.4}$$

where m denotes the term in the bracket.[2]

For sections of variable width b, such as the section of Figure 5.4,

$$M = \frac{-b_w h^2}{2}\left(\frac{1}{\varepsilon^2}\int_0^{\varepsilon} \frac{b}{b_w}\, \sigma e\, de\right) = \frac{-A_w h}{2}\, m \tag{5.3.5}$$

where b_w is the width of the web, $A_w = b_w h$, and m again denotes the parenthetical term. For sections of given shape and material, the m vs. ε curve may be calculated for design use. (Values of "m" vs. "ε" of I beams with $A_f = \frac{1}{2}A_f$ have been calculated by Hrennikoff.[2]) Since ε is directly proportional to curvature K_z, we see that m vs. ε gives the moment-curvature relation of the section.

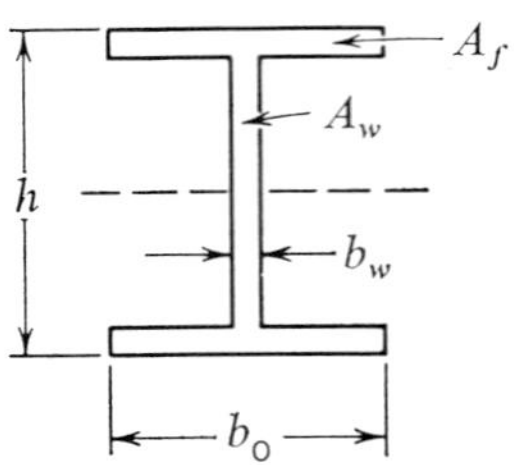

Figure 5.4. I section.

If the beam were purely elastic, the bending moment corresponding to an extreme fiber strain ε could be expressed as

$$M_E = -\int_A \sigma z \, dA = -2E \int_0^{h/2} ezb \, dz$$

With $z/(h/2) = e/\varepsilon$, and $dz = (h/2\varepsilon)\, de$, the above equation becomes

$$M_E = \frac{-b_w h^2}{2} \left(\frac{E}{\varepsilon^2} \int_0^{\varepsilon} \frac{b}{b_w} e^2 \, de \right) = \frac{-A_w h}{2} m_E \tag{5.3.6}$$

where m_E denotes the bracketed term. As shown in Figure 5.5, the loss of resisting moment due to plastic strain is given by

$$M_P = M_E - M = \frac{-A_w h}{2} (m_E - m) \tag{5.3.7}$$

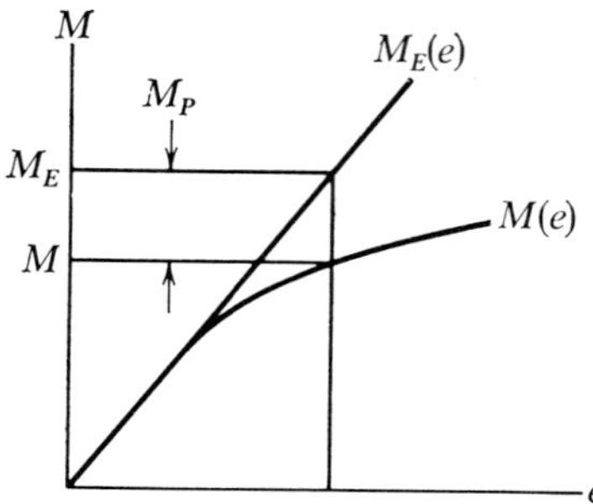

Figure 5.5. Moment relaxed owing to plastic strain.

Both m_E and m depend on the extreme fiber strain ε, which equals $K_z \cdot h/2$. Hence this gives a method for calculating the moment-curvature relation of a beam with inelastic strain.

For the bending of beams of T and other sections beyond the elastic range, see References 5 through 7.

Example. A mild-steel I beam, with cross section as shown in Figure 5.6, is loaded in bending so that its extreme fiber strain $\varepsilon = 0.0100$. The

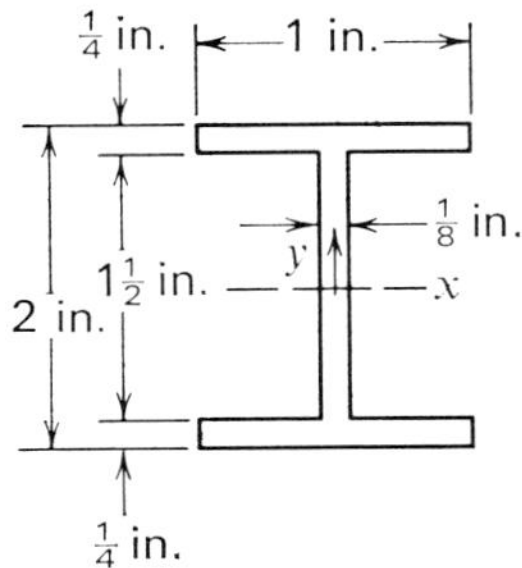

Figure 5.6

stress-strain curve is assumed to be linear up to a strain of 0.0011 and with a slope of 30×10^6 psi, as shown in Figure 5.7. In the strain range between 0.0011 and 0.0180, the stress remains constant. The bending moment resisted by the beam and the loss of moment due to plastic strain are to be determined.

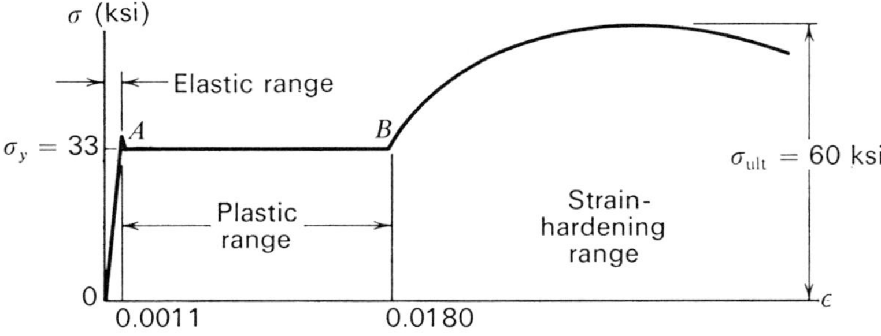

Figure 5.7. Stress-strain diagram of a mild steel.

Solution. Let y be the distance from the neutral axis of the section. Then

$$e = 0.0100 \qquad \text{at } y = 1 \text{ in.}$$

$$e = 0.0075 \qquad \text{at } y = \tfrac{3}{4} \text{ in.}$$

$$e = 0.0011 \qquad \text{at } y = 0.11 \text{ in.}$$

From Equation (5.3.5),

$$m = \left(\frac{1}{\varepsilon^2}\int_0^{\varepsilon}\frac{b}{b_w}\sigma e\, de\right)$$

For $0 \le e \le 0.0011$, we have $\sigma = Ee$ and $E = 30 \times 10^6$ psi. For $0.0011 \le e \le 0.0100$, we have $\sigma = 33{,}000$ psi. Then

$$m = \frac{1}{(0.0100)^2}\left(\int_0^{0.0011} 30 \times 10^6 e^2\, de + \int_{0.0011}^{0.0075} 33{,}000 e\, de + \int_{0.0075}^{0.0100} 8 \times 33{,}000 e\, de\right)$$

$$= \frac{1}{1 \times 10^{-4}}\left[30 \times 10^6\,\frac{(0.0011)^3}{3} + \frac{33{,}000}{2}(0.0075^2 - 0.0011^2) + \frac{8 \times 33{,}000}{2}(0.0100^2 - 0.0075^2)\right]$$

$$m = 10^4[10^7 \times 1.1^3 \times 10^{-9} + 16{,}500(7.5^2 - 1.1^2) \times 10^{-6} + 132{,}000(10^2 - 7.5^2) \times 10^{-6}]$$

$$m = 10^4(0.0133 + 0.908 + 5.78) = 67{,}000 \text{ lb in./in.}^3$$

Let $A_w = b_w h$, where h is the total height of the section. Then

$$A_w = b_w h = \tfrac{1}{8} \times 2 = \tfrac{1}{4} \text{ in.}^2$$

$$M = \frac{-A_w h}{2} \cdot m = \frac{-\frac{1}{4} \times 2}{2} \times 67{,}000 = -16{,}750 \text{ lb in.}$$

$$m_E = \frac{E}{\varepsilon^2}\int_0^{\varepsilon}\frac{b}{b_w}e^2\, de = \frac{30 \times 10^6}{(0.0100)^2}\left(\int_0^{0.0075} e^2\, de + \int_{0.0075}^{0.0100} 8e^2\, de\right)$$

$$= 30 \times 10^{10}\left[\frac{0.0075^3}{3} + \frac{8(0.01^3 - 0.0075^3)}{3}\right]$$

$$= 30 \times 10^4\left[\frac{0.422}{3} + \frac{8(1.00 - 0.422)}{3}\right]$$

$$m_E = 504{,}600 \text{ lb in./in.}^3$$

$$m_E - m = 504{,}600 - 67{,}000 = 437{,}600 \text{ lb in./in.}^3$$

From Equation (5.3.7), we have

$$M_P = \frac{-A_w h}{2}(m_E - m)$$

$$= -\tfrac{1}{4}(437{,}600) = -109{,}400 \text{ lb in.}$$

The resisting moment of the beam section, M, is 16,750 in. lb, and the loss of bending moment due to the plastic strain is 109,400 in. lb.

Note, from Figure 5.7, that for mild steel after the elastic limit is reached, strain increases to 15 times the elastic limit strain without any appreciable increase in stress. If the extreme fiber strain reaches B in the figure, the stress distribution is given as the shaded area in Figure 5.8. It is readily

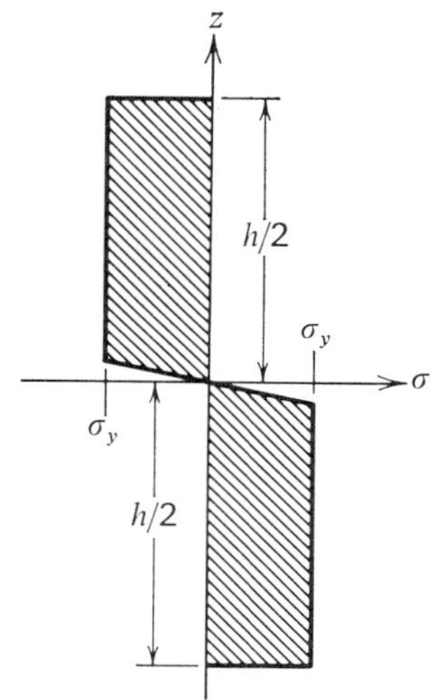

Figure 5.8. Stress distribution on the section.

seen that this stress distribution may be approximated by a rectangular stress distribution with constant compressive and tensile stresses above and below the centroidal axis. The resulting moment is readily seen to be $M = 2\sigma_y Q_m$, where

$$Q_m = \int_0^{h/2} z \, dA$$

is the static moment about the neutral axis of the cross section above the neutral axis. The longer the horizontal part of the stress-strain curve, the better will be the rectangular stress-distribution approximation.

When the resultant applied load is not parallel to one of the principal axes of the section, both M_y and M_z exist, as shown in Figure 5.9. For a section with no planes of symmetry, I_{yz} does not vanish. For this general case, the procedure to find the moment-curvature relationship is to assume a strain e_0 at the centroidal axis and two curvatures, K_z and K_y. The total strain e at any point is readily calculated from Equation (5.1.4). Corresponding to e, the plastic strain e_p is found from the stress-strain diagram of the material. From this e_p distribution, F_I, M_{y_I}, and M_{z_I} are evaluated by Equations (5.1.10) and (5.1.14). From these quantities, F, M_y, and M_z are obtained from Equations (5.1.8), (5.1.12), and (5.1.13).

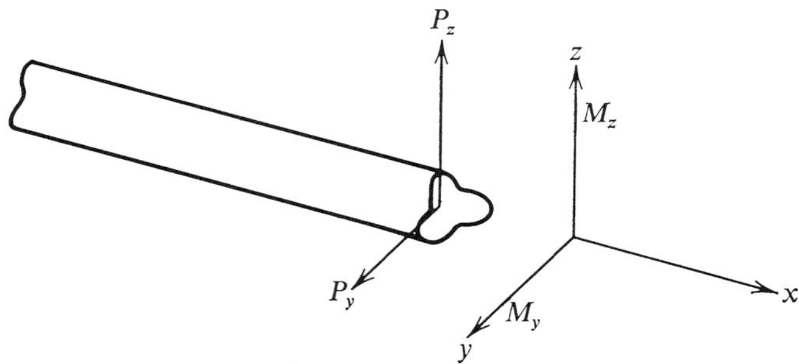

Figure 5.9. Beam with applied loads.

For each set of values of K_z, K_y, and e_0, the corresponding values of F, M_y, and M_z are obtained. For the inverse problem, where F, M_y, and M_z are given, the corresponding values of e_0, K_y, and K_z are found by trial and error. These calculations are generally lengthy. In this analysis it is assumed that local buckling or crippling does not occur. Hence the stress must be less than the local crippling stress of the compression flange.

5.4 Determinate Beams with Creep

The last section considered beams with plastic strain. We now consider beams subject to creep strain, which occurs when a beam is loaded at an elevated temperature. In many cases the initial transient creep strain is negligible in comparison to the steady creep strain, and may be ignored. For steady-state creep at constant temperature, several empirical formulas have been proposed for relating stress to steady-state creep rate. One such proposed formula is the power law $\dot{e}_c = B\sigma^n$, where $\dot{e}_c$ is the steady-state creep rate, σ is the stress, and n and B are temperature-dependent material creep constants. Following the Bernoulli-Euler assumption, the steady-state creep rate must vary linearly from the neutral axis across the section. This gives

$$\frac{z}{h/2} = \frac{\dot{e}_{c_z}}{\dot{e}_{c_{h/2}}} = \frac{B\sigma_z{}^n}{B\sigma^n_{\max}} \qquad \sigma_z = \left(\sigma^n_{\max}\frac{2z}{h}\right)^{1/n} = \sigma_{\max}\left(\frac{2z}{h}\right)^{1/n} \tag{5.4.1}$$

where subscripts z and $h/2$ denote the value of the variable at distances z and $h/2$ from the neutral axis, and $\sigma_{\max}$ is the stress in the extreme fiber at $z = h/2$.

Consider a beam of rectangular section with width b and height h, of a material whose tensile and compressive creep properties are identical. The bending moment M at any section of the beam is given by

$$\begin{aligned} M = -2\int_0^{h/2} \sigma z b\, dz &= -2\int_0^{h/2} \sigma_{\max}\left(\frac{2}{h}\right)^{1/n} b z^{1/n} z\, dz \\ &= -2b\sigma_{\max}\left(\frac{2}{h}\right)^{1/n} \int_0^{h/2} z^{(n+1)/n}\, dz \\ &= -2b\sigma_{\max}\left(\frac{2}{h}\right)^{1/n} \left[\frac{n}{2n+1} z^{(2n+1)/n}\right]_0^{h/2} \\ &= -2b\sigma_{\max}\left(\frac{2}{h}\right)^{1/n} \frac{n}{2n+1}\left(\frac{h}{2}\right)^{(2n+1)/n} \\ &= -\frac{bh^3}{12}\frac{6}{h}\frac{n}{2n+1}\sigma_{\max} \end{aligned} \tag{5.4.2}$$

The stress in any fiber during the steady-state creep becomes[5]

$$\sigma_z = \left(\frac{2z}{h}\right)^{1/n} \cdot \sigma_{\max} = \frac{-M}{I}\frac{h}{6}\frac{2n+1}{n} \cdot \left(\frac{2z}{h}\right)^{1/n} = -\left(\frac{2z}{h}\right)^{1/n} \frac{Mh}{I}\frac{2n+1}{6n} \tag{5.4.3}$$

where $I = bh^3/12$ is the moment of inertia of the cross-sectional area of the beam. Similar solutions may be obtained for other expressions of steady creep.

In those cases in which the loading time is comparatively short, the transient creep may not be negligible, and should be considered. For such beams, the increase of deflection with time is calculated as follows. At zero time, the creep strain is zero, and therefore the stress at various points of the beam may be obtained by an elastic solution. To calculate the incremental creep strain at different time intervals, the smooth stress-time curve is approximated by steps, each of which consists of a vertical segment of instantaneous change of stress followed by a horizontal segment of constant stress as shown in Figure 3.9, and explained in Chapter 3. From the creep curves of the material at different stresses, the creep strains at different points in the beam may be found for the first time interval. From these creep strains at the various points of the beam, F_I, M_{y_I}, and M_{z_I} are calculated by Equations (5.1.10) and (5.1.14). Inserting these values of F_I, M_{y_I}, and M_{z_I} into Equation (5.1.18) or (5.1.21), we compute the stress distribution at the end of the first time interval. The curvatures K_y and K_z are obtained from Equations (5.1.16) and (5.1.17) or (5.1.19) and (5.1.20). This computed stress distribution is assumed to remain constant in the second time interval. If the material may be assumed to follow the equation of state $\dot{e}_c = f(\sigma, e_c)$, then the change in creep strain Δe_c in the second time interval may readily be obtained for the various points of the beam. By adding this Δe_c to the e_c at the beginning of the second time interval, e_c at the end of the second time interval at each point is obtained. From this e_c at each point, F_I, M_{y_I}, and M_{z_I} are obtained as before. By repeating this process, the stress distribution and the curvatures K_y and K_z are obtained at different time instants. The stress-time curves at different points are thus obtained; the deflection curves of the beam at different instants are readily calculated.

Numerical Example.[5] Consider a rectangular beam 2 in. wide by 6 in. deep subjected to end moments of 162,000 in./lb. The beam is of oxygen-free copper with an elastic modulus $E = 14.1 \times 10^6$ psi. The tension-test creep curves at 165° C of this material may be approximated by the following expression.

$$e_c = 26.8 \times 10^{-6}\left[\exp\left(\frac{\sigma}{7500}\right) - 1\right]t^{0.372} \qquad \text{in./in.} \qquad (5.4.4)$$

where exp is the base of the natural logarithm and t is time in hours. With this data and assuming the existence of an equation of state $\dot{e}_c = f(\sigma, e_c)$, calculations of the stress distribution of the beam were carried out for 1, 3, 10, 50, 100, 200, 400, 1000, and 2000 hours. The time intervals were gradually increased as the creep curves flattened out and the fiber stresses became more stabilized. In the calculations, the beam cross-section was subdivided vertically into $\frac{1}{2}$-in. increments from the neutral axis. Simpson's rule was used to obtain M_I. The calculated stress distributions at the ends of the beam at different instants are shown in Figure 5.10. The stress distribution at 2000 hours is quite close to the steady-state stress distribution. For this example, it is seen that as time increases the outermost fibers, which were at first highly stressed, become somewhat relieved of stress, while the interior fibers become more highly stressed, thereby picking up the moment lost by the outermost fibers. This is known as stress relaxation. Eventually a state of stress distribution is reached in which the creep rates in the various fibers become proportional to their respective distances from the neutral axis. Any other distribution is transient in character. The approach to the steady-state stress distribution with time is shown in Figure 5.10.

In this and the two following sections, the solutions of fiber stresses of

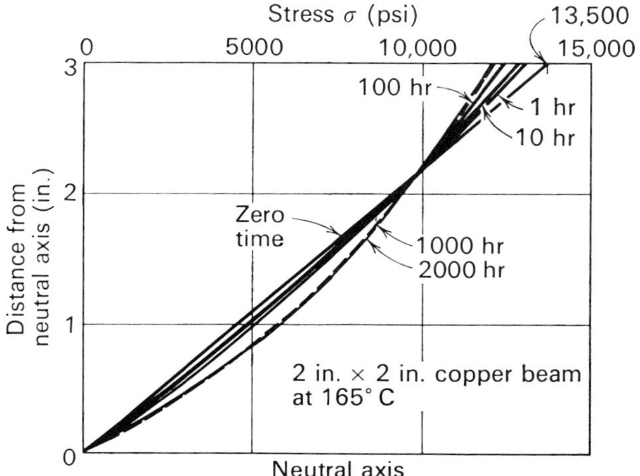

Figure 5.10. Stress distribution in a beam at different instants.

determinate beams, with thermal strain, plastic strain and creep strain are given. For beams with a combination of two or three of these inelastic strains, a similar procedure may be used.

5.5 Shearing Stresses in Inelastic Beams[2]

In general, the shearing stress in rectangular beams is small and need not be considered. In I beams and channels, however, the web thickness is small and the shear stress may be considerable. Consider the upper part of the cross section of an I beam of length dx. The extreme fiber stress at the left and right sides of this section are σ and $(\sigma + d\sigma)$, respectively, as shown in Figure 5.11. The total normal force on the left half-section, from Equation (5.1.6) and $2dz/h = de/\varepsilon$, is

$$\int_{A/2} \sigma\, dA = \int_0^{h/2} \sigma b\, dz = \int_0^{\varepsilon} \sigma b \frac{h}{2} \frac{de}{\varepsilon} = \frac{b_w h}{2} \left(\frac{1}{\varepsilon} \int_0^{\varepsilon} \frac{b}{b_w} \sigma\, de \right) = \frac{b_w h}{2} q$$

where q denotes the term in the bracket. The total normal force on the right half-section is

$$\int_{A/2} (\sigma + d\sigma)\, dA = \frac{b_w h}{2} (q + dq)$$

From the condition of equilibrium of longitudinal forces as shown in

Figure 5.11, we have

$$\tau_0 b_w \, dx = \frac{b_w h}{2} [(q + dq) - q]$$
$$\tau_0 = \frac{h}{2} \frac{dq}{dx} \tag{5.5.1}$$

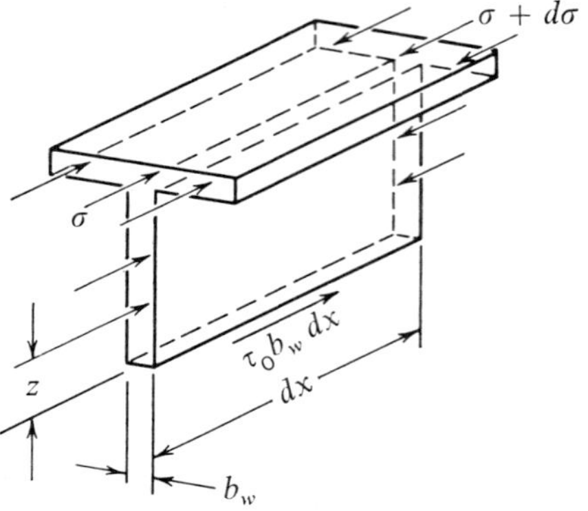

Figure 5.11. A segment of an I beam.

Since shear is equal to the slope of the moment curve, we have

$$V = \frac{dM}{dx}$$

or

$$dx = \frac{dM}{V} = \frac{1}{V} \frac{A_w h}{2} dm \tag{5.5.2}$$

where, from Equation (5.3.4), $M = (A_w h/2)m$. Substitution of Equation (5.5.2) into (5.5.1) yields

$$\tau_0 = \frac{h}{2} \frac{dq}{A_w h} \frac{2V}{dm} = \frac{V}{A_w} \frac{dq}{dm} \tag{5.5.3}$$

The function dq/dm may be computed for various values of ε by taking increments of q and m corresponding to the same increment of ε. Hence the shearing stress in the inelastic range depends not only on the shearing force V, but also on the bending moment, since the bending moment determines the strain ε, and the derivative dq/dm is a function of ε.

5.6 Indeterminate Beams with Thermal Strain

For determinate beams, the moment at a section is determined by conditions of equilibrium and is independent of the deformation in the beam. For indeterminate beams, however, the reactions and hence the moments at different sections depend not only on the external loads, but also on the distribution of the inelastic curvature along the beam. In the sections which follow, the analysis of indeterminate beams with thermal strain, creep strain, and plastic strain will be demonstrated.

Consider an indeterminate beam with an arbitrary temperature distribution T, where T is the increase in temperature above the stress-free condition. Thermal strain $e_T = \alpha T$, where α is the coefficient of expansion. Assuming that the beam is not subjected to an axial load, F in Equation (5.1.8) vanishes. From Equation (5.1.14), we obtain the inelastic moments,

$$M_{y_I} = -E \int z e_T \, dA \qquad M_{z_I} = -E \int y e_T \, dA$$

For the case with e_T varying with z and x only, $M_{z_I} = 0$. In addition, $M_z = 0$ if the beam is symmetrical about the xz plane and the applied load is in this plane. Then Equation (5.1.17) gives

$$-\frac{d^2 w}{dx^2} = -\frac{M_y + M_{y_I}}{EI_{yy}} \tag{5.1.20}$$

This equation, together with the end conditions, gives the deflection and sectional moments of the beam. This is illustrated in the following example.

Consider a rectangular beam of depth h, width b, and length l, with both ends fixed as shown in Figure 5.12. If the temperature distribution is given as $T = a[1 - (x/l)^2]z$, the thermal strain is

$$e_T = \alpha a \left[1 - \left(\frac{x}{l} \right)^2 \right] z$$

where α is the coefficient of expansion. The inelastic moment is

$$M_I = -E \int e_T z \, dA = -E\alpha a \left(1 - \frac{x^2}{l^2} \right) \int_A z^2 \, dA = -EI\alpha a \left(1 - \frac{x^2}{l^2} \right) \tag{5.6.1}$$

From equilibrium conditions, the moment at any section is

$$M = R_1 x + M_1$$

From Equation (5.1.20), we have

$$EI \frac{d^2 w}{dx^2} = R_1 x + M_1 - EI\alpha a \left(1 - \frac{x^2}{l^2} \right) \tag{5.6.2}$$

integration of which yields

$$EI\frac{dw}{dx} = R_1\frac{x^2}{2} + M_1x - EI\alpha a\left(x - \frac{x^3}{3l^2}\right) + c_1$$

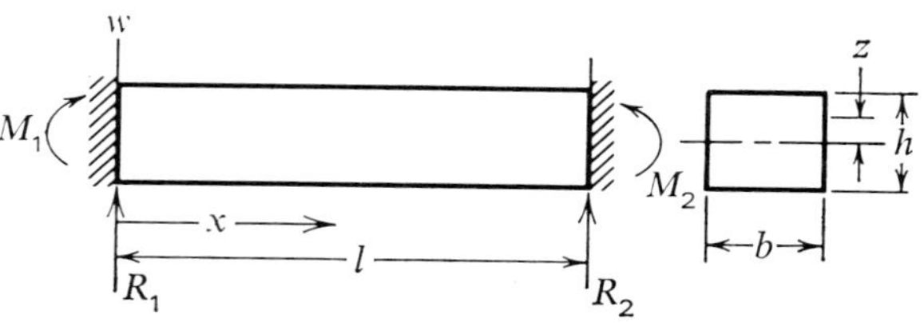

Figure 5.12

Since the ends are fixed, the end conditions are

$$\frac{dw}{dx} = 0 \qquad \text{at } x = 0$$

$$\frac{dw}{dx} = 0 \qquad \text{at } x = l$$

hence $c_1 = 0$, and

$$R_1\frac{l^2}{2} + M_1l - EI\alpha a\left(l - \frac{l}{3}\right) = 0$$

or

$$\frac{R_1l}{2} + M_1 = +EI\alpha a\frac{2}{3} \tag{5.6.3}$$

Integrating $EI\,dw/dx$ yields

$$EIw = R_1\frac{x^3}{6} + M_1\frac{x^2}{2} - EI\alpha a\left(\frac{x^2}{2} - \frac{x^4}{12l^2}\right) + c_2$$

Since the deflection is zero at the ends, we have

$$w = 0 \qquad \text{at } x = 0$$
$$w = 0 \qquad \text{at } l = 0$$

hence $c_2 = 0$, and

$$R_1\frac{l^3}{6} + M_1\frac{l^2}{2} = EI\alpha a\left(\frac{l^2}{2} - \frac{l^2}{12}\right)$$

$$\frac{R_1l}{3} + M_1 = EI\alpha a\frac{5}{6} \tag{5.6.4}$$

From Equations (5.6.3) and (5.6.4), we have

$$\frac{R_1 l}{6} = -EI\alpha a\left(-\frac{2}{3}+\frac{5}{6}\right) = -\frac{EI\alpha a}{6}$$

$$R_1 = -EI\alpha a/l$$

and

$$M_1 = EI\alpha a\left(\frac{5}{6}+\frac{1}{3}\right) = \frac{7}{6}EI\alpha a$$

Then

$$EIw = -\frac{EI\alpha a}{l}\frac{x^3}{6} + \frac{7}{6}EI\alpha a\frac{x^2}{2} - EI\alpha a\left(\frac{x^2}{2} - \frac{x^4}{12l^2}\right)$$

or

$$w = \alpha a\left(\frac{x^4}{12l^2} - \frac{x^3}{6l} + \frac{x^2}{12}\right) \tag{5.6.5}$$

This gives the deflection curve of the beam.

5.7 Indeterminate Beams with Creep

The rate of creep of a metal under uniaxial stress at an elevated temperature is a function of stress, time, creep strain, and history of loading. After following a stress-time curve, the material at some state of stress and strain has a certain strain rate under constant stress. In general, the stress-time curve at each point in the beam is a smooth curve. As explained in Section 5.4, to simplify the calculation this smooth curve is approximated by incremental horizontal and vertical steps of Δt and $\Delta\sigma$, respectively, as shown in Figure 3.9. Each step consists of a constant stress period Δt followed by an instantaneous increment of stress $\Delta\sigma$. At the initial instant of loading, the strain is purely elastic and the corresponding stress distribution σ_0 is obtained from elastic analysis. For a beam with a symmetrical section, and loaded in the plane of symmetry, Equation (5.1.20) reduces to $M/EI = d^2w_0/dx^2$, where w_0 is the initial lateral deflection. This differential equation, together with the end conditions, gives the initial moment and stress distributions in the beam. Taking a small time interval and assuming the stress remains constant in this time interval, we obtain, from the creep characteristics of the material at the given temperature, the incremental creep strain Δe_c, corresponding to the stress at a particular point in the beam. The Δe_c at different points on a section are found in the same way. The incremental inelastic sectional moment in this first time interval, $\Delta M_{I_1} = E \int \Delta e_c z \, dA$, may be obtained by numerical integration across the section, and the ΔM_{I_1}'s for the first time interval at other sections of the beam are similarly obtained. The additional lateral deflection for the first time interval Δw_1 is obtained by

$$\frac{\Delta M_{I_1}}{EI} = \frac{d^2 \Delta w_1}{dx^2} \tag{5.7.1}$$

in conjunction with the end conditions of the beam. At the end of the first time interval the deflection is $w_1 = w_0 + \Delta w_1$. Generally the creep characteristic of the material for tension is assumed to be the same as that for compression, and therefore the average creep strain over the section is zero, $\Delta F_I = 0$. At the end of the first time interval the increment of stress at any point is obtained from Equations (5.1.20) and (5.1.21) as

$$\Delta\sigma_1 = -EI \frac{d^2 \Delta w_1}{dx^2} \frac{z}{I} - E \, \Delta e_c = -E\left(z \frac{d^2 \Delta w_1}{dx^2} + \Delta e_c\right) \tag{5.7.2}$$

The stress at the end of the first time interval is $\sigma = \sigma_0 + \Delta\sigma_1$. This stress is then assumed to remain constant during the second time increment. The incremental creep strain for the second time increment, corresponding to this stress, is obtained. After the incremental strains at different points in a section are obtained, the incremental inelastic moment for the second time interval ΔM_{I_2} is obtained by numerical integration. From the ΔM_{I_2}'s at different sections, the incremental deflection curve Δw_2 is obtained as before. At the end of this second time interval, the deflection is $w_2 = w_1 + \Delta w_2$. This process is repeated for successive time intervals to obtain the stress vs. time and deflection vs. time curves. This is illustrated in the following example.

Example. Consider a fixed-end beam of 7075 aluminum alloy at 600° F with a uniformly distributed load of $q = 120$ lb/in. The beam is considered to be of an ideal I section (that is, in calculating the moment of inertia of the section, the web area is neglected and the flange area is assumed to be concentrated at the flange centroid) with dimensions as shown in Figure 5.13, and a moment of inertia $I_{zz} = 10$ in.^{4} In ideal I sections, bending moment is taken by the upper and lower flanges only. At the initial instant of loading, the fiber stresses taken by the upper and lower flanges are given by Mc/I, where c equals the distance from the centroidal axis to the flange. The initial elastic moment is given by

$$M_0 = \frac{q}{12}(6lx - l^2 - 6x^2) = 6000x - 100{,}000 - 60x^2 \tag{5.7.3}$$

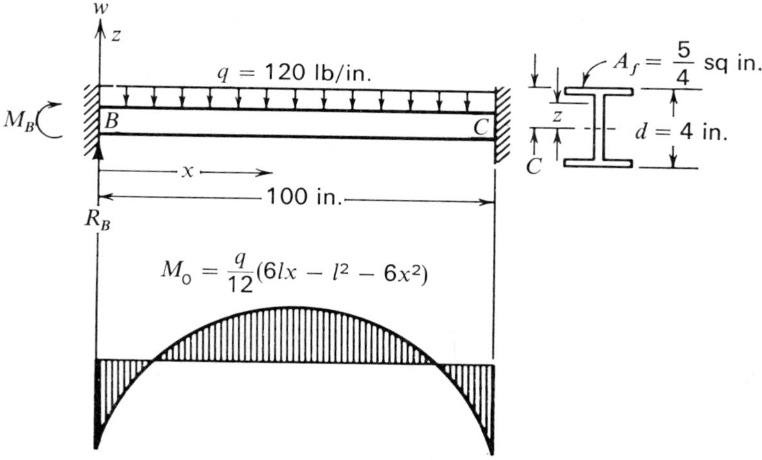

Figure 5.13. Creep bending in a beam with fixed ends.

The initial elastic stress is $\sigma_0 = -M_0 z/I$. The upper flange stresses at different sections of the beam are shown in the first row of Table 5.1.

Table 5.1 TOP FLANGE STRESS AT POINTS FOR AN INDETERMINATE BEAM SUBJECT TO CREEP†

	x (in.)						
t (hr)	1	5	11	21	31	41	49
0.00	−18,812	−14,300	−8252	−92	5668	9,028	9,988
0.5	−18,450	−13,938	−7890	270	6030	9,390	10,350
1.0	−18,288	−13,776	−7728	431	6191	9,552	10,511
1.5	−18,099	−13,587	−7539	620	6381	9,741	10,701
2.0	−17,943	−13,431	−7383	777	6537	9,897	10,857
2.5	−17,797	−13,285	−7237	923	6683	10,043	11,003
3.0	−17,666	−13,153	−7106	1054	6814	10,174	11,134
3.5	−17,545	−13,033	−6985	1175	6935	10,295	11,255
4.0	−17,435	−12,923	−6875	1285	7045	10,405	11,365
4.5	−17,334	−12,822	−6774	1386	7146	10,506	11,466
5.0	−17,242	−12,730	−6682	1478	7238	10,598	11,558
5.5	−17,157	−12,645	−6597	1563	7323	10,683	11,643

† Positive denotes compression.

The creep characteristics of the material at this temperature under constant stress σ are taken to be represented by

$$e_c = 2.64 \times 10^{-7} \exp(1.92 \times 10^{-3}\sigma)t^{0.66}$$

where exp is the base of the natural logarithm and t is in hours. From Equation (5.1.14), the inelastic moment is

$$M_I = EA_f\, d\, e_c \tag{5.7.4}$$

The increment of inelastic moment is then

$$\Delta M_I = EA_f\, d\, \Delta e_c \tag{5.7.5}$$

where d is the depth of the beam and A_f is the flange area. Referring to Figure 5.13, we see that $\Delta M = \Delta M_B + x\,\Delta R_B$. From Equation (5.1.20), we obtain

$$\Delta M_B + x\,\Delta R_B + \Delta M_I = EI\,\frac{d^2\,\Delta w}{dx^2}$$

whose integration yields

$$EI\,\frac{d\,\Delta w}{dx} = x\,\Delta M_B + \frac{x^2}{2}\,\Delta R_B + \int_0^x \Delta M_I\,dx + c_1$$

At $x = 0, l$, we have $d\,\Delta w/dx = 0$. This gives $c_1 = 0$, and

$$l\,\Delta M_B + \frac{l^2}{2}\,\Delta R_B = -\int_0^l \Delta M_I\,dx = -A \tag{5.7.6}$$

where A is the area under the ΔM_I vs. x curve. Integrating again, we obtain

$$-EI\,\Delta w = \frac{x^2}{2}\,\Delta M_B + \frac{x^3}{6}\,\Delta R_B + \int_0^x \int_0^x \Delta M_I\,dx\,dx + c_2$$

Since $\Delta w = 0$ at $x = 0$ and $x = l$, we obtain $c_2 = 0$, and

$$\frac{l^2}{2}\,\Delta M_B + \frac{l^3}{6}\,\Delta R_B = -\int_0^l \int_0^x \Delta M_I\,dx\,dx = -(l - \bar{x})A \tag{5.7.7}$$

where $\bar{x}$ is the distance from the left end of the ΔM_I vs. x curve to the centroid of A. The right-hand term of Equation (5.7.7) gives the static moment of the area of the ΔM_I vs. x curve about $x = l$.

From Equations (5.7.6) and (5.7.7), we have

$$\frac{l\,\Delta R_B}{6} = \frac{-A}{l} + \frac{2(l - \bar{x})}{l^2}\,A = \frac{-A}{l}\left(\frac{2\bar{x}}{l} - 1\right) \tag{5.7.8}$$

$$\Delta R_B = -\frac{6A}{l^2}\left(\frac{2\bar{x}}{l} - 1\right) \tag{5.7.9}$$

$$\Delta M_B = \frac{-2(l-\bar{x})A}{l^2} + \frac{2A}{l}\left(\frac{2\bar{x}}{l} - 1\right) = -\frac{2A}{l}\left(2 - \frac{3\bar{x}}{l}\right) \quad (5.7.10)$$

The left end moment and reaction are equal to the initial values plus their increments in the first time interval. With the end moment and reaction known, the flange stresses at the end of the first time increment at different sections are calculated and listed in the second row of Table 5.1. These stresses are assumed to remain constant during the second time interval (0.5 hr to 1 hr). To find the incremental creep strain in this time interval, the equation of state $\dot{e}_c = f(\sigma_1, e_c)$, shown in Chapter 3, is used. Denoting the stress in the first time interval at a particular point by σ_0 and that in the second time interval by σ_1, we see that the creep strain at the end of the first time interval is indicated by B in Figure 5.14. In the second time interval,

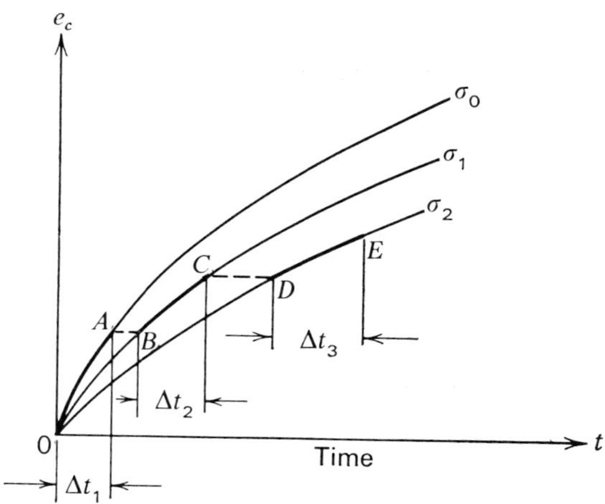

Figure 5.14

Δt_2, the stress is σ_1 instead of σ_0. The creep-strain-time relation follows the σ_1 curve, and the incremental strain is denoted by the curve BC. The vertical projection of BC gives Δe_c. From the Δe_c's at different points on a section, the section ΔM_I is obtained. From the ΔM_I's at different sections, ΔR_B, ΔM_B, and the Δw's are obtained. The incremental stresses ($\Delta\sigma$'s) are obtained from

$$\Delta\sigma = -E\left(\frac{d^2\,\Delta w}{dx^2}\,z + \Delta e_c\right)$$

From the initial stress and the incremental stresses at subsequent time intervals, the stress distributions at different time instants are readily calculated and listed in Table 5.1. This gives a procedure for calculating stresses in an indeterminate beam with nonlinear creep.

5.8 Indeterminate Beam Subject to Plastic Strain[6]

The analysis of inelastic beams was first achieved by Hrennikoff in his excellent paper, "Theory of Inelastic Bending with Reference to Limit Design."[2] The following is an alternative method for determination of the plastic-strain distributions in an inelastic beam at successive stages of loading. Consider a beam loaded beyond the elastic limit. The region in the beam stressed beyond the proportional limit undergoes plastic deformation. For a given uniaxial stress-strain curve of the material, the bending moment of any given section, expressed by Equation (5.3.4), is $M = (-Ah/2)m$ where m depends on the extreme fiber strain ε. If the strain were purely

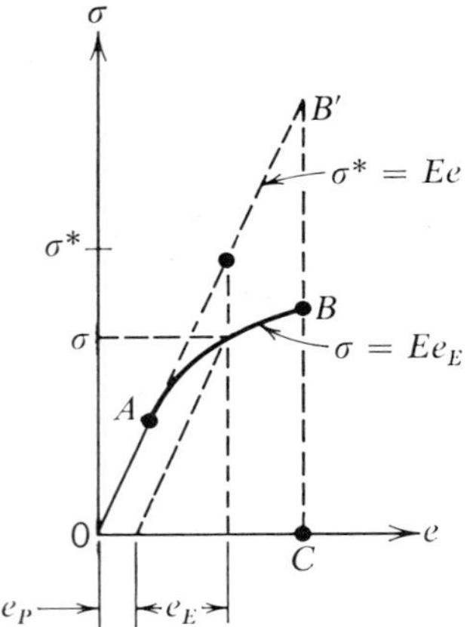

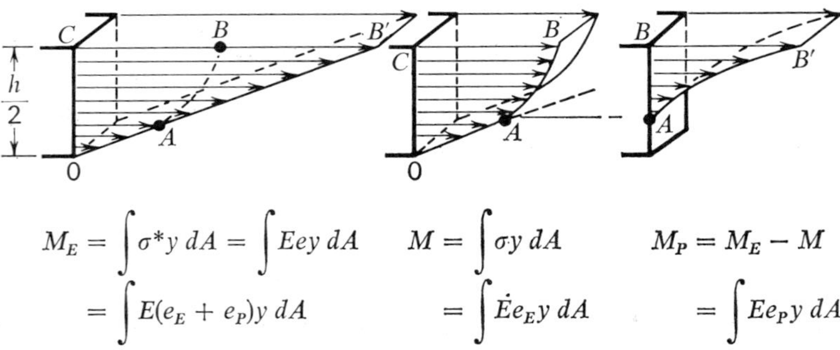

$$M_E = \int \sigma^* y \, dA = \int Eey \, dA = \int E(e_E + e_P)y \, dA$$

$$M = \int \sigma y \, dA = \int \dot{E}e_E y \, dA$$

$$M_P = M_E - M = \int Ee_P y \, dA$$

Figure 5.15. Loss of bending moment associated with plastic strain.

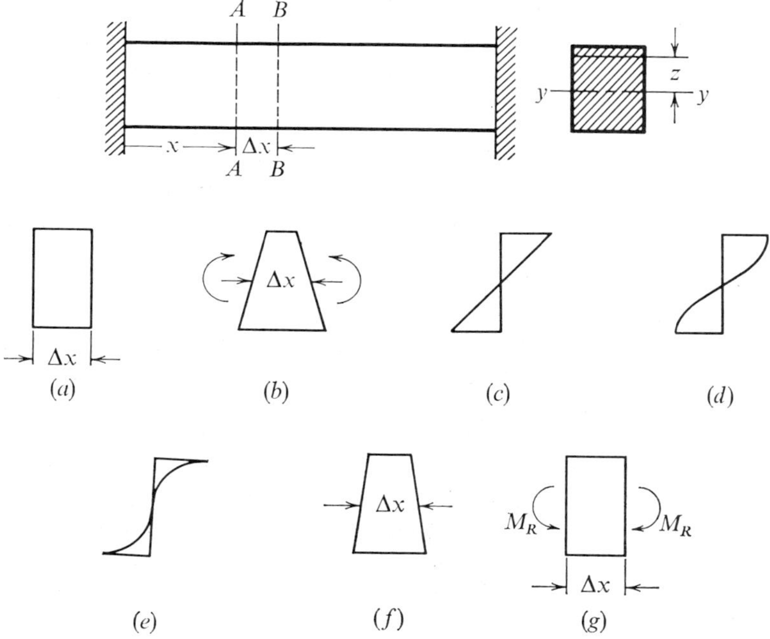

Figure 5.16. Stress and strain of a segment of an inelastic beam.[6] *(a) Segment* Δx *before loading; (b) segment* Δx *under loading; (c) strain distribution under loading; (d) bending stress distribution under loading; (e) plastic strain distribution; (f) segment* Δx *after loading and under no moment; (g) segment* Δx *under restoring moment* M_R: $M_R = M_P = +E \int e_P z\, dA$.

elastic, the moment M_E would be $E\varepsilon I/(h/2) = m_E Ah/2$ as given by Equation (5.3.5). The loss of moment, at any beam section, due to plastic strain is $M_P = M_E - M$, where M_P corresponds to M_I in Equation (5.1.14). For any given section, M_P vs. M can readily be calculated by using the stress-strain relation of the material. Using this M_P for M_I in Equation (5.1.20), we obtain the curvature and hence the deflection of the beam. Figure 5.15 shows the stress distributions, across half of a beam section, corresponding to moments M, M_P, and M_E. The equation $(M + M_P)/EI = d^2w/dx^2$ shows that M_P/EI is equivalent to an initial curvature in the beam, and that $(M_P/EI)\,dx$ gives the change of slope across dx. The M_P gives additional deflections in determinate beams but also modifies the moment distribution in indeterminate beams. Hence M_P depends on M of the same section which in turn depends on the distribution of M_P at a section throughout the beam.

Consider a beam segment of length Δx, as shown in Figure 5.16(a), stressed beyond the elastic limit. On the basis of the Bernoulli-Euler assumption of planes in a beam remaining planes during deformation, the left and right sides of this segment remain plane. Hence under loading a rectangular segment becomes trapezoidal as shown in Figure 5.16(b), and the strain is linear across every section. Because of the nonlinear stress-strain relation beyond the elastic limit, the stress distribution is not linear, as shown in Figure 5.16(d). The elastic-strain distribution is obtained by dividing the stress by Young's modulus E. The difference between the total strain and the elastic strain gives the plastic strain; its distribution is shown in Figure 5.16(e). Now imagine that this segment is cut out of the beam and the moment on the segment relieved while the left and right sides of the beam are kept plane. The elastic curvature of the segment is relieved, but the curvature caused by plastic strains, equal to M_P/EI, remains. Now imagine that a restoring moment $-M_P$ equal and opposite to M_P is applied to the segment, giving a curvature of $-M_P/EI$ and thereby eliminating the plastic curvature; the segment now has zero curvature. The beam is assumed to be cut into many segments, and each segment with plastic strain is subject to its restoring moment $-M_P$. Under such a system of restoring moments, all segments are restored to the same shape and size they had before loading. Hence they match one another perfectly.

Imagine now that they are put together and welded. Since the restoring moments for the neighboring segments may be different, an unbalanced moment $-\Delta M_{P_n} = -M_{P_{n-1}} + M_{P_n}$ exists between the nth and $(n-1)$th segments, as shown in Figure 5.17(a). Similarly, unbalanced bending moments exist between all neighboring segments. Therefore, after welding, the beam segments fit together perfectly with zero total strain under a distribution of restoring moments as shown in Figure 5.17(b). Actually, of course, no such restoring moments $-M_P$ exist, and therefore it is necessary to remove them by applying equal and opposite restoring moments at intermediate sections and $-M_{P_0}$, $-M_{P_l}$ at the left and right ends of the beam. This gives ΔM_{P_n} between the nth and $(n-1)$th segment, or $M'_P = dM_P/dx$ per unit length, and M_{P_0} and M_{P_l} at the left and right ends. Considering these moments as externally applied moments on the beam as a whole [Figure 5.17(c)], we analyze the resulting indeterminate beam as an elastic beam. Denoting the moment at any section resulting from this analysis by M_f, we see that the residual moment at any section is

$$M_{\text{res}} = M_f - M_P \tag{5.8.1}$$

where M_P exists only at those sections which experience plastic strain. Now imagine that the external load is reapplied. During this reapplication no further plastic strain develops. Since this process of cutting, relieving of moments, application of restoring moments, welding, relaxing of the restoring moments, and reloading is *all imaginary, no actual plastic strain is developed in this process.*

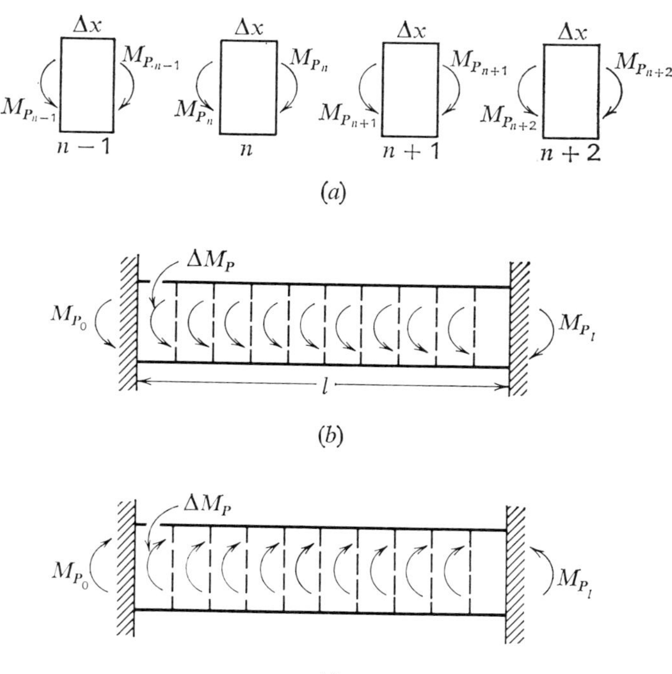

Figure 5.17. (*a*) *Restoring moments on neighboring segments*[6]; (*b*) *segments of beam welded together under restoring moments*[6]; (*c*) *relaxation of restoring moments*[6].

Let M_e be the moment in the beam that would be caused by the external loads if the beam were purely elastic (this M_e is different from M_E since M_E given previously denotes the sectional moment if the strain at the section is purely elastic). Then the moment in the beam is

$$M = M_f - M_P + M_e \tag{5.8.2}$$

where M_e can be found by elastic analysis. Finding the moment in an inelastic beam with a known distribution of M_P reduces to the problem of finding the moment in an identical elastic beam subject to an additional fictitious distributed moment M_P' on the beam plus the two end moments

M_{P_0} and M_{P_l}. This analogy is shown in the following table. The actual beam is denoted by superscript I and the fictitious elastic beam by superscript II.

Table 5.2. Analogy of Plastic Moment Gradient to Applied Distributed Moment in Beam Analysis

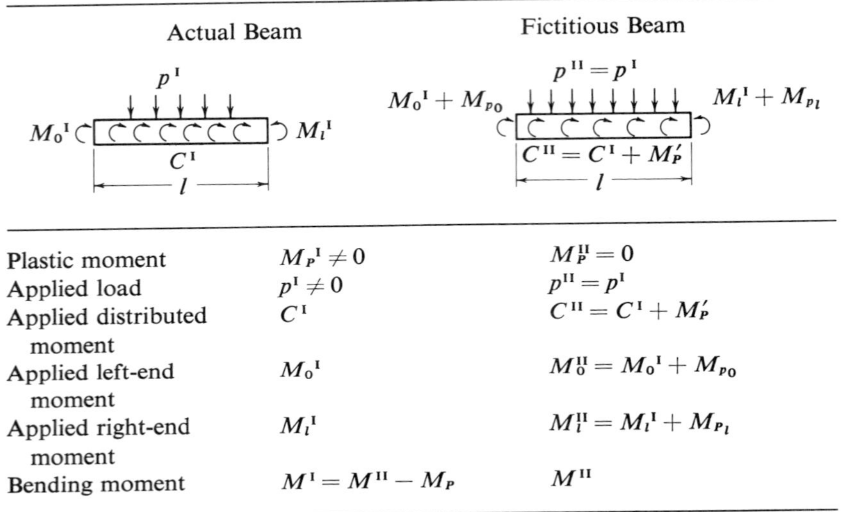

	Actual Beam	Fictitious Beam
Plastic moment	$M_P{}^{\mathrm{I}} \neq 0$	$M_P^{\mathrm{II}} = 0$
Applied load	$p^{\mathrm{I}} \neq 0$	$p^{\mathrm{II}} = p^{\mathrm{I}}$
Applied distributed moment	C^{I}	$C^{\mathrm{II}} = C^{\mathrm{I}} + M_P'$
Applied left-end moment	$M_0{}^{\mathrm{I}}$	$M_0^{\mathrm{II}} = M_0{}^{\mathrm{I}} + M_{P0}$
Applied right-end moment	$M_l{}^{\mathrm{I}}$	$M_l^{\mathrm{II}} = M_l{}^{\mathrm{I}} + M_{P_l}$
Bending moment	$M^{\mathrm{I}} = M^{\mathrm{II}} - M_P$	M^{II}

With a given distribution of plastic moment M_P, we calculate M_f as follows.[6] Let the function $g(x, x')$ denote the moment at x caused by a unit clockwise moment applied at x'. Referring to Figure 5.17(c), we see that

$$M_f(x) = M_{P_0} g(x, 0) - M_{P_l} g(x, l) + \int_{0+}^{l-} \frac{\partial M_P(x')}{\partial x'} g(x, x')\, dx' \tag{5.8.3}$$

Integrating by parts yields

$$\begin{aligned}\int_{0+}^{l-} \frac{\partial M_P(x')}{\partial x'} g(x, x')\, dx' &= M_P(x') g(x, x')\Big|_{x'=0+}^{x'=l-} - \int_{0+}^{l-} M_P(x') \frac{\partial g(x, x')}{\partial x'}\, dx' \\ &= M_{P_l} g(x, l) - M_{P_0} g(x, 0) - \int_{0+}^{x-} M_P(x') \frac{\partial g(x, x')}{\partial x'}\, dx' \\ &\quad - \int_{x-}^{x+} M_P(x') \frac{\partial g(x, x')}{\partial x'}\, dx' - \int_{x+}^{l-} M_P(x') \frac{\partial g(x, x')}{\partial x'}\, dx' \end{aligned} \tag{5.8.4}$$

where x^- and x^+ denote the point to the immediate left and right, respectively, of x. Substituting Equation (5.8.4) into Equation (5.8.3), we obtain

$$M_f = -\int_{0^+}^{x^-} M_P(x') \frac{\partial g(x, x')}{\partial x'} dx' - \int_{x^-}^{x^+} M_P(x') \frac{\partial g(x, x')}{\partial x'} dx' - \int_{x^+}^{l^-} M_P(x') \frac{\partial g(x, x')}{\partial x'} dx'$$

From the definition of the function, $g(x, x^-) = g(x, x^+) + 1$ and hence its value at x^- exceeds the corresponding value at x^+ by unity, that is

$$\int_{x^-}^{x^+} \frac{\partial g(x, x')}{\partial x'} dx' = -1$$

and therefore

$$M_f = -\int_{0^+}^{x^-} M_P(x') \frac{\partial g(x, x')}{\partial x'} dx' + M_P(x) - \int_{x^+}^{l^-} M_P(x') \frac{\partial g(x, x')}{\partial x'} dx' \tag{5.8.5}$$

From Equation (5.8.1), we have

$$\begin{aligned} M_{\text{res}}(x) &= M_f(x) - M_P(x) \\ &= -\int_{0^+}^{x^-} M_P(x') \frac{\partial g(x, x')}{\partial x'} dx' - \int_{x^+}^{l^-} M_P(x') \frac{\partial g(x, x')}{\partial x'} dx' \end{aligned} \tag{5.8.6}$$

This integral equation is used to calculate the residual moment distribution in the beam. Then

$$M(x) = M_e(x) - \int_{0^+}^{x^-} M_P(x') \frac{\partial g(x, x')}{\partial x'} dx' - \int_{x^+}^{l^-} M_P(x') \frac{\partial g(x, x')}{\partial x'} dx' \tag{5.8.7}$$

For a given beam section of a given stress-strain relation, both M and M_P are functions of the extreme fiber strain ε. Hence M is a function of M_P,

$$M(x) = F(M_P(x)) \tag{5.8.8}$$

where F is a function. Equating Equation (5.8.7) to Equation (5.8.8), we obtain

$$F(M_P(x)) = M_e(x) - \int_{0^+}^{x^-} M_P(x') \frac{\partial g(x, x')}{\partial x'} dx' - \int_{x^+}^{l^-} M_P(x') \frac{\partial g(x, x')}{\partial x'} dx' \tag{5.8.9}$$

For a given load, M_e is known. Equation (5.8.9) is an integral equation for M_P. In general $F(M_P(x))$ is nonlinear, and hence an exact solution of Equation (5.8.9) is seldom possible. An approximate solution of Equation (5.8.9) may be obtained by writing it in incremental form:

$$F'(M_P(x))\,\Delta M_P(x) = \Delta M_e(x) - \int_{0^+}^{x^-} \Delta M_P(x')\,\frac{\partial g}{\partial x'}\,dx' - \int_{x^+}^{l^-} \Delta M_P(x')\cdot\frac{\partial g}{\partial x'}\,dx' \tag{5.8.10}$$

where $F'(M_P(x))$ is the derivative of F with respect to its argument. Consider the beam to be divided into N segments and let M_{P_m} and M_{e_m} be those values at the center of the mth segment. Then Equation (5.8.10) becomes

$$F'(M_{P_m})\,\Delta M_{P_m} = \Delta M_{e_m} - \sum_{n=1}^{N} a_{mn}\,\Delta M_{P_n}\,\Delta x_n \tag{5.8.11}$$

where a_{mn} is $\partial g(x, x')/\partial x'$, with x denoting the center of the mth segment and x' the centroid of the area under the ΔM_P curve of the nth segment. For every increment of load, ΔM_e is known; also $M_P(x)$ and $F'(M_P(x))$ are known before the load increment. After the stress exceeds the elastic limit, the slope of the stress-strain curve equals the tangent modulus for increasing strain and Young's modulus for decreasing strain. In Equations (5.8.10) and (5.8.11) we write $F'(M_P(x))$ for increasing strain, and hence these two equations apply only when M and M_P increase in magnitude. Equations (5.8.10) and (5.8.11) are first applied to those sections whose fibers have reached or exceeded the elastic limit and have ΔM_e increasing in magnitude. For each of these sections, there is a ΔM_P and therefore there are as many equations (of the type 5.8.11) as the number of unknown ΔM_P's. These ΔM_P's are obtained by solution of this set of simultaneous equations. Should any of these calculated ΔM_P's be contrary in sign to the bending moment M at that section, that section is deleted from the set of simultaneous equations, and the resulting new set of equations are solved again. This process is repeated until all the calculated ΔM_p's are of the same sign as the M's. The calculated ΔM_p's are then substituted into Equation (5.8.7) to find the ΔM's and M's for the different sections. If some additional sections are found whose bending moments are increasing in magnitude in the inelastic range, these sections are added to the previous set of simultaneous equations. This results in a new set of simultaneous equations for *all* sections with increasing bending moment (in magnitude) in the inelastic range, whose solution gives all the ΔM_p's in these sections. For those sections with ΔM_p known, ΔM is readily found by

$$\Delta M = F'(M_P(x))\,\Delta M_P \tag{5.8.12}$$

This gives the distribution of moment in the beam. This procedure can be applied to inelastic indeterminate beams subject to reversed or varied loadings.

Numerical Examples: Example I. To illustrate the application of the basic concept of the analogy method, the solution of the following simple problem of Figure 5.18 is shown.

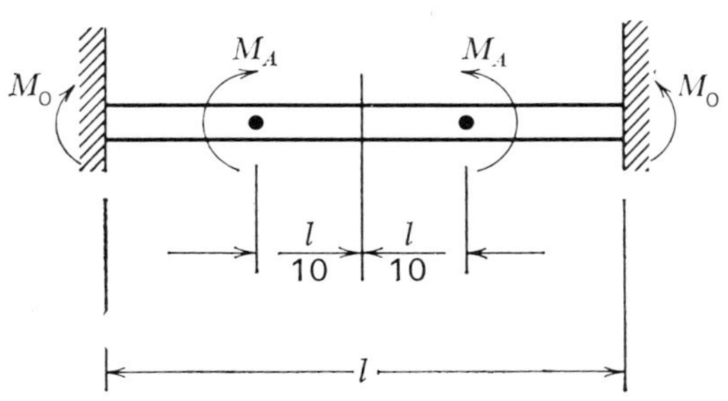

Figure 5.18

A beam of uniform I section, with fixed ends, is loaded as shown. The applied moments M_A are applied at distances $l/10$ from mid-span. The beam is assumed to be an idealized I section; i.e., the bending moment is taken by the flanges only. Hence it is assumed that the web does not take any bending moment. Let the depth of the beam be denoted by h and the flange area by A_f. The stress-strain curve of the flange is assumed to be bilinear as in Figure 5.19. Let M_y denote the moment when $\sigma = \sigma_y$; that is,

$$M_y = hA_f\sigma_y$$

The bending moments of the beam at the ends and at mid-span are to be determined for a given load $M_A = 2M_y$. Since the structure and the given loading are symmetrical with respect to the center line of the beam, the reactions at the two ends of the beam must also be symmetrical. Under symmetrically applied bending moments, the vertical reaction at each end must be zero. Under loading within the elastic limits,

$$\int_0^l \frac{M\,dx}{EI} = 0$$

and we obtain $M_0 = -(M_A/5)$. Since the $l/5$ center region of the beam has larger moment than the rest of the beam, it is this region which first attains and then exceeds M_y when the load increases. When M_y is exceeded in this

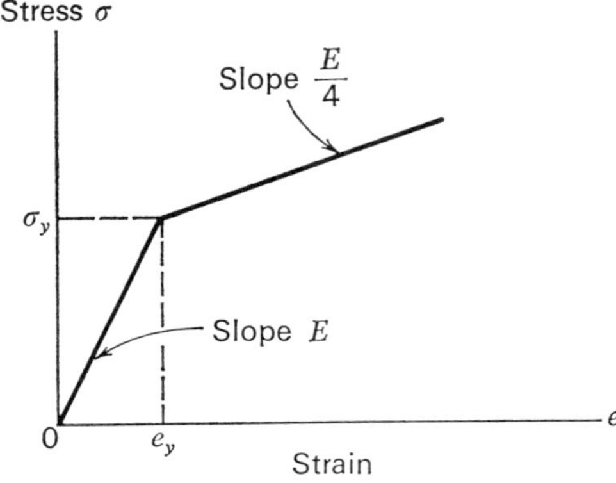

Figure 5.19. Stress-strain curve of the material.

center region, M_P occurs in this region. This gives a loading of $M_A + M_P$ on the fictitious beam at the loading points. This, in turn, causes end moments $M_0 = -\frac{1}{5}(M_A + M_P)$. From the stress-strain curve, we obtain

$$e_p = \tfrac{3}{4}(e - e_y)$$

and

$$M_P = \tfrac{3}{4}E(e - e_y)A_f h \qquad \text{for } e \geq e_y \tag{5.8.13}$$

The bending moment at the center portion of the fictitious beam is

$$\begin{aligned} M_c = M_0 + M_A + M_P &= -\tfrac{1}{5}(M_A + M_P) + M_A + M_P \\ &= \tfrac{4}{5}(M_A + M_P) \end{aligned}$$

and that of the actual beam is

$$\begin{aligned} M_c &= \tfrac{4}{5}(M_A + M_P) - M_P \\ &= \tfrac{4}{5}M_A - \tfrac{1}{5}M_P \\ &= \frac{8}{5}M_y - \frac{3E}{20}(e - e_y)A_f h \end{aligned} \tag{5.8.14}$$

From the stress-strain curve, we have

$$M_c = M_y + \frac{E}{4}(e - e_y)A_f h \qquad e \geq e_y \tag{5.8.15}$$

Equating the above two equations, we obtain

$$M_y = \frac{2E}{3}(e - e_y)A_f h \tag{5.8.16}$$

This gives

$$M_P = \tfrac{9}{8}M_y$$

$$M_0 = -\tfrac{1}{5}(2M_y + \tfrac{9}{8}M_y) = -\tfrac{5}{8}M_y$$

and

$$M_c = M_y + \tfrac{3}{8}M_y = \tfrac{11}{8}M_y$$

This gives the end and center moments for the elastoplastic beam of Figure 5.18.

Example II. To illustrate the application of the method to beams under varied and reversed loadings, a beam with fixed ends, subject to a uniform loading is analyzed.[6] The beam is taken to be of aluminum alloy 24 ST. The stress-strain relation of the material is listed in Table 5.3 and shown in Figure 5.20. It is assumed that the stress-strain relation in compression is identical to that in tension.

Table 5.3. Stress-Strain Relation of 24 ST Aluminum Alloy

Strain	Stress (psi)
0.003	31,500
0.0035	35,000
0.004	38,000
0.0045	39,500
0.005	41,000
0.0055	42,500
0.006	43,800
0.0065	44,900
0.007	45,800
0.0075	46,700
0.008	47,500

$$E = \frac{31{,}500}{0.0030} = 10{,}500{,}000 \text{ psi}$$

When a specimen is loaded beyond the elastic limit, say to point C in Figure 5.20, and then unloaded, it is assumed that the stress-strain relation will be linear (CF is a straight line) in the stress range of twice the initial elastic limit σ_E. If the reversed loading continues, the stress-strain curve will follow FH, which is assumed to be identical to CG. The beam

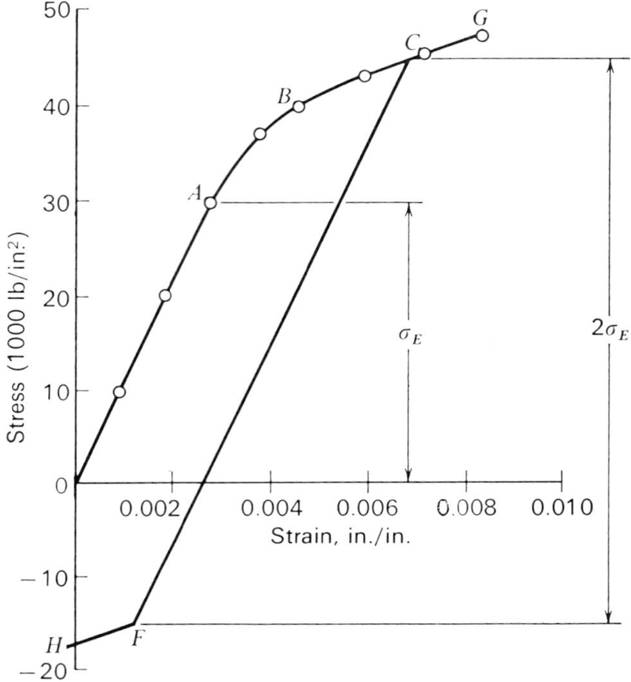

Figure 5.20. Stress-strain diagram of 24ST aluminum alloy with assumed Bauschinger effect.

section is of an idealized I section, with flange areas assumed to be concentrated at the centroids of the flanges. It is assumed that the web areas do not take any bending moment. The moment of inertia of the section is then

$$I = 2A_f\left(\frac{h}{2}\right)^2 = \frac{A_f h^2}{2} \tag{5.8.17}$$

where h is the distance between the centroids of the flanges. The bending moment is given as

$$M = \sigma A_f h$$

$$\Delta M = A_f h \, \Delta\sigma \tag{5.8.18}$$

The function $g(x, x')$ for this beam is calculated by applying a unit moment

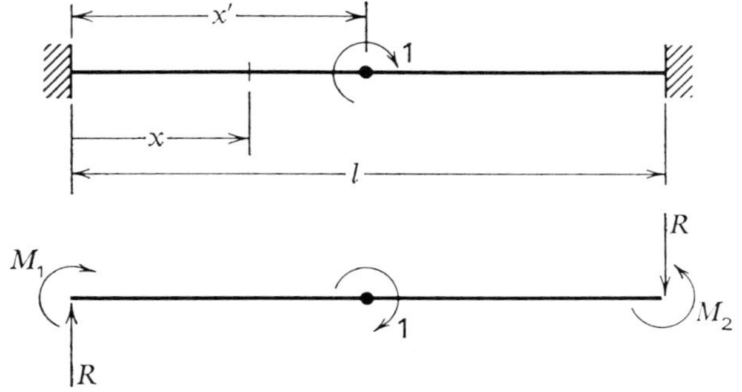

Figure 5.21. A beam with fixed ends subject to unit moment at x'.

at x' as shown in Figure 5.21. The moment and reaction at the end $x = 0$ are

$$M_1 = \frac{1}{l^2}(l - x')(3x' - l) \tag{5.8.19}$$

$$R = -\frac{6x'}{l^3}(l - x') \tag{5.8.20}$$

This gives

$$\left.\begin{aligned} g(x, x') &= M_1 + Rx & x < x' \\ g(x, x') &= M_1 + Rx + 1 & x > x' \end{aligned}\right\} \tag{5.8.21}$$

For $0 < x < x'^-$ and $x'^+ < x < l$, we have

$$\frac{\partial g}{\partial x'} = -\frac{1}{l^2}\left(6x' - 4l + 6x - \frac{12xx'}{l}\right) \tag{5.8.22}$$

Substituting this into Equation (5.8.6) gives

$$M_{\text{res}} = \int_0^l M_P(x')\frac{2}{l^2}\left(3x' - 2l + 3x - \frac{6xx'}{l}\right)dx' \tag{5.8.23}$$

Writing the last expression in incremental form yields

$$\Delta M_{\text{res}}(x) = \Delta M_P(x')\,\Delta x' \cdot \frac{2}{l^2}\left(3x' - 2l + 3x - \frac{6xx'}{l}\right)$$

$$\Delta M_{\text{res}}(x) = \Delta M_P(x')\frac{\Delta x'}{l}\,a(x, x') \tag{5.8.24}$$

where

$$a(x, x') = \frac{2}{l}\left(3x' - 2l + 3x - \frac{6xx'}{l}\right) \tag{5.8.25}$$

Under uniform loading, the maximum moment for a beam with fixed ends occurs at the two ends.

The procedure given here can be applied to beam analysis at any stage of plastic deformation. To reduce the amount of numerical calculation, the plastic deformation in the present example is limited to within $0.02l$ from both ends; that is, M_P occurs only within $0 < x < 0.02l$ and $0 < l - x < 0.02l$. From the symmetry of the beam and the loading, we have

$$M_P(x') = M_P(l - x')$$

and Equation (5.8.23) reduces to

$$\begin{aligned} M_{\text{res}} &= \int_0^{l/2} M_P(x') \frac{2}{l^2} \left\{ \left[3x' - 2l + 3x - \frac{6xx'}{l} \right] \right. \\ &\quad \left. + \left[3(l - x') - 2l + 3x - \frac{6x(l - x')}{l} \right] \right\} dx' \\ &= \int_0^{l/2} M_P(x') \frac{2}{l^2} (3l - 4l)\, dx' \\ &= -\int_0^{l/2} M_P(x') \frac{2}{l}\, dx' \end{aligned} \tag{5.8.26}$$

This shows that the residual moment is independent of x; that is, M_{res} is constant along the span. Then

$$M_{\text{res}} = -\frac{2}{l} (M_{P_0}\, \Delta x_0 + M_{P_{0.01}}\, \Delta x_{0.01} + M_{P_{0.02}}\, \Delta x_{0.02})$$

Let

$$\Delta x_0 = 0.005l \qquad \Delta x_{0.01} = \Delta x_{0.02} = 0.01l$$

Then

$$M_{\text{res}} = -0.01 M_{P_0} - 0.02 M_{P_{0.01}} - 0.02 M_{P_{0.02}} \tag{5.8.27}$$

where the subscript of M_P denotes the location of M_P in terms of l.

From the elastic analysis of a beam of uniform cross section and of length l with fixed ends, subject to a uniformly distributed load per unit length, the moments at $x = 0$, $0.01l$ and $0.02l$ are

$$M_{e_0} = -0.0833ql^2 \qquad M_{e_{0.01}} = -0.0784ql^2 \qquad M_{e_{0.02}} = -0.0735ql^2$$

The elastic limit of the material is 31.5 kips/in^2. When the maximum stress of the end sections of the beam just reach this stress,

$$\begin{aligned} M_0 &= -0.0833ql^2 = -31.5(A_f h) \qquad ql^2 = 378 A_f h \\ M_{0.01} &= -29.6 A_f h \\ M_{0.02} &= -27.8 A_f h \end{aligned}$$

where the subscript of M denotes the location in terms of l. From Equations (5.8.1), (5.8.2) and (5.8.27), we have

$$M_0 = -0.0833ql^2 - 0.010M_{P_0} - 0.020M_{P_{0.01}} - 0.020M_{P_{0.02}}$$

$$\Delta M_0 = -0.0833\Delta ql^2 - 0.010\Delta M_{P_0} - 0.020\Delta M_{P_{0.01}} - 0.020\Delta M_{P_{0.02}} \tag{5.8.28}$$

Similarly,

$$\Delta M_{0.01} = -0.0784\Delta ql^2 - 0.010\Delta M_{P_0} - 0.020\Delta M_{P_{0.01}} - 0.020\Delta M_{P_{0.02}} \tag{5.8.29}$$

$$\Delta M_{0.02} = -0.0735\Delta ql^2 - 0.010\Delta M_{P_0} - 0.020\Delta M_{P_{0.01}} - 0.020\Delta M_{P_{0.02}} \tag{5.8.30}$$

From Table 5.3, when M_0 is increased within the range $31.5A_f h$ to $35.0A_f h$, the extreme fiber strain ε increases from 0.003 to 0.0035, and

$$\frac{\Delta M}{A_f h\,\Delta\varepsilon} = \frac{3.5}{0.0005} = 7000$$

If the section is purely elastic, the resisting moment

$$M_E = E\varepsilon A_f h \qquad \Delta M_E = EA_f h\,\Delta\varepsilon$$

Since Young's modulus, E, of 24 ST aluminum alloy is 10,500 kip/in.^{2}, the previous equations become

$$\Delta M_E = 10{,}500A_f h\,\Delta\varepsilon$$

$$\Delta M_P = \Delta M_E - \Delta M = 3500A_f h\,\Delta\varepsilon$$

$$\frac{\Delta M}{\Delta M_P} = \frac{7000}{3500} = 2.00$$

Consider $M_{0.01}$ to be increased from $-29.6A_f h$ to $-31.5A_f h$; that is, the load is increased until the extreme fiber stress of the section at $x/l = 0.01$ is at the elastic limit. Then

$$\Delta M_{0.01} = -1.9A_f h$$

Substituting the above in Equations (5.8.28), (5.8.29), and (5.8.30), we obtain the system of equations

$$\Delta M_0 = 2.00\Delta M_{P_0} = -0.0833\Delta(ql^2) - 0.010\Delta M_{P_0}$$

$$\Delta M_{0.01} = -1.9A_f h = -0.0784\Delta(ql^2) - 0.010\Delta M_{P_0}$$

$$\Delta M_{0.02} = -0.0735\Delta(ql^2) - 0.010\Delta M_{P_0}$$

Solving the first two of the above equations for $\Delta(ql^2)$ and ΔM_{P_0} gives

$$\Delta(ql^2) = 24.1 A_f h \qquad \Delta M_{P_0} = -1.00 A_f h$$

$$\Delta M_0 = -2.00 A_f h \qquad \Delta M_{0.01} = -1.9 A_f h \qquad \Delta M_{0.02} = -1.76 A_f h$$

The load and moments are tabulated in column 4 of Table 5.4. Now assume the load to be increased so that $M_{0.02}$ reaches $-31.5 A_f h$, hence $\Delta M_{0.02} = -1.94 A_f h$. Since both the stresses at section $x/l = 0$, and 0.01 are between 31.5 and 35 kip/in.^{2}, we have

$$\frac{\Delta M_0}{\Delta M_{P_0}} = \frac{\Delta M_{0.01}}{\Delta M_{P_{0.01}}} = 2.00$$

Substituting these into Equations (5.8.27) to (5.8.29), we have

$$2.00 \Delta M_{P_0} = -0.0833 \Delta(ql^2) - 0.010 \Delta M_{P_0} - 0.020 \Delta M_{P_{0.01}}$$

$$2.00 \Delta M_{P_{0.01}} = -0.0784 \Delta(ql^2) - 0.010 \Delta M_{P_0} - 0.020 \Delta M_{P_{0.01}}$$

$$-1.94 A_f h = -0.0735 \Delta(ql^2) - 0.010 \Delta M_{P_0} - 0.020 \Delta M_{P_{0.01}}$$

The results of the solution of the above three simultaneous equations are listed in columns 5 and 6 of Table 5.4. The moment diagram for the beam at the end of the second stage of loading is shown in Figure 5.22.

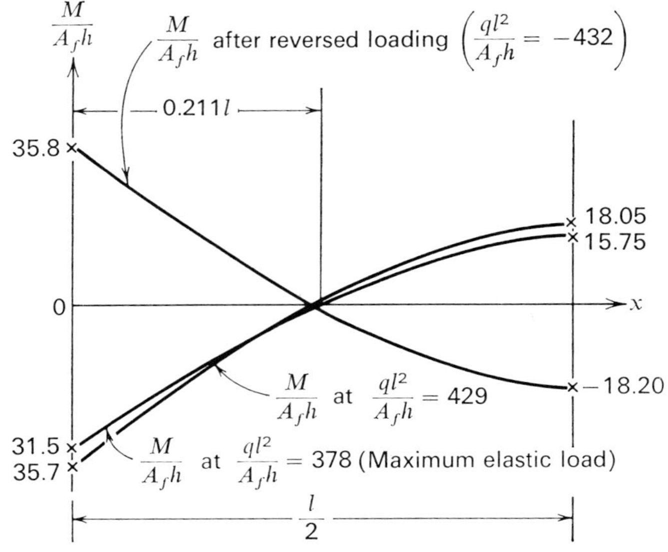

Figure 5.22. Moment diagram of the inelastic beam at $ql^2/A_f h = 378$*, 429, and* -432*.*

Now apply a reversed loading until reversed plastic deformation starts at the ends. This requires a ΔM_0 of $2 \times 31.5 A_f h = 63.0 A_f h$ and a $\Delta(ql^2)$ of $-2 \times 378 A_f h = -756 A_f h$. This gives

$$\Delta M_0 = 63.0 A_f h \qquad \Delta M_{0.01} = 59.2 A_f h \qquad \Delta M_{0.02} = 55.5 A_f h$$

$$M_0 = 27.3 A_f h \qquad M_{0.01} = 25.52 A_f h \qquad M_{0.02} = 24.0 A_f h$$

$$ql^2 = -326.95 A_f h$$

For $x/l = 0.01$ to reverse its plastic deformation, we must have

$$M_{0.01} = 63 A_f h - 33.58 A_f h = 29.42 A_f h$$

$$\Delta M_{0.01} = 29.42 A_f h - 25.52 A_f h = 3.90 A_f h$$

The stress-strain curve at the ends now follows FH, which corresponds to CG, along which σ increases from 35 to 39.5 kip/in.^{2} and ε increases from 0.0035 to 0.0045. Hence

$$\frac{\Delta M_0}{\Delta \varepsilon} = 4500 A_f h \qquad \frac{\Delta M_0}{\Delta M_{P_0}} = \frac{4500}{6000} = 0.750$$

Substituting these into Equations (5.8.28) and (5.8.29) yields

$$0.750 \Delta M_{P_0} = -0.0833 \Delta(ql^2) - 0.010 \Delta M_{P_0}$$

$$\Delta M_{0.01} = 3.9 A_f h = -0.0\ 84 \Delta(ql^2) - 0.010 \Delta M_{P_0}$$

The results of the solution of these two equations are listed in columns 9 and 10 in Table 5.4.

Now increase the reversed load until reversed plastic deformation occurs in $M_{0.02}$:

$$\Delta M_{0.02} = (31.5 - 27.65) A_f h = 3.85 A_f h$$

$$\frac{\Delta M_0}{A_f h(\Delta \varepsilon)} = \frac{43.8 - 9.5}{0.006 - 0.0045} = \frac{4.3}{0.0015} = 2870$$

$$\frac{\Delta M_0}{\Delta M_{P_0}} = \frac{2870}{7630} = 0.377$$

$$\frac{\Delta M_{0.01}}{A_f h(\Delta \varepsilon)} = \frac{38 - 31.5}{0.001} = 6500$$

$$\frac{\Delta M_{0.01}}{\Delta M_{P_{0.01}}} = \frac{6500}{4000} = 1.625$$

Table 5.4. Calculation of Moments at Different Loadnigs

1	2	3	4	5	6	7	8	9	10	11	12
1. $\frac{ql^2}{A_f h}$	378		402.1		429.05		−326.95		−377.45		−432.2
2. $\frac{\Delta(ql^2)}{A_f h}$		24.1		26.95		−756		−50.5		−54.7	
3. $\sigma_0 = \frac{M_0}{A_f h}$	−31.5		−33.5		−35.7		27.3		31.45		35.84
4. $\frac{\Delta M_0}{\Delta M_{P0}}$		−2.00		2.00				0.750		0.377	
5. $\frac{\Delta M_{P0}}{A_f h}$		−1.00		−1.10				5.53		11.63	
6. $\frac{M_{P0}}{A_f h}$	0		−1.00		−2.10		−2.10		3.43		15.06
7. $\frac{\Delta M_0}{A_f h}$		−2.00		−2.20		63.0		4.15		4.39	

8. $\sigma_{0.01} = \dfrac{M_{0.01}}{A_f h}$	−29.6		−31.5		−33.58		25.52		29.42		33.57
9. $\dfrac{\Delta M_{0.01}}{\Delta M_{P0.01}}$				2.00						1.625	
10. $\dfrac{\Delta M_{P0.01}}{A_f h}$				−1.037						2.55	
11. $\dfrac{M_{P0.01}}{A_f h}$	0		0		−1.037		−1.037		−1.037		1.51
12. $\dfrac{\Delta M_{0.01}}{A_f h}$		−1.9		−2.074		59.2		3.9		4.15	
13. $\sigma_{0.02} = \dfrac{M_{0.02}}{A_f h}$	−27.8		−29.56		−31.5		24.0		27.65		31.5
14. $\dfrac{\Delta M_{P0.02}}{A_f h}$											
15. $\dfrac{\Delta M_{0.02}}{A_f h}$		−1.76		−1.94		55.5		3.65		3.85	

Equations (5.8.28) to (5.8.30) give

$$0.377\Delta M_{P_0} = -0.0833\Delta(ql^2) - 0.010\Delta M_{P_0} - 0.020M_{P_{0.01}}$$
$$1.625\Delta M_{P_{0.01}} = -0.0784\Delta(ql^2) - 0.010\Delta M_{P_0} - 0.020M_{P_{0.01}}$$
$$3.85A_f h = -0.0735\Delta(ql^2) - 0.010\Delta M_{P_0} - 0.020M_{P_{0.01}}$$

The rrsults of the solution of these three equations are listed in columns 11 and 12 in Table 5.4. The bending-moment curves of the beam for three of the loading stages are shown in Figure 5.22, while the stress-strain history of the upper flanges of the ends of the beam is shown in Figure

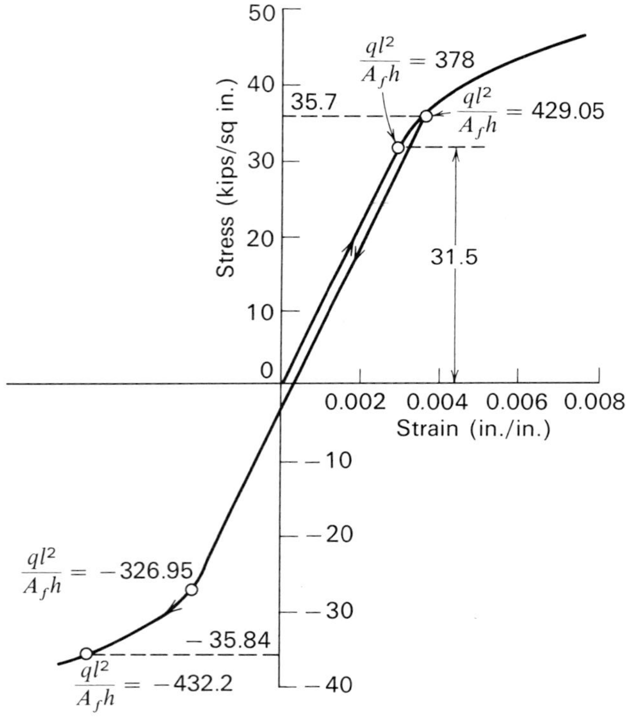

Figure 5.23. Stress-strain path of the upper flange of the ends of the beam.

5.23. Similar calculations may be made for this or other beams under arbitrary loading histories. The method given can be extended to the analysis of inelastic *rigid* frames and curved beams under an arbitrary history of loadings. For curved beams, Equation (5.1.20) is replaced by

$$\frac{d\theta}{ds} = \frac{M + M_I}{EI}$$

The preceding method of analysis for plastic beams provides the history of bending moment at all sections of a beam for strain-hardening as well as for perfectly plastic materials. However, the numerical calculations are necessarily lengthy. The stress-strain curves for mild steels may be considered to be perfectly plastic. For beams and rigid frames of such materials, the method of limit analysis has been successfully applied. Limit analysis is thoroughly treated in many excellent books,[8–13] and will not be discussed here.

REFERENCES

1. Nadai, A., *Plasticity*, McGraw-Hill, New York, p. 164, 1931.
2. Hrennikoff, A., "Theory of Inelastic Bending with Reference to Limit Design," *ASCE Trans.*, **113**, 213–246, 1948.
3. Marin, J., "Mechanics of Creep," *ASCE Trans.*, **108**, 459, 1943.
4. Shanley, F. R., *Weight-Strength Analysis of Aircraft Structures*, McGraw-Hill, New York, p. 329, 1952.
5. Popov, E. P., "Bending of Beams with Creep," *J. Appl. Phys.*, **20**, 251, 1949.
6. Lin, T. H., "Elasto-plastic Analysis of Indeterminate Beams under Reversed Loading," *J. Franklin Inst.*, Vol. 285, No. 5, pp. 364–376, 1968.
7. Ramberg, W., and W. R. Osgood, "Description of Stress-Strain Curves by Three Parameters," *NACA TN 902*, July 1943.
8. Prager, W., *An Introduction to Plasticity*, Addison-Wesley, Reading, Mass., 1959.
9. Hodge, P. G., *Plastic Analysis of Structures*, McGraw-Hill, New York, 1959.
10. Beedle, L. S., *Plastic Design of Steel Frames*, Wiley, New York, 1958.
11. Van den Broek, J. A., *Theory of Limit Design*, Wiley, New York, 1948.
12. Baker, J. F., M. R. Horne, and J. Heyman, *The Steel Skeleton*, Vols. I and II, Cambridge University Press, Cambridge, England, 1956.
13. Heyman, J., *Plastic Design of Portal Frames*, Cambridge University Press, Cambridge, England, 1957.

chapter

6

Inelastic Columns and Beam Columns

Introduction. In this chapter, the theories of elastic columns and beam-columns are briefly reviewed. Columns with thermal strain are analyzed by considering the thermal strain as an equivalent externally applied moment, as shown in Table 2.1 of Chapter 5. Then the analysis of linear viscoelastic columns is derived. For materials with nonlinear strain-time response to stress, which is common for metals, a more general method of creep analysis is presented in Section 6.7. In this method, the creep strain is shown to have the same effect as initial curvature in calculating the deflection curve. In Section 6.8, a method for analyzing beam-columns with arbitrary creep characteristics is presented. Numerical examples of columns and beam-columns are given.

In the last part of this chapter, columns with plastic strain are treated. The fallacy of the derivation of the reduced modulus theory is first given. The tangent modulus theory is then derived. A detailed discussion of the mechanism of inelastic buckling of columns is followed by numerical examples.

6.1 Elastic Columns

As in the analysis of elastic beams, the Bernoulli-Euler assumption that planes before loading remain planes after loading is applied to the analysis of inelastic columns. Two cross sections at a distance dx apart, which were parallel in the unloaded condition, have a relative angle $d\theta$ after loading. From Figure 6.1, we have

$$d\theta = \frac{(e_2 - e_1)\,ds}{h_1 + h_2} = \frac{(e_2 - e_1)\,ds}{h} \tag{6.1.1}$$

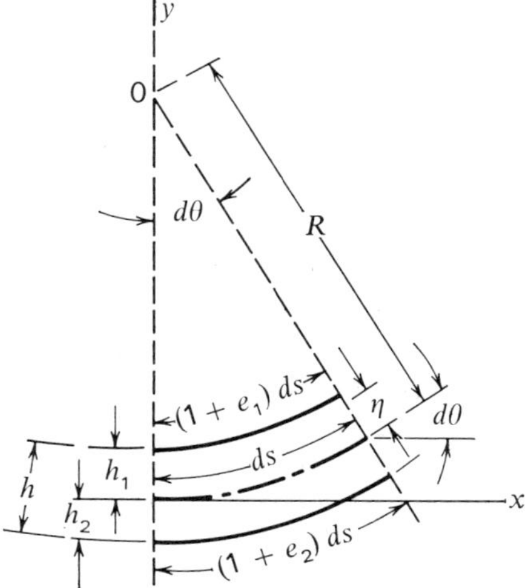

Figure 6.1

In the elastic range, stress is linear across the section, hence

$$\sigma = \frac{-M\eta}{I}$$

where η is the distance from the centroidal axis of the cross section. The maximum strains occur at $\eta = -h_1$, and $\eta = h_2$ and are

$$e_1 = \frac{\sigma_1}{E} = -\frac{Mh_1}{EI} \qquad \text{and} \qquad e_2 = \frac{\sigma_2}{E} = \frac{Mh_2}{EI} \tag{6.1.2}$$

where σ_1 is the upper extreme fiber stress and σ_2 is the stress in the lower extreme fiber. Then, from Equation (6.1.1), we obtain

$$\frac{d\theta}{ds} = \frac{Mh_2/EI + Mh_1/EI}{h} = \frac{M(h_1 + h_2)}{EIh} = \frac{M}{EI}$$

For a small lateral deflection,

$$ds \cong dx$$

$$\theta \cong \sin\theta \cong \tan\theta = \frac{dw}{dx} \tag{6.1.3}$$

and the above equation for $d\theta/ds$ becomes

$$\frac{d\theta}{ds} \cong \frac{d\theta}{dx} \cong \frac{d(dw/dx)}{dx} = \frac{d^2w}{dx^2} = \frac{M}{EI} \tag{6.1.4}$$

For a long column with hinged ends and axial forces P at both ends, as in Figure 6.2, we have, for the moment at any section,

$$M = -Pw = EI \frac{d^2w}{dx^2} \tag{6.1.5}$$

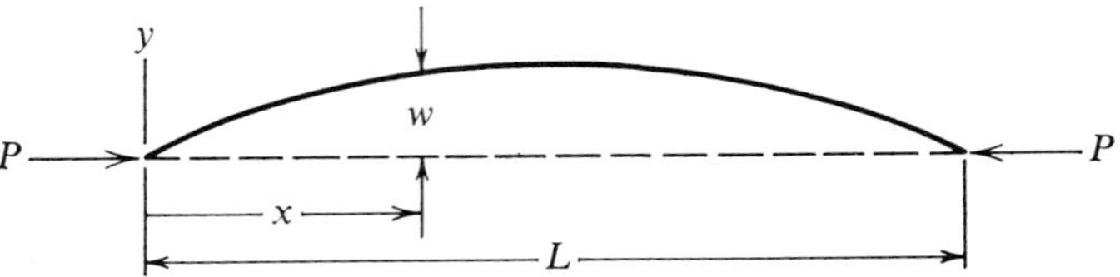

Figure 6.2. A long column with hinged ends.

For columns, P is always positive. Letting

$$K^2 = \frac{P}{EI} \tag{6.1.6}$$

we may write Equation (6.1.5) as

$$\frac{d^2w}{dx^2} + K^2w = 0 \tag{6.1.7}$$

This is a homogeneous second-order linear differential equation with the constant coefficient K^2. Let $w = e^{mx}$. Then Equation (6.1.7) gives

$$(m^2 + K^2)e^{mx} = 0$$

whose solution is $m = \pm ik$. Hence the deflection is

$$w = Ce^{iKx} + De^{-iKx} = A\frac{e^{iKx} - e^{-iKx}}{2i} + B\frac{e^{iKx} + e^{-iKx}}{2}$$

or

$$w = A \sin Kx + B \cos Kx \tag{6.1.8}$$

The end conditions for a column with hinged ends are

(1) at $x = 0$, $w = 0$, $d^2w/dx^2 = 0$, which requires that $B = 0$.
(2) at $x = L$, $w = 0$, $d^2w/dx^2 = 0$, which requires that $A \sin KL = 0$.

The second condition requires that either A or $\sin KL$ has to vanish. If $A = 0$, $w = 0$; that is, the column remains straight and no buckling occurs. Hence, for the column to buckle, $\sin KL$ must vanish. This is satisfied when KL is equal to anyone of the following values:

$$KL = 0, \pi, 2\pi, 3\pi, \ldots, n\pi$$

If $KL = 0$, then $K = 0$, and, from Equation (6.1.6), we have $P = 0$. This corresponds to the condition of a straight column with no load. As KL increases, P increases. Since the column buckles at $K = \pi$, that is, at $P = \pi^2/EI$, before reaching any of the higher buckling loads, the critical load is

$$KL = \sqrt{\frac{P}{EI}L^2} = \pi \qquad P_{cr} = \frac{\pi^2 EI}{L^2} \tag{6.1.9}$$

This load is the well-known Euler load. When $KL = \pi$, the coefficient A in Equation (6.1.8) can be any finite value. This shows that at this load, the magnitude of the lateral deflection is not unique and may be any value consistent with small deflection theory. If the column is prevented from buckling into a single sine wave corresponding to $KL = \pi$, then the column may buckle at $KL = 2\pi$. This is the case when a mid-column restraint is imposed.

6.2 Elastic Columns with Initial Curvature

Actual columns, of course, are neither absolutely straight nor homogeneous. An initial curvature is always present. In Section 6.1, the analysis of perfectly straight columns was shown. In this section the analysis of columns with initial curvature, which is present in all actual columns, will be given. Let w_0 denote the initial deflection and eccentricity of loading and w_1 the additional deflection caused by loading. The equilibrium condition gives

$$EI\frac{d^2w_1}{dx^2} + P(w_1 + w_0) = 0 \tag{6.2.1}$$

Represent w_0 as

$$w_0 = \sum_{n=1}^{\infty} A_n \sin\frac{n\pi x}{L} + b + mx \tag{6.2.2}$$

With

$$K^2 = \frac{P}{EI}$$

and using Equation (6.2.2), we see that Equation (6.2.1) becomes

$$\frac{d^2w_1}{dx^2} + K^2\left(w_1 + b + mx + \sum_{n=1}^{\infty} A_n \sin\frac{n\pi x}{L}\right) = 0$$

The solution of this differential equation is given by[1]

$$w_1 = A \sin Kx + B \cos Kx + \sum_{n=1}^{\infty} \frac{A_n}{(n^2\pi^2/K^2L^2) - 1} \sin \frac{n\pi x}{L} - b - mx \tag{6.2.3}$$

The constants of integration A and B may be obtained from the end conditions. For a column with hinged ends we have

(1) at $x = 0$, $w_1 = 0$, hence $B = b$
(2) at $x = L$, $w_1 = 0$, hence $A = [b(1 - \cos KL) + mL]/\sin KL$

Substituting these values of A and B into Equation (6.2.3) gives

$$w_1 = \frac{b(1 - \cos KL) + mL}{\sin KL} \sin Kx + b(\cos Kx - 1) - mx + \sum_{n=1}^{\infty} \frac{A_n}{(n^2\pi^2/K^2L^2) - 1} \sin \frac{n\pi x}{L} \tag{6.2.4}$$

This solution may be easily verified by substituting Equations (6.2.4) and (6.2.2) into Equation (6.2.1). For a concentrically loaded column, i.e., a column whose end loads are in line, with initial curvature,[2,3] we have $m = 0$, $b = 0$ and, from Equation (6.2.4), we see that w_1 is given by

$$w_1 = \sum_{n=1}^{\infty} \frac{A_n}{(n^2\pi^2EI/PL^2) - 1} \sin \frac{n\pi x}{L} \tag{6.2.5}$$

As KL increases and approaches π, the denominators in the first term and last terms for $n = 1$ in Equation (6.2.4) approach zero. Under this loading, w becomes large even if b, m, and A_1 are very small. The growth of w as KL approaches π is the observed phenomena of buckling in actual columns. Hence the critical load derived for perfect columns is the same as that for actual columns. At mid-span ($x = L/2$) the additional deflection for the concentrically loaded column with initial curvature is

$$(w_1)_{x=L/2} = \delta_1 = \sum_{n=1,3,\ldots}^{\infty} \frac{(-1)^{n-1}A_n}{(P_n/P) - 1}$$

where $P_n = n^2\pi^2EI/L^2$. The terms with even n do not contribute to the mid-span deflection.

The force P approaches P_1 first. As P approaches P_1, the first term,

$$\frac{A_1}{(P_1/P) - 1}$$

becomes very large, while the other terms remain small, hence

$$\delta_1 \cong \frac{A_1}{(P_1/P) - 1} = \frac{PA_1}{P_1 - P}$$

$$(P_1 - P)\delta_1 \cong PA_1 \qquad \text{or} \qquad \frac{\delta_1}{P} - \frac{1}{P_1}\delta_1 \cong \frac{A_1}{P_1} \tag{6.2.6}$$

This is a linear equation between δ_1/P and δ_1. If we plot δ_1/P vs. δ_1 when P is near P_1, the slope of this line gives the reciprocal of the critical load P_1, and the line will cut the abscissa at the point $\delta_1 = -A_1$. This method has been used by Southwell[4] to find P_1 without loading the column to failure. This also gives the first harmonic A_1 of the initial curvature.

6.3 Elastic Beam-Columns

A beam-column is a bar subject to both lateral and axial loads. Consider a beam-column with hinged ends and let the moment caused by the lateral load and its reactions be denoted by M'. A beam-column with initial deflection w_0, under both lateral and compressive axial loads, as shown in Figure 6.3, satisfies the equilibrium condition

$$M = M' - P(w_1 + w_0) = EI\frac{d^2w_1}{dx^2}$$

Rewriting the above expression, we have

$$EI\frac{d^2w_1}{dx^2} + Pw_1 = M' - Pw_0 \tag{6.3.1}$$

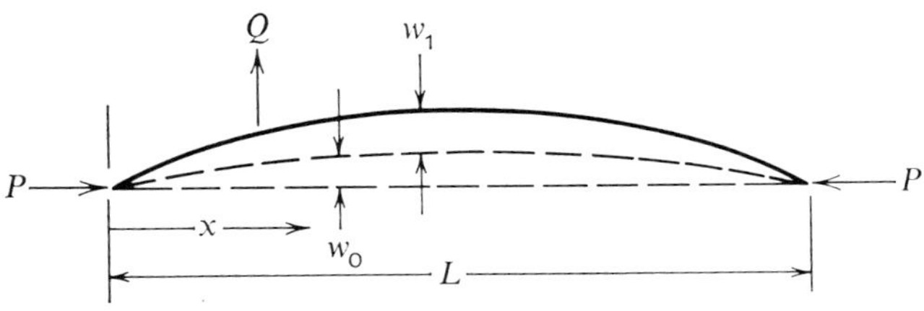

Figure 6.3

It is seen that M' is equivalent to $-Pw_0$; that is, the bending moment caused by lateral load is equivalent to an initial curvature of the column or vice versa.

Single Lateral Load on a Beam-Column. Consider a beam-column with hinged ends, subject to a single lateral load Q acting at a distance C from the right end as shown in Figure 6.4. The equilibrium condition is

$$M = \frac{QC}{L} x + Pw = -EI \frac{d^2 w}{dx^2} \qquad 0 < x < L - C$$

$$M = \frac{Q(L-C)}{L}(L-x) + Pw$$

$$= -EI \frac{d^2 w}{dx^2} \qquad L - C < x < L$$

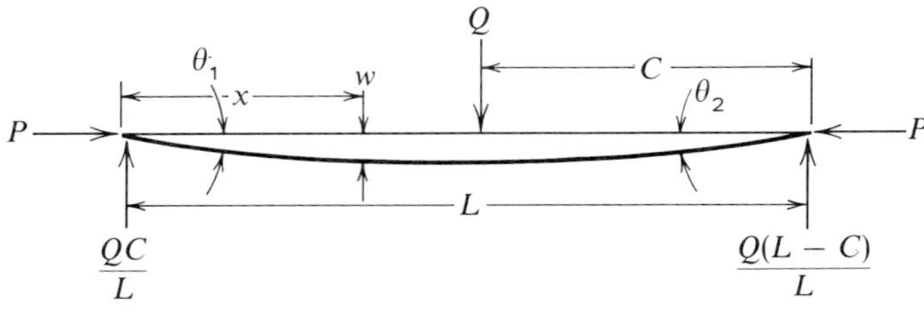

Figure 6.4

With $K^2 = P/EI$, the solution of the equilibrium equation for the left portion is

$$w = A_1 \sin Kx + A_2 \cos Kx - \frac{QC}{PL} x \qquad 0 < x < L - C$$

From the end condition $w = 0$ at $x = 0$, we obtain $A_2 = 0$, hence

$$\left.\begin{aligned} w &= A_1 \sin Kx - \frac{QC}{PL} x \\ \frac{dw}{dx} &= A_1 K \cos Kx - \frac{QC}{PL} \end{aligned}\right\} \qquad 0 < x < L - C$$

Similarly, for the right portion, we have

$$w = A_3 \sin Kx + A_4 \cos Kx - \frac{Q(L-C)(L-x)}{PL}$$

At $x = L$, $w = 0$, which gives $A_4 = -A_3 \tan KL$. Then

$$\left.\begin{aligned} w &= A_3[\sin Kx - \tan KL \cos Kx] - \frac{Q(L-C)(L-x)}{PL} \\ \frac{dw}{dx} &= A_3 K[\cos Kx + \tan KL \sin Kx] + \frac{Q(L-C)}{PL} \end{aligned}\right\} \qquad L - C < x < L$$

Continuity of deflection and slope at $x = L - C$ requires that

$$A_1 \sin K(L-C) - \frac{QC(L-C)}{PL} = A_3[\sin K(L-C) - \tan KL \cos K(L-C)] - \frac{Q(L-C)C}{PL}$$

$$A_1 K \cos K(L-C) - \frac{QC}{PL} = A_3[K \cos K(L-C) + K \tan KL \sin K(L-C)] + \frac{Q(L-C)}{PL}$$

Solving for A_1 and A_3 from these two equations yields

$$A_1 = \frac{Q \sin KC}{PK \sin KL} \qquad A_3 = -\frac{Q \sin K(L-C)}{PK \tan KL}$$

The deflection and slope are then given as

$$\left.\begin{aligned} w &= \frac{Q \sin KC}{PK \sin KL} \sin Kx - \frac{QC}{PL} x \qquad 0 < x < L - C \\ w &= \frac{Q \sin K(L-C)}{PK \sin KL} \sin K(L-x) - \frac{Q(L-C)(L-x)}{PL} \\ &\qquad L - C < x < L \end{aligned}\right\} \tag{6.3.2}$$

$$\left.\begin{aligned} \frac{dw}{dx} &= \frac{Q \sin KC}{P \sin KL} \cos Kx - \frac{QC}{PL} \qquad 0 < x < L - C \\ \frac{dw}{dx} &= \frac{Q \sin K(L-C)}{P \sin KL} \cos K(L-x) + \frac{Q(L-C)}{PL} \\ &\qquad L - C < x < L \end{aligned}\right\} \tag{6.3.3}$$

For beam-columns with more than a single concentrated load, the solutions for different loads may be superposed.[3] For distributed load q, we replace Q by $q\,dC$ and integrate with respect to C. For a uniformly distributed load q, from Equation (6.3.2), we have

$$\begin{aligned} w &= \int_0^{L-x} \left(\frac{q \sin KC}{PK \sin KL} \sin Kx - \frac{qC}{PL} x \right) dC \\ &\quad + \int_{L-x}^{L} \left[\frac{q \sin K(L-C)}{PK \sin KL} \sin K(L-x) - \frac{q(L-C)(L-x)}{PL} \right] dC \\ &= \frac{q}{PK^2} \frac{\sin K(L-x) + \sin Kx - \sin KL}{\sin KL} - \frac{q}{2P} x(L-x) \end{aligned} \tag{6.3.4}$$

From Equation (6.3.3), the end slope is

$$\left(\frac{dw}{dx}\right)_{x=0} = \int_0^L \left(\frac{q \sin C}{P \sin KL} - \frac{qC}{PL}\right) dC = \frac{q(1-\cos KL)}{PK \sin KL} - \frac{qL}{2P} \tag{6.3.5}$$

The bending moment at section x is

$$M = -EI\frac{d^2w}{dx^2} = EI\frac{q}{P}\left(\frac{\sin K(L-x) + \sin Kx}{\sin KL} - 1\right) \tag{6.3.6}$$

End Moments on a Beam-Column. Let load QC in Equation (6.3.3) be equal to a finite value of an end moment M_2 with C approaching zero and Q approaching infinity. Since C approaches zero, $\sin KC \cong KC$. From Equation (6.3.2), we have $w = (M_2/P \sin KL) \sin Kx - (M_2/PL)x$ and hence

$$\theta_1 = \left(\frac{dw}{dx}\right)_{x=0} = \frac{QCK}{P \sin KL} - \frac{QC}{PL} = \frac{M_2}{P}\left(\frac{K}{\sin KL} - \frac{1}{L}\right)$$

$$\theta_2 = -\left(\frac{dw}{dx}\right)_{x=L} = -\frac{QCK}{P \sin KL}\cos KL + \frac{QC}{PL} = -\frac{M_2}{P}\left(\frac{K \cos KL}{\sin KL} - \frac{1}{L}\right)$$

With M_1 and M_2 at both ends, we have

$$\left.\begin{aligned} w &= \frac{M_2}{P \sin KL}\sin Kx - \frac{M_2}{PL}x + \frac{M_1}{P \sin KL}\sin K(L-x) - \frac{M_1(L-x)}{PL} \\ \theta_1 &= -\frac{M_1}{P}\left(\frac{K \cos KL}{\sin KL} - \frac{1}{L}\right) + \frac{M_2}{P}\left(\frac{K}{\sin KL} - \frac{1}{L}\right) \\ \theta_2 &= \frac{M_1}{P}\left(\frac{K}{\sin KL} - \frac{1}{L}\right) - \frac{M_2}{P}\left(\frac{K \cos KL}{\sin KL} - \frac{1}{L}\right) \end{aligned}\right\} \tag{6.3.7}$$

When $M_1 = M_2 = M_0$, these equations reduce to

$$w = \frac{M_0}{P \sin KL}[\sin Kx + \sin K(L-x)] - \frac{M_0}{P} \tag{6.3.8}$$

$$\theta_1 = \theta_2 = \frac{M_0 K}{P \sin KL}(1 - \cos KL) \tag{6.3.9}$$

The moment $M = -EI(d^2w/dx^2)$ is obtained by differentiating Equation (6.3.8) twice, that is,

$$\begin{aligned} \frac{d^2w}{dx^2} &= -\frac{M_0 K^2}{P \sin KL}[\sin Kx + \sin K(L-x)] \\ M &= \frac{M_0}{\sin KL}[\sin Kx + \sin K(L-x)] \end{aligned} \tag{6.3.10}$$

For a beam-column with built-in ends, subject to a uniform lateral load q and end thrust P, the slope at either end is zero; i.e., from Equations (6.3.5) and (6.3.9), we obtain

$$\frac{M_0 K}{P \sin KL}(1 - \cos KL) + \frac{q(1 - \cos KL)}{PK \sin KL} - \frac{qL}{2P} = 0$$

Solving for the end moment M_0 yields

$$M_0 = -\frac{q}{K^2} + \frac{qL \sin KL}{2K(1 - \cos KL)} \tag{6.3.11}$$

The bending moment at section x is

$$M = M_0 + \frac{qL}{2}x - \frac{qx^2}{2} + Pw \tag{6.3.12}$$

where w is the sum of the expressions given by Equations (6.3.4) and (6.3.8) and M_0 is given by Equation (6.3.11). These results will be used in later sections. For more detailed derivations, see Reference 3.

6.4 Columns with Thermal Strain

Consider a column, with no creep or plastic strain, subject to a temperature varying across its depth and along its length. Imagine that the column is cut into elements of dx length as shown in Section 5.8. Each fiber of the element is elongated by $\alpha T\, dx$. Imagine now that a compressive stress of $E\alpha T$ is applied on the two faces of each element so that the resulting longitudinal strain just cancels the thermal longitudinal strain; each element then reverts to its original shape and length. These compressive stresses on the two cross sections of each element give a section moment $M_T = \int E\alpha T\eta\, dA$, where η is the distance from the centroidal axis of the section (Figure 6.5). Since each element has been restored to its original shape, they will now fit together perfectly, giving the original straight column. Imagine now that the elements are welded together.

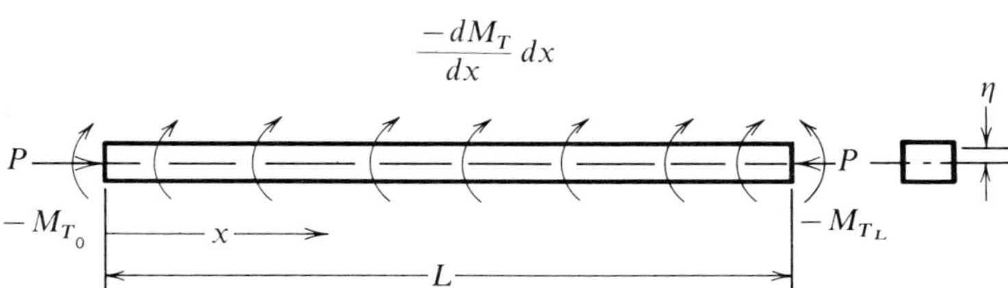

Figure 6.5

At the interface of adjacent elements, there is the unbalanced moment $(dM_T/dx)\,dx$. At the ends, $x = 0, L$, the unbalanced moments are M_{T_0} and M_{T_L}, respectively. In the actual column, no such unbalanced moments were applied, and therefore these unbalanced moments must be relaxed by applying equal and opposite moments $-(dM_T/dx)\,dx$ to the column, and $-M_{T_0}$ and $-M_{T_L}$ at the ends. This applied moment has the same effect as the moment caused by lateral load as in the case of a beam-column, hence the column with thermal gradient can be treated just like an elastic beam-column. Since this process of cutting, compressing, welding, and relaxing is entirely imaginary, no plastic strain is introduced in this process even if $E\alpha T$ is greater than the elastic limit. This is illustrated by the following example.

Example. A column, of uniform flexural rigidity (EI = constant) with pinned ends, is subject to a temperature distribution represented by

$$T = b_0\, f_0(\eta) + b_1 x f_1(\eta) + b_2\, x^2 f_2(\eta)$$

where b_0, b_1, and b_2 are constants, and f_0, f_1, and f_2 are functions. Find the expression of the lateral deflection w under axial load P, and determine the critical load.

Solution. Substituting the expression for T into the expression $M_T = \int E\alpha T\eta\, dA$ yields

$$M_T = a_0 + a_1 x + a_2\, x^2$$

where

$$a_0 = E\alpha b_0 \int f_0(\eta)\eta\, dA$$

$$a_1 = E\alpha b_1 \int f_1(\eta)\eta\, dA$$

and

$$a_2 = E\alpha b_2 \int f(\eta)\eta\, dA$$

The equivalent applied distributed moment on the column caused by thermal gradient is $-dM_T/dx = -(a_1 + 2a_2 x)$; the equivalent end moments are $-M_{T0} = -a_0$ at $x = 0$ and $-M_{T_L} = -(a_0 + a_1 L + a_2 L^2)$ at $x = L$. The load on the column is shown in Figure 6.6. The column is assumed to have no initial curvature. From the equilibrium conditions, the vertical reactions at the two ends are zero. The bending moment at section x (Figure 6.6) is

$$M = -Pw - (a_0 + a_1 x + a_2 x^2) = EI \frac{d^2 w}{dx^2} \tag{6.4.1}$$

or

$$\frac{d^2 w}{dx^2} + \frac{P}{EI} w + \frac{1}{EI}(a_0 + a_1 x + a_2 x^2) = 0 \tag{6.4.2}$$

whose solution is

$$w = A \sin Kx + B \cos Kx - \frac{1}{P}\left(a_0 + 2a_2 \frac{EI}{P} + a_1 x + a_2 x^2\right)$$

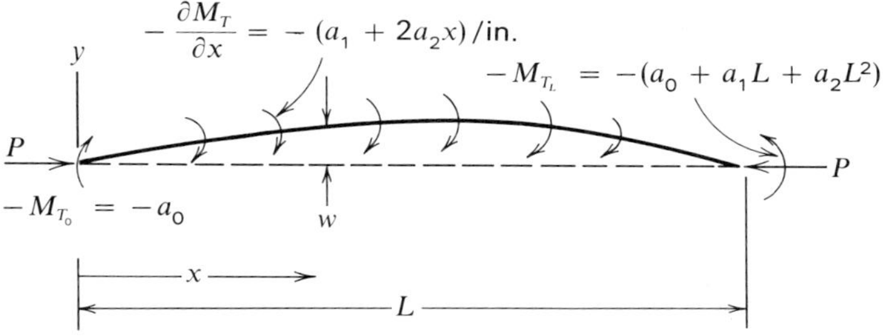

Figure 6.6

From the end conditions $w = 0$ at $x = 0$, we obtain $B = (1/P)(a_0 + 2a_2 EI/P)$ and $w = 0$ at $x = L$ gives,

$$A = -\frac{1}{P \sin KL}\left[\left(a_0 + \frac{2a_2}{K^2}\right)(1 - \cos KL) + a_1 L + a_2 L^2\right]$$

Then

$$w = -\frac{1}{P}\left[\left(a_0 + \frac{2a_2}{K^2}\right)(1 - \cos Kx) + a_1 x + a_2 x^2\right]$$
$$- \frac{\sin Kx}{P \sin KL}\left[\left(a_0 + \frac{2a_2}{K^2}\right)(1 - \cos KL) + a_1 L + a_2 L^2\right] \tag{6.4.3}$$

This is the lateral deflection under load P. It is seen that when $KL = \pi$ the second term becomes very large. Hence the critical load of a pin-ended column with thermal gradient remains $\pi^2 EI/L^2$. For columns with other temperature distributions and end conditions, similar analyses may be performed.

6.5 Linear Viscoelastic Columns

Since the stress-strain-time relations of some materials can be approximately represented by linear viscoelastic models, a number of studies have been made on viscoelastic columns. Such treatment was originally given by Freudenthal.[5] Rosenthal and Baer[6] in 1951 presented the analysis of viscoelastic columns involving only transient creep. Hilton[7] in 1951 showed the creep collapse of viscoelastic columns with initial curvature. In 1954 Kempner[8] considered the analysis for the creep bending and buckling of linearly viscoelastic columns, using a Maxwell-Kelvin model. Kempner, assuming an initial deviation of sinusoidal form, calculated the ratio of the creep deflection of the column to the deflection of the corresponding elastic column at different instants. The conclusion from Kempner's calculation showed that infinite deflection of a linearly viscoelastic column occurs only at infinite time after loading.

Column-creep analysis is based on small-deflection theory which is not valid for large deflections. However, if the initial deflection is assumed to be very small, the deflection of the purely elastic column under load will also be very small. Hence, even when the ratio of the creep deflection to the elastic deflection is large, the creep deflection may still be small. In such case small-deflection analysis is still valid. The time at which this ratio approaches infinity is here defined as the critical time of the column. Kempner's analysis shows infinite critical time for viscoelastic columns.

In 1953 the author[9] applied Laplace transformation to simplify the analysis of viscoelastic columns with initial curvature. This analysis may be extended to viscoelastic beam-columns. Suppose that the stress-strain-time relation of a material can be represented by a Maxwell-Kelvin model. The relation between the Laplace transforms of the stress and strain is given by Equation (3.6.19) as

$$\bar{\sigma} = \frac{E\left(s + \dfrac{1}{T_0}\right)s}{\left(s + \dfrac{1}{T_1}\right)\left(s + \dfrac{1}{T_2}\right)}\bar{e} \tag{3.6.19}$$

From the assumption that plane cross sections remain plane during bending, we have

$$e = e_0 + (K - K_0)\eta = e_0 + (w_{xx} - w_{0_{xx}})\eta \tag{6.5.1}$$

where K is the instantaneous curvature, K_0 is its initial value before loading, e_0 is the strain at the centroid of the geometrical section, w is the lateral deflection of the column, η is the distance from the centroidal axis, and the subscript x denotes differentiation with respect to x. Taking the Laplace transformation of Equation (6.5.1) with respect to time yields

$$\bar{e} = \bar{e}_0 + \left(\bar{w}_{xx} - \frac{1}{s} w_{0_{xx}}\right)\eta \tag{6.5.2}$$

Substituting Equation (6.5.2) into Equation (3.6.19) gives

$$\bar{\sigma} = \frac{E\left(s + \frac{1}{T_0}\right)s}{\left(s + \frac{1}{T_1}\right)\left(s + \frac{1}{T_2}\right)}\left[\bar{e}_0 + \left(\bar{w}_{xx} - \frac{1}{s} w_{0_{xx}}\right)\eta\right] \tag{6.5.3}$$

From Equation (6.5.3), we obtain

$$\bar{P} = \int \bar{\sigma}\, dA = \frac{E\left(s + \frac{1}{T_0}\right)s}{\left(s + \frac{1}{T_1}\right)\left(s + \frac{1}{T_2}\right)}\bar{e}_0 A \tag{6.5.4}$$

$$M = \int \sigma\eta\, dA$$

$$\bar{M} = \int \bar{\sigma}\eta\, dA = \frac{E\left(s + \frac{1}{T_0}\right)s}{\left(s + \frac{1}{T_1}\right)\left(s + \frac{1}{T_2}\right)}\left[\int \bar{e}_0\,\eta\, dA + \int (\bar{w}_{xx} - \frac{1}{s} w_{0_{xx}})\eta^2\, dA\right]$$

$$\bar{M} = \frac{EI\left(s + \frac{1}{T_0}\right)s}{\left(s + \frac{1}{T_1}\right)\left(s + \frac{1}{T_2}\right)}\left[\bar{w}_{xx} - \frac{1}{s} w_{0_{xx}}\right] \tag{6.5.5}$$

The external moment $M = -Pw$. If P remains constant, we have

$$-P\bar{w} = \frac{EI\left(s + \frac{1}{T_0}\right)s}{\left(s + \frac{1}{T_1}\right)\left(s + \frac{1}{T_2}\right)}[\bar{w}_{xx} - (1/s)w_{0_{xx}}] \tag{6.5.6}$$

Let

$$\phi^2 = \frac{P}{EI}\frac{\left(s + \frac{1}{T_1}\right)\left(s + \frac{1}{T_2}\right)}{\left(s + \frac{1}{T_0}\right)s} \tag{6.5.7}$$

then Equation (6.5.6) may be written

$$\bar{w}_{xx} - \frac{1}{s} w_{0_{xx}} + \phi^2 \bar{w} = 0 \tag{6.5.8}$$

Expressing the initial deflection w_0 in a sine series gives

$$w_0 = \sum_{n=1}^{\infty} A_n \sin \frac{n\pi x}{L} \tag{6.5.9}$$

and

$$w_{0_{xx}} = -\sum_{n=1}^{\infty} A_n \frac{n^2\pi^2}{L^2} \sin \frac{n\pi x}{L} \tag{6.5.10}$$

The solution of Equation (6.5.8), if we use Equation (6.5.10) for $w_{0_{xx}}$, is

$$\bar{w} = A \sin \phi x + B \cos \phi x + \sum_{n=1}^{\infty} \frac{A_n(n^2\pi^2/L^2)(1/s)}{(n^2\pi^2/L^2) - \phi^2} \sin \frac{n\pi x}{L} \tag{6.5.11}$$

For a simply supported column, the end conditions are as follows: At $x = 0$, or $x = L$, $\bar{w} = 0$, which gives $A = 0$ and $B = 0$, hence

$$\bar{w} = \sum_{n=1}^{\infty} \frac{A_n/s}{1 - \dfrac{P}{P_n} \dfrac{\left(s + \dfrac{1}{T_1}\right)\left(s + \dfrac{1}{T_2}\right)}{s\left(s + \dfrac{1}{T_0}\right)}} \sin \frac{n\pi x}{L} \tag{6.5.12}$$

where $P_n = n^2\pi^2 EI/L^2$. Equation (6.5.12) may then be written as

$$\bar{w} = \sum_{n=1}^{\infty} \frac{A_n\left(s + \dfrac{1}{T_0}\right)}{s\left(s + \dfrac{1}{T_0}\right) - \dfrac{P}{P_n}\left(s + \dfrac{1}{T_1}\right)\left(s + \dfrac{1}{T_2}\right)} \sin \frac{n\pi x}{L} \tag{6.5.13}$$

The denominator of each term of the summation is a quadratic function whose roots are denoted by a_n and b_n; that is,

$$a_n, b_n = \frac{\left(\dfrac{P}{P_n}\right)\left(\dfrac{1}{T_1} + \dfrac{1}{T_2}\right) - \dfrac{1}{T_0}}{2\left(1 - \dfrac{P}{P_n}\right)}$$

$$\pm \frac{\sqrt{\left[\left(\dfrac{P}{P_n}\right)\left(\dfrac{1}{T_1} + \dfrac{1}{T_2}\right) - \dfrac{1}{T_0}\right]^2 + 4\left(1 - \dfrac{P}{P_n}\right)\left(\dfrac{P}{P_n}\right)\left(\dfrac{1}{T_1 T_2}\right)}}{2\left(1 - \dfrac{P}{P_n}\right)} \tag{6.5.14}$$

Then $\bar{w}$ is given by

$$\bar{w} = \sum_{n=1}^{\infty} \frac{A_n\left(s + \dfrac{1}{T_0}\right)}{(s - a_n)(s - b_n)\left(1 - \dfrac{P}{P_n}\right)} \sin \frac{n\pi x}{L} \tag{6.5.15}$$

Let

$$\frac{\left(s + \dfrac{1}{T_0}\right)}{(s - a_n)(s - b_n)} = \frac{C_1}{(s - a_n)} + \frac{C_2}{(s - b_n)} \tag{6.5.16}$$

To determine C_1 we multiply both sides of the above equation by $(s - a_n)$ and let s approach a_n. Thus[10]

$$\frac{\left(a_n + \dfrac{1}{T_0}\right)}{a_n - b_n} = C_1 \tag{6.5.17}$$

Similarly,

$$\frac{\left(b_n + \dfrac{1}{T_0}\right)}{b_n - a_n} = C_2$$

Substituting these into Equation (6.5.15) yields

$$\bar{w} = \sum_{n=1}^{\infty} \frac{A_n}{\left(1 - \dfrac{P}{P_n}\right)} \left[\frac{\left(a_n + \dfrac{1}{T_0}\right)}{(a_n - b_n)(s - a_n)} + \frac{\left(b_n + \dfrac{1}{T_0}\right)}{(b_n - a_n)(s - b_n)}\right] \sin \frac{n\pi x}{L} \tag{6.5.18}$$

Since the inverse transform[10] of $1/(s - a_n)$ is $\exp(a_n t)$, the inverse transform of Equation (6.5.18) is

$$w = \sum_{n=1}^{\infty} \frac{A_n}{\left(1 - \dfrac{P}{P_n}\right)} \left[\frac{\left(a_n + \dfrac{1}{T_0}\right)}{a_n - b_n} \exp(a_n t) + \frac{\left(b_n + \dfrac{1}{T_0}\right)}{b_n - a_n} \exp(b_n t)\right] \sin \frac{n\pi x}{L} \tag{6.5.19}$$

When $t = 0$, Equation (6.5.19) reduces to

$$w = \sum_{n=1}^{\infty} \frac{A_n}{1 - (P/P_n)} \sin \frac{n\pi x}{L}$$

$$w = \sum_{n=1}^{\infty} A_n \left(\frac{1}{(P_n/P) - 1} + 1 \right) \sin \frac{n\pi x}{L}$$

Writing $w = w_0 + w_1$ and using Equation (6.5.9), we have

$$w_1 = \sum_{n=1}^{\infty} \frac{A_n}{(n^2\pi^2 EI/PL^2) - 1} \sin \frac{n\pi x}{L} \tag{6.2.5}$$

which corresponds to Southwell's results for elastic columns.

Letting α_n denote P/P_n, we see that the nth harmonic of the initial deflection of a hinged end elastic column is magnified by the factor $1/(1 - \alpha_n)$. For the viscoelastic column, the magnification factor for the nth harmonic is

$$\frac{1}{1 - \alpha_n} \left[\frac{\left(a_n + \frac{1}{T_0} \right) \exp(a_n t) - \left(b_n + \frac{1}{T_0} \right) \exp(b_n t)}{a_n - b_n} \right]$$

which varies with time. Hence the shape of the deflection curve of a viscoelastic column changes with time. Differentiating w with respect to time yields

$$\frac{dw}{dt} = \sum_{n=1}^{\infty} \frac{A_n}{1 - (P/P_n)} \times \left[\frac{\left(a_n + \frac{1}{T_0} \right) a_n \exp(a_n t) - \left(b_n + \frac{1}{T_0} \right) b_n \exp(b_n t)}{a_n - b_n} \right] \sin \frac{n\pi x}{L} \tag{6.5.20}$$

This shows that the deflection rate approaches infinity—i.e., the w vs. time curve has vertical slope—only when time approaches infinity. This result was first obtained by Kempner by a different approach.

6.6 Nonlinear Viscoelastic Columns

In Section 6.5, strain was linearly related to stress. Kempner,[11] in 1954, showed an analysis based on a model consisting of a linear spring coupled with a nonlinear dashpot, whose strain rate is proportional to a power of the applied stress. The stress-strain-time relation derived from this model closely represents the steady creep characteristics of structural materials such as aluminum or steel. Kempner applied this model to a column with idealized H section, which consists of two equal concentrated flange areas connected by a thin web of negligible bending resistance. He considered a simply supported column with an initial deflection of a single half sine wave. The deflection of the loaded column was *assumed* to remain sinusoidal in shape. Kempner has shown that the critical time of a nonlinear viscoelastic column is finite.

The assumption that the initial sinusoidal shape of deflection is maintained throughout the creep-buckling process is not compatible with the differential equation of equilibrium. However, Hoff[12] showed that a column whose stress-strain relation is given by $e_c = \lambda\sigma^n$ buckles at the critical load in a shape close to a half sine wave. For the case $n = 3$, the amplitude of the third harmonic was found to be only 1% of the amplitude of the first harmonic when the latter was equal to the radius of gyration of the section. The higher harmonics are even smaller.[13] Hence, if n is not very large, the error introduced by this assumption is not excessive.

Other simplified stress-strain-time relations have been used in the study of creep in columns. In 1947 Marin[14] presented a method of analysis involving only minimum rate creep, neglecting both elastic and transient creep strain. The analysis is valid only for cases in which transient creep and elastic strain are negligible.

6.7 Columns with Arbitrary Creep Characteristics

Libove,[15,16] in 1952 and 1953, showed the analyses of columns of H section and of rectangular section with initial curvature. The stress-strain-time relation of the material was represented by

$$e = \frac{\sigma}{E} + A \exp{(B\sigma)} t^K$$

where A, B, and K are constants. An equation of state with strain-hardening (Equation 3.4.1) was assumed. The strain rate was assumed to be a function of the stress and the creep strain, and not of the path of attaining this stress and strain. Libove also used the simplifying assumption that the deflected shape remains sinusoidal at all times.

Higgins,[17,18] in 1952, presented an iteration method for analyzing the creep deflection of columns. He used the same stress-strain-time relation and the same equation of state as Libove did. Higgins, however, removed the assumption that the deflected shape remained the same at all times. Although the particular stress-strain-time relation Higgins used does not involve the recovery effect, which usually is present in actual materials, the iteration method given can easily include this effect. The iteration method is rather lengthy. The author,[19] in 1956, showed a direct method of calculating, without iteration, stresses and deflections of columns of arbitrary creep characteristics. The method proceeds as follows.

The uniaxial strain rate under uniaxial stress at an elevated temperature is a function of the stress, time, creep strain, and the history of loading. After following a stress-time curve, the material, at some state of stress and strain, has a certain strain rate under constant stress. If the material is given an instantaneous change in stress, the instantaneous strain increment Δe will be $\Delta\sigma/E$, where E is the elastic modulus of the material at the particular temperature T. Generally the stress-time curve at each point in the column is a smooth curve. In the following analysis, this smooth curve is approximated by small horizontal and vertical steps of Δt and $\Delta\sigma$, respectively, as shown in Figure 3.9. Each step consists of a constant stress period Δt followed by an instantaneous increment of stress $\Delta\sigma$. When Δt and $\Delta\sigma$ approach zero, the step stress-time curve approaches the actual stress-time curve. Letting ϕ denote the rate of strain under constant stress, we have

$$de = \phi\, dt + \frac{d\sigma}{E} \tag{6.7.1}$$

$$\sigma = E(e - e'')$$

$$\Delta\sigma = E(\Delta e - \Delta e'') \tag{6.7.2}$$

Instead of the equilibrium conditions for the total load and total moment at each section, the equilibrium conditions for the incremental load and moment are considered,

$$\Delta P = \int \Delta\sigma\, dA = E \int (\Delta e - \Delta e'')\, dA \tag{6.7.3}$$

$$\Delta M = \int \Delta\sigma\eta\, dA = E \int (\Delta e - \Delta e'')\eta\, dA \tag{6.7.4}$$

where η is the distance from the centroidal axis of the section, that is,

$$\int \eta\, dA = 0 \tag{6.7.5}$$

From the condition that plane sections before loading remain plane sections during loading, we have

$$e = e_0 + K\eta$$

where K is the curvature and e_0 is the strain at $\eta = 0$. The incremental form of this equation is

$$\Delta e = \Delta e_0 + \eta\, \Delta K \tag{6.7.6}$$

For a column under constant load P, we have

$$\Delta P = 0 \tag{6.7.7}$$

From Equations (6.7.3) and (6.7.6), we obtain

$$\int (\Delta e - \Delta e'')\, dA = \int (\Delta e_0 + \eta\, \Delta K - \Delta e'')\, dA = 0$$

$$\Delta e_0\, A + \Delta K \int \eta\, dA = \int \Delta e''\, dA$$

$$\Delta e_0 = \frac{\int \Delta e''\, dA}{A} = \overline{\Delta e''} \tag{6.7.8}$$

where the bar denotes the average value across the section. Similarly, from Equation (6.7.4), we have

$$\Delta M = E \int [\Delta e_0 + \eta\, \Delta K - \Delta e'']\eta\, dA = EI\, \Delta K - E \int \Delta e'' \eta\, dA \tag{6.7.9}$$

where $I = \int \eta^2\, dA$ is the moment of inertia of the section. The external moment is

$$M = -P(w_1 + w_0)$$

or, in increment form with $w_1 + w_0 = w$,

$$\Delta M = -P\, \Delta w \tag{6.7.10}$$

The assumption of small curvature leads to

$$\Delta K \cong \frac{d^2\, \Delta w}{dx^2} \tag{6.7.11}$$

From Equations (6.7.9) through (6.7.11), we have

$$-P\, \Delta w = EI \frac{d^2\, \Delta w}{dx^2} - E \int \Delta e'' \eta\, dA \tag{6.7.12}$$

where the term on the left-hand side is the increment of external moment due to the increment of deflection, and the terms on the right-hand side are the increments of internal moment due to incremental curvature and a relaxation of moment due to creep in a given time interval Δt. Rewriting Equation (6.7.12) gives

$$\frac{d^2\, \Delta w}{dx^2} + \frac{P}{EI} \Delta w = -\frac{P}{EI}\left(-\frac{E}{P} \int \Delta e'' \eta\, dA\right) \tag{6.7.13}$$

where $\Delta e''$ varies from point to point in the column. Therefore the integral on the right-hand side varies from one section to another and hence is a function of x. Recall the differential equation for a column with simply supported ends subject to lateral load:

$$\frac{d^2\,\Delta w}{dx^2} + \frac{P}{EI}\,\Delta w = \frac{\Delta M'}{EI} \tag{6.3.1}$$

Comparing Equations (6.3.1) and (6.7.13), we see that the term $\int E\,\Delta e''\,\eta\,dA$ in Equation (6.7.13) plays the same role as $\Delta M'$ in Equation (6.3.1). Recall the differential equation for elastic columns with initial curvature,

$$\frac{d^2 w}{dx^2} + \frac{P}{EI}\,w = -\frac{P}{EI}\,w_0 \tag{6.2.1}$$

where w_0 is the initial deflection. By comparing Equation (6.7.13) with Equation (6.2.1), it is readily seen that the bracketed term of Equation (6.7.13) is equivalent to the initial deflection w_0 in elastic columns. Let

$$\left[-\frac{E}{P}\int \Delta e''\eta\,dA\right]$$

be represented by the Fourier series $\sum_{n=1}^{\infty} \Delta A_n \sin(n\pi x/L)$. The solution of Δw obtained from Equation (6.2.5) is

$$\Delta w = \sum_{n=1}^{\infty} \frac{\Delta A_n}{(P_n/P) - 1} \sin\frac{n\pi x}{L} \tag{6.7.14}$$

where

$$P_n = n^2\pi^2\,\frac{EI}{L^2}$$

In the calculation of the deflection-time curve of the column, the initial stress distributions at different sections are calculated as for elastic columns with initial curvature [Equation (6.2.5)]. From the stress-strain-time relation of the material at the particular temperature, the $\Delta e''$'s for the first time interval are calculated for various points on different sections. Obviously, the more points and sections used, the more accurate the results. A suitable number of points and sections, consistent with the accuracy of creep data and the desired accuracy of the results, may be chosen. Once the $\Delta e''$'s are known, $\int \Delta e''\,dA$ and $\int \Delta e''\,\eta\,dA$ for different sections are obtained by either graphical or numerical integration. The Fourier coefficients ΔA_n of the equivalent initial curvature $[-(E/P)\int \Delta e''\,\eta\,dA]$ vs. x are obtained by harmonic analysis[20] or other methods. Then Δw is obtained from Equation (6.7.14), and ΔK from Equation (6.7.11). The incremental total strain, Δe's, at various points of the column are then calculated from Equations (6.7.6) and (6.7.8).

The incremental total strain Δe minus the incremental creep strain $\Delta e''$ gives the incremental elastic strain $\Delta e'$, from which the increment of stress at the point is obtained. With the stress and the history of the stress-strain-time path known, the incremental creep strain $\Delta e''$ for the next time interval can be obtained either empirically by stressing the material following the same stress-time history, or by using some stress-strain-time relation. With $\Delta e''$ for the second time interval known, the process is repeated. By repeating the same process for successive time intervals, the stress distribution and deflections of the column at different sections at different instants can be calculated. This method is shown in the following example.

Example. A column of the same material and dimensions as calculated by Higgins[17] is analyzed by the present method in order to compare the results. The column is of aluminum alloy 75S-T6 at 600° F and having the following dimensions:

Column length $L = 8$ in. depth $h = \frac{1}{4}$ in. width $b = \frac{3}{8}$ in.

The column is pin-ended and its initial curvature is taken as a single half sine curve with an amplitude of 0.002 in.

$$w_0 = 0.002 \sin \frac{\pi x}{L}$$

$$P_{cr} = \frac{\pi^2 EI}{L^2} \qquad I = \frac{1}{12} bh^3$$

$$E = 5.2 \times 10^6 \text{ psi at } 600^\circ \text{ F}$$

$$\sigma_{cr} = \frac{P_{cr}}{bh} = \frac{\pi^2 \times 5.2 \times 10^6 \times (1/12)(1/4)^2}{8^2} = 4170 \text{ psi}$$

The applied compressive column stress is 2500 psi. The initial deflection is magnified [Equation (6.2.4)] under load at zero time to an amplitude of

$$\frac{0.002}{1 - 2500/4170} = 0.005 \text{ in.}$$

$$\sigma = \frac{P}{A} \pm \frac{(Pw)\eta}{I} = \frac{P}{A}\left(1 \pm \frac{w\eta}{\rho^2}\right) \tag{6.7.15}$$

where ρ is the radius of gyration;

$$\rho^2 = \frac{I}{A} = \frac{1}{12} h^2 = \frac{1}{12}\left(\frac{1}{4}\right)^2 = \frac{1}{192}$$

Table 6.1. CALCULATION OF CREEP STRESSES AND DEFLECTIONS OF A 75S-Tb COLUMN AT 600° F

	$x = \frac{L}{6}$					$x = \frac{L}{3}$					$x = \frac{L}{2}$				
	$\eta = \frac{h}{2}$	$\eta = \frac{h}{4}$	$\eta = 0$	$\eta = \frac{-h}{4}$	$\eta = \frac{-h}{2}$	$\eta = \frac{h}{2}$	$\eta = \frac{h}{4}$	$\eta = 0$	$\eta = \frac{-h}{4}$	$\eta = \frac{-h}{2}$	$\eta = \frac{h}{2}$	$\eta = \frac{h}{4}$	$\eta = 0$	$\eta = \frac{-h}{4}$	$\eta = \frac{-h}{2}$
	$\Delta a_1 = 0.001543$ in. $\Delta a_3 = 0.0000156$ in.										$\Delta b_1 = 0.0023$ in. $\Delta b_3 = 0.0000011$ in.				
	$\Delta a_5 = -0.0000046$ in.					Time 0–30 min					$\Delta b_5 = -0.000001133$ in. $\delta_{30} = 0.0074$ in.				
(1.1) σ	2350	2425	2500	2575	2650	2240	2370	2500	2630	2760	2200	2350	2500	2650	2800
(1.2) $\Delta\epsilon \times 10^5$	2.18	2.52	2.90	3.36	3.86	1.75	2.27	2.90	3.72	4.77	1.63	2.17	2.90	3.87	5.16
(1.3) $\Delta\epsilon_0 \times 10^5$			2.955					3.04					3.05		
(1.4) $-\frac{E}{P}\int \eta\Delta\epsilon dA$			0.00063 in.					0.00120 in.					0.00132 in.		
(1.5) $\Delta e \times 10^5$*	0.71	1.82	2.955	4.04	5.17	−0.80	1.12	3.04	5.00	6.92	1.38	0.76	3.05	5.30	7.48
(1.6) (1.5) − (1.4)	−1.47	−0.70	0.055	0.68	1.31	−2.55	−1.05	0.14	1.28	2.15	−3.01	−1.41	0.15	1.43	2.32
(1.7) $\frac{E(1.6)}{10^5}$	−76.5	−36.5	2.9	35.4	68	−132.5	−54.6	7.3	66.6	112	−1565	−73.3	7.8	74.4	120.6
	$\Delta a_1 = 0.001646$ in. $\Delta a_3 = 0.0000065$ in.										$\Delta b_1 = 0.00245$ in. $\Delta b_3 = 0$				
	$\Delta a_5 = 0.0000845$ in.					Time 30–60 min					$\Delta b^5 = 0.00000208$ in. $\delta_{60} = 0.00975$ in.				
(2.1) σ = (1.1) + (1.7)	2274	2389	2503	2610	2718	2107	2315	2507	2697	2872	2043	2277	2508	2724	2921
(2.2) $\Sigma\epsilon \times 10^5$	2.18	2.52	2.90	3.36	3.86	1.75	2.27	2.90	3.72	4.77	1.63	2.17	2.90	3.87	5.16
(2.3) $\Delta\epsilon \times 10^5$	1.00	1.26	1.70	2.18	2.85	0.62	1.09	1.70	2.60	3.74	0.50	1.00	1.70	2.96	4.42
(2.4) $\Delta\epsilon_0 \times 10^5$			1.77					1.87					3.035		
(2.5) $-\frac{E}{P}\int \eta\Delta\epsilon dA$			0.00087 in.					0.00135 in.					0.00173 in.		
(2.6) $\Delta e \times 10^5$	−0.64	0.565	1.77	2.975	4.18	−2.31	−0.21	1.87	3.96	6.05	−2.785	−0.375	2.035	4.445	6.855
(2.7) (2.6) − (2.3)	−1.64	−0.695	0.07	0.795	1.33	−2.93	−1.30	0.17	1.36	2.31	−3.285	−1.375	0.335	1.485	2.435
(2.8) $\frac{E(2.7)}{10^5}$	−85.3	−36.3	3.6	41.3	69.2	−152	−67.6	8.8	70.7	120	−171	−71.5	17.4	77.2	126.4
	$\Delta a_1 = 0.00274$ in. $\Delta a_3 = -0.000125$ in.										$\Delta b_1 = 0.0041$ in. $\Delta b_3 = -0.00000781$ in.				
	$\Delta a_5 = -0.000073$ in.					Time 60–120 min					$\Delta b_5 = -0.00000084$ in. $\delta = 0.01385$ in.				
(3.1) σ = (2.1) + (2.8)	2189	2353	2507	2651	2787	1955	2247	2516	2768	2992	1872	2206	2525	2891	3047
(3.2) $\Sigma\epsilon \times 10^5$	3.18	3.78	4.60	5.54	6.71	2.37	3.36	4.60	6.32	8.51	2.13	3.17	4.60	6.83	9.58
(3.8) $\Delta\epsilon \times 10^5$	0.74	1.20	1.76	2.62	3.61	0.50	0.90	1.84	3.66	6.4	0.28	0.80	1.85	3.58	6.88
(3.4) $\Delta\epsilon_0 \times 10^5$			1.91					2.45					2.37		
(3.5) $-\frac{E}{P}\int \eta\epsilon dA$			0.00145 in.					0.00244 in.					0.00254 in.		
(3.6) $\Delta e \times 10^5$	−2.00	−0.04	1.91	3.86	5.82	−4.33	−0.94	2.45	5.84	9.24	−5.44	−1.53	2.37	6.27	10.18
(3.7) (3.6) − (3.3)	−2.74	−1.24	0.15	1.24	2.21	−4.73	−1.84	0.61	7.18	2.84	−5.72	−2.33	0.52	2.69	3.30
(3.8) $\frac{E(3.7)}{10^5}$	−142	−64	7.8	65	115	−246	−96	32	113	147	−297	−122	27	240	172
	$\Delta a_1 = 0.00286$ in. $\Delta a_3 = 0.000200$ in.										$\Delta b_1 = 0.00427$ in. $\Delta b_3 = 0.0000143$ in				
	$\Delta a_5 = -0.000095$ in.					Time 120–180 min					$\Delta b_5 = -0.0000048$ in. $\delta_{180} = .01813$				
(4.1) (3.1) + (3.8)	2047	2289	2515	2715	2902	1709	2151	2548	2881	3139	1569	2084	2552	2941	3219
(4.2) $\Sigma\epsilon \times 10^5$	3.92	4.98	6.36	8.16	10.32	2.77	4.26	6.44	9.98	14.91	2.41	3.97	6.45	1041	16.46
(4.3) $\Delta\epsilon \times 10^5$	0.39	0.64	1.27	2.10	3.35	0.13	0.48	1.42	3.40	6.30	0.10	0.36	1.57	4.09	7.40
(4.4) $\Delta\epsilon_0 \times 10^5$			1.46					2.05							

													2.36		
(4.5) $-\frac{E}{P}\int \eta\Delta\epsilon\,dA$	0.00122 in.						0.00250 in.				0.00304 in.				
(4.6) $\Delta e \times 10^5$	−2.66	−0.60	1.46	3.52	5.58	−5.08	−1.52	2.05	5.62	9.18	−5.88	−1.76	2.36	6.48	10.60
(4.7) (4.6) − (4.3)	−3.05	−1.24	0.19	1.42	2.23	−5.21	2.00	0.63	2.22	2.88	−5.98	−2.12	0.79	2.39	3.20
(4.8) $\frac{E(4.7)}{10^5}$	−159	−645	9.9	74	116	−271	−104	32.8	115	150	−311	−110	41.1	124	166
	Δa_1 = 0.00352 in. Δa_3 = 0.000052 in. Δa_5 = −0.000143 in.						Time 180–240 min				Δb_1 = 0.00525 in. Δb_3 = 0.00000372 in. Δb_5 = −0.00000352 in. δ_{240} = 0.02338 in.				
(5.1) σ = (4.1) + (4.8)	1888	2225	2525	2789	3018	1438	2047	2581	2996	3289	1258	1974	2593	3065	3385
(5.2) $\Sigma\epsilon \times 10^5$	4.41	5.62	7.63	10.26	13.67	2.90	4.74	7.86	13.38	21.21	2.51	4.33	8.02	16.89	27.06
(5.3) $\Delta\epsilon \times 10^5$	0.14	0.50	1.11	2.40	4.10	0.044	0.23	1.33	4.0	7.5	0.034	0.17	1.35	4.05	8.55
(5.4) $\Delta e_0 \times 10^5$			1.52					2.36					2.37		
(5.5) $-\frac{E}{P}\int \eta\Delta\epsilon\,dA$	0.00164 in.						0.00317 in.				0.00343 in.				
(5.6) $\Delta e \times 10^5$	−3.26	−0.87	1.52	3.94	6.30	−6.61	−2.13	2.36	6.85	11.33	−7.73	−2.68	2.37	7.42	12.47
(5.7) (5.6) − (5.3)	−3.40	−1.37	0.41	1.51	2.20	−6.65	−2.36	1.03	2.85	3.83	−8.07	−2.85	1.02	3.37	3.92
(5.8) $\frac{E(5.7)}{10^5}$	−177	−71	21	78	114	−346	−123	54	148	199	−419	−148	53	175	204
	Δa_1 = 0.00525 in. Δa_3 = 0.000494 in. Δa_5 = 0.000150 in.						Time 240–300 min				Δb_1 = 0.0082 in. Δb_3 = 0.0000353 in. Δb_5 = 0.00000267 in. δ_{300} = 0.02792 in.				
(6.1) (5.1) + (5.8)	1711	2154	2546	2867	3132	1092	1924	2635	3144	3488	839	1826	2646	3240	4289
(6.2) $\Sigma\epsilon \times 10^5$	4.55	6.12	8.74	12.66	17.77	2.94	4.97	9.19	0.738	2871	2.54	4.50	9.37	20.94	35.61
(6.3) $\Delta\epsilon \times 10^5$	0.075	0.415	1.4	2.70	4.95	0.023	0.14	1.45	5.5	9.7	0.01	0.093	1.6	6.9	14.3
(6.4) $\Delta e_0 \times 10^5$			1.73					3.01					3.79		
(6.5) $-\frac{E}{P}\int \eta\Delta\epsilon\,dA$	0.0022 in.						0.00442 in.				0.00590 in.				
(6.6) $\Delta e \times 10^5$	−5.62	−1.95	1.73	5.41	9.08	−10.49	−3.74	3.01	9.76	16.51	−12.76	−4.48	3.79	12.06	20.34
(6.7) (6.6) − (6.3)	−5.70	−2.37	0.33	2.71	4.13	−10.51	−3.88	1.56	4.26	6.81	−12.77	−4.57	2.19	5.16	6.04
(6.8) $\frac{E(6.7)}{10^5}$	−296	−126	17	141	215	−546	−202	81	221	354	−663	−238	114	268	314
	Δa_1 = 0.0108 in. Δa_3 = 0.000468 in. Δa_5 = −0.000351 in.						Time 300–360 min				Δb_1 = 0.0188 in. Δb_3 = 0.0000334 in. Δb_5 = −0.0000086 in. δ = 0.04680 in.				
(7.1) σ = (6.1) + (6.9)	1415	2028	2563	3008	3347	546	1722	2716	3365	3842	176	1588	2760	3508	3903
(7.2) $\Sigma\epsilon \times 10^5$	4.63	6.54	10.14	15.36	22.72	2.96	5.11	10.64	22.88	38.41	2.55	4.59	10.97	27.84	49.91
(7.3) $\Delta\epsilon \times 10^5$	0.0425	0.140	1.20	3.70	9.00	0.008	0.076	1.60	9.0	31.4	0.004	0.059	2.00	12.6	35.4
(7.4) $\Delta e_0 \times 10^5$			2.16					5.80					7.60		
(7.5) $-\frac{E}{P}\int \eta\Delta\epsilon\,dA$	0.00312 in.						0.00966 in.				0.01258 in.				
(7.6) $\Delta e \times 10^5$	−15.19	−6.52	2.16	10.84	19.51	−25.9	−10.1	5.8	21.7	37.5	−28.6	−10.5	7.60	25.7	43.8
(7.7) (7.6) − (7.3)	−15.23	−6.66	0.96	7.14	10.51	−25.9	−10.2	4.2	12.7	6.1	−28.6	−10.6	5.60	13.1	8.4
(7.8) $\frac{E(7.7)}{10^5}$	−792	−246	50	372	546	−1246	−530	218	660	317	−1488	−551	291	681	436
							Time = 360 min								
(8.1) σ = (7.1) + (7.8)	623	1682	2613	3380	3893	−800	1192	2934	4025	4159	−1312	1037	3051	4189	4339

$* \; \Delta K = \Sigma - \frac{n^2\pi^2}{L^2}\Delta b_n \sin\frac{n\pi x}{L}$

$\Delta e = \Delta e_0 + \eta\Delta K$

The same stress-strain-time relation $e = (\sigma/E) + A \exp(B\sigma)t^k$ and equation of state with strain-hardening used by Higgins[17] and Libove,[15,16] are used here. For this stress-strain-time relation, $A = 2.64 \times 10^7$, $B = 1.92 \times 10^{-3}$, $K = 0.66$, σ is in psi, and t is in hours. Taking a time interval of $\frac{1}{2}$ hr, with constant stress within this interval, the creep strain $\Delta e''$ at $\eta = 0, \pm h/4$ and $\pm h/2$ on three sections $x = L/6, L/3$ and $L/2$ are evaluated. Then $\int \Delta e''\, dA$ and $\int \Delta e'' \eta\, dA$ are obtained. Obviously these integrals can be more accurately determined by calculating more points on the section. Then $[-(E/P) \int \Delta e'' \eta\, dA]$ for these sections were found. Using three terms in the Fourier series, we obtain the amplitudes of the first three harmonics Δa_1, Δa_2, Δa_3 and then the first three harmonics of the incremental deflection Δw from Equation (6.7.14). Then ΔK and Δe are found at these three sections. From the Δe's and the $\Delta e''$'s, the stress increments at these sections are calculated. With the stresses known at these sections, the process is repeated. The detail calculations are summarized in Table 6.1. The calculated deflection vs. time curve of this column is shown in Figure 6.7(a). Higgins' calculated and test results are shown in the same figure for comparison. Heller's and Finnie's[21] plot of the same results in linear scale is shown in Figure 6.7(b). The present method agrees better with the test results for $t < 350$ minutes than Higgins' method.

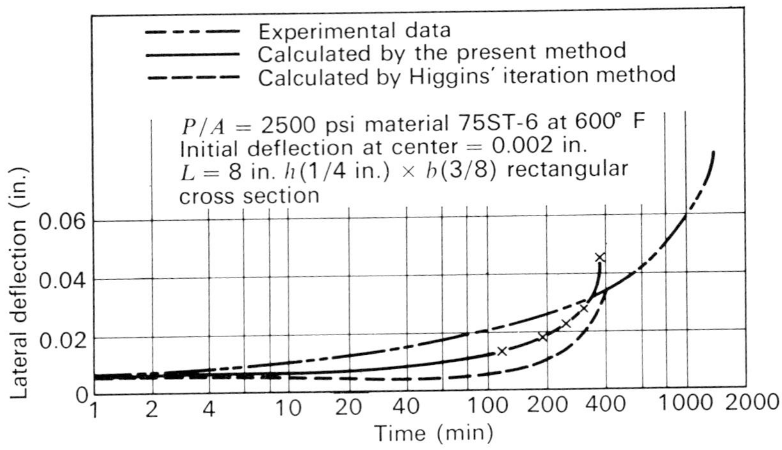

Figure 6.7(a)

The discrepancy between the calculated and test results is mainly caused by (a) the error involved in the stress-strain-time relation of the material due to an appreciable range of scatter of the test data (wide scatter of test data have been reported even from coupons cut from the same sheet); (b) the error in the determination of the amplitude of the first harmonic of the initial curvature of the column; and (c) the error in evaluating the first harmonic of the relaxation moments at different time intervals. Owing to the unavoidable imperfection of the material, the geometrical shape of the column does not give the total effective initial curvature, which determines the deflection-time curve of the column.

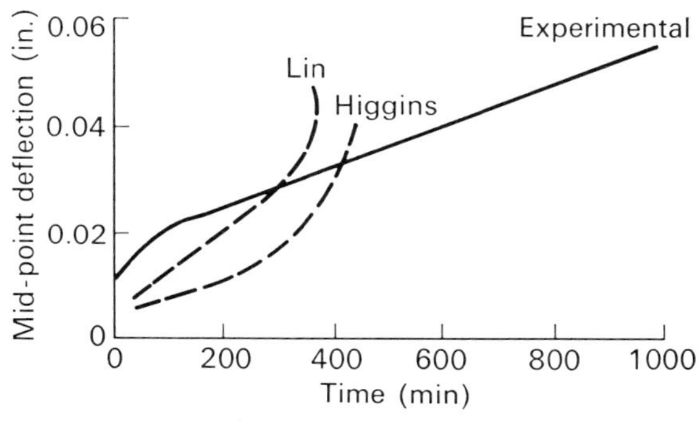

Figure 6.7(b)

For a more accurate study, the effective initial curvature may be determined by measuring the deflection curves under different loads within the elastic range at room temperature; that is, a_1 of the initial curvature may be determined from the observed w_1 curves at different loads (Section 6.2).

The stress distributions at the center section at different time instants are shown in Figure 6.8. It is seen that the distribution deviates considerably from being linear as the deflection grows. With the deflection-time curve of the column known, the life of the column can be determined.

6.8 Beam-Columns with Creep

It has been shown that the moment due to lateral load on an elastic beam-column is equivalent to the initial curvature of an elastic column (Section 6.3). In this section the method for analyzing creep stresses in columns with initial curvature is extended to beam-columns with arbitrary creep characteristics. Consider a beam-column of uniform section with restrained ends, subject to an arbitrary lateral loading. Referring to Figure 6.9, we denote by M' and R' the moment and reactions on a simply supported beam due to the same lateral loads. Let $R + R'$ be the reactions of the beam-column with restrained ends. The end moments are denoted

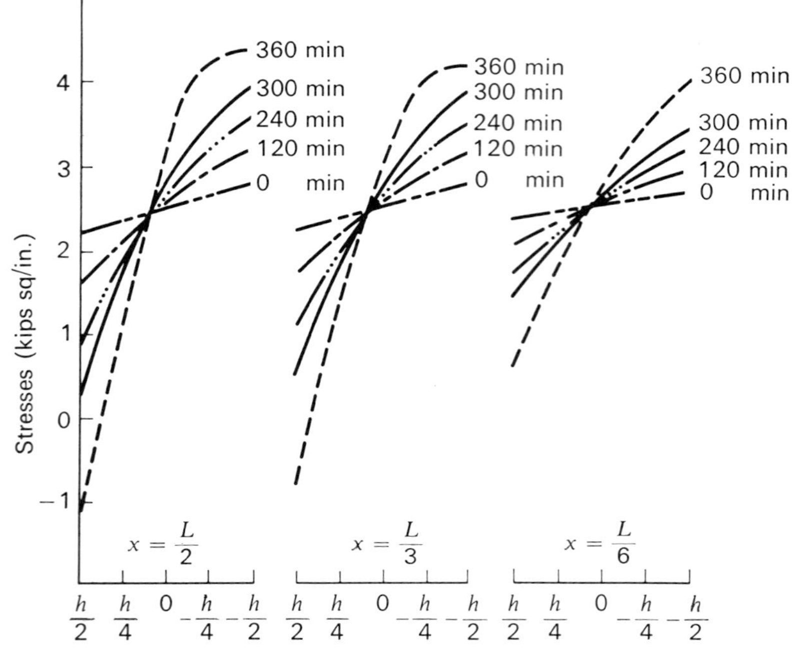

Figure 6.8

by M_1 and M_2 as shown in Figure 6.10. The external bending moment at section x is

$$M = M' + M_1 + R_1 x + Pw \tag{6.8.1}$$

where

$$R_1 = \frac{M_2 - M_1}{L}$$

Figure 6.9

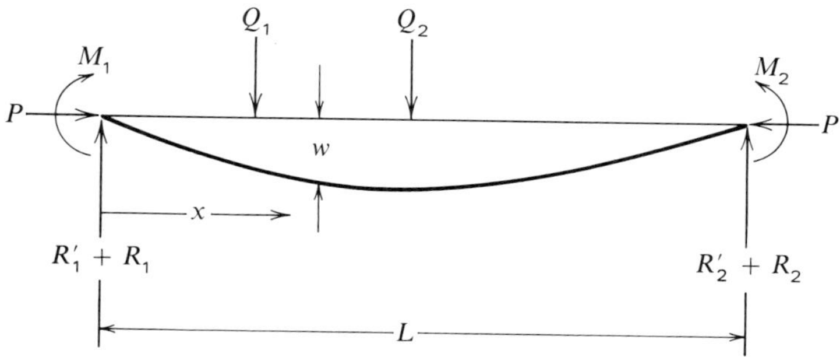

Figure 6.10

The incremental external moment in a time interval Δt is

$$\Delta M = \Delta M_1 + \Delta R_1 x + P\,\Delta w \tag{6.8.2}$$

From Equations (6.7.9) and (6.7.11), the incremental resisting moment is

$$\Delta M = -EI\,\frac{d^2\,\Delta w}{dx^2} - E\int \Delta e''\,\eta\,dA \tag{6.8.3}$$

Equating the incremental external moment to the incremental resisting moment gives

$$\Delta M_1 + x\,\Delta R_1 + P\,\Delta w = -EI\,\frac{d^2\,\Delta w}{dx^2} - E\int \Delta e''\,\eta\,dA \tag{6.8.4}$$

where the terms on the right-hand side represent the increment of internal moment due to the increment of curvature and a relaxation of moment due to creep in a given time interval Δt. Rewriting this equation, we obtain

$$\frac{d^2\,\Delta w}{dx^2} + \frac{P}{EI}\,\Delta w = -\frac{1}{I}\int \Delta e''\,\eta\,dA - \frac{\Delta M_1 + x\,\Delta R_1}{EI}$$

The integral on the right-hand side varies from one section to another and is a function of x. Let

$$-\frac{1}{I}\int \Delta e''\,\eta\,dA = F(x) \tag{6.8.5}$$

then

$$\frac{d^2\,\Delta w}{dx^2} + K^2\,\Delta w = F(x) - \frac{\Delta M_1 + x\,\Delta R_1}{EI} \tag{6.8.6}$$

where $K^2 = P/EI$. This differential equation may be resolved into two parts by letting

$$\Delta w = \Delta y_1 + \Delta y_2 \tag{6.8.7}$$

Then

$$\frac{d^2\,\Delta y_1}{dx^2} + K^2\,\Delta y_1 = F(x) \tag{6.8.8}$$

and

$$\frac{d^2\,\Delta y_2}{dx^2} + K^2\,\Delta y_2 = -\frac{\Delta M_1 + x\,\Delta R_1}{EI} \tag{6.8.9}$$

The solution of the last equation is

$$\Delta y_2 = A\sin Kx + B\cos Kx - \frac{\Delta M_1 + x\,\Delta R_1}{EIK^2}$$

To satisfy the end condition $\Delta y_2 = 0$ at $x = 0$, we must have

$$B = \frac{\Delta M_1}{K^2 EI} = \frac{\Delta M_1}{P} \tag{6.8.10}$$

Substituting this into Equation (6.8.10) and applying the end condition at $x = L$, $\Delta y_2 = 0$, we obtain

$$A = \frac{\Delta R_1}{P}\frac{L}{\sin KL} + \frac{\Delta M_1}{P}\frac{(1-\cos KL)}{\sin KL}$$

$$\Delta y_2 = \left[\frac{\Delta R_1}{P}\frac{L}{\sin KL} + \frac{\Delta M_1}{P}\frac{(1-\cos KL)}{\sin KL}\right]\sin Kx + \frac{\Delta M_1}{P}\cos Kx - \frac{\Delta M_1 + x\,\Delta R_1}{P} \tag{6.8.11}$$

$$\left(\frac{d\,\Delta y_2}{dx}\right)_{x=0} = \Delta\theta_{2_0} = \left[\frac{\Delta R_1}{P}\frac{KL}{\sin KL} + \frac{K\,\Delta M_1}{P}\frac{(1-\cos KL)}{\sin KL}\right] - \frac{\Delta R_1}{P} \tag{6.8.12}$$

$$\begin{aligned}\left(\frac{d\,\Delta y_2}{dx}\right)_{x=L} &= -\Delta\theta_{2_L} \\ &= \left[\frac{\Delta R_1}{P}\frac{KL}{\sin KL} + \frac{K\,\Delta M_1}{P}\frac{(1-\cos KL)}{\sin KL}\right]\cos KL \\ &\quad - \frac{K\,\Delta M_1}{P}\sin KL - \frac{\Delta R_1}{P} \\ &= \frac{\Delta R_1}{P}\left(\frac{KL}{\tan KL} - 1\right) + \frac{K\,\Delta M_1}{P}\left(\cot KL - \frac{1}{\sin KL}\right)\end{aligned} \tag{6.8.13}$$

$F(x)$ generally does not vanish at the ends, and its effect is the same as initial curvature and eccentricity in elastic columns as in Equation (6.2.4). Expressing $F(x)$ in the series

$$F(x) = -K^2\left[b + mx + \sum_{n=1}^{\infty} A_n \sin\frac{n\pi x}{L}\right]$$

we see that the solution of Δy_1 follows Equation (6.2.4), where use was made of $P_n = n^2\pi^2 EI/L^2$ and $K^2 = P/EI$. Then

$$\Delta y_1 = \frac{b(1-\cos KL) + mL}{\sin KL}\sin Kx + b(\cos Kx - 1) - mx + \sum_{n=1}^{\infty}\frac{A_n}{(P_n/P)-1}\sin\frac{n\pi x}{L} \tag{6.8.14}$$

and

$$\frac{d\,\Delta y_1}{dx} = \frac{b(1-\cos KL) + mL}{\sin KL}K\cos Kx - bK\sin Kx - m + \sum_{n=1}^{\infty}\frac{A_n(n\pi/L)}{(P_n/P)-1}\cos\frac{n\pi x}{L} \tag{6.8.15}$$

The slopes at the ends are

$$\left(\frac{d\,\Delta y_1}{dx}\right)_{x=0} = \Delta\theta_{1_0} = \frac{b + mL - b\cos KL}{\sin KL}K - m + \sum_{n=1}^{\infty}\frac{A_n(n\pi/L)}{(P_n/P)-1} \tag{6.8.16}$$

$$\left(\frac{d\,\Delta y_1}{dx}\right)_{x=L} = -\Delta\theta_{1_L} = K(b+mL)\cot KL - \frac{bK}{\sin KL} - m + \sum_{n=1}^{\infty}(-1)^n\frac{A_n(n\pi/L)}{(P_n/P)-1} \tag{6.8.17}$$

$$\Delta\theta_0 = \Delta\theta_{1_0} + \Delta\theta_{2_0} \qquad \Delta\theta_L = \Delta\theta_{1_L} + \Delta\theta_{2_L} \tag{6.8.18}$$

For restrained ends,

$$\Delta M_1 = -\beta_0\,\Delta\theta_0 \qquad \Delta M_2 = -\beta_L\,\Delta\theta_L \tag{6.8.19}$$

where β_0 and β_L are the moments required to rotate the left and right ends respectively by one radian. With β_0 and β_L known, substitution of Equations (6.8.11), (6.8.13), (6.8.16), (6.8.17), and (6.8.18) into Equations (6.8.19) yields two equations in the two unknowns ΔM_1 and ΔR_1, which may then be solved. In this way beam-columns with restrained ends and arbitrary creep characteristics may be analyzed. For built-in ends, $\Delta\theta_0 = \Delta\theta_L = 0$.

For simply supported ends $\beta_0 = \beta_L = 0$. With Δy_1, ΔM_1, and ΔR_1 known, Δy_2 and then Δy and ΔK at different sections may be determined. From Δe_0 and ΔK, we obtain Δe from Equation (6.7.6). With Δe and $\Delta e''$ known, $\Delta\sigma$ at different points in the beam-column are readily found by using Equation (6.7.2). The smooth stress-time curve at a point in the beam-column is approximated by small horizontal and vertical steps as shown in Figure 3.9. With the stress and the history of the stress-strain-time path known, the incremental creep strain $\Delta e''$ for the next time interval of constant stress can be obtained either empirically by testing the material following the same stress-time history or by using some stress-strain-time relation. By repeating this process for successive time intervals, the stress distribution and the deflections of the beam-column at different instants can be calculated. The above method may be applied to eccentrically loaded columns.

An Alternative Method of Solution. The method of analysis just discussed, for the solution of the differential equation $(d^2y/dx^2) + K^2y = f(x)$, is based upon first finding the solution of the homogeneous differential equation $(d^2y/dx^2) + K^2y = 0$ and then finding the particular solution of the nonhomogeneous differential equation. The complete solution is then given by the sum of the two solutions. Finally the end conditions are applied to determine the constants of integration. An alternative method of solving this differential equation is the method of determining the Green's function,[22,23] $g(x, a)$ satisfying the end conditions, for the differential operator $(d^2y/dx^2) + K^2y$. Once the Green's functions are obtained, we will show that the solution for the nonhomogeneous differential equation $(d^2y/dx^2) + K^2y = f(x)$ is $y = \int g(x, a)f(x)\,dx$. To obtain the Green's function, we proceed as follows. The column is divided at x_1 into the two parts $0 \le x < x_1$ and $x_1 < x \le L$. A solution of the homogeneous differential equation for the left part which satisfies the boundary conditions at $x = 0$, $y = 0$, and $d^2y/dx^2 = 0$ is

$$y_l = C_1 \sin Kx \qquad 0 < x < x_1$$

Similarly a solution for the right part of the beam which satisfies the boundary conditions at $x = L$, $y = 0$, and $d^2y/dx^2 = 0$ is

$$y_r = C_2 \sin K(L - x) \qquad x_1 < x < L$$

where the subscripts l and r denote the left and right parts, respectively. If these two solutions are to be continuous at $x = x_1$, we must have

$$C_2 = C_1 \frac{\sin Kx_1}{\sin K(L - x_1)} \tag{6.8.20}$$

The slopes of the y_l and y_r curves at $x = x_1$ are

$$\left(\frac{dy_l}{dx}\right)_{x_1} = C_1 K \cos Kx_1 \qquad \left(\frac{dy_r}{dx}\right)_{x_1} = -C_2 K \cos K(L - x_1) \tag{6.8.21}$$

Let the change of slope across $x = x_1$ be made unity; then

$$-C_2 K \cos K(L - x_1) - C_1 K \cos Kx_1 = 1 \tag{6.8.22}$$

From Equations (6.8.20) and (6.8.22), we have

$$C_1 = -\frac{\sin K(L - x_1)}{K \sin KL}$$

$$C_2 = -\frac{\sin Kx_1}{K \sin KL}$$

$$y_l = -\frac{\sin K(L - x_1) \sin Kx}{K \sin KL} \qquad y_r = -\frac{\sin Kx_1 \sin K(L - x)}{K \sin KL} \tag{6.8.23}$$

Thus y_l and y_r give a solution to the differential equation $(d^2y/dx^2) + K^2y = 0$ throughout the column length except at $x = x_1$, at which point there is a discontinuity of slope $(dy/dx)_{x+} - (dy/dx)_{x-} = 1$, thus causing d^2y/dx^2 to be infinite at x_1. The left and right end conditions are satisfied by y_l and y_r, respectively. The solutions y_l and y_r are the Green's function of the differential equation with the given pinned end conditions. Let this be denoted by $g(x, x_1)$. Replacing x_1 by "a" gives

$$\begin{aligned} g(x, a) = g_l(x, a) &= -\frac{\sin K(L - a) \sin Kx}{K \sin KL} \qquad 0 < x < a \\ g(x, a) = g_r(x, a) &= -\frac{\sin Ka \sin K(L - x)}{K \sin KL} \qquad a < x < L \end{aligned} \tag{6.8.24}$$

The above equations are seen to satisfy the condition given by Equation (6.8.22),

$$\left[\frac{\partial g_r(x, a)}{\partial x} - \frac{\partial g_l(x, a)}{\partial x}\right]_{x=a} = 1$$

and the conditions

$$\frac{\partial^2 g_r(x, a)}{\partial x^2} + K^2 g_r(x, a) = 0$$

$$\frac{\partial^2 g_l(x, a)}{\partial x^2} + K^2 g_l(x, a) = 0$$

It may be verified that $y = \int_0^l g(x, a)f(a)\,da$ gives the solution of the non-homogeneous differential equation

$$\frac{d^2y}{dx^2} + K^2y = f(x)$$

This can be shown by substitution:

$$\frac{dy}{dx} = \int_0^l \frac{\partial g(x, a)}{\partial x} f(a)\,da$$

$$= \int_0^x \frac{\partial g_r(x, a)}{\partial x} f(a)\,da + \int_x^l \frac{\partial g_l(x, a)}{\partial x} f(a)\,da$$

$$\frac{d^2y}{dx^2} = \frac{\partial}{\partial x}\int_0^x \frac{\partial g_r(x, a)}{\partial x} f(a)\,da + \frac{\partial}{\partial x}\int_x^l \frac{\partial g_l(x, a)}{\partial x} f(a)\,da$$

$$= \int_0^x \frac{\partial^2 g_r(x, a)}{\partial x^2} f(a)\,da + \left[\frac{\partial g_r(x, a)}{\partial x} - \frac{\partial g_l(x, a)}{\partial x}\right] f(x)$$

$$+ \int_x^l \frac{\partial^2 g_l(x, a)}{\partial x^2} f(a)\,da$$

$$= \int_0^x \frac{\partial^2 g_r(x, a)}{\partial x^2} f(a)\,da + \int_x^l \frac{\partial g_l(x, a)}{\partial x} f(a)\,da + f(x)$$

Then

$$\frac{d^2y}{dx^2} + K^2y = \int_0^x \left[\frac{\partial^2 g_r(x, a)}{\partial x^2} + K^2 g_r(x, a)\right] f(a)\,da$$

$$+ \int_x^l \left[\frac{\partial^2 g_l(x, a)}{\partial x^2} + K^2 g_l(x, a)\right] f(a)\,da + f(x) = f(x)$$

since the integrands vanish. For a given $f(x)$ on the right-hand side of Equation (6.8.8), we have

$$\Delta y_1 = \int_0^L g(x, a)f(a)\,da$$

$$= \int_0^x g_r(x, a)f(a)\,da + \int_x^L g_l(x, a)f(a)\,da$$

$$= -\sin K(L - x)\int_0^x \frac{\sin Ka}{K \sin KL} f(a)\,da$$

$$-\sin Kx \int_x^L \frac{\sin K(L - a)}{K \sin KL} f(a)\,da \qquad (6.8.25)$$

Differentiation of Equation (6.8.25) gives

$$\frac{d\,\Delta y_1}{dx} = g_r(x, x)f(x) + \int_0^x \frac{dg_r(x, a)}{dx} f(a)\,da - g_l(x, x)f(x) + \int_x^L \frac{dg_l(x, a)}{dx} f(a)\,da$$

Since $g_r(x, x) = g_l(x, x)$, we obtain

$$\frac{d\,\Delta y_1}{dx} = \int_0^x \frac{dg_r(x, a)}{dx} f(a)\,da + \int_x^L \frac{dg_l(x, a)}{dx} f(a)\,da \tag{6.8.26}$$

where $g_r(x, a)$ and $g_l(x, a)$ are given by Equation (6.8.24). This may also be used to find θ_{1_0} and θ_{1_L}:

$$\theta_{1_0} = -\int_0^L \frac{\sin K(L-a)}{\sin KL} f(a)\,da$$
$$\theta_{1_L} = \int_0^L \frac{\sin Ka}{\sin KL} f(a)\,da \tag{6.8.27}$$

For Δy_2, we see from Equation (6.8.10) that $f(x)$ is a linear function. Substitution of this linear function for $f(a)$ in $y = \int g(x, a)f(a)\,da$ yields θ_{2_0} and θ_{2_L}. This method is illustrated in the following example.

Example. A 75ST aluminum alloy beam-column with built-in ends is subjected to an axial load of 625 lb and a uniform lateral load of 2 lb/in. The column is 16 in. long and has two flanges of area 0.125 sq in. each. The distance between the centroids of the flanges is 0.289 in., and the flange area is assumed to be concentrated at its centroid. This gives a moment of inertia equivalent to a square section of $\frac{1}{2}$ in. $\times$ $\frac{1}{2}$ in. The deflection of the column is assumed to be due to the flanges only. The temperature is 600° F, and the Young's modulus E is 5.2×10^6 psi. The stress-strain-time relation for constant stress is taken to be represented by

$$e = \frac{\sigma}{E} + A \exp(B\sigma)t^K$$

where $A = 2.64 \times 10^{-7}$, $B = 1.92 \times 10^{-3}$, $K = 0.66$, σ is in psi, t is in hr, and exp is the base of the natural logarithm. The moment of inertia is $I = 2 \times \frac{1}{8} \times 0.1445^2 = 1/192$ in.^{4} and $\pi^2 EI/L^2 = 1046$ lb. With $P = 625$ lb, we have

$$K^2 = \frac{P}{EI} = \frac{625}{5.2 \times 10^6 \times (1/192)}$$

From Equation (6.3.11), we have

$$M_0 = -\frac{q}{K^2} + \frac{qL}{2K}\frac{\sin KL}{(1-\cos KL)} = -47.6 \text{ in. lb}$$

$$M = M_0 + 16x - x^2 + 625w$$

$$\sigma = \frac{P}{A} + \frac{M}{I}\eta \tag{6.8.28}$$

The stress σ is calculated for the flanges at a number of sections. For the first time interval Δt, we obtain $\Delta e''$ from the empirical relation

$$\Delta e'' = A \exp(B\sigma)(\Delta t)^K$$

The values of

$$F(x) = -\frac{1}{I}\int \Delta e''\, \eta\, dA \tag{6.8.29}$$

at different sections are obtained by graphical integration. For the given symmetrical loading, $M_1 = M_2 = M_0$, and hence from Equation (6.8.1), we see that R_1 vanishes. Equation (6.8.9) then reduces to

$$\frac{d^2\,\Delta y_2}{dx^2} + K^2\,\Delta y_2 = -\frac{\Delta M_0}{EI} \tag{6.8.30}$$

Using Equation (6.8.10), we may find Δy_2. Applying Equation (6.8.27) with

$$f(a) = -\frac{\Delta M_0}{EI} \tag{6.8.31}$$

we obtain

$$\left(\frac{d\,\Delta y_2}{dx}\right)_{x=0} = \frac{\Delta M_0}{EI}\int_0^L \frac{\sin K(L-a)}{\sin KL}\,da = \Delta\theta_{2_0} \tag{6.8.32}$$

In solving Equation (6.8.8), we see that $f(a)$ of Equation (6.8.27) corresponds to $F(a)$ given by Equation (6.8.5). $\Delta\theta_{1_0}$ is then obtained from Equation (6.8.27):

$$\Delta\theta_{1_0} = \left(\frac{d\,\Delta y_1}{dx}\right)_{x=0} = \int_0^L \frac{\sin K(L-a)}{\sin KL}\,F(a)\,da \tag{6.8.33}$$

For built-in ends, we have

$$\Delta\theta_{1_0} + \Delta\theta_{2_0} = 0$$

From Equations (6.8.30) and (6.8.31), we have

$$\frac{\Delta M_0}{KEI} \tan \frac{KL}{2} = \int_0^L \frac{\sin K(L-a)}{\sin KL} F(a)\, da \qquad (6.8.34)$$

The calculation of the deflection and moment of an elastic beam-column under different loadings is briefly shown in Section 6.3 and is treated in detail in several books.[3,24] For the beam-column under uniform distributed load, the expression of bending moment is given by Equation (6.3.12). From the stress-strain-time relation of the material at the particular temperature, the creep strain $\Delta e''$ for the first time interval is calculated at the two flanges on different sections. The more sections taken and the smaller the time interval, the more accurate the results. A suitable number of section and time intervals may be chosen for the desirable degree of accuracy. From the incremental creep strain, $\Delta e''$, at various points across a section, $F(x)$ is found from Equation (6.8.29) for different sections. The values of ΔM_0 are obtained from Equation (6.8.34), Δy_2 from Equation (6.8.30), and Δy_1 from Equation (6.8.25). Having Δy_1, and Δy_2, we may determine Δw from $\Delta w = \Delta y_1 + \Delta y_2$. With Δw known, ΔK is obtained from Equation (6.7.11). The value of $\Delta e_0''$ is found from Equation (6.7.8), and then Δe is obtained from Equation (6.7.6). From $\Delta e''$ and Δe, the value of $\Delta\sigma$ is readily found. This gives the stress σ for the subsequent time interval. In the previous analysis, it has been assumed that the equation of state is such that the strain rate is a function of the stress and the creep strain, and not of the path of attaining the stress. This equation of state was discussed in Section 3.4. The smooth stress-time curve at a point in the beam-column is approximated by small horizontal and vertical steps of constant stress and constant time as shown in Section 6.7. During each horizontal step, the stress is constant. On the basis of the equation of state, $\Delta e''$'s are calculated for different points in the subsequent interval as in Section 6.7. Then $\Delta e_0''$, $F(x)$, ΔM_0, and Δw are calculated as before. By repeating the same process to successive time intervals, the stresses and deflections at different sections at different instants are calculated. The center deflection vs. time curve, as calculated in the problem just considered, is shown in Figure 6.11. The variation of flange stresses with time is shown in Figure 6.12. Additional details of the problem may be found in Reference 36.

6.9 Creep Bending of Tubes under Internal Pressure and Axial Load

Circular tubes at elevated temperatures under internal pressure and axial load are commonly used in nuclear and chemical plants. This section shows how the change of stress and deflection with time may be calculated.

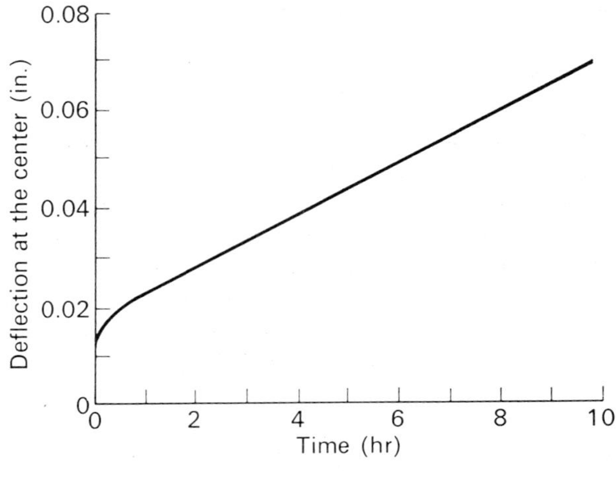

Figure 6.11

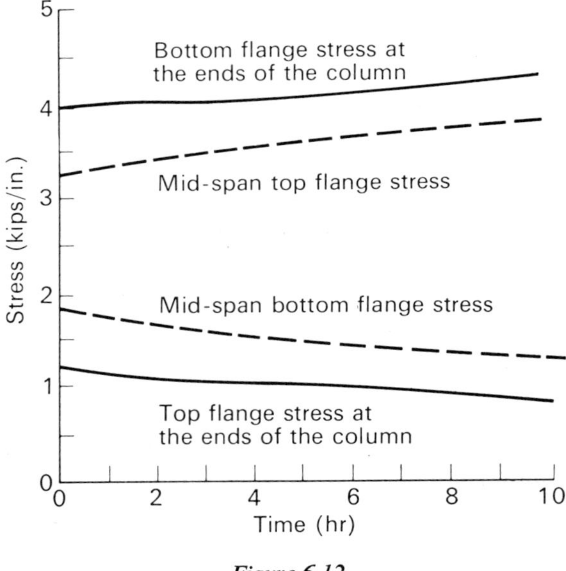

Figure 6.12

When a tube subject to internal pressure bends, the convex side lengthens while the concave side shortens. Hence the lateral area of the convex side is larger than that of the concave side. The excess of lateral area times the internal pressure gives an induced lateral load which causes further bending of the tube. Thus a tube under pressure bows just like a column under a compressive load.

Figure 6.13 shows an elemental length of tube with a bending radius of

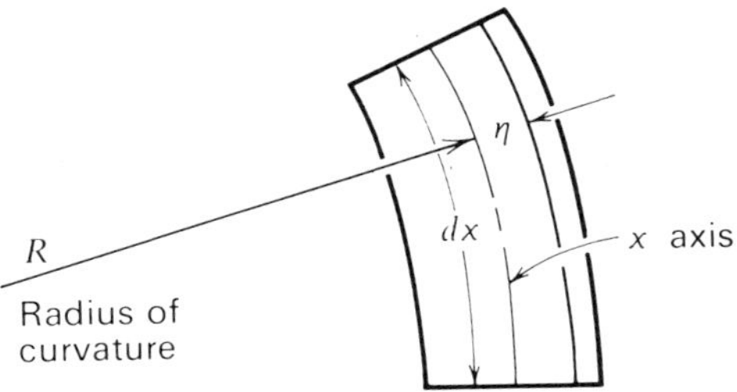

Figure 6.13. Element of tube under bending.

curvature R. The length of a fiber element at a distance η from the centroidal axis of the tube section is

$$dx' = \frac{R+\eta}{R}\,dx$$

A cross section of the tube is shown in Figure 6.14, where "a" is the inner radius. Then at the inner surface $z = a\sin\theta$, $\eta = a\cos\theta$, and $dz = a\,d\theta \cdot \cos\theta$. An elemental area $dx\,dz$ before bending becomes (Figure 6.14)

$$dx'\,dz = dx\left(1 + \frac{a\cos\theta}{R}\right)a\cos\theta\,d\theta$$

after bending, giving an increase of lateral area $[(a^2\cos^2\theta)/R]\,d\theta\,dx$.

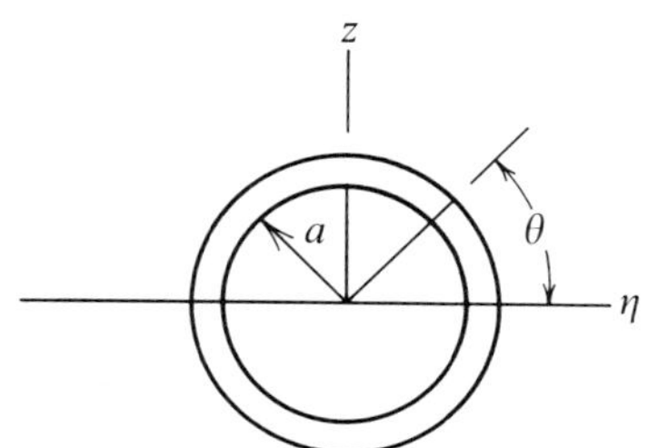

Figure 6.14. Cross-section of the tube.

This increase of area times the internal pressure p gives the induced lateral force

$$q\,dx = \int_0^{2\pi} p\,\frac{a^2\cos^2\theta}{R}\,d\theta\,dx$$

This yields a lateral load per unit length of tube as

$$q = \frac{\pi a^2}{R} p \tag{6.9.1}$$

With w denoting the lateral deflection of the tube, we obtain

$$\frac{1}{R} = -\frac{d^2w/dx^2}{[1 + (dw/dx)^2]^{3/2}} \cong -\frac{d^2w}{dx^2}$$

and Equation (6.9.1) becomes

$$q = -\pi a^2 p \frac{d^2w}{dx^2} \tag{6.9.2}$$

This is the induced lateral load on the tube caused by internal pressure associated with bending curvature.

The radial and longitudinal shear stresses in the tube are small and may be neglected. The longitudinal stress σ_x and the circumferential stress σ_θ are the two principal stresses. Assuming the radial stress σ_r to be zero, we see that the effective stress σ^* given by Equation (4.6.6) reduces to

$$\sigma^* = \frac{1}{\sqrt{2}}[(\sigma_x - \sigma_\theta)^2 + {\sigma_x}^2 + {\sigma_\theta}^2]^{1/2}$$

With e_x'', e_θ'', and e_r'' denoting the longitudinal, circumferential, and radial creep strains, respectively, the effective creep strain rate of Equation (4.7.5) becomes

$$\dot{e}^{*\prime\prime} = \frac{\sqrt{2}}{3}[(\dot{e}_x'' - \dot{e}_\theta'')^2 + (\dot{e}_\theta'' - \dot{e}_r'')^2 + (\dot{e}_r'' - \dot{e}_x'')^2]^{1/2}$$

where the dot denotes differentiation with respect to time. From Equation (4.7.7), we have

$$\dot{e}_x'' = \frac{\dot{e}^{*\prime\prime}}{\sigma^*}\left(\sigma_x - \frac{\sigma_\theta}{2}\right)$$

$$\dot{e}_\theta'' = \frac{\dot{e}^{*\prime\prime}}{\sigma^*}\left(\sigma_\theta - \frac{\sigma_x}{2}\right)$$

In creep analysis, stresses are commonly assumed to remain constant during a small time increment Δt. The increase of creep strains during a Δt time increment is given by

$$\begin{aligned} \Delta e_x'' &= \frac{\dot{e}^{*\prime\prime}}{\sigma^*}\left(\sigma_x - \frac{\sigma_\theta}{2}\right)\Delta t \\ \Delta e_\theta'' &= \frac{\dot{e}^{*\prime\prime}}{\sigma^*}\left(\sigma_\theta - \frac{\sigma_x}{2}\right)\Delta t \end{aligned} \tag{6.9.3}$$

Expressing the elastic strain as the difference between the total strain and the creep strain, we have the elastic stress-strain relations

$$\begin{aligned} \Delta e_x - \Delta e_x'' &= \frac{1}{E}(\Delta\sigma_x - \nu\,\Delta\sigma_\theta) \\ \Delta e_\theta - \Delta e_\theta'' &= \frac{1}{E}(\Delta\sigma_\theta - \nu\,\Delta\sigma_x) \end{aligned} \tag{6.9.4}$$

where ν is Poisson's ratio for elastic strain and Δ denotes the increment occurring in Δt. Solving Equations (6.9.4) for stress increments yields

$$\begin{aligned} \Delta\sigma_x &= \frac{E}{1-\nu^2}(\Delta e_x + \nu\,\Delta e_\theta - \Delta e_x'' - \nu\,\Delta e_\theta'') \\ \Delta\sigma_\theta &= \frac{E}{1-\nu^2}(\Delta e_\theta + \Delta e_x - \Delta e_\theta'' - \nu\,\Delta e_x'') \end{aligned} \tag{6.9.5}$$

Since the internal pressure remains constant, the change in hoop stress must be zero, that is, $\Delta\sigma_\theta = 0$. Then

$$\Delta e_\theta - \Delta e_\theta'' = -\nu(\Delta e_x - \Delta e_x'') \tag{6.9.6}$$

Substituting this into Equation (6.9.5), we obtain

$$\Delta\sigma_x = E(\Delta e_x - \Delta e_x'') \tag{6.9.7}$$

Suppose that, in addition to the internal pressure p, there is a constant axial compressive load P. Then

$$\Delta P = \int \Delta\sigma_x\,dA = E\int(\Delta e_x - \Delta e_x'')\,dA = 0 \tag{6.9.8}$$

From the Euler-Bernoulli assumption that planes before bending remain planes during bending, we have

$$\Delta e_x = \Delta e_{x_0} - \eta\,\frac{d^2\,\Delta w}{dx^2} \tag{6.7.6}$$

where x_0 refers to the neutral axis. From Equations (6.7.6) and (6.9.8), we have

$$\Delta e_{x_0} = \frac{1}{A}\int \Delta e_x''\,dA = \overline{\Delta e_x''}$$

where $\overline{\Delta e_x''}$ is the average $\Delta e_x''$ over the cross section. Then the increments of stress and resisting moment are

$$\Delta\sigma_x = E\left(-\eta\,\frac{d^2\,\Delta w}{dx^2} + \overline{\Delta e_x''} - \Delta e_x''\right) \tag{6.9.9}$$

$$\Delta M = \int \Delta\sigma_x \eta\,dA = -EI\,\frac{d^2\,\Delta w}{dx^2} - E\int \Delta e_x''\eta\,dA \tag{6.9.10}$$

Equation (6.9.10) is of the same form as Equation (6.7.9) for columns with no internal pressure. Equating the incremental resisting moment to the incremental external moment due to Δw, we obtain the differential equation of equilibrium. This is used for calculating the incremental deflection curve Δw.

From Equation (6.9.10), the increment of shear force is equal to

$$\Delta V = \frac{d\,\Delta M}{dx} = -EI\frac{d^3\,\Delta w}{dx^3} - E\frac{d}{dx}\int \Delta e_x''\,\eta\,dA$$

from which the increment of the distributed lateral load is obtained as

$$\Delta q = -\frac{d\,\Delta V}{dx} = EI\frac{d^4\,\Delta w}{dx^4} + E\frac{d^2}{dx^2}\int \Delta e_x''\,\eta\,dA \qquad (6.9.11)$$

The increment of lateral load due to the internal pressure, which is associated with the increase of curvature, is

$$-\pi a^2 p\,\frac{d^2\,\Delta w}{dx^2}$$

The increase of bending moment due to axial compressive load associated with the increment of deflection Δw is $-P\,\Delta w$. Taking the second derivative with respect to x gives an increment of lateral load of $-P(d^2\,\Delta w/dx^2)$. Equating the sum of the above two terms to the right side of Equation (6.9.11), we obtain the governing equilibrium equation as

$$EI\frac{d^4\,\Delta w}{dx^4} + E\frac{d^2}{dx^2}\int \Delta e_x''\,\eta\,dA = -(\pi a^2 p + P)\frac{d^2\,\Delta w}{dx^2} \qquad (6.9.12)$$

Integrating this equation twice with respect to x, assuming EI to be constant, we obtain

$$\frac{d^2\,\Delta w}{dx^2} + \frac{P + \pi a^2 p}{EI}\Delta w + \frac{1}{I}\int \Delta e_x''\,\eta\,dA = Ax + B \qquad (6.9.13)$$

where A and B are the constants of integration.

For a simply supported tube, the deflections and bending moments at the two ends are zero. The moment increment ΔM and the deflection increment Δw vanish at the ends. From Equation (6.9.10) and the conditions $\Delta M = 0$, $\Delta w = 0$ at $x = 0$ and $x = l$, we have $A = B = 0$. Equation (6.9.13) then reduces to

$$\frac{d^2\,\Delta w}{dx^2} + \frac{P + \pi a^2 p}{EI}\Delta w + \frac{1}{I}\int \Delta e_x''\,\eta\,dA = 0 \qquad (6.9.14)$$

This equation is the same as Equation (6.7.13), with P of Equation (6.7.13) replaced by $(P + \pi a^2 p)$. Therefore creep analysis of tubes under internal pressure and axial compressive load may be obtained by the method of Section 6.7.

A detailed discussion with a numerical analysis of a tube of varying cross section subject to nonuniform temperature and internal pressure distributions and a constant compressive axial load is given in Reference 37.

6.10 Critical Loads of Perfect Columns with Creep

Actual columns always have initial curvature. Methods of calculating the increase of lateral deflection with time under load for columns with initial curvature subject to creep have been shown in detail. In this section the creep of an initially straight column is considered.

The critical load of a column without creep has been taken to be the load at which the column would deflect further if it were given an impulse of slight disturbance. Rabotnov,[31] in 1957, applied this criterion to columns and plates with creep. If at a certain instant a deviation from straightness (i.e., lateral bending) is given to the compressed column, the deflection will either increase or decrease during the subsequent short interval of time. The load on the compressed column is considered stable or unstable according to whether the deflection decreases or increases. Using this criterion, Rabotnov[31] determined the stability of columns with strain-hardening creep. The creep characteristics were assumed to have the equation of state shown in Section 3.4. The creep strain rate is assumed to depend on the stress and the creep strain, and not on the history of loading. This condition is represented by

$$f(\sigma, e'', \dot{e}'') = 0 \qquad \text{or} \qquad \dot{e}'' = F(\sigma, e'') \tag{6.10.1}$$

where f and F are functions, e'' is the creep strain, and $\dot{e}''$ is the creep-strain rate. From Equation (6.10.1), we have

$$\frac{\partial f}{\partial \sigma}\,\delta\sigma + \frac{\partial f}{\partial e''}\,\delta e'' + \frac{\partial f}{\partial \dot{e}''}\,\delta\dot{e}'' = 0 \tag{6.10.2}$$

Let

$$\lambda = \frac{\partial f}{\partial \sigma} \qquad \mu = \frac{\partial f}{\partial e''} \qquad \text{and} \qquad u = \frac{\partial f}{\partial \dot{e}''} \tag{6.10.3}$$

Recall that

$$e'' = e - \frac{\sigma}{E} \qquad \dot{e}'' = \dot{e} - \frac{\dot{\sigma}}{E} \tag{6.10.4}$$

where e is the total strain and E is Young's modulus. Then

$$\lambda\,\delta\sigma + \mu\,\delta e'' + u\,\delta\dot{e}'' = 0 \tag{6.10.5}$$

$$\lambda\,\delta\sigma + \mu\left(\delta e - \frac{\delta\sigma}{E}\right) + u\left(\delta\dot{e} - \frac{\delta\dot{\sigma}}{E}\right) = 0$$

$$(E\lambda - \mu)\delta\sigma - u\,\delta\dot{\sigma} + E(\mu\,\delta e + u\,\delta\dot{e}) = 0 \tag{6.10.6}$$

Let the $\delta\sigma$, $\delta\dot{\sigma}$, δe, and $\delta\dot{e}$ correspond to the lateral bending. From the Bernoulli-Euler assumption of plane sections remaining plane, we have

$$\delta e = K\eta + \delta e_0 \tag{6.10.7}$$

where K is the curvature, η is the distance from the centroidal axis of the section, and e_0 is the axial strain at the centroidal axis. Then

$$(E\lambda - \mu)\,\delta\sigma - u\,\delta\dot{\sigma} + E(\mu K\eta + u\dot{K}\eta) + E(\mu\,\delta e_0 + u\,\delta\dot{e}_0) = 0$$

Multiplying the above expression by $\eta\,dA$ and integrating over the cross section yields

$$(E\lambda - \mu)M - u\dot{M} + EI(\mu K + u\dot{K}) = 0 \tag{6.10.8}$$

where $I = \int \eta^2\,dA$, and $M = \int \eta\,\delta\sigma\,dA$.

Let the mass density of the column material be ρ (Figure 6.15). The upward acceleration is d^2w/dt^2. The downward inertia force per unit

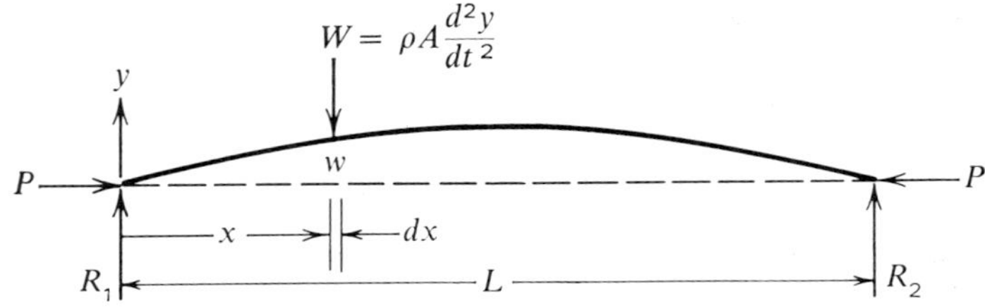

Figure 6.15

length of the column is $\rho A(d^2w/dt^2)$, where A is the cross-sectional area. The moment at a section due to the lateral load $\rho A(d^2w/dt^2)$ and the axial load P is

$$M = -Pw + R_1 x - \int_0^x \int_0^x \rho A\,\frac{\partial^2 w}{\partial t^2}\,dx\,dx$$

or

$$\frac{\partial^2 M}{\partial x^2} = -P\,\frac{\partial^2 w}{\partial x^2} - \rho A\,\frac{\partial^2 w}{\partial t^2} \tag{6.10.9}$$

where R is the reaction at the left end support. Differentiating Equation (6.10.8) with respect to x twice, substituting $-\partial^2 w/\partial x^2$ for K, and using Equation (6.10.9) gives

$$\left(\mu - E\lambda + u\frac{\partial}{\partial t}\right)\left(P\frac{\partial^2 w}{\partial x^2} + \rho A\frac{\partial^2 w}{\partial t^2}\right) + \frac{\partial^2}{\partial x^2}\left[EI\left(\mu\frac{\partial^2 w}{\partial x^2} + u\frac{\partial^3 w}{\partial x^2\,\partial t}\right)\right] = 0 \tag{6.10.10}$$

For a column of uniform section with hinged ends, this equation and the end conditions are satisfied by letting

$$w = \tau(t)\sin\frac{\pi x}{L} \tag{6.10.11}$$

Substituting Equation (6.10.11) into Equation (6.10.10), we obtain

$$\dddot{\tau} + \frac{\mu - E\lambda}{u}\ddot{\tau} + \frac{\pi^2}{\rho AL^2}(P_1 - P)\dot{\tau} + \frac{\pi^2}{\rho AuL^2}[E\lambda P + \mu(P_1 - P) - u\dot{P}]\tau = 0 \tag{6.10.12}$$

where $P_1 = \pi^2 EI/L^2$ and the dot denotes the derivative with respect to t. Letting $\tau = C\exp(\alpha t)$, we see that Equation (6.10.12) reduces to

$$\alpha^3 + \frac{\mu - E\lambda}{u}\alpha^2 + \frac{\pi^2}{\rho AL^2}(P_1 - P)\alpha + \frac{\pi^2}{\rho AuL^2}[E\lambda P + \mu(P_1 - P) - u\dot{P}]\tau = 0 \tag{6.10.13}$$

For constant-stress creep tests, $\delta\sigma = 0$, and Equation (6.10.6) reduces to

$$(E\lambda - \mu)\,\delta\sigma + E(\mu\,\delta e + u\,\delta\dot{e}) = 0$$

For a given creep strain, $\delta e = 0$, and the above expression becomes $(B\lambda - \mu)\,\delta\sigma + Eu\,\delta\dot{e} = 0$. Since creep-strain rate is expected to increase with stress, $\delta\dot{e}/\delta\sigma > 0$, hence $(\mu - E\lambda)/u$ is positive. In Equation (6.10.13), we see that P is less than the Euler load P_1, and the coefficients of α^2 and α are positive. Positive real roots exist only if the last term is negative. Hence the function τ will be decreasing if

$$\frac{E\lambda P}{P_1 - P} + \mu - u\frac{P}{P_1 - P} > 0 \tag{6.10.14}$$

For a small rate of loading, the last term is negligible, and hence the maximum load for stability is

$$\frac{P_{cr}}{P_1} = \frac{1}{1 - (E\lambda/\mu)} \tag{6.10.15}$$

Since λ and μ vary with creep strain e'', we see that P_{cr}/P_1 varies with time after loading. Consider a particular equation of strain-hardening creep, say

$$f(\sigma, e'', \dot{e}'') = \dot{e}''e''^{\beta} - A\sigma^n = 0$$

$$\dot{e}'' = \frac{1}{(e'')^{\beta}} A\sigma^n \qquad (6.10.16)$$

$$\lambda = \frac{\partial f}{\partial \sigma} = -An\sigma^{n-1} = \frac{-n\dot{e}''e''^{\beta}}{\sigma}$$

$$\mu = \frac{\partial f}{\partial e''} = \dot{e}''\beta e''^{\beta-1}$$

$$\lambda/\mu = -\frac{ne''}{\beta\sigma}$$

then

$$\frac{P_{cr}}{P_1} = \frac{1}{1 + (nEe''/\beta\sigma)} \qquad (6.10.17)$$

At zero time, $e'' = 0$, and the critical load is $P_{cr} = P_1$. After e'' occurs, P_{cr} decreases with constant stress for increasing e''. With the equation of state known, the critical load P_{cr} may be found for a given creep strain. The time to reach this amount of creep strain can be computed. Following this criterion of stability, Hoff[32] computed the stability characteristics of a simplified model of a perfect column. Gerard[33] has applied the same concept in his study of creep buckling of plates.

Shanley[18] has suggested a method to estimate the life of a column with creep. In Figure 3-10(a), creep strain is plotted vs. time at different stress levels. At a constant value of time, different stresses correspond to different values of strain. These stresses may be plotted against the corresponding strains to give a stress-strain curve called an isochronous stress-strain curve of the material. In this way a family of isochronous curves for different values of time may be obtained. Denoting the slope of the isochronous curve as the time-dependent tangent modulus E^*, Shanley suggested that a perfect column with simply supported ends would begin to deflect laterally after a slight lateral disturbance, when the load becomes

$$P_{cr} = \frac{\pi^2 E^* I}{L^2}$$

where E^* has the value of Young's modulus at the instant of load application and thereafter decreases with time. Shanley clearly recognized the fact that the time at which the column is rendered useless for structural purposes due to excessive deflection may be quite different from the time at which the isochronous tangent modulus load becomes equal to the applied load. Shanley's critical time for the perfect column differs from that determined on the basis of rigorous dynamic analysis by Rabotnov and Shesterikov. Numerical comparisons have been made of the two critical times by Hoff.[32] Shanley's method was checked with both experiments and calculations by Carson[34,35] and his associates. They showed that the error introduced by using Shanley's *isochronous* curves leads to conservative design. Experiments conducted at Batelle Institute indicate the lifetime calculated from the time dependent tangent modulus is shorter, much shorter in a number of cases, than the experimental lifetime. The time to reach e'' for a given critical load P_{cr} in Equation (6.10.17), given by Rabotnov and Shesterikov, is the time at which a perfect column starts to deflect; it does not include the time required for the deflection to develop from start to an extent that is excessive for structural purposes. Hence the critical time calculated by this method is only a part of the life of a perfect column. For estimating the life of an actual column for deflection of a given amount, the methods given in Section 6.7 are suggested.

6.11 Columns with Time-Independent Plastic Strains

Reduced Modulus Load: Consider a column which is restrained from bending during compressive loading. Suppose that after the column is loaded beyond the elastic limit, to a certain load P, the lateral bending restraint is removed. If the column tends to bend, the load P is said to be greater than the critical load; on the other hand, if the column does not tend to bend, the load P is said to be less than the critical load. Hence the critical load of such a column is taken to be the smallest load which causes lateral bending, at slight disturbance, after the restraint is removed.

At the critical load, the column bends upon removal of the restraint. During bending, P remains constant, and the compressive stress is increased on the concave side and decreased on the convex side. Referring to Figure 6.16, we see that the stress-strain relation on the concave side of the column has the slope CD, called the tangent modulus E_t, for a small incremental strain, while on the convex side of the column, where the compressive strain decreases, the stress-strain relation has the same slope as the initial modulus E (the elastic modulus): CF is parallel to OB. Plane cross-sections are assumed to remain plane. For small bending, the distribution of bending stress superposed on the uniform compressive stress before bending is shown in Figure 6.17, where h_1 denotes the part

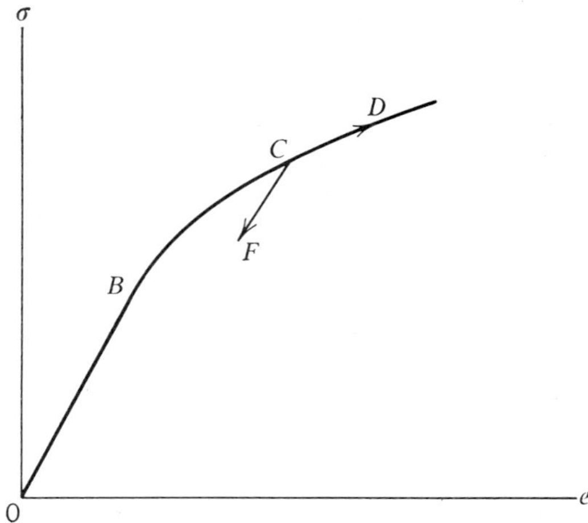

Figure 6.16

in which the compressive strain is increasing, while h_2 denotes the part in which the compressive strain is decreasing.

On the concave side of the neutral axis of bending, the tangent modulus is effective, while Young's modulus is effective on the remaining part of the section. From the condition of equilibrium, the increase of total compressive stress on the concave side must equal the decrease of compressive stress on the convex side. For a column of rectangular section, with height h and width b, we have $E_t h_1{}^2 = Eh_2{}^2$. From this and the condition $h_1 + h_2 = h$, we obtain h_1 and h_2 as

$$h_1 = \frac{h\sqrt{E}}{\sqrt{E} + \sqrt{E_t}} \qquad h_2 = \frac{h\sqrt{E_t}}{\sqrt{E} + \sqrt{E_t}}$$

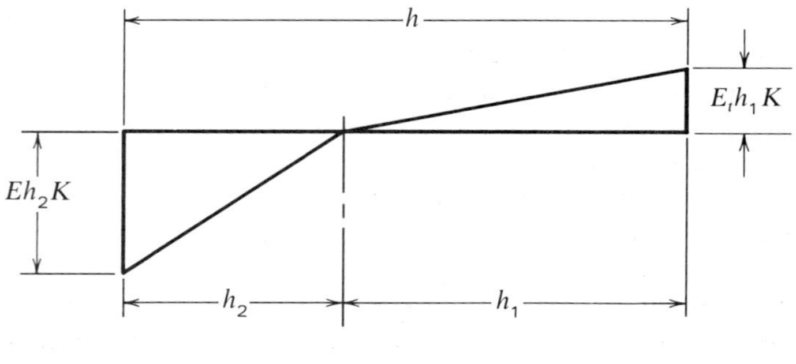

Figure 6.17

The moment given by this bending is

$$M = EKh_2 \frac{bh_2}{2} \frac{2}{3} h = \frac{bh^3}{12} K \frac{4EE_t}{(\sqrt{E} + \sqrt{E_t})^2} \tag{6.11.1}$$

where K is the curvature.

Let

$$E_r = \frac{4EE_t}{(\sqrt{E} + \sqrt{E_t})^2} \tag{6.11.2}$$

then

$$M = E_r IK \tag{6.11.3}$$

where E_r is called the reduced modulus. The external moment caused by bending is $-Pw$. For a small deflection, K is replaced by d^2w/dx^2. Equating the external moment to the resisting moment gives

$$E_r I \frac{d^2w}{dx^2} = -Pw$$

The solution of the above equation for a column with pinned ends yields the critical load

$$P_{\text{cr}} = \frac{\pi^2 E_r I}{L^2} \tag{6.11.4}$$

The reduced modulus, also known as the double modulus, was first set forth by A. Considere,[25] in 1889, for an ideal column stressed beyond the elastic limit (without any artificial lateral restraint to prevent bending during axial loading) and was subsequently used by Von Karman,[26] in 1910, in his study of inelastic columns. The lateral restraint does not exist in actual columns, and therefore the behavior predicted by the reduced modulus theory is not valid for actual columns. Test data on actual columns give lower values than that predicted by this reduced modulus. The reduced modulus load had been considered to be the theoretical critical load, and the low test values, it had been thought, were due to unavoidable eccentricities, testing technique, and so on.

Tangent Modulus Theory. In the previous section, the lateral restraint on the column prevents it from bending during axial loading. Since this restraint is not present in actual columns, it is possible for the columns to bend simultaneously with increasing axial load. This was first pointed out by Shanley[18,27,28] in 1947. With increasing axial load, bending may proceed with no reversal of strain. Without reversal of strain, the stress-strain relation over the entire section follows CD in Figure 6.16, so that the

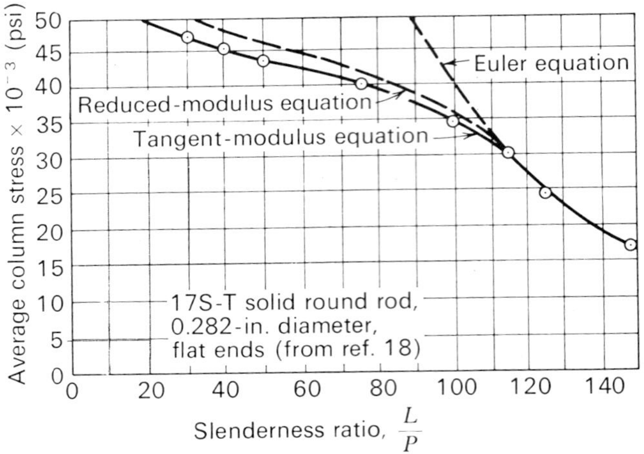

Figure 6.18

tangent modulus is effective for the entire section. Hence E_r in Equation (6.10.4) is replaced by E_t, and the critical load for pin-ended columns is

$$P_{\mathrm{cr}} = \frac{\pi^2 E_t I}{L^2} \tag{6.11.5}$$

Engesser[18] presented his tangent modulus theory in 1889, and later in 1895 acknowledging that this theory was in error, favored instead the double modulus theory. Although the reduced modulus load had been considered to be the theoretical value, the tangent modulus load had been

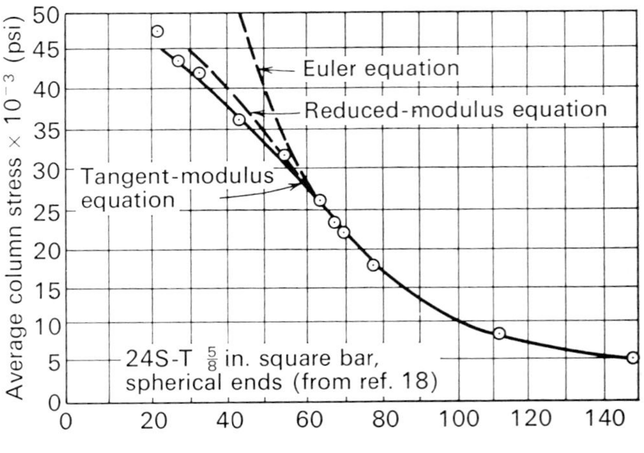

Figure 6.19

widely used in practice for its simplicity. The tangent modulus load is lower than the reduced modulus one, and hence has been considered to give more safety.

Because of the simultaneous bending and axial compression, as pointed out by Shanley, the tangent modulus load is the theoretical load at which the initially straight column may start to bend. Two sets of test data are shown in Figures 6.18 and 6.19. It is seen that the test data are closer to the tangent modulus loads than to the reduced modulus loads. To check the mechanism of buckling, Shanley tested a pin-ended column of rectangular section (2 in. × $\frac{1}{4}$ in.), which was designed to fail in the plastic range. Electric strain gages were put on both sides of the column to record the strain distribution at different stages of loading. The results are shown in Figures 6.20 and 6.21. From these figures, we see that the strain distribution remained substantially constant up to about 40,000 lb of loading.

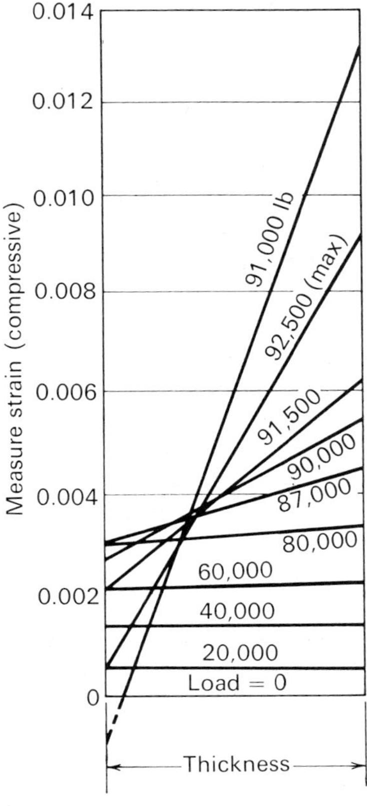

Figure 6.20

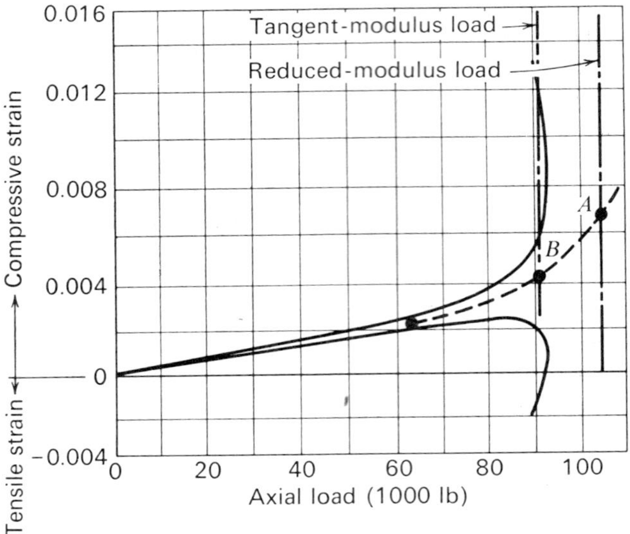

Figure 6.21

Beyond this loading, the strain varies from one side to the other indicating bending. Up to a loading of 87,000 lb no reversal of strain took place. At the maximum loading of 92,500 lb an appreciable amount of strain reversal had taken place. The small difference in the strains on the opposite sides in the early stage of loading (Figure 6.20) is due to the presence of a small initial curvature or eccentricity of the column. (The initial curvature and eccentricity of the column are magnified under loading, as shown in Section 6.2.) It is seen that the maximum load is just a little more than the tangent modulus load. Reversal of strain does not start at the beginning of bending, but at a later stage of lateral bowing. The maximum load of a perfect column lies between the tangent modulus load and the reduced modulus load. The maximum load is accurately determined and the details of the mechanism of inelastic column buckling is shown in the following analysis of an inelastic column.[29]

Mechanism of Inelastic Buckling. The basic assumptions used for the following analysis are (1) the Bernoulli-Euler assumption that plane cross sections remain planes; and (2) the assumption that the longitudinal fibers of the column act as if they were separate fibers and follow the same stress-strain curve as in simple compression tests.

Actual columns always have initial curvature, eccentricity of loading, and nonhomogeneity of material; therefore the perfect column does not exist. In the following analysis, an initial curvature is assumed. During the initial stage of loading, the column is purely elastic. Let w_0 be the initial deflection of the column's centroidal axis from the line of thrust, w the deflection at any stage of loading, and x the distance along the line of thrust measured from one end of the column. In the elastic range, the condition of equilibrium is

$$EI\left(\frac{d^2w}{dx^2} - \frac{d^2w_0}{dx^2}\right) + Pw = 0 \tag{6.11.6}$$

Letting $w_0 = \sum_{n=1}^{\infty} A_n \sin(n\pi x/L)$ and expressing w as $\sum_{n=1}^{\infty} B_n \sin(n\pi x/L)$, we have, upon substituting in Equation (6.11.6),

$$\sum_{n=1}^{\infty}\left(-\frac{n^2\pi^2}{L^2}B_n + \frac{n^2\pi^2}{L^2}A_n + B_n\frac{P}{EI}\right)\sin\frac{n\pi x}{L} = 0 \tag{6.11.7}$$

Multiplying Equation (6.11.7) by $\sin(m\pi x/L)\,dx$ and integrating from $x = 0$ to $x = L$, we see that the terms for which $n \neq m$ give

$$\begin{aligned}\int_0^L \sin\frac{m\pi x}{L}\sin\frac{n\pi x}{L}\,dx &= \frac{1}{2}\int\left(\cos\frac{(m-n)\pi x}{L} - \cos\frac{(m+n)\pi x}{L}\right)dx\\ &= \frac{1}{2}\left[\frac{L}{(m-n)\pi}\sin\frac{(m-n)\pi x}{L} - \frac{L}{(m+n)\pi}\sin\frac{(m+n)\pi x}{L}\right]_0^L = 0\end{aligned} \tag{6.11.8}$$

Only the term for which $n = m$ remains, and Equation (6.11.7) reduces to

$$\int_0^L\left(-\frac{m^2\pi^2}{L^2}B_m + \frac{m^2\pi^2}{L^2}A_m + B_m\frac{P}{EI}\right)\sin^2\frac{m\pi x}{L}\,dx = 0$$

This gives

$$\left[\left(-\frac{m^2\pi^2}{L^2} + \frac{P}{EI}\right)B_m + \frac{m^2\pi^2}{L^2}A_m\right]\frac{L}{2} = 0$$

from which we obtain

$$B_m = \frac{-A_m}{(PL^2/m^2\pi^2EI) - 1} = \frac{-A_m}{(P/P_m) - 1} \tag{6.11.9}$$

where

$$P_m = \frac{m^2\pi^2EI}{L^2} \tag{6.11.10}$$

For $m = 1$, we have

$$B_1 = \frac{-A_1}{(P/P_{cr}) - 1} \tag{6.11.11}$$

where P_{cr} is the Euler load $\pi^2 EI/L^2$. Using Equation (6.11.9), we see that the deflection $w = \sum_{n=1}^{\infty} B_n \sin(n\pi x/L)$ becomes

$$w = \sum_{n=1}^{\infty} \frac{-A_n}{(P/P_n) - 1} \sin \frac{n\pi x}{L} \tag{6.11.12}$$

This gives the increase of lateral deflection within the elastic range.

When the column is loaded beyond the elastic range, the flexural rigidities of the different sections of the column vary with the load. At a particular section, the axial load and resisting moment about a reference axis are

$$P = \int \sigma \, dA$$

$$M = \int \sigma \eta \, dA$$

where η is the distance from the reference axis. For a small increment of load ΔP which induces an increment of moment ΔM, we have

$$\begin{aligned} \Delta P &= \int \Delta\sigma \, dA \\ \Delta M &= \int \Delta\sigma \eta \, dA \end{aligned} \tag{6.11.13}$$

The concave side of the column exceeds the elastic limit first. In the region where the elastic limit is exceeded, the tangent modulus is effective. Then

$$\Delta\sigma = E_t \, \Delta e \tag{6.11.14}$$

In the region either within the elastic range or where the compressive strain decreases, $\Delta\sigma = E \, \Delta e$. In general $\Delta\sigma = E_e \, \Delta e$, where E_e is the effective modulus and equals either E_t or E, as the case may be.

From the Bernoulli-Euler assumption,

$$e = e_0 + K\eta \tag{6.11.15}$$

where e_0 is the strain at the reference axis and K is the curvature of the section. Then

$$\Delta e = \Delta e_0 + \eta \, \Delta K \tag{6.11.16}$$

and

$$\Delta P = \int E_e(\Delta e_0 + \eta \, \Delta K) b \, d\eta = E \int (\Delta e_0 + \eta \, \Delta K) \frac{E_e}{E} b \, d\eta \tag{6.11.17}$$

$$\Delta M = \int E_e(\Delta e_0 + \eta \, \Delta K)\eta b \, d\eta = E \int (\Delta e_0 + \eta \, \Delta K)\eta \frac{E_e}{E} b \, d\eta \tag{6.11.18}$$

where b is the width of the section at η. For a rectangular section, b is constant. Let $(E_e/E)b = b'$, hereafter called the modified width. Then $A' = \int b'\, d\eta$ is called the modified area of the section, and Equation (6.11.17) becomes

$$\Delta P = E \int (\Delta e_0 + \eta \,\Delta K)\, dA' \tag{6.11.19}$$

Choose the reference axis so that $\int \eta \, dA' = 0$; that is, let the reference axis pass through the centroid of the modified area. Then

$$\Delta P = E\, \Delta e_0\, A' \tag{6.11.20}$$

$$\begin{aligned} \Delta M &= E \int (\Delta e_0 + \eta\, \Delta K)\eta \, dA' \\ &= E\, \Delta K \int \eta^2 \, dA' = EI'\, \Delta K \end{aligned} \tag{6.11.21}$$

where $I' = \int \eta^2 \, dA'$ is the modified moment of inertia.

The concept of modified sections is not new. It corresponds to the transformed section of reinforced concrete used by civil engineers for a long time. For reinforced concrete, the cross section of steel is multiplied by the ratio of modulus of steel to that of concrete to get the equivalent section in concrete. Here the modulus at the region stressed beyond the elastic limit equals the tangent modulus. The elemental area dA in such a region is multiplied by E_t/E to get the equivalent area of elastic modulus. The increment of strain corresponding to an increment of P and M is

$$\Delta e = \frac{\Delta P}{EA'} + \frac{\Delta M}{EI'}\eta \tag{6.11.22}$$

where A' and I' are assumed to remain constant during ΔP and ΔM. This implies that E_t is constant during the increment of P. As P increases, σ increases and E_t changes. For accurate results, the incremental load ΔP should be small.

When the column load P increases to $P + \Delta P$, we see that w increases to $w + \Delta w$. Referring to the geometrical centroidal axis, we have

$$\Delta M = (P + \Delta P)(w + \Delta w) - Pw \tag{6.11.23}$$

Let the centroidal axis of the modified section be ξ to the right of the geometrical centroidal axis (Figure 6.22). Referring to the centroidal axis of the modified section, we see that the incremental external moment is

$$\begin{aligned} \Delta M &= (P + \Delta P)(w + \xi + \Delta w) - P(w + \xi) \\ &= (P + \Delta P)\, \Delta w + \Delta P(w + \xi) \end{aligned} \tag{6.11.24}$$

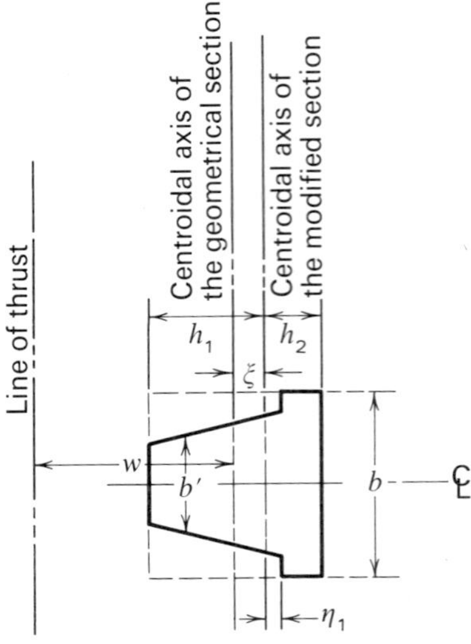

Figure 6.22

and the incremental internal moment is

$$\Delta M = EI' \, \Delta K = -EI' \frac{d^2 \, \Delta w}{dx^2} \tag{6.11.25}$$

Equating the incremental external and internal moments yields

$$\frac{d^2 \, \Delta w}{dx^2} + \frac{P + \Delta P}{EI'} \Delta w + \frac{\Delta P}{EI'} (w + \xi) = 0 \tag{6.11.26}$$

The value of EI' varies along the column length; hence this equation corresponds to an elastic column of variable cross section with initial curvature represented by the last term.

Strain Reversal. When de/dP changes sign, strain starts to reverse. When strain reversal occurs, the tangent modulus changes to Young's modulus. From Equations (6.11.22) and (6.11.24), we obtain

$$\Delta e = \frac{\Delta P}{EA'} + \frac{(P + \Delta P) \, \Delta w + \Delta P(w + \xi)}{EI'} \eta \tag{6.11.27}$$

$$\frac{\Delta e}{\Delta P} = \frac{1}{EA'} + \frac{P(\Delta w/\Delta P) + w + \Delta w + \xi}{EI'} \eta \tag{6.11.28}$$

In the limit as ΔP, Δe, and Δw approach zero, Equation (6.11.28) becomes

$$\frac{de}{dP} = \frac{1}{EA'} + \frac{P(dw/dP) + w + \xi}{EI'}\eta$$

To find the value η_1 beyond which compressive strain starts to decrease, let $de/dP = 0$, and denote I'/A' by ρ'^2, where ρ' is the radius of gyration of the modified section. This gives

$$\eta_1 = -\frac{\rho'^2}{P(dw/dP) + w + \xi} \tag{6.11.29}$$

The point η_1 is the point on the convex side beyond which the compressive strain decreases and E_t changes back to E. When η_1 is greater than h_2, there is no region of reversal of strain. Up to $\eta_1 = h_2$, the tangent modulus is effective across the whole section. Any further load and deflection would introduce a double modulus; i.e., Young's modulus E would extend over a portion of the section from the convex side and the tangent modulus would cover the remaining portion. As load and deflection increased, there would be an increase of the region in which Young's modulus is effective. This reversal of strain increases both the value of the modified moment of inertia I' and the shift of the centroidal axis ξ of the modified section toward the convex side.

The value of EI' in Equation (6.11.26) may vary along the column length. If the variation is small, which is generally the case before strain reversal occurs, EI' may be taken as being constant and equal to that at the mid-section, EI_1'. At the two pinned ends of the column, w, Δw, and ξ are zero. Let

$$w = \sum_{n=1}^{\infty} \delta_n \sin\frac{n\pi x}{L}$$

$$\Delta w = \sum_{n=1}^{\infty} \Delta\delta_n \sin\frac{n\pi x}{L}$$

$$\xi = \sum_{n=1}^{\infty} a_n \sin\frac{n\pi x}{L}$$

Substituting these into Equation (6.11.26) gives

$$-\sum_{n=1}^{\infty} \frac{n^2\pi^2}{L^2}\Delta\delta_n \sin\frac{n\pi x}{L} + \frac{P + \Delta P}{EI'}\sum_{n=1}^{\infty} \Delta\delta_n \sin\frac{n\pi x}{L}$$
$$+ \frac{\Delta P}{EI_1'}\left(\sum_{n=1}^{\infty} \delta_n \sin\frac{n\pi x}{L} + \sum_{n=1}^{\infty} a_n \sin\frac{n\pi x}{L}\right) = 0$$

Multiplying this equation by sin $(m\pi x/L)\,dx$ and integrating from $x = 0$ to $x = L$, as shown in Equations (6.11.7) and (6.11.8), we see that all the terms $n \neq m$ vanish (Section 6.2) and the terms $m = n$ give

$$-\frac{n^2\pi^2}{L^2}\Delta\delta_n + \frac{P + \Delta P}{EI'}\Delta\delta_n + \frac{\Delta P}{EI'_1}(\delta_n + a_n) = 0$$

$$\left[-\frac{n^2\pi^2 EI'_1}{L^2} + (P + \Delta P)\right]\Delta\delta_n + \Delta P(\delta_n + a_n) = 0$$

Let

$$P'_n = \frac{n^2\pi^2 EI'_1}{L^2}$$

then

$$(P'_n - P - \Delta P)\,\Delta\delta_n = \Delta P(\delta_n + a_n)$$

from which

$$\frac{\Delta\delta_n}{\Delta P} = \frac{\delta_n + a_n}{P'_n - P - \Delta P}$$

$$\Delta w = \sum_{n=1}^{\infty} \frac{(\delta_n + a_n)\,\Delta P}{P'_n - P - \Delta P} \sin\frac{n\pi x}{L}$$

This gives the solution of the differential equation. As in the elastic case, the first harmonic ($n = 1$) increases much more than the other harmonics as P approaches $P_1{}'$. *Neglecting increments of harmonics greater than* 1 *gives*

$$\frac{\Delta\delta}{\Delta P} = \frac{\delta_1 + a_1}{P'_1 - P - \Delta P}$$

As ΔP approaches zero, $\Delta\delta/\Delta P = d\delta/dP$, and we obtain

$$\frac{d\delta}{dP} = \frac{\delta_1 + a_1}{P'_1 - P} \qquad \frac{dw}{dP} = \frac{\delta_1 + a_1}{P'_1 - P}\sin\frac{\pi x}{L}$$

Then, from Equation (6.11.29), we have

$$\eta_1 = -\frac{\rho'^2}{[P/(P'_1 - P)](\delta_1 + a_1)\sin(\pi x/L) + w + \xi} \tag{6.11.30}$$

At the mid-span we have

$$\eta_1 = -\frac{\rho'^2}{(\delta_1 + a_1)[P/(P'_1 - P)] + \delta_1 + \xi} \tag{6.11.31}$$

These two equations give η_1 at different sections. When $\eta_1 > h_2$, there is no reversal of strain. If $\eta_1 < h_2$, reversal of strain occurs in the region $\eta > \eta_1$ to $\eta = h_2$.

After Δw is obtained, we find the curvature increment,

$$\Delta K = \frac{\Delta M}{EI'} = \frac{(P + \Delta P)\,\Delta w + (w + \xi)\,\Delta P}{EI'} \tag{6.11.32}$$

From the relations

$$\Delta K = \frac{\Delta(e_1 - e_2)}{h} \tag{6.11.33}$$

where e_1 and e_2 are the extreme fiber strains on the concave and convex sides, respectively, we obtain the Δe's. From the Δe's at different points on different sections, the new modified sections are obtained. The analysis is continued by adding another increment of load and repeating the procedure.

If EI' varies appreciably along the column length, the differential Equation (6.11.26) may be solved by a numerical method of finite differences. Divide the column into equal segments (Figure 6.23). The differential Equation (6.11.26) at section n, written in finite difference form, is as follows:

$$\frac{\Delta w_{n+1} + \Delta w_{n-1} - 2\Delta w_n}{\Delta x^2} + (P + \Delta P)\frac{\Delta w_n}{EI'_n} + \frac{\Delta P}{EI'_n}(w_n + \xi_n) = 0 \tag{6.11.34}$$

At the column end points, $\Delta M = 0$, $\Delta w = 0$, $w = 0$, $\xi = 0$, and the differential equation is automatically satisfied. For each of the intermediate sections, one equation of the above form may be written. The values w_n and ξ_n are those values before ΔP is added. There is one unknown Δw_n at each intermediate section. There are as many equations as unknowns,

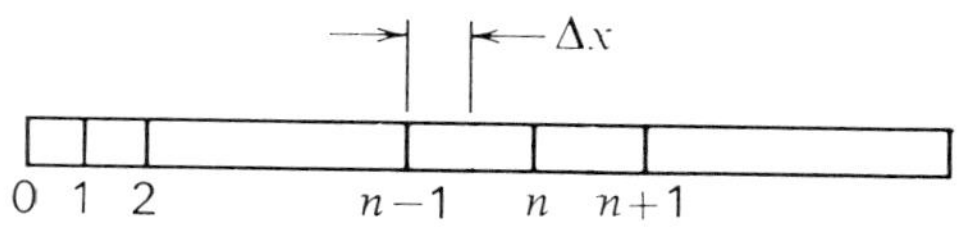

Figure 6.23

and the Δw_n's at different sections may be obtained by solving these simultaneous equations. This can be readily done with the aid of computers. With Δw_n's and ΔP, and hence the new w_n and P known, η_1's at different sections are calculated from Equation (6.11.29). Then the new ξ's are computed. Equation (6.11.34) is applied again for an additional load increment. This process is repeated until the increase of deflection becomes very large, indicating that the buckling load is being approached.

The load deflection curve of a steel column tested by Karman has been calculated by the method shown.[29] The column is of rectangular section with an initial deflection of $0.001h$ at the center and a slenderness ratio of 75. The stress-strain curve of the material and its tangent modulus vs. strain are plotted in Figure 6.24. The elastic modulus is 2,170,000 kg/sq cm. From tests, the additional lateral deflection at the center is $0.0016h$ at an average compressive stress of 2400 kg/sq cm. The modified sections of the

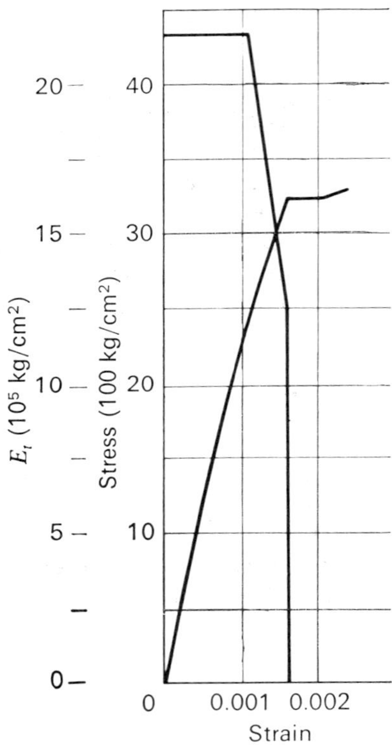

Figure 6.24

mid-section of the column at different loads are shown in Figure 6.25. The modified sections at different sections of the column near the maximum load are shown in Figure 6.26. The calculated load deflection curve of the column is shown in Figure 6.27. It is seen that the calculated curve agrees well with the test curve obtained by Karman. The detail calculation is given in Reference 29.

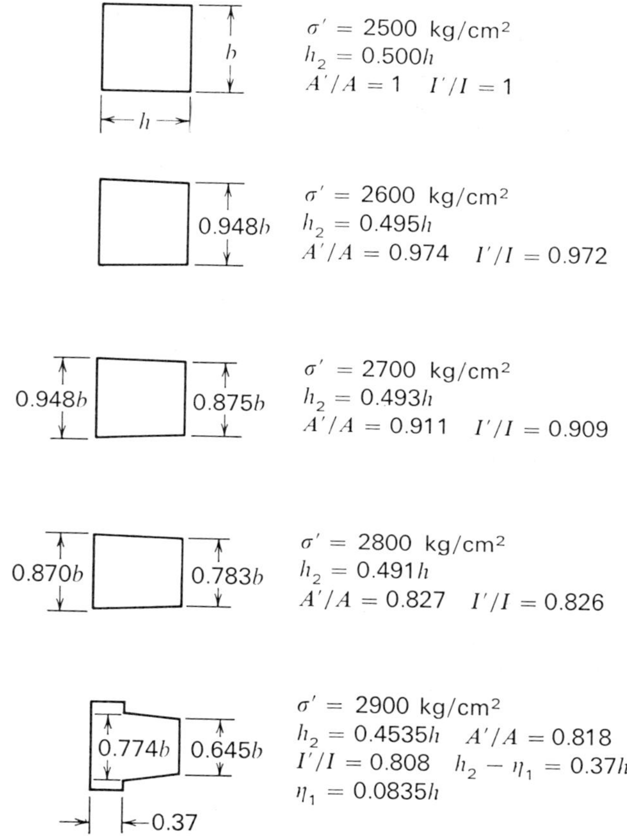

Figure 6.25.
Modified sections at the midspan of the column.

Transition of Tangent Modulus to Double Modulus of a Perfect Column. The previous section gave a detailed account of the mechanism of an inelastic column with initial curvature. It may also be of interest to see the mechanism of the transition from tangent modulus to double modulus of a perfect column. A perfect column will remain straight up to the tangent modulus load. If there is no lateral disturbance, the column may remain straight even beyond the tangent modulus load. For actual columns a slight disturbance is assumed to be present at all times. The slight disturbance is here represented by an infinitesimal lateral deflection given by $\varepsilon \sin(n\pi x/L)$, with ε approaching zero. Before the lateral deflection becomes

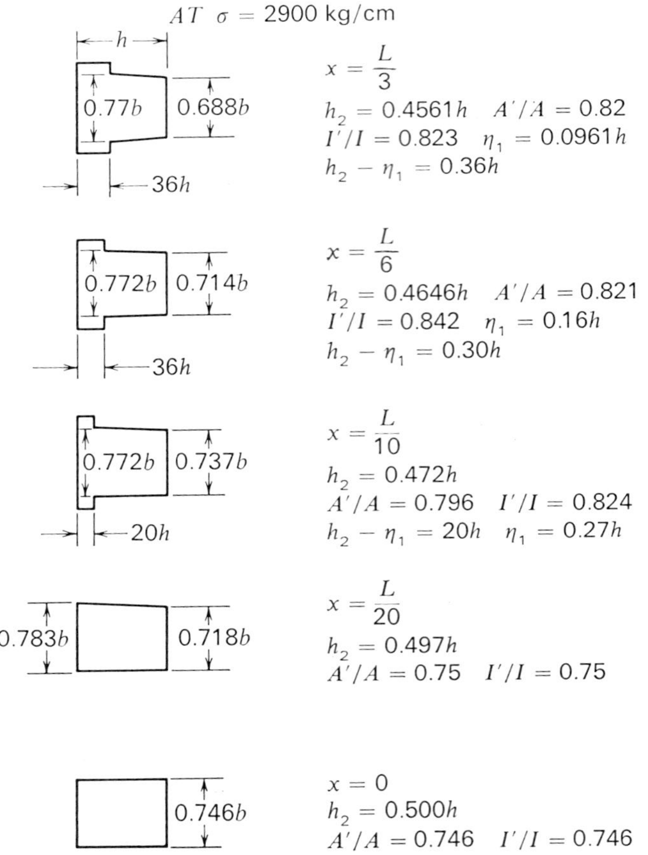

Figure 6.26

finite, EI' is constant along the column and equal to $E_t I$. From Equation (6.11.5), we have

$$P_1' = \frac{\pi^2 E_t I}{L^2} \qquad y = \varepsilon \sin \frac{\pi x}{L} \qquad \delta_1 = \varepsilon \qquad \xi = 0 \qquad a_1 = 0$$

and Equation (6.11.30) reduces to

$$\eta_1 = \frac{-\rho^2}{[P_1'\varepsilon/(P_1' - P)] \sin(\pi x/L)} \tag{6.11.35}$$

When $P_1' - P$ is finite, η_1 is infinitely large in magnitude, and strain reversal does not occur in the section. However, when P approaches P_1',

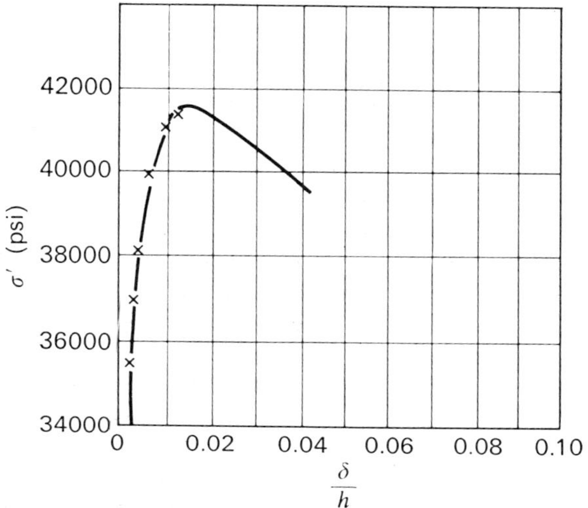

Figure 6.27

giving $P_1' - P$ the same order of magnitude as ε, say $n\varepsilon$, where n is a finite number, η_1 becomes finite,

$$\eta_1 = \frac{-n\rho^2}{P_1' \sin(\pi x/L)} \tag{6.11.36}$$

and reversal of strain starts on the convex side at mid-span, when

$$\eta_1 = \frac{-n\rho^2}{P_1'} = -h_2 \tag{6.11.37}$$

We obtain

$$n = \frac{h_2 P_1'}{\rho^2} = \frac{h_2 E_t I \pi^2}{L^2 \rho^2} = \frac{h_2 \pi^2 E_t A}{L^2}$$

Initially, as P approaches P_1', we see that n is large and $\eta_1 > h_2$. When the ratio $(P_1' - P)/\varepsilon$ reaches the value shown in Equation (6.11.37), we have $\eta_1 = h_2$. As P approaches nearer to P_1', further decrease of n occurs, and η_1 becomes less than h_2; therefore reversal of strain occurs between η_1 and h_2. In this region, the tangent modulus reverts back to the elastic modulus, and hence the double-modulus behavior occurs. The restoring of the elastic modulus in the region with reversal of strain causes the modified section and the critical load P_1' to increase and the centroidal axis of the modified section to move away from that of the geometrical section; that is, ξ grows rapidly to a finite value. Thereafter the lateral deflection increases with additional load according to Equation (6.11.26), mainly because of ξ. When δ increases to a finite value, the variation of tangent modulus across the section will also increase ξ of the modified section. The η_1 decreases according to Equation (6.11.31), thereby giving further increase of ξ. Hence the lateral deflection increases at a higher rate with further additional load. With increase of deflection, the compressive strain on the concave side increases more than the other part of the section. The tangent modulus on the concave side decreases more than that at the centroidal axis. This decrease of tangent modulus reduces the moment of inertia of the modified section.

Before bending occurs, the tangent modulus is uniform across the section. After bending occurs, the tangent modulus decreases from the convex to the concave side. After strain reversal appears, the elastic modulus returns to the convex side. The modified section changes because of the combined effects of strain reversal and the decrease of tangent modulus on the concave side with the increase of load. The modified sections along the column determine the column critical load, and therefore the critical load varies with the stage of loading.

Lateral deflection may start before reversal of strain, i.e., at $\eta_1 > h_2$ and not at $\eta_1 = 0$ as assumed in the reduced-modulus theory. This explains why a perfect column starts to bend at the tangent-modulus load and not at the reduced-modulus load. After lateral deflection increases and strain reversal occurs, η_1 decreases with load. The decrease of η_1, or the increase of the strain-reversal region, increases the flexural rigidity EI' of the modified section. On the other hand, with increasing deflection, the compressive strain on the concave side is more than the average of the whole section. This causes the tangent modulus on the concave side to be less than the tangent modulus at tangent-modulus load. The excess of the effect on EI' caused by reversal of strain over that caused by the decrease of tangent modulus on the concave side causes the critical load to be greater than the tangent-modulus load. However, the reduced-modulus load is never reached. The difference between the tangent modulus on the concave side and that corresponding to the average compressive stress on the column depends on the rate of change in the slope of the stress-strain curve in the neighborhood of the maximum compressive stress. Duberg and Wilder[30] have demonstrated this fact, in their paper on the study of idealized H-section columns of various materials, with a systematic variation of their stress-strain curves. They conclude that, for columns whose material stress-strain curves depart gradually from the initial elastic slope, as in the case of stainless steels, the maximum column load may be appreciably above the tangent-modulus load. If the departure is more abrupt, as in high-strength aluminum or magnesium alloys, the maximum column load is only slightly above the tangent-modulus load.

REFERENCES

1. Lin, T. H., "Shortening of Column with Initial Curvature and Eccentricity and its Influence on the Stress Distribution in Indeterminate Structures," *Proc. First U.S. Nat. Congr. Appl. Mech.*, 449, 1951.
2. Southwell, R. V., *An Introduction to the Theory of Elasticity*, Clarendon Press, Oxford, England, pp. 425–429, 1936.
3. Timoshenko, S., and J. Gere, *Theory of Elastic Stability*, McGraw-Hill, New York, pp. 3–18, 31–33, 1961.
4. Southwell, R. V., "On the Analysis of Experimental Observations in Problems of Elastic Stability," *Proc. Royal Soc. A*, **135**, 601–616, 1932.
5. Freudenthal, A. M., *The Inelastic Behavior of Engineering Materials and Structures*, New York, p. 518, 1950.
6. Rosenthal, D., and H. W. Baer, "An Elementary Theory of Creep of Columns," *Proc. First U.S. National Cong. Appl. Mech.*, 1951.
7. Hilton, H. H., "Creep Collapse of Visco-Elastic Columns with Initial Curvature," *J. Aerospace Sci.*, **19**, 844–846, 1952.
8. Kempner, J., *Creep Bending and Buckling of Linearly Viscoelastic Columns*, NACA Tech. Note 3136, January 1954.
9. Lin, T. H., "Stresses in Columns with Time-Dependent Elasticity," *Proc. First Midwestern Conf. on Solid Mech.*, 1953.
10. Churchill, R. V., *Modern Operational Mathematics in Engineering*, McGraw-Hill, New York, pp. 44–45, 294–295, 1944.
11. Kempner, J., *Creep Bending and Buckling of Nonlinearly Viscoelastic Columns*, NACA Tech. Note 3137, January 1954.
12. Hoff, N. J., "Buckling and Stability," Forty-First Wilbur Wright Memorial Lecture, *J. Roy. Aeron. Soc.*, **58**, 30, January 1954.
13. Patel, S. A., and J. Kempner, "Effect of Higher Harmonics Deflection Components on the Creep Buckling of Columns," *The Aeronautical Quarterly*, **8**, 215, August 1957.
14. Marin, J., "Creep Deflections in Columns," *J. Appl. Phys.*, **18**, 103–109, 1947.
15. Libove, C., "Creep Buckling of Columns," *J. Aerospace Sci.*, **19**, 459–467, 1952.
16. Libove, C., *Creep Buckling Analysis of Rectangular-Section Columns*, NACA Tech. Note 2956, 1953.
17. Higgins, T. P., Jr., "Effect of Creep on Column Deflection," Preprint No. 385, Inst. Aer. Sc., 1952.
18. Shanley, F. R., *Weight-Strength Analysis of Aircraft Structures*, McGraw-Hill, New York, pp. 323–342, 359–385, 1952.

19. Lin, T. H., "Creep Stresses and Deflections of Columns," *J. Appl. Mech.*, **78**, 214–218, 1956.
20. Sokolnikoff, I. S., and R. M. Redhefer, *Mathematics of Physics and Modern Engineering*, McGraw-Hill, New York, pp. 711–715, 1958.
21. Finnie, I., and W. R. Heller, *Creep of Engineering Materials*, McGraw-Hill, New York, p. 157, 1959.
22. Ince, E. L., *Ordinary Differential Equations*, Dover, New York, pp. 256–258, 1956.
23. Murnaghan, F. D., *Introduction to Applied Mathematics*, Vol. 2, Wiley, New York, Chap. 14, 1943.
24. Niles, A. S., and J. S. Newell, *Airplane Structures*, Vol. 2, Wiley, New York, Chap. 14, 1943.
25. Osgood, W. R., "The Double Modulus Theory of Column Action," *Civil Engineering*, **5**, 173–175, March 1935.
26. Karman, Th. Von, *Unteasuchgen über Knickfestigkeif Mitteilungen über Forschungsarbeiten Verlin dentscher Ingeniene Heft 81*, Springer, Berlin, 1910.
27. Shanley, F. R., "The Column Paradox," *J. Aerospace Sci.*, **13**, 678, December 1946.
28. Shanley, F. R., "Inelastic Column Theory," *J. Aerospace Sci.*, **14**, 261–267, May 1947.
29. Lin, T. H., "Inelastic Column Buckling," *J. Aerospace Sci.*, **17**, 159–172, 1950.
30. Duberg, J. E., and T. W. Wilder, "Column Behavior in the Plastic Stress Range," *J. Aeron. Soc.*, **17**, 323–327, June 1950.
31. Rabotnov, G. N., and S. A. Shesterikov, "Creep Stability of Columns and Plates," *J. Mech. Phys. Solids*, **6**, 27–34, 1957.
32. Hoff, N. J., "A Survey of the Theories of Creep Buckling," *Proc. 3rd U.S. Nat. Congr. Appl. Mech.*, 1958.
33. Gerard, E., and R. Papinno, "Classical Columns and Creep," *J. Aerospace Sci.*, **29**, 680–688, June 1962.
34. Carlson, R. L., "Time-dependent Tangent Modulus Applied to Column Creep Buckling," *J. Appl. Mech.*, **23**, 390, September 1956.
35. Carlson, R. L., E. E. Bodine, and G. K. Manning, *Investigation of Compressive Creep Properties of Aluminum Columns at Elevated Temperatures. Part 4, Additional Studies*, WADC Tech. Report 52–251, April 1956.
36. Lin, T. H., "Creep Deflections and Stresses of Beam-Columns," *J. Appl. Mech.*, **25**, 75–78, March 1958.
37. Anderson, D. R., W. F. Anderson, and T. H. Lin, "Creep Bowing of the Process Tube," NAA-SR TDR No. 11451, June 30, 1965.

chapter

7

Inelastic Hollow Spheres, Thick Circular Cylinders, and Rotating Disks

Introduction. In this chapter analyses of inelastic hollow spheres and inelastic long circular cylinders, under internal and external uniform pressures, are presented. The analysis of inelastic rotating disks of constant or variable thickness is also treated. These structures are among the most commonly used engineering structures. Since the sphere under pressure has spherical symmetry in both geometry and loading, the resulting stress and deformation are also spherically symmetric. Similarly, both the circular cylinder under pressure and the rotating disk are axially symmetric in geometry and loading, hence their stress and strain fields also are axially symmetric. Because of this spherical or axial symmetry, the stress and strain vary only with the radial coordinate of these bodies. The stress analyses of these bodies, therefore, is considerably simplified.

Equations for stresses, strains and displacements are derived for the inelastic problems. The elastic analyses are obtained by allowing the inelastic strains to vanish. Thermal elastic analysis is obtained by equating the inelastic strain to thermal strain. Creep analysis of a rotating disk is discussed. Finally, elastoplastic analyses for perfectly plastic materials and strain-hardening materials are shown.

7.1 Inelastic Thick Hollow Spheres[1-3]

Consider a thick-walled hollow sphere with inner radius "a" and outer radius "b" subject to an internal pressure p_i and an external pressure p_0.

Because the sphere and the load have spherical symmetry, the displacement must also have this symmetry. Referring to a system of spherical coordinates as shown in Figure 7.1, we see that the displacement, denoted

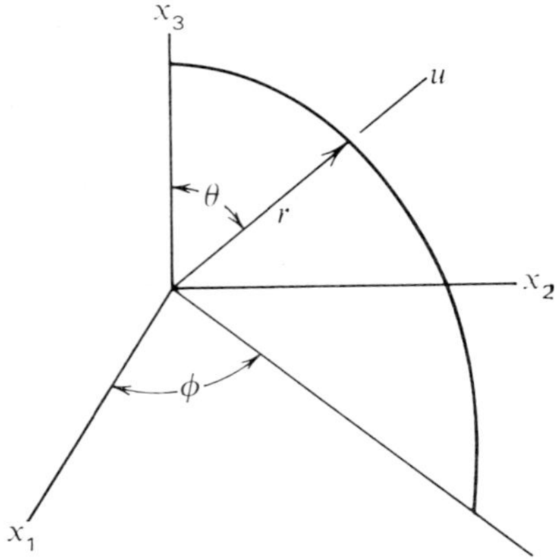

Figure 7.1. Spherical coordinates.

by u, of any point in the sphere must be radial. The symmetry of the load and geometry requires that all dependent variables be independent of θ and ϕ. Hence we have a one-dimensional problem in the independent variable r. The principal strains are

$$e_r = \frac{du}{dr}$$

and (7.1.1)

$$e_\theta = e_\phi = \frac{u}{r}$$

and the volumetric strain is

$$e = \frac{du}{dr} + 2\frac{u}{r} \tag{7.1.2}$$

With the double-prime superscript above strain denoting the inelastic part of strain, we have, from Equation (2.2.3),

$$\tau_{ij} = \lambda\, \delta_{ij}(e - e'') + 2\mu(e_{ij} - e''_{ij}) \tag{2.2.3}$$

The principal stresses are then given by

$$\sigma_r = \lambda(e - e'') + 2\mu(e_r - e_r'') \tag{7.1.3}$$

$$\sigma_\theta = \lambda(e - e'') + 2\mu(e_\theta - e_\theta'') \tag{7.1.4}$$

where the inelastic strains e'', e_r'', and e_θ'' may be due to any combination of thermal, creep, and plastic strains.

Condition of Equilibrium. Consider a differential volume element bounded by two concentric spherical surfaces with radii r and $r + dr$ and by four meridian planes as shown in Figure 7.2. Spherical symmetry causes

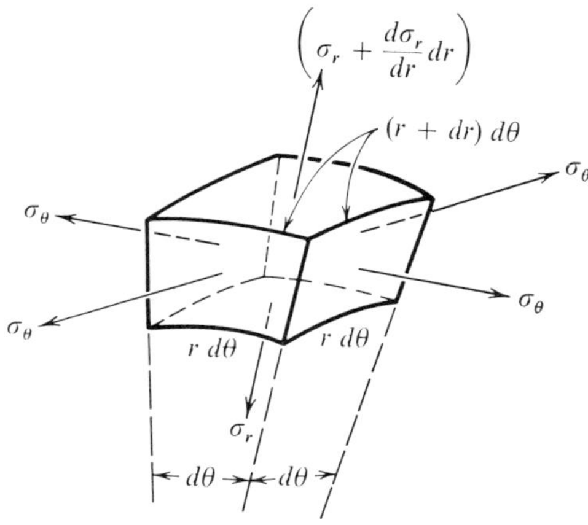

Figure 7.2. Forces acting on a differential element.

the shear stresses on the faces of this element to vanish. The normal stresses σ_r and σ_θ acting on the faces are shown. The normal stress σ_θ acting on each meridian plane gives a force component along the radial direction of $\sigma_\theta r\, dr\, d\theta(d\theta/2)$. The total radial force due to σ_θ on the four meridian planes of the differential element is $2\sigma_\theta r\, dr\,(d\theta)^2$. From the condition of equilibrium of forces along the radial direction, we have

$$\left(\sigma_r + \frac{d\sigma_r}{dr} dr\right)(r + dr)^2(d\theta)^2 - \sigma_r r^2(d\theta)^2 - 2\sigma_\theta r\, dr(d\theta)^2 = 0$$

Neglecting differentials of higher order, we obtain the *equilibrium equation*,

$$\frac{d\sigma_r}{dr} + 2\,\frac{\sigma_r - \sigma_\theta}{r} = 0 \tag{7.1.5}$$

Substituting Equations (7.1.1) and (7.1.2) into Equations (7.1.3) and (7.1.4), which in turn, are substituted into Equation (7.1.5), we obtain

$$\lambda \frac{d}{dr}\left(\frac{du}{dr} + 2\frac{u}{r}\right) + 2\mu \frac{d^2u}{dr^2} + 4\mu \frac{1}{r}\left(\frac{du}{dr} - \frac{u}{r}\right) = \lambda \frac{de''}{dr} + 2\mu \frac{de_r''}{dr} + 4\mu \frac{e_r'' - e_\theta''}{r}$$

Noting that $\dfrac{d(u/r)}{dr} = (1/r)(du/dr - u/r)$, we see that the above reduces to

$$(\lambda + 2\mu)\frac{d}{dr}\left[\frac{1}{r^2}\frac{d}{dr}(r^2u)\right] = \lambda \frac{de''}{dr} + 2\mu \frac{de_r''}{dr} + 4\mu \frac{e_r'' - e_\theta''}{r} \tag{7.1.6}$$

Integrating with respect to r yields

$$(\lambda + 2\mu)\frac{d}{dr}(r^2u) = \lambda r^2 e'' + 2\mu r^2 e_r'' + 4\mu r^2 \int_a^r \frac{e_r'' - e_\theta''}{r}\,dr + c_1 r^2$$

Integrating with respect to r again gives

$$(\lambda + 2\mu)u = \frac{\lambda}{r^2}\int_a^r r^2 e''\,dr + \frac{2\mu}{r^2}\int_a^r r^2 e_r''\,dr + \frac{4\mu}{r^2}\int_a^r r^2 \int_a^r \frac{e_r'' - e_\theta''}{r}\,dr\,dr + \frac{c_1 r}{3} + \frac{c_2}{r^2} \tag{7.1.7}$$

where c_1 and c_2 are constants of integration. Substituting this into Equations (7.1.1) yields

$$(\lambda + 2\mu)e_r = (\lambda + 2\mu)\frac{du}{dr} = -\frac{2\lambda}{r^3}\int_a^r r^2 e''\,dr + \lambda e'' - \frac{4\mu}{r^3}\int_a^r r^2 e_r''\,dr + 2\mu e_r'' - \frac{8\mu}{r^3}\int_a^r r^2 \int_a^r \frac{e_r'' - e_\theta''}{r}\,dr\,dr + 4\mu \int_a^r \frac{e_r'' - e_\theta''}{r}\,dr + \frac{c_1}{3} - \frac{2c_2}{r^3} \tag{7.1.8}$$

and

$$(\lambda + 2\mu)e_\theta = (\lambda + 2\mu)\frac{u}{r} = \frac{\lambda}{r^3}\int_a^r r^2 e''\,dr + \frac{2\mu}{r^3}\int_a^r r^2 e_r''\,dr + \frac{4\mu}{r^3}\int_a^r r^2 \int_a^r \frac{e_r'' - e_\theta''}{r}\,dr\,dr + \frac{c_1}{3} + \frac{c_2}{r^3} \tag{7.1.9}$$

From Equation (7.1.3), we have

$$\sigma_r = (\lambda + 2\mu)e_r + 2\lambda e_\theta - (\lambda + 2\mu)e_r'' - 2\lambda e_\theta'' \tag{7.1.10}$$

At $r = a$, Equations (7.1.8) and (7.1.9) yield

$$(\lambda + 2\mu)e_r = \frac{c_1}{3} - \frac{2c_2}{a^3} + \lambda e'' + 2\mu e_r''$$

and

$$(\lambda + 2\mu)e_\theta = \frac{c_1}{3} + \frac{c_2}{a^3}$$

From the boundary condition $\sigma_r = -p_i$ at $r = a$, Equations (7.1.8) to (7.1.10) give

$$\frac{c_1}{3}\left(\frac{3\lambda + 2\mu}{\lambda + 2\mu}\right) - \frac{c_2}{a^3}\frac{4\mu}{\lambda + 2\mu} = -p_i \qquad (7.1.11)$$

Let

$$I_1 = \frac{2\lambda}{b^3}\int_a^b r^2 e''\, dr$$

$$I_2 = \frac{4\mu}{b^3}\int_a^b r^2 e_r''\, dr$$

$$I_3 = \frac{8\mu}{b^3}\int_a^b r^2 \int_a^r \frac{e_r'' - e_\theta''}{r}\, dr\, dr$$

$$I_4 = 4\mu \int_a^b \frac{e_r'' - e_\theta''}{r}\, dr \qquad (7.1.12)$$

The other boundary condition, $\sigma_r = -p_0$ at $r = b$, then gives

$$-I_1 - I_2 - I_3 + I_4 + \frac{c_1}{3} - \frac{2c_2}{b^3} + \frac{2\lambda}{\lambda + 2\mu}\left(\frac{I_1}{2} + \frac{I_2}{2} + \frac{I_3}{2} + \frac{c_1}{3} + \frac{c_2}{b^3}\right) = -p_0$$

$$\frac{c_1}{3}\frac{3\lambda + 2\mu}{\lambda + 2\mu} - \frac{c_2}{b^3}\frac{4\mu}{\lambda + 2\mu} - \frac{2\mu}{\lambda + 2\mu}(I_1 + I_2 + I_3) + I_4 = -p_0 \qquad (7.1.13)$$

Solving c_1 and c_2 from Equations (7.1.11) and (7.1.13), we obtain

$$c_2 = \frac{a^3 b^3}{b^3 - a^3}\left[\frac{I_1 + I_2 + I_3}{2} - \frac{\lambda + 2\mu}{4\mu}(I_4 - p_i + p_0)\right] \qquad (7.1.14)$$

and

$$\frac{c_1}{3} = \frac{2\mu}{3\lambda + 2\mu}\frac{b^3}{b^3 - a^3}\left[I_1 + I_2 + I_3 - \frac{\lambda + 2\mu}{2\mu}\left(I_4 - \frac{a^3}{b^3}p_i + p_0\right)\right] \qquad (7.1.15)$$

From Equations (7.1.3), (7.1.4), (7.1.8), and (7.1.9), the radial and tangential stresses are

$$\sigma_r = -\frac{4\mu\lambda}{(\lambda+2\mu)r^3}\int_a^r r^2 e''\,dr - \frac{8\mu^2}{(\lambda+2\mu)}\frac{1}{r^3}\int_a^r r^2 e_r''\,dr - \frac{16\mu^2}{(\lambda+2\mu)}\frac{1}{r^3}\int_a^r r^2\int_a^r \frac{e_r''-e_\theta''}{r}\,dr\,dr + \frac{3\lambda+2\mu}{\lambda+2\mu}\frac{c_1}{3} - \frac{4\mu}{\lambda+2\mu}\frac{c_2}{r^3} \tag{7.1.16}$$

and

$$\begin{aligned}\sigma_\theta &= \lambda(e_r+2e_\theta)+2\mu e_\theta - \lambda(e_r''+2e_\theta'')-2\mu e_\theta'' \\ &= 2(\lambda+\mu)e_\theta+\lambda e_r - 2(\lambda+\mu)e_\theta''-\lambda e_r'' \\ &= \frac{2\lambda\mu}{\lambda+2\mu}\frac{1}{r^3}\int_a^r r^2 e''\,dr + \frac{4\mu^2}{\lambda+2\mu}\frac{1}{r^3}\int_a^r r^2 e_r''\,dr \\ &\quad + \frac{8\mu^2}{\lambda+2\mu}\frac{1}{r^3}\int_a^r r^2\int_a^r\frac{e_r''-e_\theta''}{r}\,dr\,dr + \frac{4\mu\lambda}{\lambda+2\mu}\int_a^r\frac{e_r''-e_\theta''}{r}\,dr \\ &\quad + \frac{3\lambda+2\mu}{\lambda+2\mu}\frac{c_1}{3} + \frac{2\mu}{\lambda+2\mu}\frac{c_2}{r^3} - \frac{(3\lambda+2\mu)\mu}{\lambda+2\mu}e_\theta''\end{aligned} \tag{7.1.17}$$

This gives the relations between stresses and the distribution of inelastic strains for a thick hollow sphere under internal and external pressures.

Elastic Hollow Spheres. For the elastic case $e_r'' = e_\theta'' = e'' = 0$, and $I_1 = I_2 = I_3 = I_4 = 0$, then, c_1 and c_2, from Equations (7.1.14) and (7.1.15), are

$$c_1 = -\frac{3(\lambda+2\mu)}{3\lambda+2\mu}\frac{b^3 p_o - a^3 p_i}{b^3-a^3} \qquad c_2 = \frac{\lambda+2\mu}{4\mu}\frac{a^3b^3}{b^3-a^3}(p_i-p_o) \tag{7.1.18}$$

Equations (7.1.7), (7.1.16), (7.1.17), and (7.1.2) then reduce to

$$\begin{aligned}u &= -\frac{r}{(3\lambda+2\mu)}\frac{b^3p_o-a^3p_i}{(b^3-a^3)} - \frac{1}{4\mu}\frac{a^3b^3(p_o-p_i)}{(b^3-a^3)r^2} \\ \sigma_r &= -\frac{b^3p_o-a^3p_i}{b^3-a^3} + \frac{a^3b^3}{b^3-a^3}(p_o-p_i)\frac{1}{r^3} \\ \sigma_\theta &= -\frac{b^3p_o-a^3p_i}{b^3-a^3} - \frac{a^3b^3(p_o-p_i)}{2(b^3-a^3)r^3} \\ e &= \frac{du}{dr}+2\frac{u}{r} = -\frac{3}{3\lambda+2\mu}\frac{b^3p_o-a^3p_i}{b^3-a^3}\end{aligned} \tag{7.1.19}$$

Hence, in the elastic range, the volumetric strain e is constant. Figure 7.3 shows the stress distributions in an elastic spherical shell.

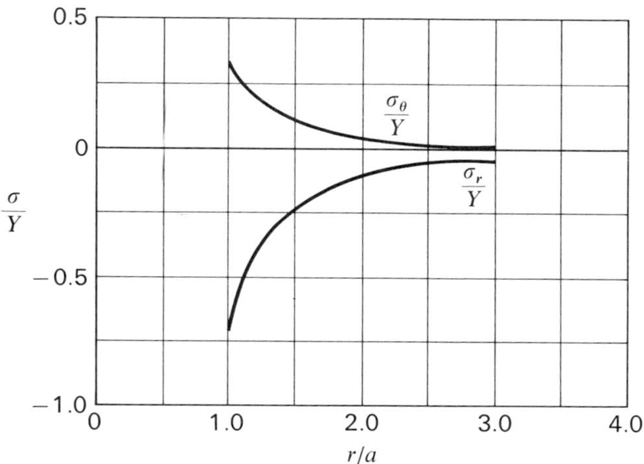

Figure 7.3. Stress distribution in elastic spherical shell (b/a = 3, ρ/a = 1, $p_i/p_o = 17.7$).

Hollow Spheres with Radial Thermal Gradients.[4] For the case with only radial thermal gradients, $e_r'' = e_\theta'' = \alpha T$ and $e'' = 3\alpha T$, where α is the coefficient of expansion and T is the increase in temperature above which the pressure-less sphere is stress-free. In this case, Equations (7.1.12), (7.1.13), and (7.1.14) become

$$I_1 = \frac{2\lambda}{b^3}\int_a^b 3r^2\alpha T\,dr \qquad I_2 = \frac{4\mu}{b^3}\int_a^b r^2\alpha T\,dr \qquad I_3 = I_4 = 0$$

$$I_1 + I_2 + I_3 = \frac{6\lambda + 4\mu}{b^3}\int_a^b r^2\alpha T\,dr$$

$$\frac{c_1}{3} = \frac{\lambda + 2\mu}{3\lambda + 2\mu}\,\frac{a^3 p_i - b^3 p_o}{b^3 - a^3} + \frac{4\mu}{b^3 - a^3}\int_a^b r^2\alpha T\,dr$$

$$c_2 = -\frac{\lambda + 2\mu}{4\mu}(p_o - p_i)\frac{a^3 b^3}{b^3 - a^3} + \frac{(3\lambda + 2\mu)a^3}{b^3 - a^3}\int_a^b r^2\alpha T\,dr$$

Using the above equations, we obtain, from Equation (7.1.16),

$$\sigma_r = \frac{4\mu(3\lambda + 2\mu)}{\lambda + 2\mu}\left[\frac{1}{b^3 - a^3}\left(1 - \frac{a^3}{r^3}\right)\int_a^b \alpha T r^2\,dr - \frac{1}{r^3}\int_a^r \alpha T r^2\,dr\right] - \frac{b^3 p_o - a^3 p_i}{b^3 - a^3} + (p_o - p_i)\frac{a^3 b^3}{(b^3 - a^3)r^3} \tag{7.1.20}$$

From the relations of the different elastic constants, we have

$$E = \frac{\mu(3\lambda + 2\mu)}{\lambda + \mu} \qquad \nu = \frac{\lambda}{2(\lambda + \mu)} \qquad \frac{2E}{1 - \nu} = \frac{4\mu(3\lambda + 2\mu)}{\lambda + 2\mu} \tag{7.1.21}$$

and Equation (7.1.20) becomes

$$\sigma_r = \frac{2E\alpha}{1 - \nu}\left[\frac{1}{b^3 - a^3}\left(1 - \frac{a^3}{r^3}\right)\int_a^b Tr^2\,dr - \frac{1}{r^3}\int_a^r Tr^2\,dr\right] - \frac{b^3 p_o - a^3 p_i}{b^3 - a^3} + (p_o - p_i)\frac{a^3 b^3}{(b^3 - a^3)r^3} \tag{7.1.22}$$

Similarly, we find

$$\sigma_\theta = \frac{E\alpha}{1 - \nu}\left[\frac{1}{b^3 - a^3}\left(2 + \frac{a^3}{r^3}\right)\int_a^b Tr^2\,dr + \frac{1}{r^3}\int_a^r Tr^2\,dr\right] - \frac{b^3 p_o - a^3 p_i}{b^3 - a^3} - \frac{1}{2r^3}(p_o - p_i)\frac{a^3 b^3}{b^3 - a^3} - \frac{E\alpha T}{2(1 - \nu)} \tag{7.1.23}$$

and the displacement is

$$\frac{u}{r} = \frac{1 + \nu}{1 - \nu}\frac{\alpha}{r^3}\int_a^r Tr^2\,dr + \frac{\alpha}{(b^3 - a^3)(1 - \nu)}\left[2(1 - 2\nu) + (1 + \nu)\frac{a^3}{r^4}\right]\int_a^b Tr^2\,dr + \frac{1}{(b^3 - a^3)E}\left[(1 - 2\nu)(a^3 p_i - b^3 p_o) - \frac{(1 + \nu)a^3 b^3}{2r^3}(p_o - p_i)\right] \tag{7.1.24}$$

For a given distribution of temperature, the stresses and displacement of the sphere are readily found from Equations (7.1.22), (7.1.23), and (7.1.24).

Hollow Spheres of Perfectly Plastic Materials. For perfectly plastic materials, plastic strain occurs when the effective stress σ^* reaches the tensile yield stress Y. In this case the effective stress is

$$\sigma^* = \frac{1}{\sqrt{2}}[(\sigma_1 - \sigma_2)^2 + (\sigma_2 - \sigma_3)^2 + (\sigma_3 - \sigma_1)^2]^{1/2} = \frac{1}{\sqrt{2}}[2(\sigma_\theta - \sigma_r)^2]^{1/2} = \sigma_\theta - \sigma_r = Y \tag{7.1.25}$$

From Equation (7.1.19), we obtain

$$\sigma^* = \sigma_\theta - \sigma_r = \frac{(p_i - p_o)a^3}{b^3 - a^3}\frac{3b^3}{2r^3}$$

The limiting pressure for elastic deformation $(p_i - p_o)^*$ is obtained by equating σ^* to Y at $r = a$; this gives

$$(p_i - p_o)^* = \frac{2Y}{3}\frac{b^3 - a^3}{b^3} \tag{7.1.26}$$

When the difference in pressure increases beyond $(p_i - p_o)^*$, plastic deformation spreads from the inner surface.

From Equations (7.1.1) and (7.1.2), the deviatoric strain components are

$$\varepsilon_\theta = e_\theta - \frac{e}{3} = \frac{1}{3}\left(\frac{u}{r} - \frac{du}{dr}\right)$$

and

$$\varepsilon_r = e_r - \frac{e}{3} = -\frac{2}{3}\left(\frac{u}{r} - \frac{du}{dr}\right)$$

Hence we have, for the deviatoric strains,

$$\varepsilon_r = -2\varepsilon_\theta \tag{7.1.27}$$

In the region where plastic deformation occurs,

$$\sigma_\theta - \sigma_r = S_\theta - S_r = Y$$

where S_θ and S_r are the deviatoric stress components, given by

$$S_\theta = \sigma_\theta - \frac{2\sigma_\theta + \sigma_r}{3} = \frac{\sigma_\theta - \sigma_r}{3} = \frac{Y}{3}$$

$$S_r = -Y + S_\theta = -\frac{2Y}{3}$$

Therefore, for the case of a thick-walled hollow sphere subject to internal and external pressure, we have

$$S_r = -2S_\theta \tag{7.1.28}$$

For a Prandtl-Reuss material [Equation (4.5.1)], we have

$$\frac{\Delta\varepsilon_r''}{S_r} = \frac{\Delta\varepsilon_\theta''}{S_\theta} \tag{7.1.29}$$

where $\Delta\varepsilon_r''$ and $\Delta\varepsilon_\theta''$ are the incremental plastic deviatoric strain components. From Equations (7.1.27) through (7.1.29), we see that S_r/S_θ, $\varepsilon_r/\varepsilon_\theta$, and $\varepsilon_r''/\varepsilon_\theta''$ all have the same ratio. This satisfies the elastic relations

$$S_r = 2\mu(\varepsilon_r - \varepsilon_r'') \qquad \text{and} \qquad S_\theta = 2\mu(\varepsilon_\theta - \varepsilon_\theta'')$$

From the equilibrium condition, Equation (7.1.5) in the plastic region is

$$\frac{d\sigma_r}{dr} - \frac{2Y}{r} = 0 \tag{7.1.30}$$

$$\sigma_r = 2Y \ln r + c$$

By the boundary condition $\sigma_r = -p_i$ at $r = a$, we obtain

$$c = -p_i - 2Y \ln a$$

Then

$$\sigma_r = -p_i + 2Y \ln \frac{r}{a} \qquad a \le r \le \rho \tag{7.1.31}$$

and

$$\sigma_\theta = Y + 2Y \ln \frac{r}{a} - p_i \qquad a \le r \le \rho$$

where $r = \rho$ is the interface between the elastic and plastic regions. At $r = \rho$, we have

$$\sigma_r = -p_i + 2Y \ln \frac{\rho}{a} \qquad \sigma_\theta = -p_i + 2Y \ln \frac{\rho}{a} + Y$$

For the elastic part, the constants in Equations (7.1.16) and (7.1.17), with $e_r'' = e_\theta'' = e'' = 0$, are now determined by the following new boundary conditions:

$$\sigma_r = -p_0 \qquad at\ r = b$$

$$\sigma_\theta - \sigma_r = Y \qquad \text{at } r = \rho$$

From this and Equations (7.1.16) and (7.1.17), the new constants are

$$\frac{c_1}{3} = \frac{\lambda + 2\mu}{3\lambda + 2\mu}\left(\frac{2\rho^3}{3b^3} Y - p_0\right) \qquad c_2 = \frac{(\lambda + 2\mu)\rho^3 Y}{6\mu}$$

The stresses and displacement for $r > \rho$, that is, in the elastic region, are then

$$\begin{aligned} \sigma_r &= \frac{2Y}{3}\frac{\rho^3}{b^3}\left(1 - \frac{b^3}{r^3}\right) - p_0 \qquad && \rho \le r \le b \\ \sigma_\theta &= \frac{2Y}{3}\frac{\rho^3}{b^3}\left(1 + \frac{b^3}{2r^3}\right) - p_0 \qquad && \rho \le r \le b \end{aligned} \tag{7.1.32}$$

$$u = \frac{2Y\rho^3}{3b^3}\left(\frac{r}{3\lambda + 2\mu} + \frac{b^3}{4\mu}\frac{1}{r^2}\right) - \frac{p_0 r}{3\lambda + 2\mu} \qquad \rho \le r \le b \tag{7.1.33}$$

Equating σ_r at $r = \rho$ of Equation (7.1.32) to that of Equation (7.1.31) yields

$$p_i - p_0 = \frac{2Y}{3}\left(\frac{b^3 - \rho^3}{b^3}\right) + 2Y \ln \frac{\rho}{a} \tag{7.1.34}$$

This gives the relation between $p_i - p_o$ and ρ. The minimum pressure $(p_i - p_o)^{**}$ for the shell to be completely plastic is found by letting $\rho = b$, giving

$$(p_i - p_o)^{**} = 2Y \ln \frac{b}{a} \tag{7.1.35}$$

The stress distribution in an elastoplastic spherical shell is shown in Figure 7.4.

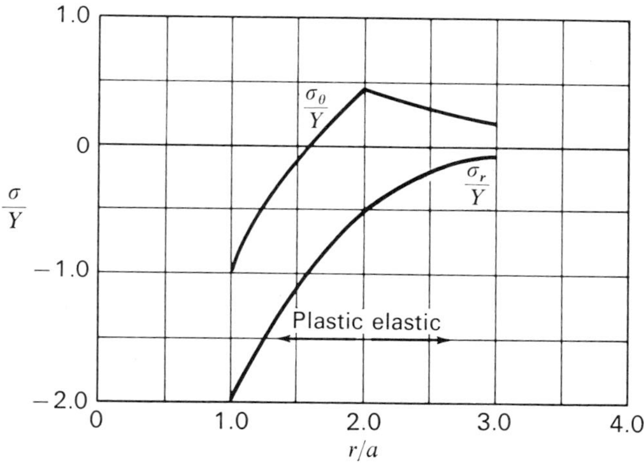

Figure 7.4. Stress distribution in elastic-plastic spherical shell ($b/a = 3$, $\rho/a = 2$, $p_o = Y/9$, $p_i/p_o = 17.7$).

Hollow Spheres of Strain-Hardening Materials. Consider strain-hardening spheres loaded beyond the elastic limit. Since plastic strain produces no dilatation, $e'' = 0$. Following the same reasoning as before, the conditions given by Equations (7.1.27) through (7.1.29) hold also for the present case; that is, $e_r'' = \varepsilon_r'' = -2\varepsilon_\theta'' = -2e_\theta''$, and Equations (7.1.12) give

$$\begin{aligned} I_1 &= 0, \\ I_2 &= \frac{4\mu}{b^3} \int_a^b r^2 e_r'' \, dr \\ I_3 &= \frac{8\mu}{b^3} \int_a^b r^2 \int_a^r \frac{3e_r''}{2r} \, dr \, dr \\ I_4 &= 4\mu \int_a^b \frac{3e_r''}{2r} \, dr \end{aligned} \tag{7.1.36}$$

From Equations (7.1.16), (7.1.17), and (7.1.25), we have

$$\sigma^* = \frac{12\mu^2}{(\lambda+2\mu)}\frac{1}{r^3}\int_a^r r^2 e_r''\,dr + \frac{24\mu^2}{(\lambda+2\mu)}\frac{1}{r^3}\int_a^r r^2\int_a^r \frac{e_r''-e_\theta''}{r}\,dr\,dr + \frac{4\mu\lambda}{\lambda+2\mu}\int_a^r \frac{e_r''-e_\theta''}{r}\,dr + \frac{6\mu}{\lambda+2\mu}\frac{c_2}{r^3} - \frac{(3\lambda+2\mu)\mu}{\lambda+2\mu}e_\theta'' \tag{7.1.37}$$

From Equations (4.5.7) and (4.5.8), noting that $e_r'' = -2e_\theta''$, and denoting the initial yielding stress in tension by Y_0, we have

$$e''^* = \begin{cases} \frac{2}{3}(e_\theta'' - e_r'') = 2e_\theta'' = f(\sigma^*) & \text{when } \sigma^* \geqq Y_0 \\ 0 & \text{when } \sigma^* \leqq Y_0 \end{cases} \tag{7.1.38}$$

The constant c_2 is given by Equation (7.1.14) and depends on $(p_i - p_0)$, and the e_r'''s and e_θ'''s along the radius. Since $e_r'' = -2e_\theta''$, we may express σ^* [given by Equation (7.1.37)] in terms of $(p_i - p_0)$ and the e_θ'''s at different values of r. Let $r = \rho$ denote the interface between the elastic and plastic regions. At $r = \rho$, we have $\sigma^* = Y_0$ and for $r > \rho$, we have $\sigma^* < Y_0$. This condition, together with Equations (7.1.37) and (7.1.38), gives the relation between the e_θ'''s at different points under a given p_i and p_0 pressure. Writing Equation (7.1.38) in incremental form to facilitate numerical calculation, we obtain

$$2\Delta e_\theta'' = f'(\sigma^*)\,\Delta\sigma^* \qquad \text{for } r \leq \rho \tag{7.1.39}$$

where $f'(\sigma^*)$ is the derivative of $f(\sigma^*)$ with respect to its argument. The derivative $f'(\sigma^*)$ is taken to be constant for the increment $\Delta\sigma^*$. From Equation (7.1.25), we have $\Delta\sigma^* = \Delta\sigma_\theta - \Delta\sigma_r$, and from Equation (7.1.37), we have

$$\Delta\sigma^* = \frac{-24\mu^2}{(\lambda+2\mu)}\frac{1}{r^3}\int_a^r r^2\,\Delta e_\theta''\,dr - \frac{72\mu^2}{\lambda+2\mu}\frac{1}{r^3}\int_a^r r^2\int_a^r \frac{\Delta e_\theta''}{r}\,dr\,dr - \frac{12\mu\lambda}{\lambda+2\mu}\int_a^r \frac{\Delta e_\theta''}{r}\,dr + \frac{6\mu}{\lambda+2\mu}\frac{\Delta c_2}{r^3} - \frac{(3\lambda+2\mu)\mu}{\lambda+2\mu}\Delta e_\theta'' \tag{7.1.40}$$

where, from Equation (7.1.14), we have Δc_2 given by

$$\Delta c_2 = \frac{a^3b^3(\lambda+2\mu)}{(b^3-a^3)4\mu}\Delta(p_i - p_0) - \frac{4a^3\mu}{(b^3-a^3)}\left[\int_a^b r^2\,\Delta e_\theta''\,dr - \frac{3(\lambda+2\mu)b^3}{4\mu}\int_a^b \frac{\Delta e_\theta''}{r}\,dr + 3\int_a^b r^2\int_a^r \frac{\Delta e_\theta''}{r}\,dr\right] \tag{7.1.41}$$

Divide $(\rho - a)$ into N equal segments. The value ρ may then be determined by trial and error. Let e''_{θ_n} be the e_θ'' at the center of the nth segment. The integrals in Equations (7.1.40) and (7.1.41) may be written as linear combinations of the $\Delta e''_{\theta_n}$'s. There are N unknown $\Delta e''_{\theta_n}$'s. Applying Equation (7.1.39) to the center point of each segment, we have N linear equations to solve for N unknowns. After $\Delta e_\theta''$, and hence e_θ'', of each point is known, σ^* at the end of this load increment $\Delta(p_i - p_0'')$ is obtained. Using this σ^* to evaluate $f'(\sigma^*)$ for the next load increment, the process is repeated until the final pressure is reached. This gives the values of the e_r'''s, e_θ'''s, and σ^*'s at different values of r for different loadings. Since, from Equations (7.1.5) and (7.1.25), we have

$$2\frac{\sigma^*}{r} = 2\frac{\sigma_\theta - \sigma_r}{r} = \frac{d\sigma_r}{dr}$$

the σ_r's may be obtained by numerical integration along r from the inner surface at which $\sigma_r = -p$. This gives a numerical method of calculating the stresses σ_r and σ_θ at different points of a hollow sphere of a strain-hardening material subject to internal and external pressure.

Residual Stresses. If after a portion of the sphere is stressed in the plastic range, either for perfectly plastic or strain-hardening materials, the sphere is unloaded, residual stresses remain. If no additional plastic strain has occurred during this unloading process, the residual stress can be calculated by superposing the stresses from an elastic analysis caused by a negative pressure $-(p_i - p_0)$ on the stresses due to $(p_i - p_0)$ obtained in the elasto-plastic solution.

The discussions in this section were for cases in which the strains are small, and hence in which the change in internal radius is not considered in calculating the stresses. For strains that are large, see Reference 3.

7.2 Inelastic Thick Circular Cylinders[1–3]

Consider a long, thick circular cylinder of inner radius "a" and outer radius "b" subject to a uniform internal pressure p_i and an outer pressure p_0. Let r, θ, z be a set of cylindrical coordinates, as shown in Figure 7.5, and u, v, and w, respectively, be the displacements along these three axes. Displacement w is assumed to be zero throughout; that is,

$$e_z = 0 \tag{7.2.1}$$

Because of the axial symmetry of structure and loading, v is zero, and the three principal stresses are σ_r, σ_θ, and σ_z. Hence the shear stresses and shear strains on these planes are zero. As in the previous section,

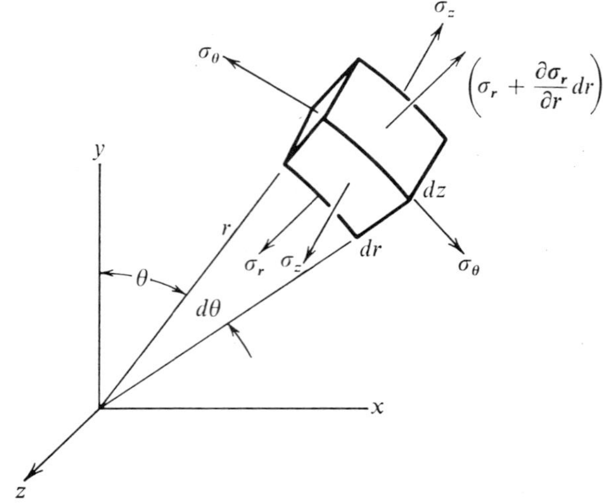

Figure 7.5. Stresses acting on a differential element of the cylinder.

$$e_r = \frac{du}{dr} \qquad \text{and} \qquad e_\theta = \frac{u}{r} \tag{7.2.2}$$

$$e = e_r + e_\theta = \frac{du}{dr} + \frac{u}{r} \tag{7.2.3}$$

The stress-strain relations are

$$\begin{aligned} \sigma_r &= \lambda(e - e'') + 2\mu(e_r - e_r'') \\ \sigma_\theta &= \lambda(e - e'') + 2\mu(e_\theta - e_\theta'') \\ \sigma_z &= \lambda(e - e'') + 2\mu(e_z - e_z'') \end{aligned} \tag{7.2.4}$$

where

$$e'' = e_r'' + e_\theta'' + e_z''$$

Substituting Equations (7.2.2) and (7.2.3) into Equation (7.2.4) yields

$$\begin{aligned} \sigma_r &= (\lambda + 2\mu)\left(\frac{du}{dr} - e_r''\right) + \lambda\left(\frac{u}{r} - e_\theta''\right) - \lambda e_z'' \\ \sigma_\theta &= (\lambda + 2\mu)\left(\frac{u}{r} - e_\theta''\right) + \lambda\left(\frac{du}{dr} - e_r''\right) - \lambda e_z'' \\ \sigma_z &= -(\lambda + 2\mu)e_z'' + \lambda\left(\frac{du}{dr} - e_r''\right) + \lambda\left(\frac{u}{r} - e_\theta''\right) \end{aligned} \tag{7.2.5}$$

Condition of Equilibrium. Referring to Figure 7.5, and equating the sum of the radial forces acting on this differential element of the cylinder to zero, we obtain the governing equation of equilibrium,

$$\frac{d\sigma_r}{dr} + \frac{\sigma_r - \sigma_\theta}{r} = 0 \tag{7.2.6}$$

Substituting Equation (7.2.5) into Equation (7.2.6) yields

$$(\lambda + 2\mu)\frac{d}{dr}\left[\frac{1}{r}\frac{d(ru)}{dr}\right] = \lambda\frac{d}{dr}(e_r'' + e_\theta'' + e_z'') + 2\mu\frac{d}{dr}e_r'' + \frac{2\mu}{r}(e_r'' - e_\theta'') \tag{7.2.7}$$

Integrating with respect to r yields

$$(\lambda + 2\mu)\frac{d(ru)}{dr} = \lambda(e_r'' + e_\theta'' + e_z'')r + 2\mu e_r'' r + 2\mu r\int_a^r \frac{e_r'' - e_\theta''}{r}\,dr + c_1 r$$

Integrating with respect to r again, we obtain

$$(\lambda + 2\mu)u = \frac{\lambda}{r}\int_a^r (e_r'' + e_\theta'' + e_z'')r\,dr + \frac{2\mu}{r}\int_a^r re_r''\,dr + \frac{2\mu}{r}\int_a^r r\int_a^r \frac{e_r'' - e_\theta''}{r}\,dr\,dr + \frac{c_1 r}{2} + \frac{c_2}{r} \tag{7.2.8}$$

Now differentiating Equation (7.2.8) with respect to r gives

$$(\lambda + 2\mu)\frac{du}{dr} = -\frac{\lambda}{r^2}\int_a^r (e_r'' + e_\theta'' + e_z'')r\,dr + \lambda(e_r'' + e_\theta'' + e_z'') - \frac{2\mu}{r^2}\int_a^r re_r''\,dr + 2\mu e_r'' - \frac{2\mu}{r^2}\int_a^r r\int_a^r \frac{e_r'' - e_\theta''}{r}\,dr\,dr + 2\mu\int_a^r \frac{e_r'' - e_\theta''}{r}\,dr + \frac{c_1}{2} - \frac{c_2}{r^2}$$

From Equation (7.2.5), we have

$$\sigma_r = -\frac{2\mu\lambda}{(\lambda + 2\mu)}\frac{1}{r^2}\int_a^r (e_r'' + e_\theta'' + e_z'')r\,dr - \frac{4\mu^2}{\lambda + 2\mu}\frac{1}{r^2}\int_a^r re_r''\,dr - \frac{4\mu^2}{(\lambda + 2\mu)r^2}\int_a^r r\int_a^r \frac{e_r'' - e_\theta''}{r}\,dr\,dr + 2\mu\int_a^r \frac{e_r'' - e_\theta''}{r}\,dr + \left(\frac{\lambda + \mu}{\lambda + 2\mu}\right)c_1 - \frac{2\mu c_2}{(\lambda + 2\mu)r^2} \tag{7.2.9}$$

From the boundary conditions, $\sigma_r = -p_i$ at $r = a$ and $\sigma_r = -p_0$ at $r = b$, we obtain, from Equation (7.2.9), the following two equations:

$$-p_i = \frac{\lambda + \mu}{\lambda + 2\mu} c_1 - \frac{2\mu}{(\lambda + 2\mu)a^2} c_2 \tag{7.2.10}$$

$$-p_0 = -I_1 - I_2 - I_3 + I_4 + \frac{\lambda + \mu}{\lambda + 2\mu} c_1 - \frac{2\mu c_2}{(\lambda + 2\mu)b^2} \tag{7.2.11}$$

where

$$\begin{aligned}
I_1 &= \frac{2\mu\lambda}{(\lambda + 2\mu)b^2} \int_a^b e''r\,dr \\
I_2 &= \frac{4\mu^2}{(\lambda + 2\mu)b^2} \int_a^b re_r''\,dr \\
I_3 &= \frac{4\mu^2}{(\lambda + 2\mu)b^2} \int_a^b r \int_a^r \frac{e_r'' - e_\theta''}{r}\,dr\,dr \\
I_4 &= 2\mu \int_a^b \frac{e_r'' - e_\theta''}{r}\,dr
\end{aligned} \tag{7.2.12}$$

The solution for the constants c_1 and c_2 from the above two equations is

$$c_1 = \frac{\lambda + 2\mu}{\lambda + \mu} \left[\frac{a^2 p_i - b^2 p_0 + b^2(I_1 + I_2 + I_3 - I_4)}{b^2 - a^2} \right] \tag{7.2.13}$$

and

$$c_2 = (p_i - p_0 + I_1 + I_2 + I_3 - I_4) \frac{\lambda + 2\mu}{2\mu} \frac{a^2 b^2}{(b^2 - a^2)} \tag{7.2.14}$$

Substituting Equations (7.2.13) and (7.2.14) into Equation (7.2.9) yields

$$\begin{aligned}
\sigma_r = &-\frac{2\mu\lambda}{\lambda + 2\mu} \frac{1}{r^2} \int_a^r e''r\,dr - \frac{4\mu^2}{\lambda + 2\mu} \frac{1}{r^2} \int_a^r re_r''\,dr \\
&- \frac{4\mu^2}{(\lambda + 2\mu)r^2} \int_a^r r \int_a^r \frac{e_r'' - e_\theta''}{r}\,dr\,dr + 2\mu \int_a^r \frac{e_r'' - e_\theta''}{r}\,dr \\
&+ \frac{1}{b^2 - a^2} [a^2(p_i - p_0) + b^2(I_1 + I_2 + I_3 - I_4)] - p_0 \\
&- (p_i - p_0 + I_1 + I_2 + I_3 - I_4) \frac{a^2 b^2}{(b^2 - a^2)r^2}
\end{aligned} \tag{7.2.15}$$

Similarly, σ_θ and σ_z, from Equations (7.2.5), are

$$\sigma_\theta = \frac{2\mu\lambda}{(\lambda + 2\mu)r^2}\int_a^r e''r\,dr + \frac{4\mu^2}{(\lambda + 2\mu)r^2}\int_a^r re_r''\,dr$$

$$+ \frac{4\mu^2}{(\lambda + 2\mu)r^2}\int_a^r r\int_a^r \frac{e_r'' - e_\theta''}{r}\,dr\,dr + \frac{2\mu\lambda}{\lambda + 2\mu}\int_a^r \frac{e_r'' - e_\theta''}{r}\,dr$$

$$- \frac{2\mu\lambda}{\lambda + 2\mu}(e_\theta'' + e_z'') - 2\mu e_\theta''$$

$$+ \frac{1}{b^2 - a^2}[a^2(p_i - p_0) + b^2(I_1 + I_2 + I_3 - I_4)]$$

$$- p_0 + (p_i - p_0 + I_1 + I_2 + I_3 - I_4)\frac{a^2b^2}{(b^2 - a^2)r^2} \tag{7.2.16}$$

$$\sigma_z = \frac{2\mu\lambda}{\lambda + 2\mu}\int_a^r \frac{e_r'' - e_\theta''}{r}\,dr + \frac{\lambda}{\lambda + \mu}\cdot\frac{1}{b^2 - a^2}[a^2p_i - b^2p_0$$

$$+ b^2(I_1 + I_2 + I_3 - I_4)] - \frac{2\mu\lambda}{\lambda + 2\mu}(e_\theta'' + e_z'') - 2\mu e_z''$$

and, from Equation (7.2.8), we have

$$u = \frac{\lambda}{(\lambda + 2\mu)}\frac{1}{r}\int_a^r e''r\,dr + \frac{2\mu}{(\lambda + 2\mu)}\frac{1}{r}\int_a^r re_r''\,dr$$

$$+ \frac{2\mu}{(\lambda + 2\mu)}\frac{1}{r}\int_a^r r\int_a^r \frac{e_r'' - e_\theta''}{r}\,dr\,dr$$

$$+ \frac{r}{2(\lambda + \mu)}\left\{\frac{1}{(b^2 - a^2)}[a^2(p_i - p_0) + b^2(I_1 + I_2 + I_3 - I_4)] - p_0\right\}$$

$$+ (p_i - p_o + I_1 + I_2 + I_3 - I_4)\frac{1}{2\mu}\frac{a^2b^2}{(b^2 - a^2)r} \tag{7.2.17}$$

The above expressions give the relations between stresses, displacements, and the distribution of inelastic strains for thick circular cylinders under uniform pressure.

Elastic Cylinder. When the entire cylinder is stressed within the elastic limit, we have

$$e_r'' = e_\theta'' = e_z'' = e'' = 0 \qquad I_1 = I_2 = I_3 = I_4 = 0$$

and the above equations reduce to

$$\sigma_r = \frac{a^2}{b^2 - a^2}(p_i - p_o) - \frac{a^2 b^2 (p_i - p_o)}{(b^2 - a^2) r^2} - p_o = \frac{(p_i - p_o) a^2}{(b^2 - a^2)}\left(1 - \frac{b^2}{r^2}\right) - p_o \tag{7.2.18}$$

$$\sigma_\theta = \frac{a^2 (p_i - p_o)}{b^2 - a^2} + \frac{(p_i - p_o) a^2 b^2}{(b^2 - a^2) r^2} - p_o = \frac{(p_i - p_o) a^2}{b^2 - a^2}\left(1 + \frac{b^2}{r^2}\right) - p_o \tag{7.2.19}$$

$$\sigma_z = \nu(\sigma_r + \sigma_\theta) = 2\nu\left[\frac{(p_i - p_o) a^2}{b^2 - a^2} - p_o\right] \tag{7.2.20}$$

$$u = \frac{a^2 r}{2(b^2 - a^2)} \frac{(p_i - p_o)}{(\lambda + \mu)} - \frac{p_o r}{2(\lambda + \mu)} + \frac{p_i - p_o}{2\mu} \frac{a^2 b^2}{(b^2 - a^2) r} \tag{7.2.21}$$

This gives the elastic analysis of circular cylinders.

Cylinder with Radial Thermal Gradients. When the cylinder is subject to radial thermal gradients only, $e_z'' = e_\theta'' = e_z'' = \alpha T$, where α and T are as defined in Section 7.1. Then, from Equations (7.2.12), we have

$$I_1 = \frac{6\mu\lambda}{\lambda + 2\mu} \frac{1}{b^2} \int_a^b \alpha T r \, dr$$

$$I_2 = \frac{4\mu^2}{\lambda + 2\mu} \frac{1}{b^2} \int_a^b \alpha T r \, dr$$

$$I_3 = I_4 = 0$$

$$I_1 + I_2 = \frac{2\mu(2\mu + 3\lambda)}{\lambda + 2\mu} \frac{1}{b^2} \int_a^b \alpha T r \, dr$$

Substitution of the above expressions for I_1, I_2, I_3, and I_4 into Equation (7.2.15) gives

$$\sigma_r = \frac{a^2}{b^2 - a^2}(p_i - p_o)\left(1 - \frac{b^2}{r^2}\right) - p_o - \frac{2\mu(3\lambda + 2\mu)}{\lambda + 2\mu} \frac{1}{r^2} \int_a^r \alpha T r \, dr + \frac{2\mu(2\mu + 3\lambda)}{(\lambda + 2\mu)(b^2 - a^2)}\left(1 - \frac{a^2}{r^2}\right) \int_a^b \alpha T r \, dr$$

From the relations between the elastic constants in Equation (7.1.21), we obtain

$$\sigma_r = \frac{a^2}{b^2 - a^2}(p_i - p_o)\left(1 - \frac{b^2}{r^2}\right) - p_o - \frac{E}{1-\nu}\frac{1}{r^2}\int_a^r \alpha T r\, dr$$
$$+ \frac{E}{(1-\nu)(b^2 - a^2)}\left(1 - \frac{a^2}{r^2}\right)\int_a^b \alpha T r\, dr \qquad (7.2.22)$$

Similarly, we have

$$\sigma_\theta = \frac{(p_i - p_o)a^2}{b^2 - a^2}\left(1 + \frac{b^2}{r^2}\right) - p_o + \frac{E}{(1-\nu)}\left[\frac{1}{(b^2 - a^2)}\left(1 + \frac{a^2}{r^2}\right)\int_a^b \alpha T r\, dr\right.$$
$$\left. + \frac{1}{r^2}\int_a^r \alpha T r\, dr\right] - \frac{E}{1-\nu}\alpha T \qquad (7.2.23)$$
$$\sigma_z = -\frac{E\alpha T}{1-\nu} + \frac{2\nu}{b^2 - a^2}\left(a^2 p_i - b^2 p_o + \frac{E}{1-\nu}\int_a^r \alpha T r\, dr\right)$$

This gives the thermal stresses in a circular cylinder under a radial temperature distribution.

Cylinders of Perfectly Plastic Materials. Equations (7.2.15) and (7.2.16) hold for cylinders of ideally plastic materials, with no thermal gradients. Since plastic deformation here is taken to have no dilatation, $e'' = 0$, and I_1 in Equation (7.2.12) vanishes. We assume the material to follow Tresca's yield criterion. From Equations (7.2.18) through (7.2.20), we have $|\sigma_\theta - \sigma_r| > |\sigma_\theta - \sigma_z|$, so the yield condition is given by $\sigma_\theta - \sigma_r = Y$. From this and Equation (7.2.6), we obtain

$$\frac{d\sigma_r}{dr} = \frac{Y}{r} \qquad (7.2.24)$$

In the plastic region,

$$\sigma_r = Y \ln r + c_3 \qquad a \le r \le \rho \qquad (7.2.25)$$

where $r = \rho$ denotes the interface between the elastic and plastic regions. The boundary condition, $\sigma_r = -p_i$ at $r = a$, determines the constant c_3,

$$-p_i = Y \ln a + c_3$$
$$c_3 = -p_i - Y \ln a$$

Therefore, from Equation (7.2.25), we have

$$\sigma_r = -p_i + Y \ln \frac{r}{a} \qquad a \le r \le \rho \qquad (7.2.26)$$

In the elastic region, from Equations (7.2.8) and (7.2.9), we have

$$(\lambda + 2\mu)u = \frac{c_1 r}{2} + \frac{c_2}{r}$$

$$\sigma_r = \frac{\lambda + \mu}{\lambda + 2\mu} c_1 - \frac{2\mu c_2}{(\lambda + 2\mu)r^2}$$

$$\sigma_\theta = \frac{(\lambda + \mu)}{\lambda + 2\mu} c_1 + \frac{2\mu c_2}{(\lambda + 2\mu)r^2}$$

$$\sigma_\theta - \sigma_r = \frac{4\mu c_2}{(\lambda + 2\mu)r^2}$$

Equating the elastic solutions to the plastic solutions at the elastic-plastic interface, $r = \rho$, gives

$$\frac{\lambda + u}{\lambda + 2\mu} c_1 - \frac{2\mu c_2}{(\lambda + 2\mu)\rho^2} = -p_i + Y \ln \frac{\rho}{a}$$

$$\frac{4\mu c_2}{(\lambda + 2\mu)\rho^2} = Y \qquad c_2 = \frac{\lambda + 2\mu}{4\mu} \rho^2 Y$$

$$\frac{\lambda + \mu}{\lambda + 2\mu} c_1 - \frac{Y}{2} = -p_i + Y \ln \frac{\rho}{a} \qquad c_1 = \frac{\lambda + 2\mu}{\lambda + \mu} \left(\frac{Y}{2} + Y \ln \frac{\rho}{a} - p_i \right)$$

In the elastic region, $\rho \le r \le b$, we have, using the values of c_1 and c_2 determined above,

$$\sigma_r = \left(\frac{Y}{2} + Y \ln \frac{\rho}{a} - p_i \right) - \frac{Y}{2} \frac{\rho^2}{r^2}$$

$$\sigma_\theta = \left(\frac{Y}{2} + Y \ln \frac{\rho}{a} - p_i \right) + \frac{Y}{2} \frac{\rho^2}{r^2}$$

$$\sigma_z = \frac{\lambda}{\lambda + \mu} \left(\frac{Y}{2} + Y \ln \frac{\rho}{a} - p_i \right)$$

At $r = b$, we have $\sigma_r = -p_0$. This is the condition to determine ρ. In the plastic region $a \le r \le \rho$, we have

$$\sigma_r = -p_i + Y \ln \frac{r}{a}$$

$$\sigma_\theta = Y - p_i + Y \ln \frac{r}{a}$$

For a cylinder of strain-hardening materials stressed beyond the elastic limit, a numerical solution of Equations (7.2.15) and (7.2.16) similar to the one used in the treatment of hollow spheres of strain-hardening materials discussed at the end of the last section may be applied. The detail solution of a cylinder of perfectly plastic material with the Von Mises criterion of yielding has been discussed in detail by Prager and Hodge,[5] and is not repeated here.

7.3 Inelastic Rotating Disks of Uniform Thickness

For a uniformly thin, flat circular disk of inner radius "a" and outer radius "b" rotating about its polar axis at an angular velocity ω, the forces acting on a differential element of the disk, including the inertia force, are shown in Figure 7.6, where cylindrical coordinates are again used. Equating

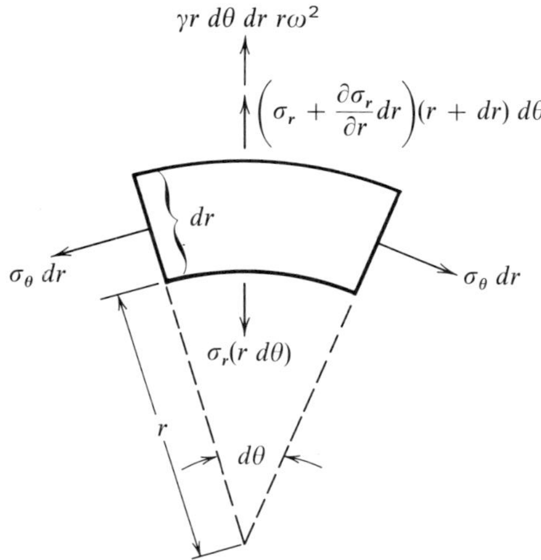

Figure 7.6. Stresses acting on an element of the disk.

the sum of radial forces acting on the element to zero yields the *equilibrium equation*

$$\frac{d\sigma_r}{dr} - \frac{\sigma_\theta - \sigma_r}{r} + \gamma\omega^2 r = 0 \tag{7.3.1}$$

where γ is the density of the material, and higher-order terms are neglected.

Since the disk is thin, the stresses on planes parallel to the plane of the disk are taken to be zero; the stress-strain relations of the disk are

$$\sigma_r = \lambda(e - e'') + 2\mu(e_r - e_r'')$$
$$\sigma_\theta = \lambda(e - e'') + 2\mu(e_\theta - e_\theta'')$$
$$\sigma_z = \lambda(e - e'') + 2\mu(e_z - e_z'') = 0$$

Eliminating $(e_z - e_z'')$ from the above, we find

$$\sigma_r = \frac{4\mu(\lambda + \mu)}{\lambda + 2\mu}(e_r - e_r'') + \frac{2\mu\lambda}{\lambda + 2\mu}(e_\theta - e_\theta'')$$

$$\sigma_\theta = \frac{4\mu(\lambda + \mu)}{\lambda + 2\mu}(e_\theta - e_\theta'') + \frac{2\mu\lambda}{\lambda + 2\mu}(e_r - e_r'')$$

In terms of Young's modulus E and Poisson's ratio v, the above equations are

$$\begin{aligned}\sigma_\theta &= \frac{E}{1 - v^2}[(e_\theta - e_\theta'') + v(e_r - e_r'')] \\ \sigma_r &= \frac{E}{1 - v^2}[(e_r - e_r'') + v(e_\theta - e_\theta'')]\end{aligned} \tag{7.3.2}$$

From Equation (7.2.2), we have $e_r = du/dr$, $e_\theta = u/r$, and Equations (7.3.2) become

$$\begin{aligned}\sigma_\theta &= \frac{E}{1 - v^2}\left[\frac{u}{r} + v\frac{du}{dr} - e_\theta'' - ve_r''\right] \\ \sigma_r &= \frac{E}{1 - v^2}\left[\frac{du}{dr} + v\frac{u}{r} - e_r'' - ve_\theta''\right]\end{aligned} \tag{7.3.3}$$

Substituting Equations (7.3.3) into (7.3.1) yields

$$\frac{d}{dr}\left(\frac{1}{r}\frac{d(ru)}{dr}\right) = -\frac{1 - v^2}{E}\gamma\omega^2 r + \frac{d}{dr}(e_r'' + ve_\theta'') - (1 - v)\frac{e_\theta'' - e_r''}{r} \tag{7.3.4}$$

Integrating with respect to "r" twice, we obtain

$$\begin{aligned}u = &-\frac{1 - v^2}{E}\gamma\omega^2\frac{r^3}{8} + \frac{1}{r}\int_a^r (e_r'' + ve_\theta'')r\,dr \\ &- (1 - v)\frac{1}{r}\int_a^r r\int_a^r \frac{e_\theta'' - e_r''}{r}\,dr\,dr + c_1\frac{r}{2} + \frac{c_2}{r}\end{aligned} \tag{7.3.5}$$

Differentiating u with respect to r yields

$$\frac{du}{dr} = -\frac{1-v^2}{E}\frac{3}{8}\gamma\omega^2 r^2 - \frac{1}{r^2}\int_a^r (e_r'' + ve_\theta'')r\,dr + e_r'' + ve_\theta''$$

$$+ (1-v)\frac{1}{r^2}\int_a^r r\int_a^r \frac{e_\theta'' - e_r''}{r}\,dr\,dr - (1-v)\int_a^r \frac{e_\theta'' - e_r''}{r}\,dr + \frac{c_1}{2} - \frac{c_2}{r^2}$$

From Equation (7.3.3), we have

$$\sigma_\theta = \frac{E}{1-v^2}\left[-\frac{1-v^2}{E}\gamma\omega^2\left(\frac{1}{8}+\frac{3v}{8}\right)r^2 + \frac{(1-v)}{r^2}\int_a^r (e_r'' + ve_\theta'')r\,dr\right.$$

$$-\frac{(1-v)^2}{r^2}\int_a^r r\int_a^r \frac{e_\theta'' - e_r''}{r}\,dr\,dr + \frac{c_1}{2}(1+v) + \frac{c_2}{r^2}(1-v)$$

$$\left.- v(1-v)\int_a^r \frac{e_\theta'' - e_r''}{r}\,dr - (1-v^2)e_\theta''\right] \tag{7.3.6}$$

$$\sigma_r = \frac{E}{1-v^2}\left[-\frac{1-v^2}{E}\gamma\omega^2 r^2\left(\frac{3}{8}+\frac{v}{8}\right) - \frac{1-v}{r^2}\int_a^r (e_r'' + ve_\theta'')r\,dr\right.$$

$$+\frac{(1-v)^2}{r^2}\int_a^r r\int_a^r \frac{e_\theta'' - e_r''}{r}\,dr\,dr - (1-v)\int_a^r \frac{e_\theta'' - e_r''}{r}\,dr$$

$$\left.+\frac{c_1}{2}(1+v) - \frac{c_2}{r^2}(1-v)\right]$$

$$= -\gamma\omega^2 r^2\frac{3+v}{8} - \frac{E}{(1+v)}\frac{1}{r^2}\int_a^r (e_r'' + ve_\theta'')r\,dr$$

$$+\frac{E(1-v)}{(1+v)}\frac{1}{r^2}\int_a^r r\int_a^r \frac{e_\theta'' - e_r''}{r}\,dr\,dr - \frac{E}{1+v}\int_a^r \frac{e_\theta'' - e_r''}{r}\,dr$$

$$+\frac{E}{2(1-v)}c_1 - \frac{E}{1+v}\frac{c_2}{r^2} \tag{7.3.7}$$

The disk is taken to be subject to no forces at the inner and outer radii. From these boundary conditions, $\sigma_r = 0$ at $r = a, b$, constants c_1 and c_2 may be determined:

$$0 = \frac{E}{2(1-v)}c_1 - \frac{E}{1+v}\frac{c_2}{a^2} - \gamma\omega^2 a^2\frac{3+v}{8}$$

$$0 = -\gamma\omega^2 b^2\frac{3+v}{8} - I_1 + I_2 - I_3 + \frac{E}{2(1-v)}c_1 - \frac{E}{1+v}\frac{c_2}{b^2}$$

Substracting the first equation from the second gives

$$\frac{E}{1+\nu}\left(\frac{1}{a^2}-\frac{1}{b^2}\right)c_2+\gamma\omega^2\frac{3+\nu}{8}(a^2-b^2)-I_1+I_2-I_3=0$$

where

$$I_1=\frac{E}{1+\nu}\frac{1}{b^2}\int_a^b(e_r''+\nu e_\theta'')r\,dr$$

$$I_2=\frac{E(1-\nu)}{1+\nu}\frac{1}{b^2}\int_a^b r\int_a^r\frac{e_\theta''-e_r''}{r}\,dr\,dr$$

$$I_3=\frac{E}{1+\nu}\int_a^b\frac{e_\theta''-e_r''}{r}\,dr$$

$$c_2=\frac{1+\nu}{E}\frac{a^2b^2}{b^2-a^2}\left[I_1-I_2+I_3+\gamma\omega^2\frac{3+\nu}{8}(b^2-a^2)\right] \quad (7.3.8)$$

$$\frac{E}{2(1-\nu)}c_1=\gamma\omega^2a^2\frac{3+\nu}{8}+\frac{b^2}{b^2-a^2}\left[I_1-I_2+I_3+\gamma\omega^2\frac{3+\nu}{8}(b^2-a^2)\right]$$

$$=\frac{b^2}{b^2-a^2}(I_1-I_2+I_3)+\gamma\omega^2\frac{3+\nu}{8}(a^2+b^2)$$

$$c_1=\frac{2(1-\nu)}{E}\frac{b^2}{b^2-a^2}(I_1-I_2+I_3)+\frac{(1-\nu)(3+\nu)}{4E}\gamma\omega^2(a^2+b^2) \quad (7.3.9)$$

Substituting these values of c_1 and c_2 into Equations (7.3.6) and (7.3.7) gives

$$\sigma_\theta=-\gamma\omega^2\left(\frac{1+3\nu}{8}\right)r^2+\frac{E}{1+\nu}\frac{1}{r^2}\int_a^r(e_r''+\nu e_\theta'')r\,dr$$

$$-\frac{(1-\nu)E}{1+\nu}\frac{1}{r^2}\int_a^r r\int_a^r\frac{e_\theta''-e_r''}{r}\,dr\,dr-Ee_\theta''-\frac{E\nu}{1+\nu}\int_a^r\frac{e_\theta''-e_r''}{r}\,dr$$

$$+\frac{b^2}{b^2-a^2}(I_1-I_2+I_3)+\frac{3+\nu}{8}\gamma\omega^2(a^2+b^2)$$

$$+\frac{a^2b^2}{(b^2-a^2)}\left[I_1-I_2+I_3+\gamma\omega^2\frac{3+\nu}{8}(b^2-a^2)\right]\frac{1}{r^2} \quad (7.3.10)$$

or, rearranging, we have

$$\sigma_\theta = -\gamma\omega^2\left(\frac{1+3\nu}{8}\right)r^2 + \frac{E}{1+\nu}\frac{1}{r^2}\int_a^r (e_r'' + \nu e_\theta'')r\,dr$$

$$-\frac{(1-\nu)E}{1+\nu}\frac{1}{r^2}\int_a^r r\int_a^r \frac{e_\theta'' - e_r''}{r}\,dr\,dr - Ee_\theta'' - \frac{E\nu}{1+\nu}\int_a^r \frac{e_\theta'' - e_r''}{r}\,dr$$

$$+\frac{b^2}{b^2-a^2}(I_1 - I_2 + I_3)\left(1 + \frac{a^2}{r^2}\right)$$

$$+\frac{3+\nu}{8}\gamma\omega^2\left(a^2 + b^2 + \frac{a^2b^2}{r^2}\right) \tag{7.3.11}$$

and

$$\sigma_r = -\gamma\omega^2 r^2\frac{3+\nu}{8} - \frac{E}{1+\nu}\frac{1}{r^2}\int_a^r (e_r'' + \nu e_\theta'')r\,dr$$

$$+\frac{E(1-\nu)}{1+\nu}\frac{1}{r^2}\int_a^r r\int_a^r \frac{e_\theta'' - e_r''}{r}\,dr\,dr - \frac{E}{1+\nu}\int_a^r \frac{e_\theta'' - e_r''}{r}\,dr$$

$$+\frac{b^2}{b^2-a^2}(I_1 - I_2 + I_3)\left(1 - \frac{a^2}{r^2}\right) + \frac{3+\nu}{8}\gamma\omega^2\left(a^2 + b^2 - \frac{a^2b^2}{r^2}\right) \tag{7.3.12}$$

This gives the relation between stresses and inelastic strain distribution in a rotating circular disk.

Elastic Disks. For elastic disks, the above equations reduce to

$$\begin{aligned} \sigma_\theta &= \frac{3+\nu}{8}\gamma\omega^2\left(a^2 + b^2 + \frac{a^2b^2}{r^2}\right) - \gamma\omega^2\frac{1+3\nu}{8}r^2 \\ \sigma_r &= \frac{3+\nu}{8}\gamma\omega^2\left(a^2 + b^2 - \frac{a^2b^2}{r^2}\right) - \gamma\omega^2\frac{3+\nu}{8}r^2 \end{aligned} \tag{7.3.13}$$

Disks with Radial Thermal Gradients. For rotating disks with radial thermal gradients only, $e_r'' = e_\theta'' = \alpha T$, and

$$I_1 = \frac{E}{b^2}\int_a^b \alpha Tr\,dr \qquad I_2 = I_3 = 0$$

Then Equations (7.3.11) and (7.3.12) give

$$\sigma_\theta = -\gamma\omega^2\left(\frac{1+3\nu}{8}\right)r^2 + \frac{E}{r^2}\int_a^r \alpha T r\,dr - E\alpha T$$
$$+ \frac{E}{b^2-a^2}\left(1+\frac{a^2}{r^2}\right)\int_a^b \alpha T r\,dr + \frac{3+\nu}{8}\gamma\omega^2\left(a^2+b^2+\frac{a^2b^2}{r^2}\right) \quad (7.3.14)$$

and

$$\sigma_r = -\frac{3+\nu}{8}\gamma\omega^2 r^2 - \frac{E}{r^2}\int_a^r \alpha T r\,dr$$
$$+ \frac{E}{b^2-a^2}\left(1-\frac{a^2}{r^2}\right)\int_a^r \alpha T r\,dr + \frac{3+\nu}{8}\gamma\omega^2\left(a^2+b^2-\frac{a^2b^2}{r^2}\right) \quad (7.3.15)$$

This gives the relation between stresses and temperature distribution in a rotating circular disk.

Rotating Disks of Perfectly Plastic Materials. Consider a rotating disk of a perfectly plastic material with no thermal gradients but stressed beyond the elastic limit. From Equation (7.3.13), we have

$$\sigma_r = \frac{3+\nu}{8}\gamma\omega^2\frac{(r^2-a^2)(b^2-r^2)}{r^2} \geq 0$$

hence

$$\sigma_\theta \geq \sigma_r \geq \sigma_z = 0 \quad (7.3.16)$$

Using Tresca's criterion of yielding, from Equation (7.1.25), we have

$$\sigma_\theta - \sigma_z = \sigma_\theta = Y \quad (7.3.17)$$

in the plastic region. Substituting this into Equation (7.3.1), we find

$$r\frac{d\sigma_r}{dr} - (Y - \sigma_r) + \gamma\omega^2 r^2 = 0$$

or

$$\frac{d}{dr}(r\sigma_r) = Y - \gamma\omega^2 r^2$$

Integration of the above equilibrium equation yields

$$r\sigma_r = Yr - \frac{\gamma\omega^2 r^3}{3} + c_3$$

$$\sigma_r = Y - \frac{\gamma\omega^2 r^2}{3} + \frac{c_3}{r}$$

The boundary condition $\sigma_r = 0$ at $r = a$ determines c_3 as

$$c_3 = -a\left(Y - \frac{\gamma\omega^2}{3}a^2\right)$$

Then

$$\sigma_r = Y\left(1 - \frac{a}{r}\right) - \frac{\gamma\omega^2}{3}\left(r^2 - \frac{a^3}{r}\right) \tag{7.3.18}$$

In the elastic region Equation (7.3.5) reduces to

$$u = -\frac{1 - v^2}{E}\gamma\omega^2\frac{r^3}{8} + c_1\frac{r}{2} + \frac{c_2}{r}$$

$$\frac{du}{dr} = -\frac{1 - v^2}{E}\gamma\omega^2\frac{3r^2}{8} + \frac{c_1}{2} - \frac{c_2}{r^2}$$

Substituting the above expressions for u and du/dr in Equations (7.3.3) yields

$$\sigma_r = \frac{E}{1 - v^2}\left[-\frac{1 - v^2}{E}\gamma\omega^2\frac{3r^2}{8} + \frac{c_1}{2} - \frac{c_2}{r^2} - \frac{1 - v^2}{E}v\gamma\omega^2\frac{r^2}{8} + v\frac{c_1}{2} + v\frac{c_2}{r^2}\right]$$

$$= -\gamma\omega^2\frac{3 + v}{8}r^2 + \frac{E}{1 - v}\frac{c_1}{2} - \frac{E}{1 + v}\frac{c_2}{r^2}$$

$$\sigma_\theta = \frac{E}{1 - v^2}\left[-\frac{1 - v^2}{E}\gamma\omega^2\frac{r^2}{8} + \frac{c_1}{2} + \frac{c_2}{r^2} - \frac{1 - v^2}{E}v\gamma\omega^2\frac{3r^2}{8} + v\frac{c_1}{2} - v\frac{c_2}{r^2}\right]$$

$$= -\gamma\omega^2 r^2\frac{1 + 3v}{8} + \frac{c_1}{2}\frac{E}{1 - v} + \frac{c_2}{r^2}\frac{E}{1 + v}$$

At $r = b$, we have $\sigma_r = 0$, and the above equation for σ_r gives

$$-\gamma\omega^2\frac{3 + v}{8}b^2 + \frac{E}{(1 - v)}\frac{c_1}{2} - \frac{E}{(1 + v)}\frac{c_2}{b^2} = 0$$

At the elastic-plastic interface, $r = \rho$, $\sigma_\theta = Y$, and the above equation for σ_θ gives

$$-\gamma\omega^2\rho^2\frac{1 + 3v}{8} + \frac{c_1}{2}\frac{E}{1 - v} + \frac{c_2}{\rho^2}\frac{E}{1 + v} = Y$$

Substracting the first of the above equations from the second yields

$$\gamma\omega^2\left[\frac{3+\nu}{8}b^2 - \frac{1+3\nu}{8}\rho^2\right] + \frac{Ec_2}{1+\nu}\left[\frac{1}{\rho^2}+\frac{1}{b^2}\right] = Y$$

from which we obtain

$$\frac{Ec_2}{1+\nu} = \left[Y - \gamma\omega^2\left(\frac{3+\nu}{8}b^2 - \frac{1+3\nu}{8}\rho^2\right)\right]\frac{\rho^2 b^2}{b^2+\rho^2}$$

Then, solving for c_1, we obtain

$$\frac{Ec_1}{2(1-\nu)} = \gamma\omega^2\frac{3+\nu}{8}b^2 + \frac{\rho^2}{b^2+\rho^2}\left[Y - \gamma\omega^2\left(\frac{3+\nu}{8}b^2 - \frac{1+3\nu}{8}\rho^2\right)\right]$$

Using these values for c_1 and c_2, we see that σ_θ and σ_r become

$$\begin{aligned}
\sigma_\theta &= -\gamma\omega^2 r^2\frac{1+3\nu}{8} + \gamma\omega^2\frac{3+\nu}{8}b^2 + \frac{\rho^2}{b^2+\rho^2} \\
&\quad\times\left[Y - \gamma\omega^2\left(\frac{3+\nu}{8}b^2 - \frac{1+3\nu}{8}\rho^2\right)\right] \\
&\quad - \frac{\rho^2 b^2}{(b^2+\rho^2)r^2}\left[Y - \gamma\omega^2\left(\frac{3+\nu}{8}b^2 - \frac{1+3\nu}{8}\rho^2\right)\right] \\
&= \frac{\gamma\omega^2}{8}\left[(3+\nu)b^2 - (1+3\nu)r^2\right] \\
&\quad + \frac{\rho^2}{b^2+\rho^2}\left[Y - \gamma\omega^2\left(\frac{3+\nu}{8}b^2 - \frac{1+3\nu}{8}\rho^2\right)\right]\left(1+\frac{b^2}{r^2}\right)
\end{aligned} \tag{7.3.19}$$

$$\begin{aligned}
\sigma_r &= -\gamma\omega^2\frac{3+\nu}{8}r^2 + \gamma\omega^2\frac{3+\nu}{8}b^2 + \frac{\rho^2}{b^2+\rho^2} \\
&\quad\times\left[Y - \gamma\omega^2\left(\frac{3+\nu}{8}b^2 - \frac{1+3\nu}{8}\rho^2\right)\right] \\
&\quad - \frac{\rho^2 b^2}{(b^2+\rho^2)r^2}\left[Y - \gamma\omega^2\left(\frac{3+\nu}{8}b^2 - \frac{1+3\nu}{8}\rho^2\right)\right] \\
&= \gamma\omega^2\frac{3+\nu}{8}(b^2 - r^2) + \frac{\rho^2}{b^2+\rho^2} \\
&\quad\times\left[Y - \gamma\omega^2\left(\frac{3+\nu}{8}b^2 - \frac{1+3\nu}{8}\rho^2\right)\right]\left(1-\frac{b^2}{r^2}\right)
\end{aligned} \tag{7.3.20}$$

To find ρ, we equate σ_r from Equation (7.3.20) to that from Equation (7.3.18) at $r = \rho$, thereby obtaining

$$Y\left(1 - \frac{a}{\rho}\right) - \frac{\gamma\omega^2}{3}\left(\rho^2 - \frac{a^3}{\rho}\right) = \gamma\omega^2 \frac{3 + \nu}{8}(b^2 - \rho^2)$$
$$- \left[Y - \gamma\omega^2\left(\frac{3 + \nu}{8} b^2 - \frac{1 + 3\nu}{8} \rho^2\right)\right]\left(\frac{b^2 - \rho^2}{b^2 + \rho^2}\right) \quad (7.3.21)$$

This gives the relation between ρ and ω. When $\rho = b$, the disk is completely plastic.

7.4 Inelastic Rotating Disks of Variable Thickness

Rotating disks of variable thickness have been widely used in machines. For such disks, the equilibrium Equation (7.3.1) becomes

$$\frac{d}{dr}(rh\sigma_r) - h\sigma_\theta + \gamma\omega^2 r^2 h = 0 \quad (7.4.1)$$

where h is the variable thickness.

The stress-strain relations are given by Equations (7.3.2), and the strain displacement relations are given by Equation (7.2.2). From Equation (7.2.2), we have

$$e_r = \frac{d}{dr}(re_\theta) \quad (7.4.2)$$

Substituting this into Equation (7.3.2), we have

$$\sigma_r = \frac{E}{1 - \nu^2}\left[\frac{d}{dr}(re_\theta) - e_r'' + \nu(e_\theta - e_\theta'')\right]$$
$$\sigma_\theta = \frac{E}{1 - \nu^2}\left\{e_\theta - e_\theta'' + \nu\left[\frac{d}{dr}(re_\theta) - e_r''\right]\right\} \quad (7.4.3)$$

From this and Equation (7.4.1), we obtain

$$\frac{d}{dr}\left\{\frac{Erh}{1 - \nu^2}\left[\frac{d}{dr}(re_\theta) + \nu e_\theta - e_r'' - \nu e_\theta''\right]\right\}$$
$$- \frac{Eh}{1 - \nu^2}\left[e_\theta + \nu\frac{d}{dr}(re_\theta) - e_\theta'' - \nu e_r''\right] + \gamma\omega^2 r^2 h = 0 \quad (7.4.4)$$

At both inner boundary, $r = a$, and outer boundary, $r = b$, we have $\sigma_r = 0$. If e_r'' and e_θ'' are known, these two boundary conditions, together with the differential equation (7.4.4), are sufficient to determine the variation of e_θ with r. This can be solved numerically by the method of finite difference. For disks with known radial thermal gradients $e_r'' = e_\theta'' = \alpha T$, the distribution of e_θ and e_r, and hence of σ_r and σ_θ, can be readily calculated by solving Equation (7.4.4) with the boundary conditions.[6]

Rotating Disk with Radial Thermal Gradient and Creep. If the disk is subject to both thermal gradients and creep, the stress distribution due to thermal gradients alone is first found. This is the stress distribution at the initial instant of time. From these σ_r and σ_θ stress components, the deviatoric stress components S_θ and S_r are then found. Assuming the existence of the mechanical equation of state and using the Von Mises criterion and its flow rule, we have creep rate given by Equation (4.8.3):

$$\dot{e}_c^* = f(e_c^*, \sigma^*) \tag{7.4.5}$$

where $\dot{e}_c^*$ is given by Equation (4.8.5), and the effective stress σ^* is

$$\sigma^* = \sqrt{\tfrac{1}{2}}[(\sigma_r - \sigma_\theta)^2 + \sigma_r^2 + \sigma_\theta^2]^{1/2} \tag{7.4.6}$$

The rate of creep-strain components is given by Equation (4.8.7). Referring to Figure 3.9, we see that to calculate the incremental creep strain for an incremental time interval, the stresses are assumed to remain constant in this interval. The time interval Δt is chosen sufficiently small to keep the change of stress due to creep strain in this time interval small. From the creep rate and Δt, we obtain Δe_{r_c} and Δe_{θ_c} where the lower subscript "c" denotes the creep part of strain. Writing Equation (7.4.4) in incremental form by keeping ω constant, we have

$$\frac{d}{dr}\left\{\frac{Erh}{1-\nu^2}\left[\frac{d}{dr}(r\,\Delta e_\theta) + \nu\,\Delta e_\theta - \Delta e_{r_c} - \nu\,\Delta e_{\theta_c}\right]\right\}$$

$$-\frac{Eh}{1-\nu^2}\left[\Delta e_\theta + \nu\frac{d}{dr}(r\,\Delta e_\theta) - \Delta e_{\theta_c} - \nu\,\Delta e_{r_c}\right] = 0 \tag{7.4.7}$$

Substituting the values of Δe_{r_c} and Δe_{θ_c} into the above, the above equation may be written in finite difference form as a set of simultaneous equations, from which Δe_θ's at different points are solved. From these Δe_{r_c}'s, Δe_{θ_c}'s, and Δe_θ's the incremental stresses, $\Delta\sigma_\theta$'s and $\Delta\sigma_r$'s, are obtained. The process is repeated for successive time increments until the desired life of the disk is obtained.

Rotating Disks with Plastic Strain. For a disk stressed beyond the elastic limit, Equations (4.5.1) and (4.5.9) may be used, giving

$$\frac{de_r^p}{S_r} = \frac{de_\theta^p}{S_\theta} = \frac{de_z^p}{S_z} = \frac{3de^{*p}}{2\sigma^*} = \frac{3(de^{*p}/d\sigma^*)\,d\sigma^*}{2\sigma^*} \tag{7.4.8}$$

where the superscript p denotes the plastic part of strain. The deviatoric stresses are

$$S_r = \frac{2\sigma_r - \sigma_\theta}{3} \qquad S_\theta = \frac{2\sigma_\theta - \sigma_r}{3} \qquad S_z = -\frac{\sigma_\theta + \sigma_r}{3}$$

and $de^{*p}/d\sigma^*$ is obtained from a tensile test, with σ^* corresponding to the tensile stress τ_{11}, and e^{*p} to the *plastic* tensile strain e_{11}^p. We defined σ^* and e^{*p} in Equations (4.5.6) and (4.5.7). Writing Equations (7.4.4) and (7.4.5) in incremental form, we have

$$\frac{d}{dr}\left\{\frac{Erh}{1-\nu^2}\left[\frac{d}{dr}(r\,\Delta e_\theta) + \nu\,\Delta e_\theta - \Delta e_r{}^p - \nu\,\Delta e_\theta{}^p\right]\right\}$$
$$-\frac{Eh}{1-\nu^2}\left[\Delta e_\theta + \nu\frac{d}{dr}(r\,\Delta e_\theta) - \Delta e_\theta{}^p - \nu\,\Delta e_\theta{}^p\right] + \gamma r^2 h\,\Delta(\omega^2) = 0 \tag{7.4.9}$$

$$\frac{3\Delta e_r{}^p}{2\sigma_r - \sigma_\theta} = \frac{3\Delta e_\theta{}^p}{2\sigma_\theta - \sigma_r} = \frac{3\Delta e_z{}^p}{-(\sigma_r + \sigma_\theta)} = \frac{3(\Delta e^{*p}/\Delta\sigma^*)\,\Delta\sigma^*}{2\sigma^*} \tag{7.4.10}$$

From Equations (4.5.6) and (7.4.3), we obtain

$$\sigma^*\,\Delta\sigma^* = (\sigma_\theta - \sigma_r)(\Delta\sigma_\theta - \Delta\sigma_r) + \sigma_r\,\Delta\sigma_r + \sigma_\theta\,\Delta\sigma_\theta \tag{7.4.11}$$

$$\Delta\sigma_r = \frac{E}{1-\nu^2}\left[\frac{d}{dr}(r\,\Delta e_\theta) - \Delta e_r{}^p + \nu(\Delta e_\theta - \Delta e_\theta{}^p)\right]$$
$$\Delta\sigma_\theta = \frac{E}{1-\nu^2}\left\{\Delta e_\theta - \Delta e_\theta{}^p + \nu\left[\frac{d}{dr}(r\,\Delta e_\theta) - \Delta e_r{}^p\right]\right\} \tag{7.4.12}$$

Substituting Equation (7.4.12) into Equation (7.4.11) yields

$$\sigma^*\,\Delta\sigma^* = -(\sigma_\theta - \sigma_r)\frac{E}{1+\nu}\left[\frac{d}{dr}(r\,\Delta e_\theta) - \Delta e_r{}^p - (\Delta e_\theta - \Delta e_\theta{}^p)\right]$$
$$+\sigma_r\frac{E}{1-\nu^2}\left[\frac{d}{dr}(r\,\Delta e_\theta) - \Delta e_r{}^p + \nu(\Delta e_\theta - \Delta e_\theta{}^p\right]$$
$$+\sigma_\theta\frac{E}{1-\nu^2}\left\{\Delta e_\theta - \Delta e_\theta{}^p + \nu\left[\frac{d}{dr}(r\,\Delta e_\theta) - \Delta e_r{}^p\right]\right\} \tag{7.4.13}$$

For numerical calculation, the radial distance $(b - a)$ is divided into N segments. The nth segment is denoted by subscript n; that is, σ_{θ_n} denotes σ_θ at the middle of the nth segment. Expressing $\Delta e_r{}^p$ in terms of $\Delta e_\theta{}^p$ by Equation (7.4.10) and eliminating $\Delta\sigma^*$, by Equations (7.4.10) and (7.4.13), we obtain a set of equations in the $\Delta e_{\theta_n}^p$'s and Δe_{θ_n}'s. Hence the $\Delta e_{\theta_n}^p$'s and

$\Delta e_{r_n}^p$'s are expressed as a linear combination of the Δe_{θ_n}'s. Substituting these expressions into Equation (7.4.9), we obtain a system of simultaneous equations in the unknown Δe_{θ_n}'s. These Δe_{θ_n}'s are readily solved for a given $\Delta(\omega^2)$ with the aid of a computer. From the Δe_{θ_n}'s, the $\Delta e_\theta{}^p$'s and $\Delta e_r{}^p$'s may be determined. Then $\Delta\sigma_r$, $\Delta\sigma_\theta$, and $\Delta\sigma^*$ are obtained. Adding these to the σ_r, σ_θ, and σ^* which existed before the increment of speed $\Delta(\omega^2)$, we obtain the new set of stresses σ_I, σ_θ, and σ^*. This process is repeated to obtain the successive incremental stresses until the final desirable rotating speed is obtained. Details of numerical methods for the determination of stresses in rotating disks of variable thickness subjected to plastic flow or creep, or both were given by Millenson and Manson,[6] Wu,[7] Wahl,[8–10] and Ma,[11–12] and are not repeated here.

REFERENCES

1. Timoshenko, S., and J. N. Goodier, *Theory of Elasticity*, McGraw-Hill, New York, pp. 416–421, 1951.
2. Johnson, W., and P. B. Mellor, *Plasticity for Mechanical Engineers*, Van Nostrand, Princeton, N.J., pp. 141–154, 1962.
3. Hill, R., *The Mathematical Theory of Plasticity*, Clarendon Press, Oxford, England, pp. 97–106, 1950.
4. Whalley, E., "The Design of Pressure Vessels Subject to Thermal Stress," *Can. J. Tech.*, **34**, 268–303, 1956.
5. Prager, W., and P. G. Hodge, *Theory of Perfectly Plastic Solids*, Wiley, New York, pp. 95–122, 1951.
6. Millenson, M. B., and S. S. Manson, "Determination of Stresses in Gas Turbine Disks Subject to Plastic Flow and Creep," NACA Tech. Note No. 1636, 1948.
7. Wu, M. H. Lee, "General Plastic Behavior and Approximate Solutions of Rotating Disk in Strain-Hardening Range," NACA Tech. Note 2367, 1951.
8. Wahl, A. M., "Stress Distributions in Rotating Disks Subject to Creep Including Effects of Variable Thickness and Temperature," *J. Appl. Mech. Trans., ASME*, **79**, 299–305, 1957.
9. Wahl, A. M., "Analysis of Creep in Rotating Disks Based on the Tresca Criterion and Associated Flow Rule, *J. Appl. Mech. Trans. ASME*, **78**, 231–238, 1956.
10. Wahl, A. M., "Further Studies of Stress Distribution in Rotating Disks and Cylinders Under Elevated-Temperature Creep Conditions," *J. Appl. Mech.*, **25**, 243–250, 1958.
11. Ma, B. M., "A Creep Analysis of Rotating Solid Disks," *J. Franklin Inst.*, **267**, 150–165, 1959.
12. Ma, B. M., "A Further Creep Analysis for Rotating Solid Disks of Variable Thickness," *J. Franklin Inst.*, **269**, 408–419, 1960.

chapter

8

Inelastic Torsion of Prismatic Bars

Introduction. This chapter begins the analysis of two-dimensional inelastic problems by considering the torsion problem of prismatic bars. The derivation, by St. Venant in 1853, of warping functions for elastic noncircular bars is first shown. This is followed by a description of Prandtl's membrane analogy, which gives an experimental method for obtaining the warping functions. Then the analyses of inelastic torsional problems with creep and plastic strains, for both perfectly plastic and strain-hardening materials, are presented. It is shown that the inelastic strain has the equivalent effect of a nonuniformly distributed load on the membrane of Prandtl's membrane analogy.

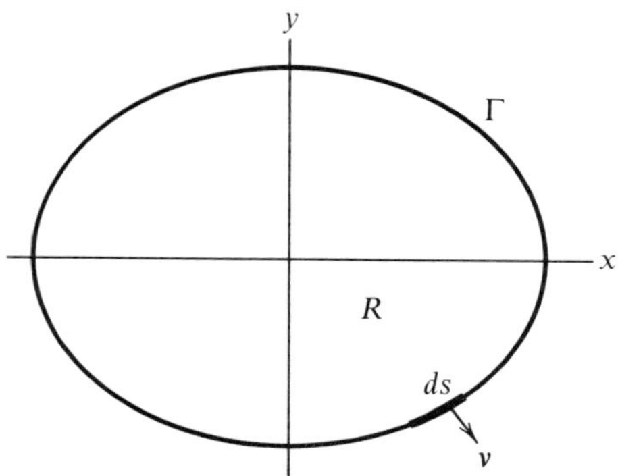

Figure 8.1. A cross section of the bar.

8.1 Elastic Torsion[1,2]

Consider a prismatic bar twisted by equal and opposite couples of magnitude M at the two ends. Let x, y, and z be a set of rectangular coordinates, with the z axis along the length of the bar and the x and y axes lying in the plane of the cross section. The displacements along the x, y, and z axes are denoted by u, v, and w, respectively. A cross section of the bar is shown in Figure 8.1. If the body forces within the bar are zero, then within R we have the equilibrium equations

$$\frac{\partial \tau_{xx}}{\partial x} + \frac{\partial \tau_{yx}}{\partial y} + \frac{\partial \tau_{zx}}{\partial z} = 0$$

$$\frac{\partial \tau_{xy}}{\partial x} + \frac{\partial \tau_{yy}}{\partial y} + \frac{\partial \tau_{zy}}{\partial z} = 0$$

$$\frac{\partial \tau_{xz}}{\partial x} + \frac{\partial \tau_{yz}}{\partial y} + \frac{\partial \tau_{zz}}{\partial z} = 0 \tag{8.1.1}$$

Let ν denote the normal to the boundary Γ of the cross section (Figure 8.1). Since there is no traction on the lateral surface of the bar, $\tau_{\nu z} = 0$. This gives

$$\tau_{\nu z} = \tau_{z\nu} = \tau_{zx} \cos(\nu, x) + \tau_{zy} \cos(\nu, y) = 0 \qquad \text{on } \Gamma \tag{8.1.2}$$

Assume the displacements of the form[1,2]

$$u = -\theta z y \qquad v = \theta z x \qquad w = \theta \phi(x, y) \tag{8.1.3}$$

where θ is the twist per unit length of the bar. The original plane cross sections of the bar are distorted into surfaces given by the function $\phi(x, y)$, known as the *warping function*. The function ϕ must be so determined as to satisfy the equilibrium conditions, Equations (8.1.1) and (8.1.2). From Equation (8.1.3) and the strain-displacement relations, we obtain

$$e_{xx} = e_{yy} = e_{zz} = e_{xy} = 0$$

$$e_{xz} = \frac{1}{2}\left(\frac{\partial u}{\partial z} + \frac{\partial w}{\partial x}\right) = \frac{\theta}{2}\left(\frac{\partial \phi}{\partial x} - y\right) \tag{8.1.4}$$

$$e_{yz} = \frac{1}{2}\left(\frac{\partial v}{\partial z} + \frac{\partial w}{\partial y}\right) = \frac{\theta}{2}\left(\frac{\partial \phi}{\partial y} + x\right)$$

At the start of loading, M is small and the deformation is *elastic*. Then Equation (8.1.4), together with the elastic stress-strain relations, gives

$$\tau_{xx} = \tau_{yy} = \tau_{zz} = \tau_{xy} = 0$$

$$\tau_{xz} = \mu\theta\left(\frac{\partial\phi}{\partial x} - y\right) \qquad \tau_{yz} = \mu\,\theta\left(\frac{\partial\phi}{\partial y} + x\right) \tag{8.1.5}$$

Substituting Equations (8.1.5) into Equations (8.1.1) and (8.1.2) yields

$$\nabla^2\phi = \frac{\partial^2\phi}{\partial x^2} + \frac{\partial^2\phi}{\partial y^2} = 0 \qquad \text{in } R \tag{8.1.6}$$

and

$$\left(\frac{\partial\phi}{\partial x} - y\right)\cos(x, v) + \left(\frac{\partial\phi}{\partial y} + x\right)\cos(y, v) = 0 \qquad \text{on } \Gamma$$

or

$$\frac{\partial\phi}{\partial x}\cos(x, v) + \frac{\partial\phi}{\partial y}\cos(y, v) = \frac{d\phi}{dv} = y\cos(x, v) - x\cos(y, v) \qquad \text{on } \Gamma \tag{8.1.7}$$

From Equations (8.1.5), it is seen that the first two of Equations (8.1.1) are identically satisfied. Any function which satisfies Equation (8.1.6) is called a harmonic function; hence ϕ is a harmonic function. Therefore the solution of the elastic torsion of a prismatic-bar problem reduces to the determination of a harmonic function ϕ in a given region R whose normal derivative $d\phi/dv$ is prescribed on the boundary Γ of the region.

If $f(z)$ is a complex function of the complex variable $z = x_1 + ix_2$, then $f(z)$ may be expressed as $f(z) = \zeta(x_1, x_2) + i\eta(x_1, x_2)$. Then

$$\frac{\partial f}{\partial x_1} = \frac{df}{dz}\frac{\partial z}{\partial x_1} = \frac{df}{dz} = \frac{\partial\zeta}{\partial x_1} + i\,\frac{\partial\eta}{\partial x_1}$$

$$\frac{\partial f}{\partial x_2} = \frac{df}{dz}\frac{\partial z}{\partial x_2} = i\,\frac{df}{dz} = \frac{\partial\zeta}{\partial x_2} + i\,\frac{\partial\eta}{\partial x_2}$$

Equating the above two expressions for df/dz, we have

$$\frac{\partial\zeta}{\partial x_1} + i\,\frac{\partial\eta}{\partial x_1} = -i\,\frac{\partial\zeta}{\partial x_2} + \frac{\partial\eta}{\partial x_2}$$

Equating the real and imaginary parts of the above relation gives

$$\frac{\partial\zeta}{\partial x_1} = \frac{\partial\eta}{\partial x_2} \qquad \frac{\partial\eta}{\partial x_1} = -\frac{\partial\zeta}{\partial x_2}$$

These are called the Cauchy- Riemann conditions. Continuous functions whose real and imaginary parts satisfy the Cauchy-Riemann conditions are called analytic functions. Differentiating the above first equation with respect to x_1 and the second with respect to x_2, we obtain

$$\frac{\partial^2 \zeta}{\partial {x_1}^2} = \frac{\partial^2 \eta}{\partial x_1 \partial x_2} \qquad \frac{\partial^2 \eta}{\partial x_1 \partial x_2} = -\frac{\partial^2 \zeta}{\partial {x_2}^2}$$

This gives $\partial^2 \zeta/\partial {x_1}^2 + \partial^2 \zeta/\partial {x_2}^2 = \nabla^2 \zeta = 0$. Similarly, $\nabla^2 \eta = 0$. Hence if $f(t) = \zeta + i\eta$ is analytic, ζ and η satisfy the conditions $\nabla^2 \zeta = 0$ and $\nabla^2 \eta = 0$.

Since ϕ is harmonic within R, we can construct an analytic function $F(z) = \phi + i\psi$, where z is the complex variable $x + iy$. By the Cauchy-Riemann conditions for analytic functions,

$$\frac{\partial \phi}{\partial x} = \frac{\partial \psi}{\partial y} \qquad \frac{\partial \phi}{\partial y} = -\frac{\partial \psi}{\partial x} \tag{8.1.8}$$

Substituting Equation (8.1.8) into Equation (8.1.5), we obtain

$$\tau_{xz} = \mu\theta\left(\frac{\partial \psi}{\partial y} - y\right) \qquad \tau_{yz} = \mu\theta\left(-\frac{\partial \psi}{\partial x} + x\right) \tag{8.1.9}$$

From Figure 8.1, we have

$$\cos(v, x) = \frac{dy}{ds} \quad \text{and} \quad \cos(v, y) = -\frac{dx}{ds} \tag{8.1.10}$$

and therefore, from Equations (8.1.7), (8.1.8), and (8.1.10), we find

$$\frac{d\psi}{ds} = \frac{\partial \psi}{\partial y}\frac{dy}{ds} + \frac{\partial \psi}{\partial x}\frac{dx}{ds} = y\frac{dy}{ds} + x\frac{dx}{ds} = \frac{1}{2}\frac{d}{ds}(x^2 + y^2) \quad \text{on } \Gamma \tag{8.1.11}$$

$$\psi = \tfrac{1}{2}(x^2 + y^2) + c \quad \text{on } \Gamma \tag{8.1.12}$$

Hence the solution of the problem of elastic torsion of a prismatic bar reduces to the determination of a harmonic function ψ in a given region R with prescribed value on the boundary Γ of the region.

To show that the distribution of stresses given by Equation (8.1.9) corresponds to the torsional couples applied, we will first show the vanishing of the resultant force for each cross section of the bar. The resultant force in the x direction is

$$X = \iint_R \mu\theta\left(\frac{\partial\psi}{\partial y} - y\right) dx\, dy$$

$$= \mu\theta \iint_R \left\{\frac{\partial}{\partial x}\left[x\left(\frac{\partial\psi}{\partial y} - y\right)\right] - \frac{\partial}{\partial y}\left[x\left(\frac{\partial\psi}{\partial x} - x\right)\right]\right\} dx\, dy$$

By Green's theorem* for the plane, the area integral is related to the line integral, and the above equation becomes

$$X = \mu\theta \int_\Gamma x\left(\frac{\partial\psi}{\partial x} - x\right) dx + x\left(\frac{\partial\psi}{\partial y} - y\right) dy$$

$$= \mu\theta \int_\Gamma x\left[\left(\frac{\partial\psi}{\partial x}\frac{dx}{ds} + \frac{\partial\psi}{\partial y}\frac{dy}{ds}\right) ds - \left(x\frac{dx}{ds} + y\frac{dy}{ds}\right) ds\right]$$

From Equation (8.1.11) the bracket term vanishes; hence $X = 0$. It can similarly be shown that

$$Y = \iint_R \tau_{zy}\, dx\, dy = 0$$

Therefore the resultant force over each cross section vanishes. The resultant moment is

$$M = \iint (x\tau_{zy} - y\tau_{zx})\, dx\, dy \tag{8.1.13}$$

$$= \mu\theta \iint \left[x\left(-\frac{\partial\psi}{\partial x} + x\right) - y\left(\frac{\partial\psi}{\partial y} - y\right)\right] dx\, dy$$

$$= \mu\theta \iint \left[(x^2 + y^2) - x\frac{\partial\psi}{\partial x} - y\frac{\partial\psi}{\partial y}\right] dx\, dy \tag{8.1.14}$$

The torsional moment M is proportional to the angle θ of twist per unit length, where the constant of proportionality

$$D = \mu \iint \left[x^2 + y^2 - x\frac{\partial\psi}{\partial x} - y\frac{\partial\psi}{\partial y}\right] dx\, dy$$

* Green's theorem for the plane:[1] If $P(x, y)$ and $Q(x, y)$, $\partial P/\partial y$ and $\partial Q/\partial x$ are single-valued, continuous functions over a closed region R in a plane bounded by the boundary curve Γ, then

$$\iint_R \left(\frac{\partial P}{\partial y} - \frac{\partial Q}{\partial x}\right) dx\, dy = -\int_\Gamma (P\, dx + Q\, dy)$$

is called the torsional rigidity of the bar. Hence the harmonic function ψ gives the elastic solution of a prismatic bar.

From the above discussions, we can see that a harmonic function $\psi(x, y)$ equated to $\frac{1}{2}(x^2 + y^2)$ on the boundary curve Γ of a prismatic bar gives the elastic torsional stress distribution as expressed by Equation (8.1.9). For example, consider the harmonic function

$$\psi = k_1{}^2(x^2 - y^2) + k_2{}^2 \tag{8.1.15}$$

To find the boundary curve Γ of the bar for this ψ, we let

$$\psi = k_1{}^2(x^2 - y^2) + k_2{}^2 = \tfrac{1}{2}(x^2 + y^2)$$

Then

$$(\tfrac{1}{2} - k_1{}^2)x^2 + (\tfrac{1}{2} + k_1{}^2)y^2 = k_2{}^2 \qquad \text{on } \Gamma \tag{8.1.16}$$

Letting $k_1{}^2 < \frac{1}{2}$, we may write the above as

$$\frac{x^2}{a^2} + \frac{y^2}{b^2} = 1$$

where

$$a = \frac{k_2}{\sqrt{\frac{1}{2} - k_1{}^2}} \qquad b = \frac{k_2}{\sqrt{\frac{1}{2} + k_1{}^2}}$$

$$k_1{}^2 = \frac{1}{2}\frac{a^2 - b^2}{a^2 + b^2} \qquad k_2{}^2 = \frac{a^2 b^2}{a^2 + b^2}$$

Substituting these into Equation (8.1.14) yields

$$\psi = \frac{1}{2}\frac{a^2 - b^2}{(a^2 + b^2)}(x^2 - y^2) + \frac{a^2 b^2}{a^2 + b^2} \tag{8.1.17}$$

From this and Equation (8.1.9), we obtain

$$\tau_{xz} = \frac{-2\mu\theta a^2 y}{a^2 + b^2} \qquad \tau_{yz} = \frac{2\mu\theta b^2 x}{a^2 + b^2} \tag{8.1.18}$$

This gives the stresses in an elastic prismatic bar of elliptic section under torsion.

8.2 Membrane Analogy

From Equations (8.1.1) and (8.1.5), we have

$$\frac{\partial \tau_{xz}}{\partial x} + \frac{\partial \tau_{yz}}{\partial y} = 0 \tag{8.2.1}$$

This differential equation is automatically satisfied by expressing the stress components in terms of a stress function[3] $F(x, y)$ defined by

$$\tau_{xz} = \frac{\partial F}{\partial y} \qquad \text{and} \qquad \tau_{yz} = -\frac{\partial F}{\partial x} \tag{8.2.2}$$

From Equation (8.1.5), we have

$$\frac{\partial F}{\partial y} = \mu\theta\left(\frac{\partial \phi}{\partial x} - y\right) \qquad -\frac{\partial F}{\partial x} = \mu\theta\left(\frac{\partial \phi}{\partial y} + x\right) \tag{8.2.3}$$

Differentiating the first of Equations (8.2.3) with respect to y and the second of these equations with respect to x, and then subtracting the second from the first, we find

$$\nabla^2 F = \frac{\partial^2 F}{\partial x^2} + \frac{\partial^2 F}{\partial y^2} = -2\mu\theta \tag{8.2.4}$$

This is Poisson's differential equation in two dimensions. Substituting Equation (8.2.2) into Equation (8.1.2) yields

$$\frac{\partial F}{\partial y}\cos(v, x) - \frac{\partial F}{\partial x}\cos(v, y) = 0 \qquad \text{on } \Gamma$$

Substituting Equation (8.1.10) into the above, we obtain

$$\frac{\partial F}{\partial x}\frac{dx}{ds} + \frac{\partial F}{\partial y}\frac{dy}{ds} = \frac{dF}{ds} = 0 \qquad \text{on } \Gamma \tag{8.2.5}$$

which requires that

$$F = \text{const.} \qquad \text{on } \Gamma$$

Since the stress components depend only on the derivative of the stress function F, the constant may be taken as zero. The stress function F must satisfy Poisson's equation, Equation (8.2.4), in the region R and must equal zero on the boundary Γ. The resultant force on the cross section along the x direction is given by

$$X = \iint_R \frac{\partial F}{\partial y}\,dx\,dy = \int [F(y_2, x) - F(y_1, x)]\,dx = 0 \tag{8.2.6}$$

where y_1 and y_2 are the lower and upper y coordinates of the section at a particular x, and both (y_2, x) and (y_1, x) are on Γ; hence $F(y_2, x) - F(y_1, x)$ is zero. Similarly, the resultant force along the y direction is

$$Y = 0$$

and the moment at a section is given by

$$M = \iint_R (\tau_{zy} x - \tau_{zx} y)\, dx\, dy$$

$$= -\iint_R \left(\frac{\partial F}{\partial x} x + \frac{\partial F}{\partial y} y\right) dx\, dy$$

$$= -\int \left(xF - \int F\, dx\right) dy + \int \left(yF - \int F\, dy\right) dx$$

$$= 2 \iint_R F\, dx\, dy \tag{8.2.7}$$

Hence the stress function F gives a resultant couple to the prismatic bar. Equations (8.2.6) and (8.2.7) hold for both elastic and plastic torsions.

Consider a homogeneous membrane stretched under uniform tension T per unit length over an opening made in a rigid plate where the opening has the same shape as the cross section of the twisted bar. Let q be the pressure per unit area applied on the membrane, and w the small deflection of the membrane. The net downward vertical force acting on the left and right sides of the differential element $dx\, dy$ as shown in Figure 8.2 is

$$T \frac{\partial^2 w}{\partial x^2}\, dx\, dy$$

and that on the front and rear of this element is

$$T \frac{\partial^2 w}{\partial y^2}\, dx\, dy$$

Equating the sum of the vertical forces acting on this element gives

$$T \frac{\partial^2 w}{\partial x^2}\, dx\, dy + T \frac{\partial^2 w}{\partial y^2}\, dx\, dy + q\, dx\, dy = 0$$

or

$$\nabla^2 w = \frac{\partial^2 w}{\partial x^2} + \frac{\partial^2 w}{\partial y^2} = -\frac{q}{T} \tag{8.2.8}$$

Hence the lateral deflection w satisfies Poisson's differential equation as given by Equation (8.2.8) within the cross section R and equals zero on the boundary curve Γ. If q/T is replaced by $2\mu\theta$, then w has the same differential equation and boundary condition as F. From this analogy and Equation (8.2.7), we have

$$F = \frac{2\mu\theta T}{q} w$$

and

$$M = \frac{4\mu\theta T}{q} \iint w \, dx \, dy \tag{8.2.9}$$

Therefore the stress function F for elastic torsion may be obtained experimentally by measuring the lateral deflection of the membrane. This is called Prandtl's membrane analogy.

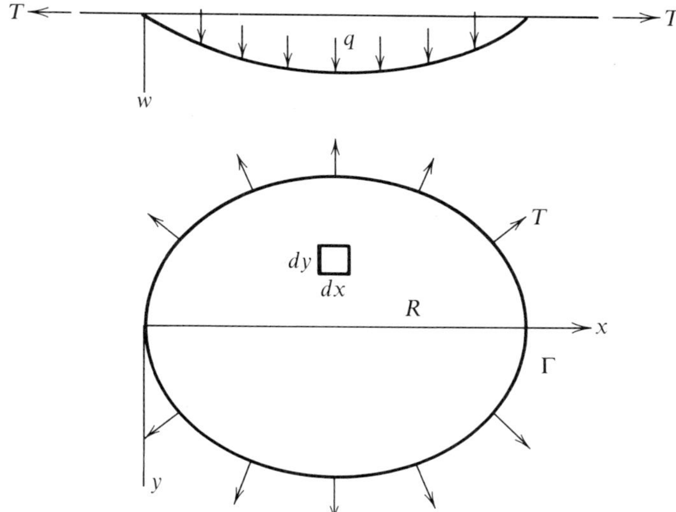

Figure 8.2. Membrane deflection under edge tension and lateral pressure.

8.3 Torsion with Creep

If a prismatic bar is twisted at an elevated temperature, creep strain occurs. The stress-strain-time relation of the material is assumed to follow Equations (4.8.3) and (4.8.7). The initial stress distribution is obtained by an elastic analysis. During creep the stress component vs. time curve may be approximated by horizontal and vertical segments as shown in Figure 3.9. Within each horizontal Δt segment, the stress is assumed constant. Among the six stress components, only τ_{zx} and τ_{zy} exist at the start of loading. For the first Δt time interval, the incremental creep strain obtained by multiplying the creep strain rate [given by Equation (4.8.7)] by Δt has only two nonvanishing components, Δe_{zx_c} and Δe_{zy_c}, given by

$$\frac{\Delta e_{zx_c}}{\tau_{zx}} = \frac{\Delta e_{zy_c}}{\tau_{zy}} = \frac{3\Delta e_c{}^*}{2\sigma^*} \tag{8.3.1}$$

where

$$\Delta e_c{}^* = \frac{2}{\sqrt{3}}\,(\Delta e_{zx_c}^2 + \Delta e_{zy_c}^2)^{1/2} \tag{8.3.2}$$

and

$$\sigma^* = \sqrt{3}\,(\tau_{zx}^2 + \tau_{zy}^2)^{1/2} \tag{8.3.3}$$

The creep rate of metals at a given temperature is commonly assumed to be a function of stress and the current amount of creep strain. This function is experimentally determined from uniaxial tests as $\dot{e}_{11_c} = f(\tau_{11}, e_{11_c})$. This is generalized to

$$\dot{e}_c{}^* = f(\sigma^*, e_c{}^*)$$

for multiaxial creep. With σ^* and $e_c{}^*$ known, the creep rate $\dot{e}_c{}^*$ is determined. The displacements are assumed to be of the form given by Equation (8.1.3). For the first time interval, $\Delta e_{zx_c} = e_{zx_c}$ and $\Delta e_{zy_c} = e_{zy_c}$. We have two nonvanishing stress components, given by

$$\tau_{zx} = \frac{\partial F}{\partial y} = \mu\left[\theta\left(\frac{\partial \phi}{\partial x} - y\right) - 2e_{zx_c}\right] \tag{8.3.4}$$

$$\tau_{zy} = -\frac{\partial F}{\partial x} = \mu\left[\theta\left(\frac{\partial \phi}{\partial y} + x\right) - 2e_{zy_c}\right] \tag{8.3.5}$$

Differentiating Equation (8.3.4) with respect to y, and Equation (8.3.5) with respect to x, and subtracting the latter from the former, we find

$$\frac{\partial^2 F}{\partial x^2} + \frac{\partial^2 F}{\partial y^2} = \nabla^2 F = -2\mu\left[\theta + \frac{\partial(e_{zx_c})}{\partial y} - \frac{\partial(e_{zy_c})}{\partial x}\right] \tag{8.3.6}$$

This is of the form of Poisson's differential equation. The boundary condition is $F = 0$ on Γ. The second and third terms in the bracketed term of Equation (8.3.6) vary from point to point on the section. Comparing Equation (8.3.6) with Equation (8.2.8), with F replaced by w, we see that the inelastic creep strain has the equivalent effect of a nonuniform pressure on the membrane. This shows that a membrane under a uniform edge tension T and a nonuniformly distributed load may be used to find the stress distribution of a prismatic bar with creep strain under torsion. The solution of F for rectangular sections may be obtained by expanding F and the bracketed term of Equation (8.3.6) in terms of trigonometric series.[4] For sections of arbitrary form, a finite difference method may be used. If the twist is held constant, that is, $\theta = \text{const.}$, the torsional couple M decreases with time. If M is kept constant,

$$2 \iint F \, dx \, dy = M = \text{const.} \tag{8.3.7}$$

and θ will increase with time. In either case, F is solved from the differential Equation (8.3.6), with boundary condition $F = 0$ on Γ. Under constant moment, θ varies to satisfy Equation (8.3.7). Then τ_{zx} and τ_{zy} are obtained from F by Equations (8.2.2). The other stress components at the end of the first time interval are again zero. During the second time interval the stress is again assumed to be constant, and again only Δe_{zx_c} and Δe_{zy_c} occur. Again F is solved in the same way. This process is repeated for successive time increments for any desired life of the bar. Hence, among the six creep strain components, only e_{zx_c} and e_{zy_c} exist at all instants of time. If we let e_{zx_c} and e_{zy_c} be the creep strain components at the beginning of a time interval, and Δe_{zx_c}, Δe_{zy_c} the incremental creep strain components for that interval, Δe^* in Equation (8.3.1) becomes

$$\Delta e_c^* = \frac{2}{\sqrt{3}} [(e_{zx_c} + \Delta e_{zx_c})^2 + (e_{zy_c} + \Delta e_{zy_c})^2]^{1/2} - \frac{2}{\sqrt{3}} [e_{zx_c}^2 + e_{zy_c}^2]^{1/2} \tag{8.3.8}$$

This reduces to Equation (8.3.2) for the first time interval, at the beginning of which $e_{zx_c} = e_{zy_c} = 0$.

For a bar under constant twisting moment M, Equation (8.3.6) may be written in the following form:

$$\nabla^2 F_1 = -2\mu\theta \tag{8.3.9}$$

$$\nabla^2 F_2 = -2\mu\left(\frac{\partial(e_{zx_c})}{\partial y} - \frac{\partial(e_{zy_c})}{\partial x}\right) \tag{8.3.10}$$

$$F = F_1 + F_2 \tag{8.3.11}$$

The torsional moment due to F_1 is denoted by M_1, which is the same as the elastic moment under a twisting angle θ per unit length. Then

$$M_1 = D\theta \tag{8.3.12}$$

$$M_2 = 2 \iint F_2 \, dx \, dy \tag{8.3.13}$$

$$M = M_1 + M_2 = D\theta + M_2 = \text{const.} \tag{8.3.14}$$

After M_2 is evaluated, θ is readily found.

At the beginning of a time interval, e_{zx_c}, e_{zy_c}, τ_{zx}, and τ_{zy} are known at all points. From the relation $\dot{e}_c{}^* = F(\sigma^*, e_c{}^*)$ and Equation (4.8.7), the $\dot{e}_{zx_c}$ and $\dot{e}_{zy_c}$ are obtained. Multiplying these creep strain rates by Δt, we obtain Δe_{zx_c} and Δe_{zy_c}. Adding these incremental creep strain components to the corresponding creep strain components at the beginning of the interval, we obtain the e_{zx_c} and e_{zy_c} at the end of the interval. Substituting these values of e_{zx_c} and e_{zy_c} into Equation (8.3.10), we may determine F_2. We then obtain M_2 by Equation (8.3.13) and θ from Equation (8.3.14). This process is repeated to any desired length of time. This gives a summary of the procedure for calculating the torsional stresses in prismatic bars with creep.

8.4 Elastoplastic Torsion

If the prismatic bar is twisted just beyond the elastic limit, plastic strain starts at the highest stressed points. The material is assumed to have the incremental stress-strain relation given by Equation (4.5.1). At the elastic limit, all stress components except τ_{zx} and τ_{zy} vanish, hence only $\Delta e''_{zx}$ and $\Delta e''_{zy}$ of the incremental plastic strain components exist. The displacements are assumed to be of the form given by Equation (8.1.3). We have the stress components, given by

$$\tau_{xx} = \tau_{yy} = \tau_{zz} = \tau_{xy} = 0$$

$$\tau_{zx} = \frac{\partial F}{\partial y} = \mu\left[\theta\left(\frac{\partial \phi}{\partial x} - y\right) - 2\,\Delta e''_{zx}\right] \tag{8.4.1}$$

$$\tau_{zy} = -\frac{\partial F}{\partial x} = \mu\left[\theta\left(\frac{\partial \phi}{\partial y} + x\right) - 2\,\Delta e''_{zy}\right] \tag{8.4.2}$$

As in Section 8.3, from these two equations, we obtain

$$\frac{\partial^2 F}{\partial x^2} + \frac{\partial^2 F}{\partial y^2} = \nabla^2 F = -2\mu\theta - 2\mu\left[\frac{\partial}{\partial y}(\Delta e''_{zx}) - \frac{\partial}{\partial x}(\Delta e''_{zy})\right] \tag{8.4.3}$$

The function F is obtained from the solution of this Poisson's differential equation (Equation 8.4.3) with the boundary condition $F = 0$ on the boundary curve Γ. Having F, we then determine τ_{zx} and τ_{zy} by Equations (8.4.1) and (8.4.2). The other stress components are again zero. As the twist is further increased, the incremental plastic strain is again confined to e''_{zx} and e''_{zy}. Hence for torsional loading on a prismatic bar, only these two plastic strain components exist. At any stage of loading, Equation (8.4.3) becomes

$$\nabla^2 F = -2\mu\theta - 2\mu\left[\frac{\partial}{\partial y}(e''_{zx}) - \frac{\partial}{\partial x}(e''_{zy})\right] \quad \text{in } R \tag{8.4.4}$$

and $F = 0$ on Γ.

For Prandtl-Reuss materials with strain-hardening, from Equations (4.5.1) and (4.5.9), we have the relation

$$\frac{de''_{zx}}{\tau_{zx}} = \frac{de''_{zy}}{\tau_{zy}} = \frac{3de''^*}{2\sigma^*} \tag{8.4.5}$$

In the present case,

$$\sigma^* = \sqrt{3}(\tau_{zx}^2 + \tau_{zy}^2)^{1/2}$$

and (8.4.6)

$$e''^* = \frac{2}{\sqrt{3}}(e''^2_{zx} + e''_{zy})^{1/2}$$

From Equation (4.5.8), we have $e''^* = f(\sigma^*)$. The function f is determined from uniaxial tests, and then generalized to

$$de''^* = f'(\sigma^*)\,d\sigma^*$$

$$\frac{3\,de''^*}{2\sigma^*} = \frac{3f'(\sigma^*)\,d\sigma^*}{2\sigma^*} = \frac{3}{2}\frac{f'(\sigma^*)(\tau_{zx}\,d\tau_{zx} + \tau_{zy}\,d\tau_{zy})}{(\tau_{zx}^2 + \tau_{zy}^2)}$$

for the multiaxial stress state. Substituting this into Equation (8.4.5) and replacing the differential by the small finite increment "Δ", we obtain

$$\frac{\Delta e''_{zx}}{\tau_{zx}} = \frac{\Delta e''_{zy}}{\tau_{zy}} = \frac{3}{2}\frac{f'(\sigma^*)(\tau_{zx}\,\Delta\tau_{zx} + \tau_{zy}\,\Delta\tau_{zy})}{(\tau_{zx}^2 + \tau_{zy}^2)} \tag{8.4.7}$$

Writing Equation (8.4.4) in incremental form, we have

$$\nabla^2(\Delta F) = -2\mu\,\Delta\theta - 2\mu\left[\frac{\partial}{\partial y}(\Delta e''_{zx}) - \frac{\partial}{\partial x}(\Delta e''_{zy})\right] \tag{8.4.8}$$

and from Equations (8.2.2), we have

$$\Delta\tau_{zx} = \frac{\partial}{\partial y}(\Delta F) \qquad \Delta\tau_{zy} = -\frac{\partial}{\partial x}(\Delta F) \tag{8.4.9}$$

From Equations (8.4.7), and (8.4.9), we find

$$\Delta e''_{zx} = \frac{3}{2}\frac{\tau_{zx} f'(\sigma^*)}{(\tau_{zx}^2 + \tau_{zy}^2)}\left[\tau_{zx}\frac{\partial}{\partial y}(\Delta F) - \tau_{zy}\frac{\partial}{\partial x}(\Delta F)\right]$$

$$\Delta e''_{zy} = \frac{3}{2}\frac{\tau_{zy} f'(\sigma^*)}{(\tau_{zx}^2 + \tau_{zy}^2)}\left[\tau_{zx}\frac{\partial}{\partial y}(\Delta F) - \tau_{zy}\frac{\partial}{\partial x}(\Delta F)\right]$$

$$\begin{aligned}\frac{\partial}{\partial y}(\Delta e''_{zx}) = {} & \frac{3}{2}\frac{\partial}{\partial y}\left(\frac{\tau_{zx}^2 f'(\sigma^*)}{\tau_{zx}^2 + \tau_{zy}^2}\right)\frac{\partial}{\partial y}(\Delta F) + \frac{3}{2}\frac{\tau_{zx}^2 f'(\sigma^*)}{\tau_{zx}^2 + \tau_{zy}^2}\frac{\partial^2}{\partial y^2}(\Delta F) \\ & - \frac{3}{2}\frac{\partial}{\partial y}\left(\frac{\tau_{zx}\tau_{zy} f'(\sigma^*)}{\tau_{zx}^2 + \tau_{zy}^2}\right)\frac{\partial}{\partial x}(\Delta F) - \frac{3}{2}\frac{\tau_{zx}\tau_{zy} f'(\sigma^*)}{\tau_{zx}^2 + \tau_{zy}^2}\frac{\partial^2}{\partial x\,\partial y}(\Delta F)\end{aligned} \tag{8.4.10}$$

$$\begin{aligned}\frac{\partial}{\partial x}(\Delta e''_{zy}) = {} & \frac{3}{2}\frac{\partial}{\partial x}\left(\frac{\tau_{zy}\tau_{zx} f'(\sigma^*)}{\tau_{zx}^2 + \tau_{zy}^2}\right)\frac{\partial}{\partial y}(\Delta F) + \frac{3}{2}\frac{\tau_{zy}\tau_{zx} f'(\sigma^*)}{\tau_{zx}^2 + \tau_{zy}^2}\frac{\partial^2}{\partial x\,\partial y}(\Delta F) \\ & - \frac{3}{2}\frac{\partial}{\partial x}\left(\frac{\tau_{zy}^2 f'(\sigma^*)}{\tau_{zx}^2 + \tau_{zy}^2}\right)\frac{\partial}{\partial x}(\Delta F) - \frac{3}{2}\frac{\tau_{zy}^2 f'(\sigma^*)}{\tau_{zx}^2 + \tau_{zy}^2}\frac{\partial^2}{\partial x^2}(\Delta F)\end{aligned} \tag{8.4.11}$$

Substituting these expressions into Equation (8.4.8) results in a linear differential equation with variable coefficients for ΔF. This equation may be solved numerically by the method of finite difference with the aid of a high speed computer. An additional $\Delta\theta$ may then be added to the bar, and the resulting additional ΔF is solved in the same way. The process may be repeated for successive increments of θ until the full amount of twist is attained.

If we assume the material to have the stress-strain relation proposed by Hencky, as given by Equation (4.5.16), in which $F(\sigma^*)$ corresponds to $3f(\sigma^*)/2\sigma^*$, then

$$\begin{aligned} e''_{zx} &= \tau_{zx}\frac{3f(\sigma^*)}{2\sigma^*} \\ e''_{zy} &= \tau_{zy}\frac{3f(\sigma^*)}{2\sigma^*} \end{aligned} \tag{8.4.12}$$

where

$$\sigma^* = \sqrt{3}[F_{,y}{}^2 + F_{,x}{}^2]^{1/2}$$

Then

$$\begin{aligned} e''_{zx} &= F_{,y}\frac{3f[\sqrt{3}(F_{,y}{}^2 + F_{,x}{}^2)^{1/2}]}{2\sqrt{3}[F_{,y}{}^2 + F_{,x}{}^2]^{1/2}} = F_{,y}\chi[(F_{,y}{}^2 + F_{,x}{}^2)^{1/2}] \\ e''_{zy} &= -F_{,x}\chi[(F_{,y}{}^2 + F_{,x}{}^2)^{1/2}] \end{aligned} \tag{8.4.13}$$

and χ is defined by

$$\chi[(F_{,y}{}^2 + F_{,x}{}^2)^{1/2}] = \frac{\sqrt{3}f[\sqrt{3}(F_{,y}{}^2 + F_{,x}{}^2)^{1/2}]}{2[F_{,y}{}^2 + F_{,x}{}^2]^{1/2}}$$

and the subscripts x and y after a comma denote differentiation with respect to x and y, respectively. Substituting Equation (8.4.13) into Equation (8.4.4) yields

$$\frac{\nabla^2 F}{2\mu} = -\theta - \{F_{,y}\chi[(F_{,y}{}^2 + F_{,x}{}^2)^{1/2}]\}_{,y} + \{F_{,x}\chi[(F_{,y}{}^2 + F_{,x}{}^2)^{1/2}]\}_{,x} \tag{8.4.14}$$

This is a nonlinear differential equation with $F = 0$ on the boundary Γ of the cross section. This equation has been solved by Onat[5] by an iteration method for prismatic bars under torsion.

Torsion of Perfectly Plastic Bars. If we assume the material to have no strain-hardening, σ^* remains constant. Let this constant be the yield stress in tension Y. Then

$$3(\tau_{xz}^2 + \tau_{yz}^2) = Y^2 \tag{8.4.15}$$

Let

$$Y^2 = 3k^2 \tag{8.4.16}$$

Then Equation (8.4.15) becomes

$$\tau_{xz}^2 + \tau_{yz}^2 = k^2 \tag{8.4.17}$$

From Equations (8.4.1), and (8.4.2), we have

$$\left(\frac{\partial F}{\partial x}\right)^2 + \left(\frac{\partial F}{\partial y}\right)^2 = k^2 \tag{8.4.18}$$

The left side of Equation (8.4.18) is the square of the absolute value of the gradient "grad. F", that is, the maximum slope of the surface F. Hence, in the plastic region,

$$|\text{grad. } F| = k \tag{8.4.19}$$

and in the elastic region,

$$\nabla^2 F = -2\mu\theta \tag{8.2.4}$$

At the boundary Γ of the cross section, $F = 0$.

Referring to Figure 8.3, we see that the shaded parts are plastic and the remaining part is elastic. On the inner boundaries of the plastic areas, *abc* and *def*, the resultant shearing stress $(\tau_{zx}^2 + \tau_{zy}^2)^{1/2} = k$. This value is approached from both the plastic side and the elastic side. This means the

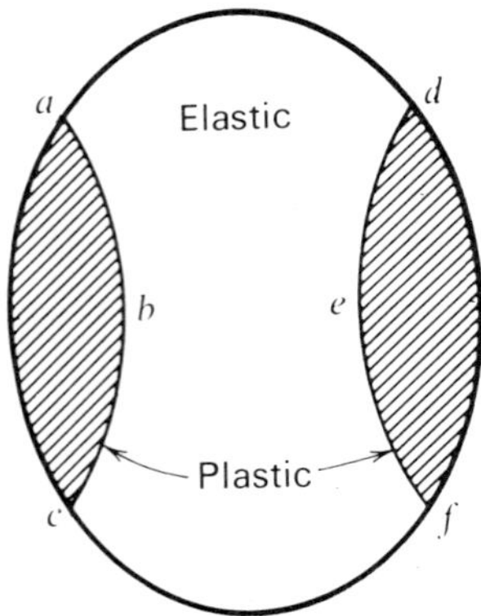

Figure 8.3. Elastic and plastic areas of a twisted bar.

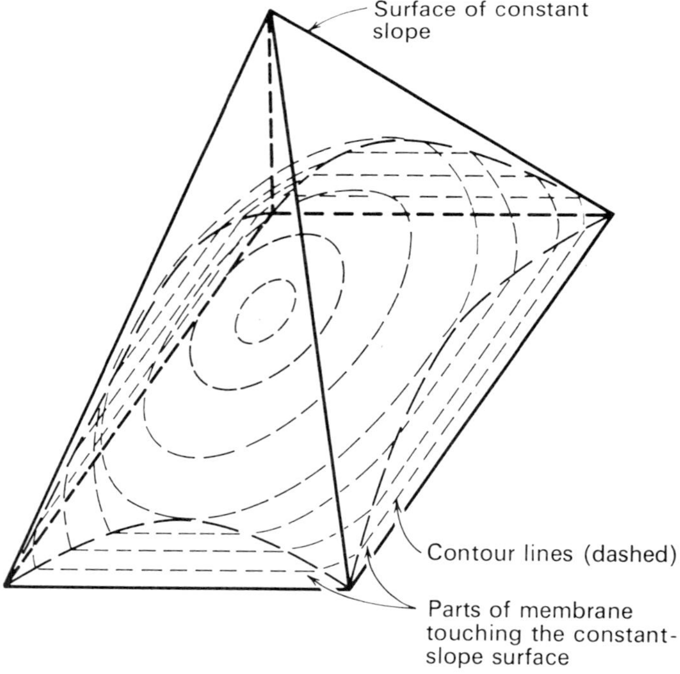

Figure 8.4. Membrane surface within a surface of constant slope.

slope of the $F(x, y)$ surface must be continuous across the boundary between the elastic and plastic regions. An experimental representation of this surface has been given by Nadai[3] as follows: A horizontal cardboard having the geometrical shape of the cross section of the prismatic bar is covered with sand so as to form a surface with constant slope. The interior surface of a roof of constant slope over the cross section of the bar is made the same as the surface of the sand heap. The base of this roof has the same shape as the section. The base is closed with a uniformly stretched membrane subject to a uniform lateral pressure. As the pressure is increased a portion of the membrane will touch the interior surface of the roof (see Figure 8.4). *The membrane, including the portion touching the roof, gives the surface, $F(x, y)$, for the elastic-perfectly plastic torsion of a prismatic bar.* The areas under the parts of the membrane touching the roof are plastic, and the remaining area of the section is elastic. The slope of the membrane surface constructed this way is continuous from the part touching the roof to the part not touching the roof. This gives the surface F, and its derivatives give the shearing stresses (Equation 8.2.2). For a more detailed discussion of the elastic-perfectly plastic torsion of prismatic bars, see the excellent discussions by Nadai,[3] by Prager and Hodge,[6] and by Hill.[7] It is seen that the solution of elastic-perfectly plastic torsion is simpler than the solution of elastic-plastic torsion with strain-hardening.

REFERENCES

1. Sokolnikoff, I. S., *Mathematical Theory of Elasticity*, McGraw-Hill, New York, p. 110, 1956.
2. Timoshenko, S., and J. N. Goodier, *Theory of Elasticity*, McGraw-Hill, New York, p. 259, 1951.
3. Nadai, A., *Theory of Flow and Fracture of Solids*, McGraw-Hill, New York, pp. 490–504, 1954.
4. Morse, P. M., and H. Feshbach, *Methods of Theoretical Physics*, McGraw-Hill, New York, pp. 797–798, 1953.
5. Onat, B. T., "Torsion of Prismatic Rods of Work-Hardening Material," Thesis, Istanbul Technical University, 1951.
6. Prager, W., and P. E. Hodge, Jr., *Theory of Perfectly Plastic Solids*, Wiley, New York, pp. 55–90, 1951.
7. Hill, R., *The Mathematical Theory of Plasticity*, Clarendon Press, Oxford, England, pp. 84–86, 1958.

chapter

9

Inelastic Plane Problems

Introduction. This chapter presents a general method for the solution of two-dimensional inelastic problems. These problems are of two physically distinct types, the plane strain problem and the plane stress problem. The former problem is characterized by a deformation state such that deformation along one coordinate axis is zero, while the deformations along the other two coordinate axes vary along these two axes only. Long cylindrical bodies subjected to external forces which do not vary along the length of the body are examples of plane strain problems. The plane stress problem is characterized by a stress state such that the components of stress associated with one of the coordinate axes are zero while the components of stress associated with the other two coordinate axes are dependent only on these axes. Thin plates, under in-plane loads which do not vary over the thickness of the plate, are examples of plane stress problems, since all stress components associated with the thickness coordinate are zero. *The mathematical formulations of these two types of elastic and inelastic problems are identical.*

Section 9.1 gives the derivation of the stress field in an elastic infinite medium under plane strain caused by a given distribution of body forces. Sections 9.2 and 9.3 review Muskhelishvili's method of complex variables for the solution of elastic plane problems. Sections 9.4 and 9.5 present the analysis of plane problems with circular and noncircular boundaries. In the final sections, a method for the solution of inelastic plane strain and plane stress problems, based on the principle of analogy described in Chapter 2, is presented.

9.1 Elastic Plane Deformation of an Infinite Medium[1]

Consider an elastic body referred to a cartesian coordinate system with axes x_1, x_2, and x_3 in which one component, say u_3, of the displacement vector is zero, and the other components u_1 and u_2 vary only with x_1 and x_2. With subscripts i, j ranging on values 1, 2, and 3, and α, β ranging on 1 and 2, we have

$$u_3 = 0 \qquad u_\alpha = u_\alpha(x_1, x_2) \tag{9.1.1}$$

$$e_{31} = e_{32} = e_{33} = 0 \qquad e_{\alpha\beta} = e_{\beta\alpha}(x_1, x_2) \tag{9.1.2}$$

Deformations satisfying Equations (9.1.1) *and* (9.1.2) *are called plane deformations.* The stress components corresponding to the plane deformation state, from Equations (1.9.7), are

$$\tau_{\alpha\beta} = \tau_{\alpha\beta}(x_1, x_2) = \lambda\, \delta_{\alpha\beta}\,\theta + \mu(u_{\alpha,\beta} + u_{\beta,\alpha}) \tag{9.1.3}$$

$$\begin{aligned} \tau_{31} &= \tau_{32} = 0 \\ \tau_{33} &= \lambda\theta \end{aligned} \tag{9.1.4}$$

where θ is the dilatation, which is equal to $(u_{1,1} + u_{2,2})$ in this case. Equation (9.1.3) yields $(\tau_{11} + \tau_{22}) = 2(\lambda + \mu)\theta$, from which we obtain, upon using Equation (9.1.4),

$$\tau_{33} = \lambda\theta = \frac{\lambda}{2(\lambda + \mu)}(\tau_{11} + \tau_{22}) = \nu(\tau_{11} + \tau_{22}) \tag{9.1.5}$$

where $\nu = \lambda/2(\lambda + \mu)$ is Poisson's ratio.

The equilibrium conditions $\tau_{ij,j} + F_i = 0$ reduce to

$$\tau_{\alpha\beta,\beta} + F_\alpha = 0 \qquad \text{and} \qquad \tau_{33,3} + F_3 = 0 \tag{9.1.6}$$

where τ_{11}, τ_{22}, and τ_{12} are independent of x_3. From Equation (9.1.5), τ_{33} is also independent of x_3. Hence F_3 for this case must vanish, and F_α must be independent of x_3. For a finite medium the equilibrium condition on the boundary is $T_\alpha = \tau_{\alpha\beta} n_\beta$, where T_α is the α component of the surface traction per unit length of boundary and n_β is the β component of the unit normal $\mathbf{n}$ of the boundary.

Substituting Equation (9.1.3) into Equation (9.1.6), we obtain

$$\mu\nabla^2 u_\alpha + (\lambda + \mu)(u_{1,1} + u_{2,2})_{,\alpha} + F_\alpha(x_1, x_2) = 0 \tag{9.1.7}$$

where

$$\nabla^2 \equiv \frac{\partial^2}{\partial {x_1}^2} + \frac{\partial^2}{\partial {x_2}^2}$$

which gives the two equations

$$\lambda\theta_{,1} + 2\mu u_{1,11} + \mu(u_{1,22} + u_{2,12}) + F_1 = 0 \tag{9.1.8}$$

and

$$\lambda\theta_{,\,2} + 2\mu u_{2,\,22} + \mu(u_{2,\,11} + u_{1,\,21}) + F_2 = 0 \tag{9.1.9}$$

The displacements u_α, together with their first, second, and third derivatives, are assumed to be continuous; that is,

$$u_{\alpha,\,\beta\gamma} = u_{\alpha,\,\gamma\beta}$$

The rotation ω about the x_3 axis is given in Equation (1.5.15) as

$$\omega = \tfrac{1}{2}(u_{2,\,1} - u_{1,\,2})$$

From the above two relations, Equation (9.1.8) may be written as

$$\lambda\theta_{,\,1} + 2\mu(u_{1,\,1} + u_{2,\,2})_{,\,1} + \mu(u_{1,\,2} - u_{2,\,1})_{,\,2} + F_1 = 0$$

or

$$(\lambda + 2\mu)\theta_{,\,1} - 2\mu\omega_{,\,2} + F_1 = 0 \tag{9.1.10}$$

Similarly, Equation (9.1.9) may be written as

$$(\lambda + 2\mu)\theta_{,\,2} + 2\mu\omega_{,\,1} + F_2 = 0 \tag{9.1.11}$$

In the regions with no body force, Equations (9.1.10) and (9.1.11) reduce to

$$\left.\begin{aligned}(\lambda + 2\mu)\frac{\partial\theta}{\partial x_1} - 2\mu\frac{\partial\omega}{\partial x_2} &= 0\\ (\lambda + 2\mu)\frac{\partial\theta}{\partial x_2} + 2\mu\frac{\partial\omega}{\partial x_1} &= 0\end{aligned}\right\} \tag{9.1.12}$$

Since the displacements u_α are continuous up to their third derivatives, θ and ω are continuous up to their second derivatives. Equation (9.1.12) shows that $(\lambda + 2\mu)\theta$ and $2\mu\omega$ satisfy the Cauchy-Riemann conditions,[2] and hence $(\lambda + 2\mu)\theta + i2\mu\omega$ is an analytic function of the complex variable $z = x_1 + ix_2$; that is,

$$f(z) = (\lambda + 2\mu)\theta + i2\mu\omega \tag{9.1.13}$$

Any analytical function $f(z)$ gives a distribution of dilatation θ and rotation ω satisfying the equilibrium conditions (9.1.7) and compatibility conditions in regions with no body force. If a function $f(z)$ gives infinite value of plane strain at specified points, such points must not be in the substance of the body, but they may be in cavities within the body. It will be useful to introduce a function

$$F(z) = \int^z f(z)\,dz = \xi + i\eta \tag{9.1.14}$$

It is well known that the integral of an analytic function is also analytic, hence $F(z)$ is an analytic function. Since both the real and the imaginary parts of an analytic function are harmonic functions, i.e., they satisfy $\nabla^2 u = 0$, we have

$$\nabla^2\xi = 0 \qquad \nabla^2\eta = 0 \tag{9.1.15}$$

$$\frac{\partial F}{\partial x_1} = \frac{\partial \xi}{\partial x_1} + i\frac{\partial \eta}{\partial x_1} = \frac{dF}{dz} = f(z) = (\lambda + 2\mu)\theta + i2\mu\omega$$

$$\frac{\partial F}{\partial x_2} = \frac{\partial \xi}{\partial x_2} + i\frac{\partial \eta}{\partial x_2} = i\frac{dF}{dz} = if(z) = i(\lambda + 2\mu)\theta - 2\mu\omega$$

By equating the real and imaginary parts of the above equations, we obtain

$$\begin{aligned} (\lambda + 2\mu)\theta &= \frac{\partial \xi}{\partial x_1} = \frac{\partial \eta}{\partial x_2} \\ 2\mu\omega &= -\frac{\partial \xi}{\partial x_2} = \frac{\partial \eta}{\partial x_1} \end{aligned} \tag{9.1.16}$$

Writing the displacement components u_1 and u_2 as

$$\left.\begin{aligned} u_1 &= \frac{\partial \phi}{\partial x_1} + \frac{\partial \psi}{\partial x_2} + u_1' \\ u_2 &= \frac{\partial \phi}{\partial x_2} - \frac{\partial \psi}{\partial x_1} + u_2' \end{aligned}\right\} \tag{9.1.17}$$

we obtain the equations

$$\theta = u_{1,1} + u_{2,2} = \nabla^2\phi + \frac{\partial u_1'}{\partial x_1} + \frac{\partial u_2'}{\partial x_2} \tag{9.1.18}$$

$$2\omega = u_{2,1} - u_{1,2} = -\nabla^2\psi + \left(\frac{\partial u_2'}{\partial x_1} - \frac{\partial u_1'}{\partial x_2}\right) \tag{9.1.19}$$

Let $\nabla^2\phi = \theta$ and $-\nabla^2\psi = 2\omega$. Then

$$\frac{\partial u_1'}{\partial x_1} + \frac{\partial u_2'}{\partial x_2} = 0 \qquad \frac{\partial u_2'}{\partial x_1} - \frac{\partial u_1}{\partial x_2} = 0 \tag{9.1.20}$$

This means that u_1' and u_2' give neither dilatation nor rotation; that is, u_1' and u_2' represent simple shear.[1] Taking $u_2' + iu_1'$ as an analytic function of $(x_1 + ix_2)$, we see that u_1' and u_2' are plane harmonic functions.

From Equations (9.1.15), (9.1.16), (9.1.18), and (9.1.20) we have

$$\theta = \nabla^2\phi = \frac{1}{\lambda + 2\mu}\frac{\partial \eta}{\partial x_2} = \frac{1}{2(\lambda + 2\mu)}\left[2\frac{\partial \eta}{\partial x_2} + x_2\nabla^2\eta\right] = \frac{1}{\lambda + 2\mu}\nabla^2\left(\frac{x_2\eta}{2}\right) \tag{9.1.21}$$

Similarly,

$$2\omega = -\nabla^2\psi = -\frac{1}{\mu}\nabla^2\left(\frac{x_2\xi}{2}\right) \tag{9.1.22}$$

From Equations (9.1.21) and (9.1.22), we find

$$\phi = \frac{1}{\lambda + 2\mu}\left(\frac{x_2\eta}{2}\right) + g \tag{9.1.23}$$

$$\psi = \frac{1}{\mu}\left(\frac{x_2\xi}{2}\right) + h$$

where g and h are plane harmonic functions; that is, $\nabla^2 g$ and $\nabla^2 h$ are zero. From Equation (9.1.17), we have

$$u_1 = \frac{\partial}{\partial x_1}\left[\frac{x_2\eta}{2(\lambda + 2\mu)}\right] + \frac{\partial}{\partial x_2}\left[\frac{x_2\xi}{2\mu}\right] + \frac{\partial g}{\partial x_1} + \frac{\partial h}{\partial x_2} + u_1'$$

$$u_2 = \frac{\partial}{\partial x_2}\left[\frac{x_2\eta}{2(\lambda + 2\mu)}\right] - \frac{\partial}{\partial x_1}\left[\frac{x_2\xi}{2\mu}\right] + \frac{\partial g}{\partial x_2} - \frac{\partial h}{\partial x_1} + u_2'$$

By calculating $(u_{1,1} + u_{2,2})$ and $(u_{2,1} - u_{1,2})$ it can be shown that $\partial g/\partial x_1$, $\partial g/\partial x_2$, $\partial h/\partial x_1$, and $\partial h/\partial x_2$ do not contribute to dilatation or rotation. They represent only simple shear. Since simple shear is already represented by u_1' and u_2', we may assume that g and h are zero. Then the above equations become

$$\begin{aligned} u_1 &= \frac{\partial}{\partial x_1}\left[\frac{x_2\eta}{2(\lambda + 2\mu)}\right] + \frac{\partial}{\partial x_2}\left[\frac{x_2\xi}{2\mu}\right] + u_1' \\ u_2 &= \frac{\partial}{\partial x_2}\left[\frac{x_2\eta}{2(\lambda + 2\mu)}\right] - \frac{\partial}{\partial x_1}\left[\frac{x_2\xi}{2\mu}\right] + u_2' \end{aligned} \tag{9.1.24}$$

Stress Field Due to a Point Force. Now consider the case $f(z) = A/z$, where z is the complex variable $(x_1 + ix_2)$ and A is a real constant. This gives zero strain and hence zero stress at infinite distance. This corresponds to a stress distribution in an infinite medium. In this case we have

$$F(z) = \int^z f(z)\,dz = A \ln z$$

$$\xi + i\eta = A(\ln r + i\theta) \tag{9.1.25}$$

where r and θ are polar coordinates in the x_1x_2 plane. Equating real and imaginary parts yields

$$\xi = A \ln r \qquad \eta = A\theta = A \tan^{-1}\frac{x_2}{x_1}$$

Substituting these expressions into Equation (9.1.24), we obtain

$$u_1 = \frac{A}{2(\lambda + 2\mu)}\left(-\frac{x_2^2}{r^2}\right) + \frac{1}{2\mu}\left(A \ln r + A\frac{x_2^2}{r^2}\right) + u_1'$$

$$= \frac{A}{2\mu}\ln r + \frac{A(\lambda+\mu)}{2\mu(\lambda+2\mu)}\frac{x_2^2}{r^2} + u_1'$$

$$u_2 = \frac{A}{2(\lambda + 2\mu)}\theta - \frac{\lambda+\mu}{2\mu(\lambda+2\mu)}A\frac{x_1x_2}{r^2} + u_2'$$

For u_2 to be single-valued, u_2' must be equal to $-A\theta/2(\lambda + 2\mu)$. Since $u_2' + iu_1'$ is a function of $x_1 + ix_2$, that is, u_1' is the conjugate function of u_2', we have $u_1' = [A/2(\lambda + 2\mu)] \ln r$. This gives the displacements as

$$u_1 = \frac{\lambda + 3\mu}{2\mu(\lambda+2\mu)}A \ln r + \frac{\lambda+\mu}{2\mu(\lambda+2\mu)}A\frac{x_2^2}{r^2} \qquad (9.1.26)$$

$$u_2 = -\frac{\lambda+\mu}{2\mu(\lambda+2\mu)}A\frac{x_1x_2}{r^2} \qquad (9.1.27)$$

From the stress-displacement relations, Equation (9.1.3), we have, upon using the above displacement relations,

$$\tau_{11} = A\frac{2(\lambda+\mu)}{\lambda+2\mu}\frac{x_1}{r^2}\left(\frac{2\lambda+3\mu}{2(\lambda+\mu)} - \frac{x_2^2}{r^2}\right)$$

$$\tau_{22} = A\frac{2(\lambda+\mu)}{\lambda+2\mu}\frac{x_1}{r^2}\left(-\frac{\mu}{2(\lambda+\mu)} + \frac{x_2^2}{r^2}\right)$$

$$\tau_{12} = A\frac{2(\lambda+\mu)}{\lambda+2\mu}\frac{x_2}{r^2}\left(\frac{\mu}{2(\lambda+\mu)} + \frac{x_1^2}{r^2}\right) \qquad (9.1.28)$$

$$\tau_{33} = \nu(\tau_{11} + \tau_{22}) = \nu A\frac{x_1}{r^2}\frac{2\lambda+2\mu}{\lambda+2\mu}$$

$$\tau_{23} = \tau_{13} = 0$$

Since the stress and strain become infinite at the origin, the origin is considered to be a cavity. The resultant of the force acting at the cavity may be found by integrating the forces acting on a circular boundary about the cavity. The x_1 component of the resultant is

$$P_1 = \int_0^{2\pi} -\left(\tau_{11}\frac{x_1}{r} + \tau_{12}\frac{x_2}{r}\right) r\, d\theta = -2A\pi$$

or

$$A = -\frac{P_1}{2\pi}$$

The x_2 component of the resultant has been found to be zero. The moment of the boundary forces about the origin also vanishes. Hence the stresses given by Equation (9.1.28) correspond to the single force ($2\pi A$) acting at the origin along the negative direction of the x_1 axis. Since $\nu = \lambda/2(\lambda + \mu)$, we have,

$$\frac{1}{1-\nu} = \frac{2(\lambda+\mu)}{\lambda+2\mu}$$

$$\frac{1-2\nu}{2} = \frac{\mu}{2(\lambda+\mu)} \qquad (9.1.29)$$

$$\frac{3-2\nu}{2} = \frac{2\lambda+3\mu}{2(\lambda+\mu)}$$

and Equations (9.1.28) may then be written as

$$\tau_{11} = \frac{-P_1 x_1}{2\pi(1-\nu)r^2}\left(\frac{3-2\nu}{2} - \frac{{x_2}^2}{r^2}\right)$$

$$\tau_{22} = \frac{-P_1 x_1}{2\pi(1-\nu)r^2}\left(-\frac{1-2\nu}{2} + \frac{{x_2}^2}{r^2}\right) \qquad (9.1.30)$$

$$\tau_{12} = \frac{-P_1 x_2}{2\pi(1-\nu)r^2}\left(\frac{1-2\nu}{2} + \frac{{x_1}^2}{r^2}\right)$$

These same results may also be obtained by the method of Fourier transforms.[3] Similarly, for a load P_2 acting at the origin along the x_2 axis, the stress distribution may be obtained by replacing P_1 by P_2 and interchanging the roles of subscripts 1 and 2. From Equations (9.1.30), we have

$$\tau_{11} = -\frac{P_2 x_2}{2\pi(1-\nu)r^2}\left(-\frac{1-2\nu}{2} + \frac{{x_1}^2}{r^2}\right)$$

$$\tau_{22} = -\frac{P_2 x_2}{2\pi(1-\nu)r^2}\left(\frac{3-2\nu}{2} - \frac{{x_1}^2}{r^2}\right) \qquad (9.1.31)$$

$$\tau_{12} = -\frac{P_2 x_1}{2\pi(1-\nu)r}\left(\frac{1-2\nu}{2} + \frac{{x_2}^2}{r^2}\right)$$

For an infinite elastic medium under plane strain and subject to a distribution of body force components $F_1(x_1', x_2')$ and $F_2(x_1', x_2')$, the stress distribution is obtained by replacing P_1 with $F_1\,dx_1'\,dx_2'$, P_2 with $F_2\,dx_1'\,dx_2'$, x_1 with $x_1 - x_1'$ and x_2 with $x_2 - x_2'$. The resulting stress distribution becomes

$$\tau_{11} = \iint \left\{ \frac{-F_1(x_1 - x_1')}{2\pi(1-\nu)r^2} \left[\frac{(3-2\nu)}{2} - \frac{(x_2 - x_2')^2}{r^2} \right] - \frac{F_2(x_2 - x_2')}{2\pi(1-\nu)r^2} \left[-\frac{1-2\nu}{2} + \frac{(x_1 - x_1')^2}{r^2} \right] \right\} dx_1' \, dx_2'$$

$$\tau_{22} = \iint \left\{ \frac{-F_1(x_1 - x_1')}{2\pi(1-\nu)r^2} \left[-\frac{1-2\nu}{2} + \frac{(x_2 - x_2')^2}{r^2} \right] - \frac{F_2(x_2 - x_2')}{2\pi(1-\nu)r^2} \left[\frac{3-2\nu}{2} - \frac{(x_1 - x_1')^2}{r^2} \right] \right\} dx_1' \, dx_2'$$

$$\tau_{12} = \iint \left\{ \frac{-F_1(x_2 - x_2')}{2\pi(1-\nu)r^2} \left[\frac{1-2\nu}{2} + \frac{(x_1 - x_1')^2}{r^2} \right] - \frac{F_2(x_1 - x_1')}{2\pi(1-\nu)r^2} \left[\frac{1-2\nu}{2} + \frac{(x_2 - x_2')^2}{r^2} \right] \right\} dx_1' \, dx_2' \tag{9.1.32}$$

where $r^2 = (x_1 - x_1')^2 + (x_2 - x_2')^2$. This gives the general solution for an infinite medium under plane strain subject to an arbitrary distributed body force $F_\alpha(x_1, x_2)$.

Consider a uniform cylinder of infinite length along the x_3 axis which is cut out of an infinite medium under plane strain and subject to a given distribution of body force F_α. The stresses or tractions on the boundary of the cylinder are found from Equation (9.1.32). This traction is denoted by $\mathbf{T}_0$, as shown in Figure 9.1(a). Now consider the same cylinder without body force F_α but subject to a boundary traction of $-\mathbf{T}_0 + \mathbf{T}_B$ as shown in Figure 9.1(b). The superposition of the stress field of Figure 9.1(a) to the stress field of Figure 9.1(b) gives the stress field in the cylinder subject to a body force F_α and a boundary force $\mathbf{T}_B$ as shown in Figure 9.1(c). The solution for the problem of an elastic body of finite uniform section under plane strain and subject to both body and boundary forces may be reduced,

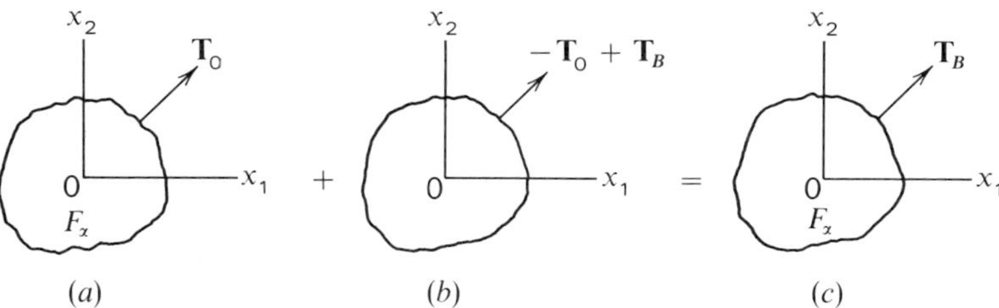

Figure 9.1

through the solution of an infinite medium under plane strain and the same body force, to the problem of determining the stress distribution in a cylinder under plane strain with no body force but subject to a certain boundary force.

9.2 Airy's Stress Function[4,5]

For the case of plane deformation with no body force, the equilibrium conditions are

$$\frac{\partial \tau_{11}}{\partial x_1} + \frac{\partial \tau_{12}}{\partial x_2} = 0 \qquad \frac{\partial \tau_{21}}{\partial x_1} + \frac{\partial \tau_{22}}{\partial x_2} = 0 \tag{9.2.1}$$

The first of Equations (9.2.1) is satisfied by letting

$$\tau_{11} = \frac{\partial \Phi}{\partial x_2} \qquad \text{and} \qquad \tau_{12} = -\frac{\partial \Phi}{\partial x_1} \tag{9.2.2}$$

The second of Equations (9.2.1) is satisfied by letting

$$\tau_{22} = \frac{\partial \Psi}{\partial x_1} \qquad \tau_{21} = -\frac{\partial \Psi}{\partial x_2} \tag{9.2.3}$$

Since $\tau_{12} = \tau_{21}$, we see that Φ and Ψ must satisfy the condition

$$\frac{\partial \Phi}{\partial x_1} = \frac{\partial \Psi}{\partial x_2} \tag{9.2.4}$$

This condition is satisfied by letting

$$\Phi = \frac{\partial U}{\partial x_2} \qquad \text{and} \qquad \Psi = \frac{\partial U}{\partial x_1} \tag{9.2.5}$$

This gives

$$\tau_{11} = U,_{22} \qquad \tau_{22} = U,_{11} \qquad \tau_{12} = -U,_{12} \tag{9.2.6}$$

The compatibility condition, Equation (1.8.13), requires that

$$\frac{\partial^2 e_{11}}{\partial {x_2}^2} + \frac{\partial^2 e_{22}}{\partial {x_1}^2} = 2\frac{\partial^2 e_{12}}{\partial x_1\, \partial x_2} \tag{9.2.7}$$

Strain is related to the stress by

$$e_{11} = \frac{1}{E}[\tau_{11} - \nu(\tau_{22} + \tau_{33})] \qquad e_{22} = \frac{1}{E}[\tau_{22} - \nu(\tau_{11} + \tau_{33})]$$

$$e_{12} = \frac{1}{2\mu}\tau_{12}$$

where E is the Young's modulus and ν is the Poisson's ratio. From Equations (9.1.5), we find

$$e_{11} = \frac{1}{E}[(1 - \nu^2)\tau_{11} - \nu(1 + \nu)\tau_{22}]$$

$$e_{22} = \frac{1}{E}[(1 - \nu^2)\tau_{22} - \nu(1 + \nu)\tau_{11}]$$

Substituting these expressions into Equation (9.2.7) and noting that $E = 2\mu(1 + \nu)$, we have

$$(1 - \nu)\frac{\partial^2\tau_{11}}{\partial x_2{}^2} - \nu\frac{\partial^2\tau_{22}}{\partial x_2{}^2} + (1 - \nu)\frac{\partial^2\tau_{22}}{\partial x_1{}^2} - \nu\frac{\partial^2\tau_{11}}{\partial x_1{}^2} = 2\frac{\partial^2\tau_{12}}{\partial x_1\,\partial x_2} \tag{9.2.8}$$

From Equation (9.2.1), we obtain

$$2\frac{\partial^2\tau_{12}}{\partial x_1\,\partial x_2} = -\frac{\partial^2\tau_{11}}{\partial x_1{}^2} - \frac{\partial^2\tau_{22}}{\partial x_2{}^2}$$

Equation (9.2.8) then reduces to

$$\nabla^2(\tau_{11} + \tau_{22}) = 0 \tag{9.2.9}$$

Substituting the expressions of Equations (9.2.6) into Equation (9.2.9), we obtain

$$\nabla^2\nabla^2 U = \nabla^4 U = 0 \tag{9.2.10}$$

The U defined by Equation (9.2.6) is called Airy's stress function and, from Equation (9.2.10), is seen to be biharmonic. The U must have derivatives up to the fourth order. For the stress to be single-valued, the second partial derivatives of U must be single-valued.

For the case in which the traction on the boundary is specified we have, from Equation (1.1.4),

$$f_{n1} = T_1 = \tau_{\alpha 1}\, l_{n\alpha} = \tau_{11} l_{n1} + \tau_{21} l_{n2} = U,_{22}\, l_{n1} - U,_{12}\, l_{n2} \tag{9.2.11}$$

$$f_{n2} = T_2 = \tau_{\alpha 2}\, l_{n\alpha} = \tau_{12}\, l_{n1} + \tau_{22}\, l_{n2} = -U,_{12}\, l_{n1} + U,_{11} l_{n2} \tag{9.2.12}$$

where T_1 and T_2 are the components of $\mathbf{T}$ along the x_1 and x_2 axis respectively. Referring to Figure 9.2, we note that

$$l_{n1} = \cos(x_1, n) = \cos(x_2, s) = \frac{dx_2}{ds}$$

$$l_{n2} = \cos(x_2, n) = -\cos(x_1 s) = -\frac{dx_1}{ds}$$

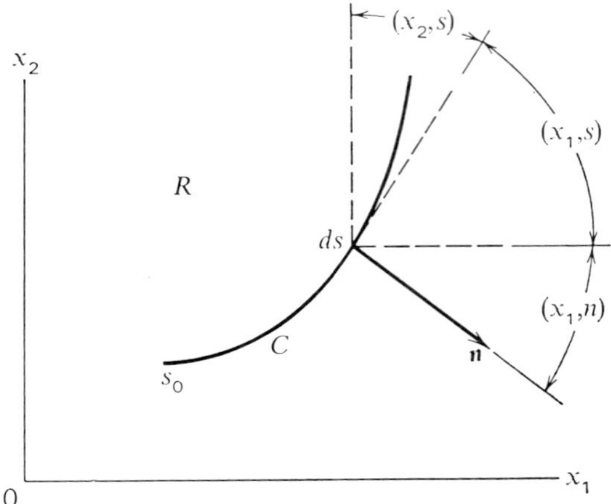

Figure 9.2. Segment of boundary C.

Substituting the above expressions for l_{n1} and l_{n2} into Equations (9.2.11) and (9.2.12), we obtain

$$U,_{22}\frac{dx_2}{ds} - U,_{21}\left(-\frac{dx_1}{ds}\right) = T_1(s)$$

or

$$\frac{d}{ds}(U,_2) = T_1(s) \tag{9.2.13}$$

and similarly,

$$-\frac{d}{ds}(U,_1) = T_2(s) \tag{9.2.14}$$

Integrating these two equations from some fixed point s_0, we obtain

$$\begin{aligned} U,_1(s) &= -\int_{s_0}^{s} T_2(s)\,ds = f_1(s) + c_1 \\ U,_2(s) &= \int_{s_0}^{s} T_1(s)\,ds = f_2(s) + c_2 \end{aligned} \tag{9.2.15}$$

Hence the solution of stresses in a body under plane strain and subject to boundary forces is reduced to the determination of a biharmonic function U satisfying $\nabla^4 U = 0$ within the cross section R and $U,_\alpha = f_\alpha(s)$ on the boundary C.

Since the problem of a body under plane strain requires the determination of U, the general solution of the biharmonic equation is first shown. Consider $\nabla^2 U = P_1(x_1, x_2)$ and recall that the second derivatives of U correspond to stresses which are single-valued. Hence P_1 is single-valued, and since $\nabla^4 U = 0$, we see that P_1 must satisfy $\nabla^2 P_1 = 0$, which may be written as

$$\frac{\partial}{\partial x_1}\left(\frac{\partial P_1}{\partial x_1}\right) + \frac{\partial}{\partial x_2}\left(\frac{\partial P_1}{\partial x_2}\right) = 0$$

This equation is satisfied by letting

$$\frac{\partial P_1}{\partial x_1} = \frac{\partial P_2}{\partial x_2} \quad \text{and} \quad \frac{\partial P_1}{\partial x_2} = -\frac{\partial P_2}{\partial x_1}$$

Since these equations satisfy the Cauchy-Riemann conditions for an analytic function of the complex variable $z = x_1 + ix_2$, an analytical function

$$F(z) = P_1 + iP_2 \tag{9.2.16}$$

may be formed. It is well known that the integral of an analytic function is also analytic, hence the function $\phi(z)$ defined by

$$\phi(z) = \frac{1}{4}\int^z F(z)\, dz = p_1 + ip_2 \tag{9.2.17}$$

is analytic and has the derivative $\phi'(z)$. Since ϕ is an analytic function, p_1 and p_2 are harmonic in R. From Equations (9.2.16) and (9.2.17), we have

$$\frac{\partial \phi}{\partial x_1} = \frac{d\phi}{dz}\frac{\partial z}{\partial x_1} = \frac{\partial p_1}{\partial x_1} + i\frac{\partial p_2}{\partial x_1}$$

and therefore, since $\partial z/\partial x_1 = 1$, we have

$$\phi'(z) = \tfrac{1}{4}(P_1 + iP_2) = p_{1,1} + ip_{2,1}$$

Since p_1 and p_2 satisfy the Cauchy-Riemann conditions, the following equations result:

$$p_{1,1} = p_{2,2} \quad \text{and} \quad p_{1,2} = -p_{2,1} \tag{9.2.18}$$

$$\nabla^2 U = P_1 = 4p_{1,1} = 2(p_{1,1} + p_{2,2})$$

$$\nabla^2(x_1 p_1) = x_1 \nabla^2 p_1 + 2p_{1,1} = 2p_{1,1}$$

Similarly,

$$\nabla^2(x_2 p_2) = 2p_{2,2}$$

Hence, from the above three equations, we obtain

$$\nabla^2(U - x_1 p_1 - x_2 p_2) = 0 \qquad \text{in } R$$

whose solution is

$$U = x_1 p_1 + x_2 p_2 + q_1(x_1, x_2)$$

where $q_1(x_1, x_2)$ is harmonic in R.

Now let $\chi(z)$ be the function of the complex variable z, whose real part is q_1, that is,

$$\chi(z) = q_1 + iq_2$$

Since

$$\begin{aligned}\bar{z}\phi(z) + \chi(z) &= (x_1 - ix_2)(p_1 + ip_2) + q_1 + iq_2 \\ &= x_1 p_1 + x_2 p_2 + q_1 + i[x_1 p_2 - x_2 p_1 + q_2]\end{aligned}$$

where the bar on top of a quantity denotes its complex conjugate, we have

$$U = \mathscr{R}\{\bar{z}\phi(z) + \chi(z)\} \tag{9.2.19}$$

where $\mathscr{R}$ denotes the real part of the expression in the bracket. Equation (9.2.19) may be written alternatively as

$$2U = \bar{z}\phi(z) + z\,\overline{\phi(z)} + \chi(z) + \overline{\chi(z)} \tag{9.2.20}$$

This expresses Airy's stress function in terms of two analytical complex functions, ϕ and χ.

From Equation (9.2.6), we have

$$\begin{aligned}\tau_{11} + i\tau_{12} &= U,_{22} - iU,_{12} = -i(U,_1 + iU,_2),_2 \\ \tau_{22} - i\tau_{12} &= U,_{11} + iU,_{12} = (U,_1 + iU,_2),_1\end{aligned} \tag{9.2.21}$$

From Equations (9.2.20), we have

$$U,_1 = \frac{\partial U}{\partial z}\frac{\partial z}{\partial x_1} + \frac{\partial U}{\partial \bar{z}}\frac{\partial \bar{z}}{\partial x_1} = \frac{\partial U}{\partial z} + \frac{\partial U}{\partial \bar{z}}$$

$$U,_2 = \frac{\partial U}{\partial z}\frac{\partial z}{\partial x_2} + \frac{\partial U}{\partial \bar{z}}\frac{\partial \bar{z}}{\partial x_2} = i\frac{\partial U}{\partial z} - i\frac{\partial U}{\partial \bar{z}}$$

$$U,_1 + iU,_2 = 2\frac{\partial U}{\partial \bar{z}} = \phi(z) + z\,\overline{\phi'(z)} + \overline{\psi(z)} \tag{9.2.22}$$

where $\psi(z) = \chi'(z)$. Substituting Equations (9.2.22) into Equations (9.2.21) yields

$$\begin{aligned}\tau_{11} + i\tau_{12} &= \phi'(z) + \overline{\phi'(z)} - z\,\overline{\phi''(z)} - \overline{\psi'(z)} \\ \tau_{22} - i\tau_{12} &= \phi'(z) + \overline{\phi'(z)} + z\,\overline{\phi''(z)} + \overline{\psi'(z)}\end{aligned} \tag{9.2.23}$$

Writing these in a different form, we have

$$\begin{aligned}\tau_{11} + \tau_{22} &= 2[\phi'(z) + \overline{\phi'(z)}] = 4\mathscr{R}[\phi'(z)] \\ \tau_{22} - \tau_{11} + 2i\tau_{12} &= 2[\bar{z}\phi''(z) + \psi'(z)]\end{aligned} \tag{9.2.24}$$

where the second of Equations (9.2.24) is obtained by subtracting the first of Equations (9.2.23) from the second and taking the complex conjugate of the result. These give the stress components in terms of the two analytic functions ϕ and ψ.

From Equations (9.2.15), we have

$$U,_1 + iU,_2 = f_1(s) + if_2(s) + \text{const.} \qquad \text{on } C$$

where

$$f_1(s) + if_2(s) = i\int^s [T_1(s) + iT_2(s)]\, ds$$

Substituting the expression given by Equation (9.2.22) gives

$$\phi(z) + z\,\overline{\phi'(z)} + \overline{\psi(z)} = f_1 + if_2 + \text{const.} \qquad \text{on } C \tag{9.2.25}$$

For a simply connected finite region, the stress distribution of the body is not affected by replacing ϕ with ϕ + const. Therefore the constant in the above equation is arbitrary and may be omitted, and then

$$\phi(z) + z\,\overline{\phi'(z)} + \overline{\psi(z)} = f_1 + if_2 = U,_1 + iU,_2 \qquad \text{on } C \tag{9.2.26}$$

Hence

$$\int_{s_1}^{s_2} [T_1(s) + iT_2(s)]\, ds = -i\Big[\phi(z) + z\,\overline{\phi'(z)} + \overline{\psi(z)}\Big]_{s_1}^{s_2} \tag{9.2.27}$$

gives the resultant force over the boundary $s_1 s_2$. The resultant moment of the boundary tractions from s_1 to s_2 about the origin is

$$M = \int_{s_1}^{s_2} (x_1 T_2 - x_2 T_1)\, ds$$

Substituting the expressions for T_1 and T_2 from Equations (9.2.13) and (9.2.14) and integrating by parts, we obtain

$$\begin{aligned}M &= -\int_{s_1}^{s_2} \{x_1\, dU,_1 + x_2\, dU,_2\}\, ds \\ &= -\Big[x_1 U,_1 + x_2 U,_2\Big]_{s_1}^{s_2} + \Big[U\Big]_{s_1}^{s_2}\end{aligned} \tag{9.2.28}$$

Using

$$x_1 U,_1 + x_2 U,_2 = \mathscr{R}[z(U,_1 - iU,_2)] \tag{9.2.29}$$

Equation (9.2.22),

$$U,_1 - iU,_2 = \overline{\phi(z)} + \bar{z}\phi'(z) + \psi(z)$$

and Equation (9.2.19), we see that Equation (9.2.28) for M becomes

$$M = \mathscr{R}\Big[\chi(z) - z\psi(z) - z\bar{z}\phi'(z)\Big]_{s_1}^{s_2} \tag{9.2.30}$$

For a simply connected region, letting $s_1 = s_2$, we see that the resultant forces along the x_1 and x_2 directions, and the resultant moment all vanish, since $\phi(z)$, $\psi(z)$, and $\chi(z)$ are single-valued.

Expressing the stresses in terms of displacements, we have, from Equations (9.1.4) and (9.1.5),

$$\left.\begin{aligned} \tau_{11} &= U,_{22} = \lambda(u_{1,1} + u_{2,2}) + 2\mu u_{1,1} \\ \tau_{22} &= U,_{11} = \lambda(u_{1,1} + u_{2,2}) + 2\mu u_{2,2} \\ \tau_{12} &= -U,_{12} = \mu(u_{1,2} + u_{2,1}) \end{aligned}\right\} \tag{9.2.31}$$

Then, from Equations (9.2.31), we find

$$\nabla^2 U = 2(\lambda + \mu)(u_{1,1} + u_{2,2})$$

$$2\mu u_{1,1} = U,_{22} - \frac{\lambda}{2(\lambda + \mu)}\nabla^2 U = -U,_{11} + \frac{\lambda + 2\mu}{2(\lambda + \mu)}\nabla^2 U$$

$$2\mu u_{2,2} = -U,_{22} + \frac{\lambda + 2\mu}{2(\lambda + \mu)}\nabla^2 U$$

Noting

$$\nabla^2 U = P_1 = 4p_{1,1} = 4p_{2,2}$$

$$2\mu u_{1,1} = -U,_{11} + \frac{2(\lambda + 2\mu)}{\lambda + \mu}p_{1,1}$$

and integrating the last expression gives

$$2\mu u_1 = -U,_1 + \frac{2(\lambda + 2\mu)}{\lambda + \mu}p_1 + f(x_2)$$

Similarly,

$$2\mu u_2 = -U,_2 + \frac{2(\lambda + 2\mu)}{\lambda + \mu}p_2 + g(x_1)$$

where $f(x_2)$ and $g(x_1)$ are, as yet, arbitrary functions. It follows, from the two above equations and the last of Equations (9.2.31), that

$$2\tau_{12} = 2\mu(u_{1,2} + u_{2,1}) = -2U_{,12} + \frac{2(\lambda + 2\mu)}{\lambda + \mu}(p_{1,2} + p_{2,1}) + f'(x_2) + g'(x_1)$$

From Equations (9.2.18) and (9.2.31) ,we have

$$f'(x_2) + g'(x_1) = 0$$

and therefore

$$f(x_2) = ax_2 + b \qquad g(x_1) = -ax_1 + c$$

where a, b, and c are constants. The functions f and g in u_1 and u_2 give rigid body displacements and may be disregarded in strain analysis. With $f = g = 0$, we have

$$2\mu(u_1 + iu_2) = \frac{2(\lambda + 2\mu)}{\lambda + \mu}(p_1 + ip_2) - U_{,1} - iU_{,2}$$

From Equations (9.2.17) and (9.2.22), we have

$$2\mu(u_1 + iu_2) = \left(\frac{2(\lambda + 2\mu)}{\lambda + \mu} - 1\right)\phi(z) - z\overline{\phi'(z)} - \overline{\psi(z)}$$

$$= \frac{\lambda + 3\mu}{\lambda + \mu}\phi(z) - z\overline{\phi'(z)} - \overline{\psi(z)} \qquad (9.2.32)$$

If the displacement, instead of traction, is specified along the boundary, Equation (9.2.32) is used instead of (9.2.27).

In many cases, it is more convenient to express stresses and displacements in polar coordinates r and α. Referring to Figure 9.3, we see that $z = x_1 + ix_2 = r \exp(i\alpha)$. With u_r and u_α denoting the components of

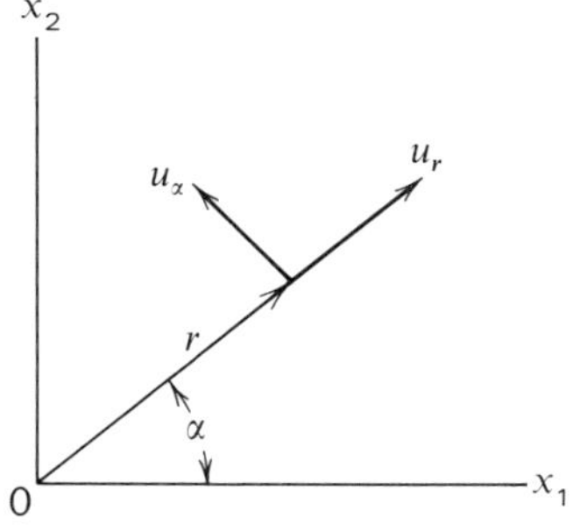

Figure 9.3

displacement along the radial and circumferential directions, we see also that

$$u_1 = u_r \cos \alpha - u_\alpha \sin \alpha$$

and

$$u_2 = u_r \sin \alpha + u_\alpha \cos \alpha$$

Hence

$$u_1 + iu_2 = (u_r + iu_\alpha) \exp(i\alpha)$$

Solving this for $(u_r + iu_\alpha)$ yields

$$u_r + iu_\alpha = (u_1 + iu_2) \exp(-i\alpha)$$

Using Equation (9.2.32), we obtain

$$2\mu(u_r + iu_\alpha) = \exp(-i\alpha)\left[\frac{\lambda + 3\mu}{\lambda + \mu}\,\phi(z) - z\,\overline{\phi'(z)} - \overline{\psi(z)}\right] \tag{9.2.33}$$

The stresses in polar coordinates are obtained from Equation (9.2.24) through the transformation of coordinates,[4] as

$$\begin{aligned} \tau_{rr} + \tau_{\alpha\alpha} &= 4\mathscr{R}[\phi'(z)] \\ \tau_{\alpha\alpha} - \tau_{rr} + 2i\tau_{r\alpha} &= 2[\bar{z}\phi''(z) + \psi'(z)] \exp(2i\alpha) \end{aligned} \tag{9.2.34}$$

Denoting $\phi'(z)$ by $\Phi(z)$ and $\psi'(z)$ by $\Psi(z)$, we have

$$\begin{aligned} \tau_{rr} + \tau_{\alpha\alpha} &= 4\mathscr{R}[\Phi(z)] \\ \tau_{\alpha\alpha} - \tau_{rr} + 2i\tau_{r\alpha} &= 2[\bar{z}\Phi'(z) + \Psi(z)] \exp(2i\alpha) \end{aligned} \tag{9.2.35}$$

9.3 Generalized Plane Stress in an Elastic Body[4,5]

Consider an elastic thin plate under forces acting in its plane. The thickness $2h$ (Figure 9.4) is small compared with the linear dimensions of the cross section. It is assumed that the faces are free of applied loads, and that the applied loads on the edges are parallel to the faces and symmetrically distributed with respect to the middle surface. The component of the body force F_3 normal to the plate is assumed to be zero and the other two components F_α are assumed to be symmetrical with respect to the middle plane. Under these conditions, the points on the middle plane undergo no displacement along the x_3 axis. If the plate is thin, u_3 will be small. The mean value of u_3 throughout the plate thickness vanishes. The mean values of the displacements along the other two axes are given as

$$\overline{u_\alpha}(x_1, x_2) = \frac{1}{2h}\int_{-h}^{h} u_\alpha(x_1, x_2, x_3)\, dx_3 \tag{9.3.1}$$

Since the faces are free of external loads, we have

$$\tau_{13}(x_1, x_2, \pm h) = \tau_{23}(x_1, x_2, \pm h) = \tau_{33}(x_1, x_2, \pm h) = 0 \tag{9.3.2}$$

Applying the equilibrium condition on the faces of the plate yields

$$\tau_{13,1} + \tau_{23,2} + \tau_{33,3} = 0$$

Since $\tau_{13,1}(x_1, x_2, \pm h) = \tau_{23,3}(x_1, x_2, \pm h) = 0$, we find

$$\tau_{33,3}(x_1, x_2, \pm h) = 0$$

Both τ_{33} and $\tau_{33,3}$ vanish on the faces, so for small h, we generally have τ_{33} approximately equal to zero. This justifies the assumption that $\tau_{33} = 0$. Consider the equilibrium conditions along the x_α axis,

$$\tau_{\alpha 1,1} + \tau_{\alpha 2,2} + \tau_{\alpha 3,3} + F_\alpha = 0$$

Multiplying this equation by dx_3 and integrating from $-h$ to $+h$ and noting

$$\bar{\tau}_{\alpha 3} = \frac{1}{2h}\int_{-h}^{h} \tau_{\alpha 3,3}\, dx_3 = \frac{1}{2h}[\tau_{\alpha 3}(x_1, x_2, h) - \tau_{\alpha 3}(x_1, x_2, -h)] = 0,$$

we obtain

$$\bar{\tau}_{\alpha 1,1} + \bar{\tau}_{\alpha 2,2} + \bar{F}_\alpha = 0 \qquad \bar{\tau}_{\alpha 3} = 0 \tag{9.3.3}$$

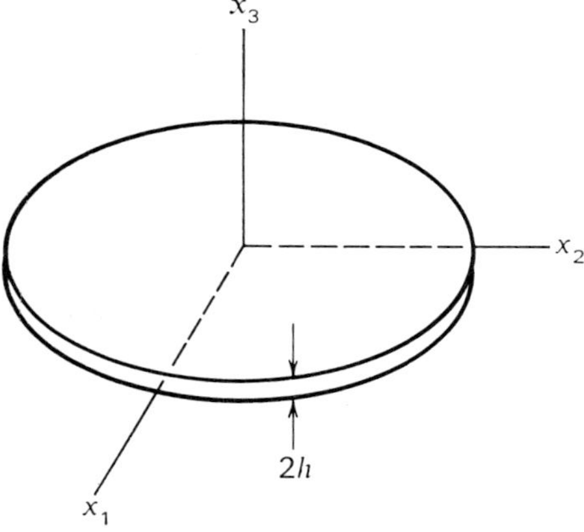

Figure 9.4

where the bar on the top denotes the average value stress through the thickness. *This state of stress is called generalized plane stress.*

$$\bar{\tau}_{33} = \lambda(\bar{u}_{1,1} + \bar{u}_{2,2} + \bar{u}_{3,3}) + 2\mu\bar{u}_{3,3} = 0$$

$$\bar{u}_{3,3} = -\frac{\lambda}{\lambda + 2\mu}(\bar{u}_{1,1} + \bar{u}_{2,2})$$

$$\left.\begin{aligned} \bar{\tau}_{11} &= \frac{2\lambda\mu}{\lambda + 2\mu}(\bar{u}_{1,1} + \bar{u}_{2,2}) + 2\mu\bar{u}_{1,1} \\ \bar{\tau}_{22} &= \frac{2\lambda\mu}{\lambda + 2\mu}(\bar{u}_{1,1} + \bar{u}_{2,2}) + 2\mu\bar{u}_{2,2} \\ \bar{\tau}_{12} &= \mu(\bar{u}_{1,2} + \bar{u}_{2,1}) \end{aligned}\right\} \tag{9.3.4}$$

This may be written as

$$\bar{\tau}_{\alpha\beta} = \bar{\lambda}\,\delta_{\alpha\beta}(\bar{u}_{1,1} + \bar{u}_{2,2}) + \mu(\bar{u}_{\alpha,\beta} + \bar{u}_{\beta,\alpha}) \tag{9.3.5}$$

where

$$\bar{\lambda} = \frac{2\lambda\mu}{\lambda + 2\mu} \tag{9.3.6}$$

Substituting Equation (9.3.5) into Equation (9.3.3), we obtain

$$\mu\nabla^2\bar{u}_\alpha + (\bar{\lambda} + \mu)(\bar{u}_{1,1\alpha} + \bar{u}_{2,2\alpha}) + \bar{F}_\alpha = 0 \tag{9.3.7}$$

Equation (9.3.7) is identical with Equation (9.1.7), (9.3.5) with (9.1.3), and the first equation of (9.3.3) with the first of (9.1.6) if λ, $\tau_{\alpha\beta}$, and u_α in the former are replaced by $\bar{\lambda}$, $\bar{\tau}_{\alpha\beta}$, and $\bar{u}_\alpha$ respectively. For plane stress problems, the stress field due to distributed forces $\bar{F}_1$ and $\bar{F}_2$ is the same as Equation (9.1.32) with ν replaced by

$$\bar{\nu} = \frac{\bar{\lambda}}{2(\bar{\lambda} + \mu)} = \frac{\lambda}{3\lambda + 2\mu} = \frac{\nu}{1 + \nu} \tag{9.3.8}$$

this gives

$$\bar{\tau}_{11} = \iint\left\{-\frac{\bar{F}_1(x_1 - x_1')}{4\pi r^2}\left[(3 + \nu) - 2(1 + \nu)\frac{(x_2 - x_2')^2}{r^2}\right]\right.$$
$$\left.-\frac{\bar{F}_2(x_2 - x_2')}{4\pi r^2}\left[-(1 - \nu) + 2(1 + \nu)\frac{(x_1 - x_1')^2}{r^2}\right]\right\} dx_1'\,dx_2'$$

$$\bar{\tau}_{22} = \iint \left\{ -\frac{\bar{F}_1(x_1 - x_1')}{4\pi r^2} \left[-(1 - \nu) + 2(1 + \nu) \frac{(x_2 - x_2')^2}{r^2} \right] \right. \tag{9.3.9}$$

$$\left. - \frac{\bar{F}_2(x_2 - x_2')}{4\pi r^2} \left[(3 + \nu) - 2(1 + \nu) \frac{(x_1 - x_1')^2}{r^2} \right] \right\} dx_1'\, dx_2'$$

$$\bar{\tau}_{12} = \iint \left\{ -\frac{\bar{F}_1(x_2 - x_2')}{4\pi r^2} \left[(1 - \nu) + 2(1 + \nu) \frac{(x_1 - x_1')^2}{r^2} \right] \right.$$

$$\left. - \frac{\bar{F}_2(x_1 - x_1')}{4\pi r^2} \left[(1 - \nu) + 2(1 + \nu) \frac{(x_2 - x_2')^2}{r^2} \right] \right\} dx_1'\, dx_2'$$

Hence the method for the solution of plane deformation problems shown in the previous two sections may be applied as well to the solution of generalized plane stress problems.

9.4 Regions with Circular Boundaries[4,5]

Let the origin of the coordinates x_1, x_2 be at the center of the circle with radius R as shown in Figure 9.5. From Section 9.2, we have

$$f_1 + if_2 = i \int_0^s (T_1 + iT_2)\, ds = iR \int_0^\theta (T_1 + iT_2)\, d\theta \tag{9.2.15}$$

On the boundary,

$$f_1 + if_2 = \phi(z) + z\,\overline{\phi'(z)} + \overline{\psi(z)}$$

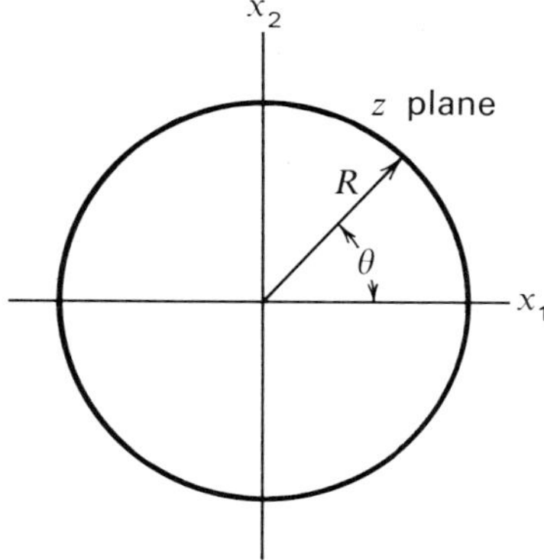

Figure 9.5

where $z = R \exp i\theta$. Representing $f_1 + if_2$ in series,[4] we have

$$f_1 + if_2 = \sum_{k=-\infty}^{\infty} A_k \exp(ik\theta) \tag{9.4.1}$$

where

$$A_k = \frac{1}{2\pi} \int_0^{2\pi} [f_1(\theta) + if_2(\theta)] \exp(-ik\theta)\, d\theta \tag{9.4.2}$$

Since $\phi(z)$ and $\psi(z)$ are analytic inside the boundary, they may be expressed in power series for $|z| < R$. From Equation (9.2.24), the stresses depend not on $\phi(z)$ but on $\phi'(z)$ and ϕ''. Hence $\phi(z)$ may be determined within an arbitrary constant. Letting $\phi(0) = 0$, we may write

$$\begin{aligned} \phi(z) &= \sum_{k=1}^{\infty} a_k z^k \qquad \psi(z) = \sum_{k=0}^{\infty} b_k z^k \\ \overline{\phi'(z)} &= \sum_{k=1}^{\infty} k\bar{a}_k \bar{z}^{k-1} \qquad \overline{\psi(z)} = \sum_{k=\infty}^{\infty} \bar{b}\bar{z}^k \end{aligned} \tag{9.4.3}$$

These series are assumed to converge in the interior and on the boundary. On the boundary $z = R\exp(i\theta)$ and $\bar{z} = R\exp(-i\theta)$. Hence on the boundary

$$\begin{aligned} z\bar{\phi}' &= z \sum_{k=1}^{\infty} k\bar{a}_k(\bar{z})^{k-1} = \sum_{k=1}^{\infty} k\bar{a}_k R^k \exp[-(k-2)i\theta] \\ &= \bar{a}_1 R \exp(i\theta) + \sum_{k=0}^{\infty} (k+2)\bar{a}_{k+2} R^{k+2} \exp(-ik\theta) \end{aligned} \tag{9.4.4}$$

Substituting Equations (9.4.1), (9.4.3), and (9.4.4) into Equation (9.2.26), we obtain

$$\begin{aligned} \sum_{k=1}^{\infty} a_k R^k \exp(ik\theta) + \bar{a}_1 R \exp(i\theta) + \sum_{k=0}^{\infty} (k+2)\bar{a}_{k+2} R^{k+2} \exp(-ik\theta) \\ + \sum_{k=0}^{\infty} \bar{b}_k R^k \exp(-ik\theta) = \sum_{k=-\infty}^{\infty} A_k \exp(ik\theta) \end{aligned} \tag{9.4.5}$$

Equating the coefficients of $\exp(in\theta)$, we obtain,

$$a_1 R + \bar{a}_1 R = A_1 \qquad a_1 + \bar{a}_1 = \frac{A_1}{R} \tag{9.4.6}$$

$$a_n R^n = A_n \quad a_n = \frac{A_n}{R^n} \qquad n > 1 \tag{9.4.7}$$

$$(n+2)\bar{a}_{n+2} R^{n+2} + R^n \bar{b}_n = A_{-n} \qquad n \geq 0$$

$$b_n = \frac{\bar{A}_{-n}}{R^n} - (n+2)a_{n+2} R^2 = \frac{\bar{A}_{-n}}{R^n} - (n+2)\frac{A_{n+2}}{R^n} \qquad n \geq 0 \tag{9.4.8}$$

As an illustration of the use of the above formulae, the following two examples are presented.

Example 1: Uniform Pressure. A uniform pressure p is acting on the boundary of a circle of radius R. The following expressions may be obtained:

$$T_1 = -p\cos\theta \qquad T_2 = -p\sin\theta$$

$$f_1(s) + if_2(s) = i\int^s (T_1 + iT_2)\,ds = -i\int^\theta p\exp(i\theta)R\,d\theta = -pR\exp(i\theta)$$

$$A_1 = -pR \quad A_k = 0 \qquad k \neq 1$$

$$a_1 + \bar{a}_1 = -p \quad a_n = 0 \qquad n \neq 1$$

$$b_n = 0$$

$$\phi(z) = \frac{-p + i\alpha}{2}z \qquad \psi(z) = 0$$

where α is a constant. From Equation (9.2.34), we have

$$\tau_{rr} + \tau_{\theta\theta} = -2p \qquad \tau_{\theta\theta} - \tau_{rr} + 2i\tau_{r\theta} = 0 \qquad \tau_{rr} = \tau_{\theta\theta} = -p \qquad \tau_{r\theta} = 0$$

This gives equal biaxial compressive stresses throughout the plate.

Example 2: Sinusoidal Pressure. A sinusoidal pressure $p = p_0 \sin 2\theta$ acts on the boundary of a circle with radius R. Then we obtain

$$T_1 = -p_0\sin 2\theta\cos\theta \qquad T_2 = -p_0\sin 2\theta\sin\theta$$

$$f_1(s) + if_2(s) = i\int^\theta -p_0(\cos\theta\sin 2\theta + i\sin\theta\sin 2\theta)R\,d\theta$$

$$= -i\int^\theta p_0\sin 2\theta\exp(i\theta)R\,d\theta$$

$$\sin 2\theta = \frac{\exp(2i\theta) - \exp(-2i\theta)}{2i}$$

$$f_1(s) + if_2(s) = \frac{-p_0 R}{2}\int^\theta (\exp(3i\theta) - \exp(-i\theta))\,d\theta$$

$$= -\frac{p_0 R}{2}\left[\frac{\exp(3i\theta)}{3i} + \frac{\exp(-i\theta)}{i}\right] = i\frac{p_0 R}{2}\left[\frac{\exp(3i\theta)}{3} + \exp(-i\theta)\right]$$

$$= \sum_{n=-\infty}^{\infty} A_k\exp(ik\theta)$$

where $A_k = 0$ except for $k = -1$ and $k = 3$. For $k = -1$ and $k = 3$ we have

$$A_{-1} = \frac{ip_0 R}{2} \qquad A_3 = \frac{ip_0 R}{6}$$

From Equations (9.4.7) and (9.4.8), we obtain

$$a_3 = \frac{ip_0}{6R^2} \qquad b_1 = -\frac{ip_0}{2} - 3\frac{ip_0}{6} = -ip_0$$

and all other a_k and b_k coefficients vanish. From these coefficients, the following expressions result:

$$\phi(z) = \frac{ip_0}{6R^2} z^3 \qquad \phi'(z) = \frac{ip_0}{2R^2} z^2 \qquad \phi''(z) = \frac{ip_0}{R^2} z$$

$$\psi(z) = -ip_0 z \qquad \psi'(z) = -ip_0$$

From Equation (9.2.34), we have

$$\tau_{rr} + \tau_{\theta\theta} = 4\mathscr{R}[\phi'(z)] = 4\mathscr{R}\left[\frac{ip_0}{2R^2} r^2 \exp(2i\theta)\right]$$

$$= \mathscr{R}\left[\frac{2ip_0 r^2}{R^2}(\cos 2\theta + i \sin 2\theta)\right] = -\frac{2p_0 r^2}{R^2} \sin 2\theta$$

$$\tau_{\theta\theta} - \tau_{rr} + 2i\tau_{r\theta} = 2\bar{z}\frac{ip_0}{R^2} z \exp(2i\theta) - 2ip_0 \exp(2i\theta)$$

$$= 2ip_0\left(\frac{r^2}{R^2} - 1\right)[\cos 2\theta + i \sin 2\theta]$$

$$= 2p_0\left(\frac{r^2}{R^2} - 1\right)[-\sin 2\theta + i \cos 2\theta]$$

$$\tau_{\theta\theta} - \tau_{rr} = -2p_0\left(\frac{r^2}{R^2} - 1\right) \sin 2\theta \qquad \tau_{r\theta} = p_0\left(\frac{r^2}{R^2} - 1\right) \cos 2\theta$$

$$\tau_{\theta\theta} = -p_0\left(\frac{2r^2}{R^2} - 1\right) \sin 2\theta \qquad \tau_{rr} = -p_0 \sin 2\theta$$

For a pressure $p = p_0 \sin n\theta$, where $n > 2$, the same procedure can be used. For other examples, see References 4 and 5.

9.5 Regions with Noncircular Boundaries

Solutions of stresses and displacements for a simply connected region with a circular boundary have been shown in the previous section. For simply connected regions R of noncircular shape in the z plane, a mapping function $z = \omega(\zeta)$ can always be found to map this noncircular region into a unit circle in the ζ plane. Both z and ζ are complex variables. If the points in the z plane and those in the ζ plane are uniquely related, $\omega'(\zeta)$ must not vanish throughout the unit circle $|\zeta| \le 1$. This requires that the boundary C of R has continuously changing curvature.[3] For such cases, with $z = 0$ in the interior of R, we can represent $\omega(\zeta)$ by the convergent series

$$z = \omega(\zeta) = \sum_{n=1}^{\infty} c_n \zeta^n \qquad |\zeta| \le 1 \tag{9.5.1}$$

Let $z = r \exp(i\alpha)$, $\zeta = \rho \exp(i\theta)$, and

$$\phi_1(\zeta) = \phi[\omega(\zeta)] \qquad \text{and} \qquad \psi_1(\zeta) = \psi[\omega(\zeta)] \tag{9.5.2}$$

then

$$\phi'(z) = \frac{d\phi_1}{d\zeta}\frac{d\zeta}{dz} = \frac{\phi_1'(\zeta)}{\omega'(\zeta)} \tag{9.5.3}$$

and Equations (9.2.22) and (9.2.32) become

$$U,_1 + iU,_2 = \phi_1(\zeta) + \frac{\omega(\zeta)}{\overline{\omega'(\zeta)}}\overline{\phi_1'(\zeta)} + \overline{\psi_1(\zeta)}, \qquad |\zeta| \le 1 \tag{9.5.4}$$

$$2\mu(u_1 + iu_2) = \frac{\lambda + 3\mu}{\lambda + \mu}\phi_1(\zeta) - \frac{\omega(\zeta)}{\overline{\omega'(\zeta)}}\overline{\phi_1'(\zeta)} - \overline{\psi_1(\zeta)}, \qquad |\zeta| \le 1 \tag{9.5.5}$$

$$dz = |dz| \exp(i\alpha) \qquad d\zeta = |d\zeta| \exp(i\theta)$$

$$\exp(i\alpha) = \frac{dz}{|dz|} = \frac{\omega'(\zeta)\, d\zeta}{|\omega'(\zeta)|\, |d\zeta|} = \frac{\omega'(\zeta)}{|\omega'(\zeta)|}\exp(i\theta) = \frac{\zeta}{\rho}\frac{\omega'(\zeta)}{|\omega'(\zeta)|}$$

$$\exp(-i\alpha) = \exp(-i\theta)\frac{\overline{\omega'(\zeta)}}{|\omega'(\zeta)|} = \frac{\bar{\zeta}}{\rho}\frac{\overline{\omega'(\zeta)}}{|\omega'(\zeta)|}$$

$$\exp(2i\alpha) = \frac{\zeta^2}{\rho^2}\frac{[\omega'(\zeta)]^2}{|\omega'(\zeta)|^2} = \frac{\zeta^2}{\rho^2}\frac{[\omega'(\zeta)]^2}{\omega'(\zeta)\overline{\omega'(\zeta)}} = \frac{\zeta^2}{\rho^2}\frac{\omega'(\zeta)}{\overline{\omega'(\zeta)}} \tag{9.5.6}$$

Substituting Equations (9.5.6) into Equation (9.2.35), we obtain

$$\tau_{\rho\rho} + \tau_{\theta\theta} = 4\mathscr{R}\Phi(\zeta) = 2[\Phi_1(\zeta) + \overline{\Phi_1(\zeta)}] \tag{9.5.7}$$

$$\tau_{\theta\theta} - \tau_{\rho\rho} + 2i\tau_{\rho\theta} = \frac{2\zeta^2}{\rho^2\overline{\omega'(\zeta)}}[\overline{\omega(\zeta)}\Phi_1'(\zeta) + \omega'(\zeta)\Psi_1(\zeta)] \tag{9.5.8}$$

where Φ_1 denotes ϕ_1' and Ψ_1 denotes ψ_1'. Subtracting Equation (9.5.8) from Equation (9.5.7) yields

$$\tau_{\rho\rho} - i\tau_{\rho\theta} = \Phi_1(\zeta) + \overline{\Phi_1(\zeta)} - \frac{\zeta^2}{\rho^2\overline{\omega'(\zeta)}}[\overline{\omega(\zeta)}\Phi_1'(\zeta) + \omega'(\zeta)\Psi_1(\zeta)] \tag{9.5.9}$$

This gives the stress components within a simply connected region in the ζ plane bounded by the unit circle.

A point on the boundary C in the z plane corresponds to a point on the circumference of the unit circle $\zeta = \exp(i\theta)$. Denoting this point by σ, we may express the equation

$$f_1(s) + if_2(s) = i\int^s [T_1(s) + iT_2(s)]\, ds$$

as a function $F(\sigma)$ of σ. Equation (9.2.26), after mapping into the ζ plane, becomes

$$\phi_1(\zeta) + \frac{\omega(\zeta)}{\overline{\omega'(\zeta)}}\overline{\phi_1'(\zeta)} + \overline{\psi_1(\zeta)} = F(\sigma) \qquad |\zeta| = 1 \tag{9.5.10}$$

Let

$$\phi_1(\zeta) = \sum_0^\infty a_k\zeta^k \qquad \psi_1(\zeta) = \sum_0^\infty b_k\zeta^k \tag{9.5.11}$$

With $\omega(\zeta)$ known, the coefficients a_k and b_k can be obtained in the same way as shown in Section 9.4. With $\phi_1(\zeta)$ and $\psi_1(\zeta)$ known, $\phi(z)$, $\psi(z)$ and hence the stress distribution are readily obtained. For illustrative examples, see References 4 and 5.

9.6 Inelastic Plane Deformation Problems

Consider the cases of plane deformation with total strain satisfying the conditions given in Section 9.1, that is,

$$e_{31} = e_{32} = e_{33} = 0 \qquad e_{\beta\gamma} = e_{\beta\gamma}(x_1 x_2) \tag{9.6.1}$$

where β and γ assume values 1 and 2. The inelastic strain components are subject to the following conditions:

$$e''_{31} = e''_{32} = 0 \qquad e''_{33} = e''_{33}(x_1, x_2) \qquad e''_{\beta\gamma} = e''_{\beta\gamma}(x_1, x_2) \tag{9.6.2}$$

hence

$$\theta = \theta(x_1, x_2) \qquad \text{and} \qquad \theta'' = \theta''(x_1, x_2)$$

From the stress-strain relationship

$$\tau_{ij} = \lambda\, \delta_{ij}(\theta - \theta'') + 2\mu(e_{ij} - e''_{ij}) \tag{4.2.2}$$

we obtain

$$\tau_{\beta\gamma} = \tau_{\beta\gamma}(x_1, x_2) \tag{9.6.3}$$

$$\tau_{31} = \tau_{32} = 0 \qquad \tau_{33} = \tau_{33}(x_1, x_2) \tag{9.6.4}$$

The equilibrium condition $\tau_{ij,j} + F_i = 0$ reduces to

$$\tau_{\beta\gamma,\gamma} + F_\beta = 0 \qquad \text{and} \qquad F_3 = 0 \tag{9.6.5}$$

in the region R, while $T_i = \tau_{ij} n_j$ on the boundary C reduces to $T_\beta = \tau_{\beta\gamma} n_\gamma$. Substituting the stress-strain relation, Equation (4.2.2), into the above equilibrium conditions, we have

$$\lambda\, \delta_{\beta\gamma}(\theta - \theta'')_{,\gamma} + 2\mu(e_{\beta\gamma,\gamma} - e''_{\beta\gamma,\gamma}) + F_\beta = 0$$

$$\lambda\theta_{,\beta} + 2\mu e_{\beta\gamma,\gamma} - \lambda\theta''_{,\beta} - 2\mu e''_{\beta\gamma,\gamma} + F_\beta = 0 \tag{9.6.6}$$

$$T_\beta = (\lambda\, \delta_{\beta\gamma}\theta + 2\mu e_{\beta\gamma})n_\gamma - (\lambda\, \delta_{\beta\gamma}\theta'' + 2\mu e''_{\beta\gamma})n_\gamma \tag{9.6.7}$$

It is seen the terms $[-\lambda\theta''_{,\beta} - 2\mu e''_{\beta\gamma,\gamma}]$ and $(\lambda\, \delta_{\beta\gamma}\theta'' + 2\mu e''_{\beta\gamma})n_\gamma$ have the same effect as the body force F_β and the surface traction T_β on the strain distribution $e_{\beta\gamma}$. Hence these terms are called *equivalent body* and *surface forces*, respectively, and are denoted by $\underline{F}_\beta$ and $\underline{T}_\beta$. With this notation Equations (9.6.6) and (9.6.7) may be written as

$$\lambda\theta_{,\beta} + 2\mu e_{\beta\gamma,\gamma} + \underline{F}_\beta + F_\beta = 0 \tag{9.6.8}$$

$$T_\beta + \underline{T}_\beta = (\lambda\, \delta_{\beta\gamma}\theta + 2\mu e_{\beta\gamma})n_\gamma \tag{9.6.9}$$

If the distributions of the inelastic strains $e''_{\beta\gamma}$ and e''_{33} are known, $\underline{F}_\beta$ and $\underline{T}_\beta$ can be calculated. Considering these $\underline{F}_\beta$ and $\underline{T}_\beta$ as an additional set of body and surface forces, we can calculate the strain distribution $e_{\beta\gamma}$ in the body as if it were purely elastic. The methods shown in the previous sections for the solution of elastic plane deformation problems are applicable. With $e_{\beta\gamma}$ known, the stresses τ^{I}_{ij} caused by the combination of actual and equivalent body and surface forces are

$$\tau^{\mathrm{I}}_{ij} = \lambda\, \delta_{ij}\theta + 2\mu e_{ij} \tag{9.6.10}$$

and the stresses in the actual body, from Equation (4.2.2), are

$$\tau_{ij} = \tau^{\mathrm{I}}_{ij} - \lambda\,\delta_{ij}\theta'' - 2\mu e''_{ij} \tag{9.6.11}$$

Hence the strain and stress distributions in the actual body can be found.

If there is only thermal strain and neither creep nor plastic strain, then for an increase in temperature T above the stress-free temperature, we have

$$\begin{aligned} e''_{ij} &= e^T_{ij} = \delta_{ij}\alpha T \qquad \theta = 3\alpha T \\ \tau_{ij} &= \lambda\,\delta_{ij}(\theta - 3\alpha T) + 2\mu(e_{ij} - \delta_{ij}\alpha T) \\ &= \lambda\,\delta_{ij}\theta + 2\mu e_{ij} - \delta_{ij}(3\lambda + 2\mu)\alpha T \end{aligned} \tag{9.6.12}$$

where α is the coefficient of thermal expansion. Putting $i = j$, summing over i from 1 to 3, and letting $\Theta = \tau_{ii}$, we obtain

$$\Theta = (3\lambda + 2\mu)(\theta - 3\alpha T)$$

$$\theta = \frac{\Theta}{3\lambda + 2\mu} + 3\alpha T \tag{9.6.13}$$

$$e_{ij} = \frac{\tau_{ij}}{2\mu} - \frac{\lambda}{2\mu}\delta_{ij}\frac{\Theta}{3\lambda + 2\mu} + \delta_{ij}\alpha T \tag{9.6.14}$$

Since $e_{33} = 0$, Equation (9.6.14) gives

$$\frac{\tau_{33}}{2\mu} - \frac{\lambda}{2\mu}\frac{(\tau_{11} + \tau_{22} + \tau_{33})}{3\lambda + 2\mu} + \alpha T = 0 \tag{9.6.15}$$

or

$$\tau_{33} = -\frac{\mu(3\lambda + 2\mu)}{\lambda + \mu}\alpha T + \frac{\lambda(\tau_{11} + \tau_{22})}{2(\lambda + \mu)} \tag{9.6.16}$$

For plane deformation, the compatibility conditions reduce to

$$e_{11,22} + e_{22,11} - 2e_{12,12} = 0 \tag{9.6.17}$$

Substituting Equation (9.6.14) into Equation (9.6.17) and eliminating Θ by Equation (9.6.15), we have

$$\tau_{11,22} + \tau_{22,11} - 2\tau_{12,12} - \nabla_1{}^2(\tau_{33}) = 0 \tag{9.6.18}$$

From the equilibrium conditions

$$\tau_{\beta\gamma,\gamma} + F_\beta = 0$$

we eliminate $\tau_{12,12}$ in Equation (9.6.18) and obtain

$$\nabla_1{}^2(\tau_{11} + \tau_{22}) + F_{1,1} + F_{2,2} - \nabla_1{}^2(\tau_{33}) = 0$$

where

$$\nabla_1{}^2 = \left(\frac{\partial^2}{\partial x_1{}^2} + \frac{\partial^2}{\partial x_2{}^2}\right)$$

From this and Equation (9.6.16), we have

$$\frac{\lambda + 2\mu}{2(\lambda + \mu)} \nabla_1{}^2(\tau_{11} + \tau_{22}) + F_{1,1} + F_{2,2} + \frac{\mu(3\lambda + 2\mu)}{\lambda + \mu} \alpha \nabla_1{}^2 T = 0 \quad (9.6.19)$$

For the case of no body force, the above Equation reduces to

$$\nabla_1{}^2\left[\frac{\lambda + 2\mu}{2(\lambda + \mu)}(\tau_{11} + \tau_{22}) + \frac{\mu(3\lambda + 2\mu)}{\lambda + \mu} T\right] = 0 \quad (9.6.20)$$

Let U be the Airy stress function given in Section 9.2. Then Equation (9.6.20) may be written as

$$\nabla_1{}^2\left[\frac{\lambda + 2\mu}{2(\lambda + \mu)} \nabla_1{}^2 U + \frac{\mu(3\lambda + 2\mu)}{\lambda + \mu} T\right] = 0 \quad (9.6.21)$$

This is the *differential equation for thermoelastic plane strain problems.*[6]

Expressing the strain in terms of displacement, Equation (9.6.12) becomes

$$\tau_{ij} = \lambda\, \delta_{ij} u_{k,k} + \mu(u_{i,j} + u_{j,i}) - \delta_{ij}(3\lambda + 2\mu)\alpha T$$

Substituting this in the equation of equilibrium $\tau_{ij,j} = 0$ yields

$$\lambda\, \delta_{ij} u_{k,kj} + \mu(u_{i,jj} + u_{j,ij}) - \delta_{ij}(3\lambda + 2\mu)\alpha T,_j = 0$$

Letting V be the displacement potential, that is,

$$u_i = V,_i \quad (9.6.22)$$

we see that the previous equation becomes

$$\lambda\, \delta_{ij} V,_{kkj} + \mu(V,_{ijj} + V,_{jij}) - \delta_{ij}(3\lambda + 2\mu)\alpha T,_j = 0$$

$$[(\lambda + 2\mu)V,_{kk} - (3\lambda + 2\mu)\alpha T],_i = 0$$

$$\left[\frac{\lambda + 2\mu}{3\lambda + 2\mu} V,_{kk} - \alpha T\right],_i = 0 \quad (9.6.23)$$

This gives a differential equation in terms of displacement potential for thermoelastic plane strain problems.[6–8] Both Equations (9.6.21) and (9.6.23) are two-dimensional Poisson's differential equations. These equations are to be solved with given boundary conditions. Equation (9.6.21) is more convenient when the tractions on the boundary are given, while Equations (9.6.23) are preferable when the displacement is prescribed. The solution of this type of differential equation is discussed in books on potential theory[9] and on mathematical methods of physics.[10]

9.7 Inelastic Generalized Plane Stress Problems

Referring to Section 9.3, for generalized plane stress as in the case of thin plates, we have $\bar{\tau}_{31} = \bar{\tau}_{32} = \bar{\tau}_{33} = 0$. Writing the stress-strain relation for average stress and strain across the thickness, we have

$$\bar{\tau}_{ij} = \lambda\, \delta_{ij}(\bar{\theta} - \bar{\theta}'') + 2\mu(\bar{e}_{ij} - \bar{e}''_{ij}) \tag{9.7.1}$$

where the bar denotes the average value through the thickness. Since in this section all stress, strain, and force components are the average values through the thickness, it will be convenient to delete the bar in the remainder of this section. Creep and plastic strain, e_{ij_c} and e_{ij_p}, give no change in volume, hence θ'' is due to thermal expansion only; that is, $\theta'' = 3\alpha T$. Then

$$\begin{aligned}
e''_{ij} &= e_{ij_c} + e_{ij_p} + \delta_{ij}\alpha T \\
\tau_{11} &= \lambda(\theta - 3\alpha T) + 2\mu(e_{11} - \alpha T - e_{11_c} - e_{11_p}) \\
\tau_{22} &= \lambda(\theta - 3\alpha T) + 2\mu(e_{22} - \alpha T - e_{22_c} - e_{22_p}) \\
\tau_{33} &= 0 = \lambda(\theta - 3\alpha T) + 2\mu(e_{33} - \alpha T - e_{33_c} - e_{33_p})
\end{aligned}$$

From the above equations, and using $e_{ii_c} = e_{ii_p} = 0$, we easily obtain

$$(\tau_{11} + \tau_{22}) = (3\lambda + 2\mu)(\theta - 3\alpha T)$$

or, solving for $(\theta - 3\alpha T)$, we have

$$\theta - 3\alpha T = \frac{\tau_{11} + \tau_{22}}{3\lambda + 2\mu} \tag{9.7.2}$$

Using this last expression, we see that the above equations become

$$\tau_{11} = \frac{\lambda(\tau_{11} + \tau_{22})}{3\lambda + 2\mu} + 2\mu(e_{11} - e''_{11})$$

$$\tau_{22} = \frac{\lambda(\tau_{11} + \tau_{22})}{3\lambda + 2\mu} + 2\mu(e_{22} - e''_{22})$$

or

$$\frac{2(\lambda + \mu)}{3\lambda + 2\mu}\tau_{11} = \frac{\lambda}{3\lambda + 2\mu}\tau_{22} + 2\mu(e_{11} - e''_{11})$$

Finally, then,

$$\tau_{11} = \frac{\lambda}{2(\lambda + \mu)} \tau_{22} + \frac{\mu(3\lambda + 2\mu)}{(\lambda + \mu)} (e_{11} - e''_{11}) \tag{9.7.3}$$

Using the relations between Young's modulus E, Poisson's ratio ν, and Lame's constants λ, μ,

$$E = \frac{\mu(3\lambda + 2\mu)}{\lambda + \mu} \qquad \nu = \frac{\lambda}{2(\lambda + \mu)} \tag{9.7.4}$$

we see that Equation (9.7.3) becomes

$$\tau_{11} = \nu\tau_{22} + E(e_{11} - e''_{11}) \tag{9.7.5}$$

Similarly, we have

$$\tau_{22} = \nu\tau_{11} + E(e_{22} - e''_{22}) \tag{9.7.6}$$

From these two equations, we obtain

$$\begin{aligned} \tau_{11} &= \frac{E}{1 - \nu^2} [(e_{11} + \nu e_{22}) - (e''_{11} + \nu e''_{22})] \\ \tau_{22} &= \frac{E}{1 - \nu^2} [(e_{22} + \nu e_{11}) - (e''_{22} + \nu e''_{11})] \end{aligned} \tag{9.7.7}$$

Writing the shear modulus μ as G, we have

$$\tau_{12} = 2G(e_{12} - e''_{12}) \tag{9.7.8}$$

Substituting the above into the equilibrium conditions

$$\tau_{\beta\gamma,\gamma} + F_\beta = 0$$

yields

$$\frac{E}{1 - \nu^2} (e_{11,1} + \nu e_{22,1}) + 2Ge_{12,2} + \left[-\frac{E}{1 - \nu^2} (e''_{11,1} + \nu e''_{22,1}) - 2Ge''_{12,2} \right] + F_1 = 0$$

$$\frac{E}{1 - \nu^2} (e_{22,2} + \nu e_{11,2}) + 2Ge_{12,1} + \left[-\frac{E}{1 - \nu^2} (e''_{22,2} + \nu e''_{11,2}) - 2Ge''_{12,1} \right] + F_2 = 0$$

In this case the equivalent body forces are

$$\begin{aligned} \underline{F}_1 &= -\frac{E}{1 - \nu^2} (e''_{11,1} + \nu e''_{22,1}) - 2Ge''_{12,2} \\ \underline{F}_2 &= -\frac{E}{1 - \nu^2} (e''_{22,2} + \nu e''_{11,2}) - 2Ge''_{12,1} \end{aligned} \tag{9.7.9}$$

From the equilibrium condition on the boundary $T_\beta = \tau_{\beta\gamma} n_\gamma$, and Equations (9.7.7), (9.7.8), we have

$$T_1 = \frac{E}{1-\nu^2}[(e_{11} + \nu e_{22}) - (e''_{11} + \nu e''_{22})]n_1 + 2G(e_{12} - e''_{12})n_2$$

Hence the equivalent surface forces are

$$\underline{T}_1 = \frac{E}{1-\nu^2}(e''_{11} + \nu e''_{22})n_1 + 2Ge''_{12}n_2$$

and (9.7.10)

$$\underline{T}_2 = \frac{E}{1-\nu^2}(e''_{22} + \nu e''_{11})n_2 + 2Ge''_{12}n_1$$

To analyze the strain and stress in an inelastic body subject to plane stress, we first consider the body as purely elastic and apply both the actual and the equivalent sets of body and surface forces. The resulting strain distribution $e_{\beta\gamma}$ can be found by the method of analysis for elastic plane stress problems. The stress obtained in this way is denoted by $\tau^{\mathrm{I}}_{\beta\gamma}$, where

$$\tau^{\mathrm{I}}_{\beta\gamma} = \lambda\,\delta_{\beta\gamma}\theta + 2\mu e_{\beta\gamma} \tag{9.7.11}$$

The actual stress in the body with inelastic strain is then given by

$$\tau_{\beta\gamma} = \tau^{\mathrm{I}}_{\beta\gamma} - \lambda\,\delta_{\beta\gamma}\theta'' - 2\mu e''_{\beta\gamma} \tag{9.7.12}$$

This gives the general procedure by which the stress and strain distributions in an elastic body under plane stress may be analyzed.

Thermostress. If the inelastic strain consists of thermal strain alone, the inelastic body is thermoelastic, and the equations become

$$e''_{11} = e''_{22} = \alpha T \qquad e''_{12} = 0$$

$$e_{11} = \frac{\tau_{11}}{E} - \nu\frac{\tau_{22}}{E} + \alpha T$$

$$e_{22} = \frac{\tau_{22}}{E} - \nu\frac{\tau_{11}}{E} + \alpha T \tag{9.7.13}$$

$$e_{12} = \frac{\tau_{12}}{2G} = \frac{\tau_{12}(1+\nu)}{E}$$

The equivalent body forces due to temperature gradient, from Equation (9.7.9), are

$$\underline{F}_1 = -\frac{E}{1-\nu}\alpha T_{,1}$$
$$\underline{F}_2 = -\frac{E}{1-\nu}\alpha T_{,2} \tag{9.7.14}$$

While the equivalent surface forces, as obtained from Equation (9.7.10), are

$$\underline{T}_1 = \frac{E}{1-\nu}\alpha T n_1$$
$$\underline{T}_2 = \frac{E}{1-\nu}\alpha T n_2 \tag{9.7.15}$$

With the equivalent body and surface forces known, the stresses are readily obtained by the method given in Sections 9.1 through 9.3, as demonstrated by the following examples.

Example. This example illustrates the use of the analogy between inelastic strain and body force for solving thermal stress problems. Consider an infinite plate subject to a uniform temperature T within a rectangular area of sides $2a$ and $2b$, as shown in Figure 9.6, and zero temperature

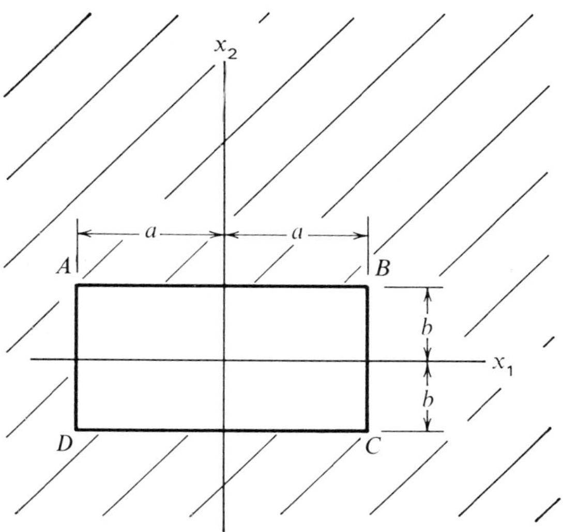

Figure 9.6

exterior to this rectangular region. Since the temperature is uniform within and exterior to the rectangular region, there are no equivalent body forces within and outside the rectangle. However, there is a discontinuity in temperature across the boundary of the rectangular area, and hence equivalent forces exist along this boundary. From Equation (9.7.14) the equivalent body forces are given by

$$\underline{F}_i = -\frac{E}{1-\nu}\alpha T,_i$$

The stress field in an infinite plate caused by a body force distribution is given by Equation (9.3.9). Across BC, we see that F_i is given by

$$\underline{F}_1 = -\frac{E}{1-\nu}\alpha\{T - T \cdot U(x_1 - a)\},_{x_1} = \frac{E\alpha T}{1-\nu}\delta(x_1 - a)$$

where $U(x_1 - a)$ is the unit step function and $\delta(x_1 - a)$ is the Dirac delta function. Substituting this into Equation (9.3.9) gives that part of the stress τ_{11} due to the temperature discontinuity across BC,

$$\tau_{11} = \frac{E\alpha T}{1-\nu}\int_{-b}^{b}\int_{0}^{\infty} -\frac{\delta(x_1' - a)}{4\pi}\left\{\frac{(3+\nu)(x_1 - x_1')}{(x_1 - x_1')^2 + (x_2 - x_2')^2}\right.$$
$$\left. - 2(1+\nu)\frac{(x_1 - x_1')(x_2 - x_2')^2}{[(x_1 - x_1')^2 + (x_2 - x_2')^2]^2}\right\} dx_1'\, dx_2$$
$$= \frac{E\alpha T}{4\pi(1-\nu)}\left[-2\tan^{-1}\left(\frac{x_2 - b}{x_1 - a}\right) + 2\tan^{-1}\left(\frac{x_2 + b}{x_1 - a}\right)\right.$$
$$\left. + (1+\nu)\left\{-\frac{(x_1 - a)(x_2 - b)}{(x_1 - a)^2 + (x_2 - b)^2} + \frac{(x_1 - a)(x_2 + b)}{(x_1 - a)^2 + (x_2 + b)^2}\right\}\right] \tag{9.7.16}$$

Similarly the contributions to the τ_{11} stress due to the equivalent force along the boundary segments AD, AB, and DC are also obtained by integration of Equation (9.3.9). Adding up the contributions to τ_{11} caused by the equivalent body forces acting along the four sides of the rectangular region, we obtain

$$\tau_{11} = -\frac{E\alpha T}{2\pi}\left[\cot^{-1}\frac{x_2 + b}{x_1 + a} + \cot^{-1}\frac{x_2 - b}{x_1 - a}\right.$$
$$\left. - \cot^{-1}\frac{x_2 + b}{x_1 - a} - \cot^{-1}\frac{x_2 - b}{x_1 + a}\right] \tag{9.7.17}$$

Similarly the τ_{22} and τ_{12} stresses due to the equivalent body forces along the sides of the rectangle are

$$\tau_{22} = -\frac{E\alpha T}{2\pi}\left[\tan^{-1}\frac{x_2+b}{x_1+a} + \tan^{-1}\frac{x_2-b}{x_1-a} - \tan^{-1}\frac{x_2+b}{x_1-a} - \tan^{-1}\frac{x_2-b}{x_1+a}\right] \tag{9.7.18}$$

$$\tau_{12} = \frac{E\alpha T}{\pi}\ln\frac{[(x_1-a)^2+(x_2-b)^2]^{1/2}[(x_1+a)^2+(x_2+b)^2]^{1/2}}{[(x_1-a)^2+(x_2+b)^2]^{1/2}[(x_1+a)^2+(x_2-b)^2]^{1/2}} \tag{9.7.19}$$

Equations (9.7.17) through (9.7.19) gives the stress field in an infinite plate caused by the given discontinuous temperature distribution.

If the thermal rectangular area of the previous example is within a circular plate of radius R whose center coincides with that of the rectangular region, as shown in Figure 9.7, and if in addition this plate is free from boundary tractions, that is, if $\mathbf{T}_B$ in Figure 9.1 vanishes, then the thermal stress field in this circular plate may be determined, using the scheme of Figure 9.1.

The traction per unit length of the circular boundary, as shown in Figure 9.1(a), is

$$\mathbf{T}_0 = T_1 + iT_2 \tag{9.7.20}$$

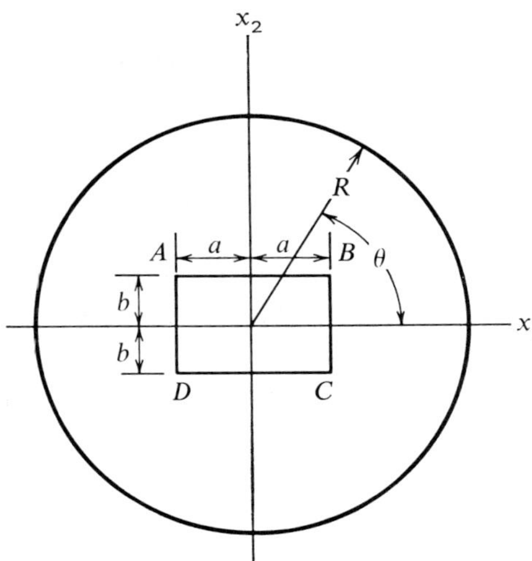

Figure 9.7

where the components of the traction force along the x_1 and x_2 axes are

$$T_1 = \tau_{1\alpha} \nu_\alpha = \tau_{11} \cos\theta + \tau_{21} \sin\theta$$
$$T_2 = \tau_{2\alpha} \nu_\alpha = \tau_{12} \cos\theta + \tau_{22} \sin\theta \tag{9.7.21}$$

where τ_{11}, τ_{12}, and τ_{22} are the stresses obtained from Equations (9.7.17) through (9.7.19) at distance R from the center of the rectangle. The stress field given by Equations (9.7.17) to (9.7.19) also gives the stress field in a circular plate of radius R with uniform temperature T in the rectangle $ABCD$ and zero temperature exterior to the rectangle when the circular plate has boundary tractions $\mathbf{T}_0$. In the present problem the circular plate has no traction forces on its boundary, hence $\mathbf{T}_0$ has to be removed by applying an equal and opposite boundary traction $-\mathbf{T}_0$. The stress field caused by this $-\mathbf{T}_0$ is found as follows. From Equation (9.2.26), we have

$$f_1(\theta) = R \int_0^\theta -(\tau_{11} \cos\theta + \tau_{21} \sin\theta)\, d\theta$$

$$f_2(\theta) = R \int_0^\theta -(\tau_{12} \cos\theta + \tau_{22} \sin\theta)\, d\theta$$

Since the stress components $\tau_{\alpha\beta}$ are complicated functions of θ [Equations (9.7.17) through (9.7.19)], it is difficult to evaluate these integrals analytically. However, the integrals may be readily integrated numerically with the aid of a computer. Then the resulting complex function $f_1(\theta) + if_2(\theta)$ may be expressed in series form as

$$f_1(\theta) + if_2(\theta) = \sum_{k=-\infty}^{\infty} A_k \exp(ik\theta) \tag{9.4.1}$$

where k's are integers and

$$A_k = \frac{1}{2\pi} \int_0^{2\pi} [f_1(\theta) + if_2(\theta)] \exp(-ik\theta)\, d\theta \tag{9.4.2}$$

Noting that $\exp(-ik\theta) = \cos k\theta - i \sin k\theta$, we may obtain the A_k's through numerical integration of Equation (9.4.2). From Equation (9.2.26), with $z = R \exp(i\theta)$, we have

$$\phi(z) + z\overline{\phi'(z)} + \psi(z) = f_1 + if_2 = \sum A_k \exp(ik\theta)$$

where

$$\phi(z) = \sum_{k=1}^{\infty} a_k z^k \qquad \psi(z) = \sum_{k=0}^{\infty} b_k z^k \tag{9.4.3}$$

and the a_k and b_k coefficients are obtained from the A_k coefficients by Equations (9.4.6) through (9.4.8). Then the stresses in a circular plate due to the boundary traction $-T_0 = -(T_1 + iT_2)$ are obtained from Equation (9.2.24) as

$$\begin{aligned} \tau_{11} + \tau_{22} &= 4\mathscr{R}[\phi'(z)] \\ \tau_{22} - \tau_{11} + 2i\tau_{12} &= 2[\bar{z}\phi''(z) + \psi'(z)] \end{aligned} \tag{9.7.22}$$

From Figure 9.1, superposing the stresses given by Equations (9.7.17) through (9.7.19) on those of Equation (9.7.22), we obtain the thermal stress field in a circular plate, subject to no boundary tractions, with uniform temperature T in a rectangular area and zero temperature in the remaining part of the plate. This illustrates a general procedure for obtaining the thermal stress distribution in a finite plate. Generally the calculations are lengthy, and a computer is required.

Creep Analysis. For the case with creep strain the initial stress is obtained by elastic analysis. The stress component versus time curve is often approximated by horizontal and vertical segments as shown in Figure 3.9. Within each horizontal segment Δt, the stress τ_{ij} and hence the deviatoric stress S_{ij} are assumed constant. For each time interval Δt, the incremental creep strain Δe_{ij_c} is obtained by multiplying Equation (4.8.7) by Δt,

$$\frac{\Delta e_{11_c}}{S_{11}} = \frac{\Delta e_{22_c}}{S_{22}} = \frac{\Delta e_{33_c}}{S_{33}} = \frac{\Delta e_{12_c}}{S_{12}} = \frac{3\,\Delta e_c^*}{2\sigma^*} \tag{9.7.23}$$

where

$$e_c^* = \frac{\sqrt{2}}{3}\,[(e_{11_c} - e_{22_c})^2 + (e_{22_c} - e_{33_c})^2 + (e_{33_c} - e_{11_c})^2 + 6e_{12_c}^2]^{1/2}$$

and

$$\sigma^* = \frac{1}{\sqrt{2}}\,[(\tau_{11} - \tau_{22})^2 + \tau_{22}^2 + \tau_{11}^2 + 6\tau_{12}^2]^{1/2}$$

The creep strain rate $\dot{e}_c^*$ is assumed to depend on the stress σ^* and the current amount of creep strain e_c^*; that is, $\dot{e}_c^* = f(e_c^*, \sigma^*)$, where the function f is determined from uniaxial creep test relation $\dot{e}_{11_c} = f(e_{11}, \tau_{11})$. At the beginning of each time interval Δt, the e_c^* and σ^* are known. From the creep data of the material at the particular temperature, $\dot{e}_c^*$ and Δe_c^* are found. Then, by Equation (9.7.23), the Δe_{ij_c} are obtained. From these Δe_{ij_c}, the incremental body and surface forces $\Delta \underline{F}_\beta$ and $\Delta \underline{S}_\beta$ for the subsequent time interval are computed. Adding these to the $\underline{F}_\beta$ and $\underline{S}_\beta$ of the previous interval, we obtain the equivalent forces for the present time interval. From the equivalent and the actual body and surface forces, the strain and stress distributions are obtained. The process is repeated for successive time intervals to get the stress versus time curve for different points in the body. The calculations are generally lengthy, and the use of a computer is required. A numerical example of creep analysis of a uniformly stretched circular annulus plate is given in Reference 8.

Elastoplastic Analysis. For the case with plastic strain, the incremental stress-strain relations, Equations (4.5.1), for plane stress, and Equations (4.5.9) give

$$\frac{\Delta e_{11_p}}{S_{11}} = \frac{\Delta e_{22_p}}{S_{22}} = \frac{\Delta e_{33_p}}{S_{33}} = \frac{\Delta e_{12_p}}{S_{12}} = \frac{3\,\Delta e_p{}^*}{2\sigma^*} \tag{9.7.24}$$

where the subscript p on strain denotes plastic strain and e^* and σ^* are the effective strain and stress given by

$$e^* = \frac{\sqrt{2}}{3}\left[(e_{11} - e_{22})^2 + (e_{22} - e_{33})^2 + (e_{33} - e_{11})^2 + 6e_{12}^2\right]^{1/2} \tag{9.7.25}$$

$$\sigma^* = \frac{1}{\sqrt{2}}\left[(\tau_{11} - \tau_{22})^2 + \tau_{22}^2 + \tau_{11}^2 + 6\tau_{12}^2\right]^{1/2} \tag{9.7.26}$$

The effective plastic strain $e_p{}^*$ is assumed to be a function of σ^*, $e_p{}^* = f(\sigma^*)$, where the function f is determined from uniaxial tests; that is, $e_{11_p} = f(\tau_{11})$. Writing Equation (9.7.24) in a different form, we have

$$\left.\begin{aligned} \Delta e_{11_p} &= \frac{2\tau_{11} - \tau_{22}}{2\sigma^*}\,\Delta e_p{}^* \\ \Delta e_{22_p} &= \frac{2\tau_{22} - \tau_{11}}{2\sigma^*}\,\Delta e_p{}^* \\ \Delta e_{12_p} &= \frac{3}{2}\frac{\tau_{12}}{\sigma^*}\,\Delta e_p{}^* \end{aligned}\right\} \tag{9.7.27}$$

From Equations (9.7.7) and (9.7.8), the loss of stress $\tau_{\beta\gamma_p}$ at a point due to plastic strain at that point is

$$\begin{aligned} \tau_{11_p} &= -\frac{E}{1-\nu^2}(e_{11_p} + \nu e_{22_p}) \qquad \tau_{22_p} = -\frac{E}{1-\nu^2}(e_{22_p} + \nu e_{11_p}) \\ \tau_{12_p} &= -2Ge_{12_p} \end{aligned} \tag{9.7.28}$$

Let $\tau_{\beta\gamma_0}$ be the stress caused by the applied loads if the body is purely elastic and $\tau'_{\beta\gamma}$ be the stress caused by the equivalent body and surface forces. From Equation (9.7.12), the stress at any point in the actual body is expressed as

$$\tau_{\beta\gamma} = \tau_{\beta\gamma_0} + \tau'_{\beta\gamma} + \tau_{\beta\gamma_p} \tag{9.7.29}$$

where $\tau_{\beta\gamma_0} + \tau'_{\beta\gamma}$ is equal to $\tau^{\mathrm{I}}_{\beta\gamma}$ in Equation (9.7.11). Writing this in incremental form, we have

$$\Delta\tau_{\beta\gamma} = \Delta\tau_{\beta\gamma_0} + \Delta\tau'_{\beta\gamma} + \Delta\tau_{\beta\gamma_p} \tag{9.7.30}$$

When the entire plate is within the elastic limit, $\tau'_{\beta\gamma} = \tau_{\beta\gamma_p} = 0$, and $\tau_{\beta\gamma} = \tau_{\beta\gamma_0}$. When the effective stress σ^* exceeds the elastic limit in some region of the plate, plastic strain, and hence plastic strain gradients, will occur and produce equivalent body and surface forces. For numerical calculation of the equivalent body force, the area with plastic strain and its immediate neighbors are divided into a rectangular grid. Let m and n denote the location of the grid point along the x_1 and x_2 axes, respectively. The equivalent body forces $\underline{F}_1$ and $\underline{F}_2$ given by Equation (9.7.9) on an elemental area $\Delta x_1\, \Delta x_2$ bounded by grid lines are here expressed in finite difference form as

$$\begin{aligned}
\underline{F}_{1_{m,n}} = \Bigg\{ &-\frac{E}{1-\nu^2}\left[\frac{\begin{aligned}&(e^p_{11_{m+1,n}} + e^p_{11_{m+1,n+1}} - e^p_{11_{m,n}} - e^p_{11_{m,n+1}})\\ &\quad + \nu(e^p_{22_{m+1,n}} + e^p_{22_{m+1,n+1}} - e^p_{22_{m,n}} - e^p_{22_{m,n+1}})\end{aligned}}{2\,\Delta x_1}\right]\\
&+ 2G\left[\frac{(e^p_{12_{m,n+1}} + e^p_{12_{m+1,n+1}} - e^p_{12_{m,n}} - e^p_{12_{m+1,n}})}{2\,\Delta x_2}\right]\Bigg\}\Delta x_1\,\Delta x_2\\
\underline{F}_{2_{m,n}} = \Bigg\{ &-\frac{E}{1-\nu^2}\left[\frac{\begin{aligned}&(e^p_{22_{m,n+1}} + e^p_{22_{m+1,n+1}} - e^p_{22_{m,n}} - e^p_{22_{m+1,n}})\\ &\quad + \nu(e^p_{11_{m,n+1}} + e^p_{11_{m+1,n+1}} - e^p_{11_{m,n}} - e^p_{11_{m+1,n}})\end{aligned}}{2\,\Delta x_2}\right]\\
&+ 2G\left[\frac{(e^p_{12_{m+1,n}} + e^p_{12_{m+1,n+1}} - e^p_{12_{m,n}} - e^p_{12_{m,n+1}})}{2\,\Delta x_1}\right]\Bigg\}\Delta x_1\,\Delta x_2
\end{aligned} \tag{9.7.31}$$

where the superscript p also denotes plastic and the comma in the subscript does not denote differentiation.

This form is written for non-uniform grid spacing. For uniform grid spacing, the expressions for $\underline{F}_1$ and $\underline{F}_2$ in finite difference form may easily be written from Equation (9.7.9). Equation (9.7.31) shows that $\underline{F}_{1_{mn}}$ and $\underline{F}_{2_{mn}}$ vary linearly with $e^p_{\beta\gamma}$ at different points. The $\Delta\tau'_{\beta\gamma}$ at any point varies linearly with the equivalent body and surface forces $\underline{F}_\beta$, $\underline{T}_\beta$, and hence also linearly with values of $\Delta e_{\beta\gamma_p}$ at different points. At any stage of loading, τ_{11}, τ_{22}, τ_{12}, and σ^* are known. From Equation (9.7.27), we see that $\Delta e_{\beta\gamma_p}$ are proportional to $\Delta e_p{}^*$. Hence at grid point (pq) we see that

$\Delta\tau'_{\beta\gamma}$ is linearly related to the $\Delta e_p{}^*$ at different points in the plate. This may be written as

$$(\Delta\tau'_{\beta\gamma})_{pq} = a_{\beta\gamma_{pqmn}}(\Delta e_p{}^*)_{mn} \tag{9.7.32}$$

in which pq and mn denote grid-point locations and $a_{\beta\gamma_{pqmn}}$ is the influence coefficient giving the stress component $\tau_{\beta\gamma}$ at grid point (pq) caused by a unit effective plastic strain $e_p{}^*$ at grid point (mn). The repetition of the subscript mn in the right side denotes summation over all values of mn. As seen from Equation (9.7.27), $a_{\beta\gamma_{pqmn}}$ depends on σ^* and the stress components $\tau_{\beta\gamma}$ and hence varies with loading.

With $\tau_{\beta\gamma}$ and σ^* known, we obtain, from Equation (9.7.28),

$$\Delta\tau_{\beta\gamma_p} = b_{\beta\gamma}\,\Delta e_p{}^* \tag{9.7.33}$$

where $b_{\beta\gamma}$ also varies with loading. From Equation (9.7.30), at any grid point (pq), we have

$$\Delta\tau_{\beta\gamma_{pq}} = \Delta\tau_{\beta\gamma o_{pq}} + a_{\beta\gamma_{pqmn}}(\Delta e_p^*)_{mn} + b_{\beta\gamma}(\Delta e_p^*)_{pq} \tag{9.7.34}$$

The incremental effective stress $\Delta\sigma^*$ at each pq grid location is obtained from Equation (9.7.26) as

$$\Delta\sigma^* = \frac{2\tau_{11} - \tau_{22}}{2\sigma^*}\,\Delta\tau_{11} + \frac{2\tau_{22} - \tau_{11}}{2\sigma^*}\,\Delta\tau_{22} + \frac{3\tau_{12}}{\sigma^*}\,\Delta\tau_{12} \tag{9.7.35}$$

where the stresses $\tau_{\beta\gamma}$'s and incremental stresses $\Delta\tau_{\beta\gamma}$'s refer to the same pq point at which $\Delta\sigma^*$ is being determined. Equation (9.7.35) may be concisely written as

$$\Delta\sigma^* = c_{\beta\gamma}\,\Delta\tau_{\beta\gamma} \qquad \beta, \gamma = 1, 2 \tag{9.7.36}$$

where it is emphasized again that $\Delta\sigma^*$, $c_{\beta\gamma}$, and $\Delta\tau_{\beta\gamma}$ are all evaluated at the same point.

From Equations (9.7.34) and (9.7.36), the increment of effective stress $\Delta\sigma^*$ at each pq location due to an increment of effective plastic strain $\Delta e_p{}^*$ at every mn location is

$$\begin{aligned}\Delta\sigma^* &= c_{\beta\gamma}\,\Delta\tau_{\beta\gamma o} + c_{\beta\gamma}\,a_{\beta\gamma_{pqmn}}\,(\Delta e_p{}^*)_{mn} + c_{\beta\gamma}\,b_{\beta\gamma}\,(\Delta e_p{}^*)_{pq} \\ &= \frac{\Delta\sigma^*}{\Delta e_p{}^*}\,\Delta e_p{}^*\end{aligned} \tag{9.7.37}$$

in which $\Delta\sigma^*/\Delta e_p^*$ is obtained from uniaxial tests. Writing Equation (9.7.37) in expanded form, we have

$$\frac{\Delta\sigma^*}{\Delta e_p^*}\Delta e_p^* = \left\{\frac{2\tau_{11}-\tau_{22}}{2\sigma^*}\Delta\tau_{11_0} + \frac{2\tau_{22}-\tau_{11}}{2\sigma^*}\Delta\tau_{22_0} + 3\frac{\tau_{12}}{\sigma^*}\Delta\tau_{12_0}\right\}$$
$$+\sum_m\sum_n\left(\frac{2\tau_{11}-\tau_{22}}{2\sigma^*}a_{11_{pqmn}} + \frac{2\tau_{22}-\tau_{11}}{2\sigma^*}a_{22_{pqmn}}\right.$$
$$\left. + 3\frac{\tau_{12}}{\sigma^*}a_{12_{pqmn}}\right)(\Delta e_p^*)_{mn}$$
$$-\frac{E}{1-\nu^2}\left\{\left(\frac{2\tau_{11}-\tau_{22}}{2\sigma^*}\right)\left[\frac{(2-\nu)\tau_{11}-(1-2\nu)\tau_{22}}{2\sigma^*}\right]\right.$$
$$\left. + \frac{2\tau_{22}-\tau_{11}}{2\sigma^*}\left[\frac{(2-\nu)\tau_{22}-(1-2\nu)\tau_{11}}{2\sigma^*}\right]\right\}\Delta e_p^*$$
$$-9G\left(\frac{\tau_{12}}{\sigma^*}\right)^2\Delta e_p^* \qquad (9.7.38)$$

where the quantities $\Delta\sigma^*$, Δe_p^*, $\Delta\tau_{\beta\gamma}$, and $\tau_{\beta\gamma}$'s without a subscript refer to these quantities at each pq location.

Taking $\Delta e^*_{p_{mn}}$ at grid points (mn) as unknowns, we see that there are as many unknowns as there are grid points with incremental plastic strain Δe_p^*, since Equation (9.7.37) holds for each of these grid points. The set of simultaneous equations (9.7.38) may be solved for values of Δe_p^* at different points. With Δe_p^* known, $\Delta\tau_{\beta\gamma}$ is readily calculated. $\tau_{\beta\gamma} + \Delta\tau_{\beta\gamma}$ gives $\tau_{\beta\gamma}$ for the next increment of applied loads, and the cycle is repeated. The rate of plastic strain component is proportional to the corresponding deviatoric stress component

$$\dot{e}_{ij_P} \propto S_{ij}$$

Equation (9.7.24) is valid if the S_{ij} remain in the same ratio during each load increment which causes Δe_{ij_p}. Hence if the change of the ratios of the different deviatoric stress components in a load increment is large, the results are not valid. In this case the load increment should be reduced until the analysis yields ratios of the S_{ij} which remain approximately the same. The accuracy of this method may be increased by increasing the fineness of the grid and by decreasing the load increments.

A numerical illustrative example of elastoplastic analysis of an infinite plate is shown in Reference 12.

REFERENCES

1. Love, A. E., *A Treatise on the Mathematical Theory of Elasticity*, Dover, New York, pp. 35, 204–210, 1944.
2. Sokolnikoff, I. S. and R. M. Redheffer, *Mathematics of Physics and Modern Engineering*, McGraw-Hill, New York, pp. 535–544, 1958.
3. Sneddon, I. S., *Fourier Transforms*, McGraw-Hill, New York, pp. 400–402, 1951.
4. Sokolnikoff, I. S., *Mathematical Theory of Elasticity*, McGraw-Hill, New York, pp. 136, 249–286, 1956.
5. Muskhelishvili, N. I., *Some Basic Problems of the Mathematical Theory of Elasticity*, Noordhoff, Holland, p. 146, 1953.
6. Boley, B. A., and J. H. Weiner, *Theory of Thermal Stresses*, Wiley, New York, Chap. 4, 1960.
7. Nowacki, W., *Thermoelasticity*, Pergamon, London, England, Chap. IX, p. 553, 1962.
8. Zudans, Z., T. C. Yen, and W. H. Steigelman, *Thermal Stress Techniques in the Nuclear Industry*, American Elsevier, New York, pp. 239–241, 1965.
9. MacMillan, W. D., *The Theory of the Potential*, Dover, New York, p. 35, 126, 1930.
10. Morse, P. M., and H. Feshbach, *Methods of Theoretical Physics*, McGraw-Hill, New York, pp. 797–799, 1953.
11. Platus, D. L., "The Plane Stress Problems for a Doubly Connected Region Undergoing Elastic and Creep Deformations," Ph.D. Dissertation, UCLA, 1962; *J. Appl. Mech.*, **31**, 54–60, 1964.
12. Lin, T. H., and L. M. Lackman, "Relief of Residual Stresses by Heating Patterns," *J. Engr. Mech.*, ASCE, June 1964.

chapter

10

Inelastic Bending of Plates

Introduction. This chapter presents a method of analysis for inelastic bending of plates. The analysis, using the analogy concept, reduces the analysis of plates with inelastic strain to the analysis of identical elastic plates with an additional set of lateral loads and edge moments.

The chapter begins with a brief review of the theory of bending of elastic plates. The analogy between inelastic strain and lateral load and edge moments of plates is derived in Section 10.2. The analysis of thermoelastic plates is given in Section 10.3. Sections 10.4 and 10.5 give the methods of analysis for viscoelastic plates and plates with nonlinear strain-hardening creep characteristics. Section 10.6 shows the method of elastoplastic analysis of plates of strain-hardening materials. Examples illustrating the various inelastic analyses are given. The numerical methods employed in the inelastic analyses are generally quite lengthy, and computation by a digital computer is required.

10.1 Elastic Bending of Thin Plates

Before discussing the inelastic bending of thin plates, the elastic bending of plates will be briefly reviewed. Let the xy plane coincide with the middle plane of the plate and the z axis be the normal to this plane with positive direction downward. Under loading, the middle plane of the plate deforms into a curved surface, called the middle surface. Kirchhoff[1] assumed in the analysis of thin plates that normals to the middle plane before loading remain normal to the middle surface during loading. This is analogous to the Bernoulli-Euler assumption that plane sections of a beam remain plane during loading. He further assumed that the normal stress on planes parallel to the middle surface may be neglected in comparison with other stresses. Assuming, then, that $\tau_{zz} = 0$, we see that the stress-strain relations for a thin plate in the elastic range are

$$\begin{aligned} e_{xx} &= \frac{\tau_{xx}}{E} - \nu \frac{\tau_{yy}}{E} \\ e_{yy} &= \frac{\tau_{yy}}{E} - \nu \frac{\tau_{xx}}{E} \\ e_{xy} &= \frac{\tau_{xy}}{2G} \end{aligned} \tag{10.1.1}$$

where E is Young's modulus, G is the shear modulus, and ν is Poisson's ratio. Expressing stresses in terms of strains, we obtain

$$\begin{aligned} \tau_{xx} &= \frac{E}{1-\nu^2}(e_{xx} + \nu e_{yy}) \\ \tau_{yy} &= \frac{E}{1-\nu^2}(e_{yy} + \nu e_{xx}) \\ \tau_{xy} &= 2Ge_{xy} \end{aligned} \tag{10.1.2}$$

Let u, v, and w be the displacement along the x, y, and z axes; and let u_0, v_0, and w_0 be the corresponding displacement of points on the middle plane. Referring to Figure 10.1, we see that, since the normal to the middle plane remains normal to the middle surface after deformation,

$$u_0 - u = z \frac{\partial w}{\partial x}$$

or (10.1.3)

$$u = u_0 - z \frac{\partial w}{\partial x}$$

Similarly, for displacements in the y direction,

$$v = v_0 - z \frac{\partial w}{\partial y} \tag{10.1.4}$$

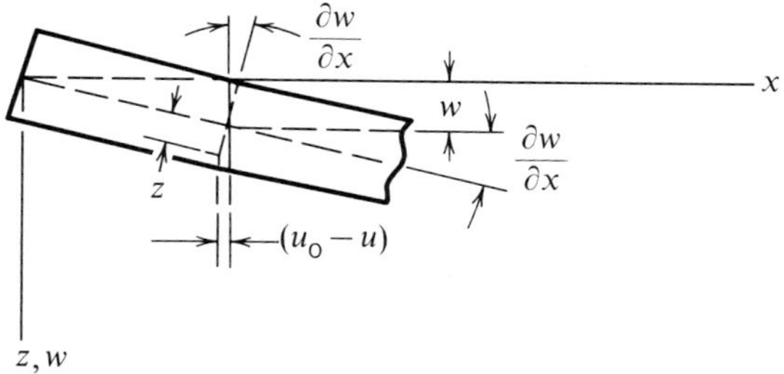

Figure 10.1

Differentiating Equations (10.1.3) and (10.1.4) with respect to z gives

$$\frac{\partial u}{\partial z} = -\frac{\partial w}{\partial x} \qquad \frac{\partial v}{\partial z} = -\frac{\partial w}{\partial y}$$

Recalling the strain-displacement relations, we find

$$e_{xz} = \frac{1}{2}\left(\frac{\partial u}{\partial z} + \frac{\partial w}{\partial x}\right) = 0 \qquad e_{yz} = \frac{1}{2}\left(\frac{\partial v}{\partial z} + \frac{\partial w}{\partial y}\right) = 0$$

Hence Kirchhoff's assumption implies zero shear strains[2] e_{xz} and e_{yz}. From Equations (10.1.3) and (10.1.4), the strain components are

$$\begin{aligned} e_{xx} &= \frac{\partial u}{\partial x} = \frac{\partial u_0}{\partial x} - z\frac{\partial^2 w}{\partial x^2} \\ e_{yy} &= \frac{\partial v}{\partial y} = \frac{\partial v_0}{\partial y} - z\frac{\partial^2 w}{\partial y^2} \\ e_{xy} &= \frac{1}{2}\left(\frac{\partial u}{\partial y} + \frac{\partial v}{\partial x}\right) = \frac{1}{2}\left(\frac{\partial u_0}{\partial y} + \frac{\partial v_0}{\partial x}\right) - z\frac{\partial^2 w}{\partial x\,\partial y} \end{aligned} \tag{10.1.5}$$

Substituting Equations (10.1.5) into Equation (10.1.2) yields

$$\begin{aligned} \tau_{xx} &= \frac{E}{1-\nu^2}\left[\frac{\partial u_0}{\partial x} - z\frac{\partial^2 w}{\partial x^2} + \nu\left(\frac{\partial v_0}{\partial y} - z\frac{\partial^2 w}{\partial y^2}\right)\right] \\ \tau_{yy} &= \frac{E}{1-\nu^2}\left[\frac{\partial v_0}{\partial y} - z\frac{\partial^2 w}{\partial y^2} + \nu\left(\frac{\partial u_0}{\partial x} - z\frac{\partial^2 w}{\partial x^2}\right)\right] \\ \tau_{xy} &= G\left(\frac{\partial u_0}{\partial y} + \frac{\partial v_0}{\partial x} - 2z\frac{\partial^2 w^x}{\partial x\,\partial y}\right) \end{aligned} \tag{10.1.6}$$

The stresses τ_{xx}, τ_{yy}, τ_{xy}, τ_{xz}, and τ_{yz} vary across the plate thickness h.

Their resultants on the sections *ABCD* and *CBEF* in Figure 10.2 are obtained by integration; that is,

$$N_x = \int_{-h/2}^{h/2} \tau_{xx}\, dz \qquad N_y = \int_{-h/2}^{h/2} \tau_{yy}\, dz \qquad N_{xy} = \int_{-h/2}^{h/2} \tau_{xy}\, dz \tag{10.1.7}$$

$$\left.\begin{aligned} Q_x &= \int_{-h/2}^{h/2} \tau_{xz}\, dz \qquad Q_y = \int_{-h/2}^{h/2} \tau_{yz}\, dz \\ M_x &= \int_{-h/2}^{h/2} \tau_{xx} z\, dz \qquad M_y = \int_{-h/2}^{h/2} \tau_{yy} z\, dz \\ M_{xy} &= \int_{-h/2}^{h/2} -\tau_{xy} z\, dz = -M_{yx} \end{aligned}\right\} \tag{10.1.8}$$

The signs of M_{xy} and M_{yx} are due to the sign conventions adopted in Figure 10.2. Substituting Equations (10.1.6) into Equations (10.1.7) and (10.1.8) and integrating, we obtain

$$\begin{aligned} N_x &= \frac{Eh}{1-\nu^2}\left(\frac{\partial u_0}{\partial x} + \nu \frac{\partial v_0}{\partial y}\right) \\ N_y &= \frac{Eh}{1-\nu^2}\left(\frac{\partial v_0}{\partial y} + \nu \frac{\partial u_0}{\partial x}\right) \\ N_{xy} &= Gh\left(\frac{\partial u_0}{\partial y} + \frac{\partial v_0}{\partial x}\right) \end{aligned} \tag{10.1.9}$$

$$\begin{aligned} M_x &= -\frac{Eh^3}{12(1-\nu^2)}\left(\frac{\partial^2 w}{\partial x^2} + \nu \frac{\partial^2 w}{\partial y^2}\right) \\ M_y &= -\frac{Eh^3}{12(1-\nu^2)}\left(\frac{\partial^2 w}{\partial y^2} + \nu \frac{\partial^2 w}{\partial x^2}\right) \\ M_{xy} &= -M_{yx} = \frac{2Gh^3}{12}\frac{\partial^2 w}{\partial x\, \partial y} = \frac{Eh^3}{12(1+\nu)}\frac{\partial^2 w}{\partial x\, \partial y} \end{aligned} \tag{10.1.10}$$

Upon recalling $G = E/2(1+\nu)$ and defining $D = Eh^3/12(1-\nu^2)$, we see that Equations (10.1.10) become,

$$\begin{aligned} M_x &= -D\left(\frac{\partial^2 w}{\partial x^2} + \nu \frac{\partial^2 w}{\partial y^2}\right) \\ M_y &= -D\left(\frac{\partial^2 w}{\partial y^2} + \nu \frac{\partial^2 w}{\partial x^2}\right) \\ M_{xy} &= D(1-\nu)\frac{\partial^2 w}{\partial x\, \partial y} \end{aligned} \tag{10.1.11}$$

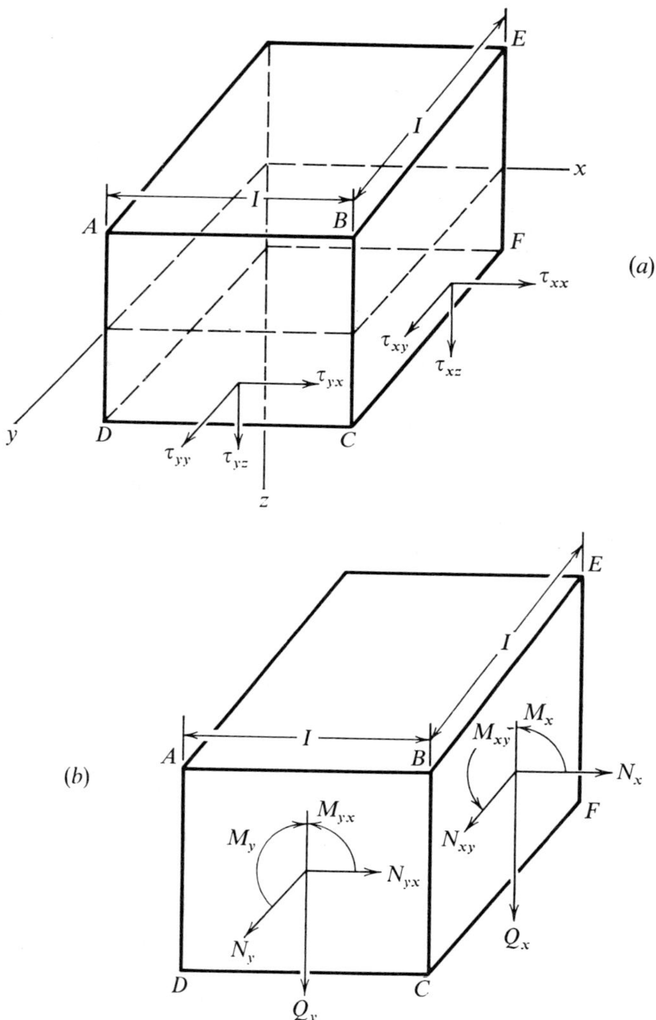

Figure 10.2

Consider a $dx\, dy$ element of plate (Figure 10.3). The conditions of equilibrium give

$$\sum F_x = 0 \qquad \frac{\partial N_x}{\partial x} + \frac{\partial N_{yx}}{\partial y} = 0$$

$$\sum F_y = 0 \qquad \frac{\partial N_{xy}}{\partial x} + \frac{\partial N_y}{\partial y} = 0 \tag{10.1.12}$$

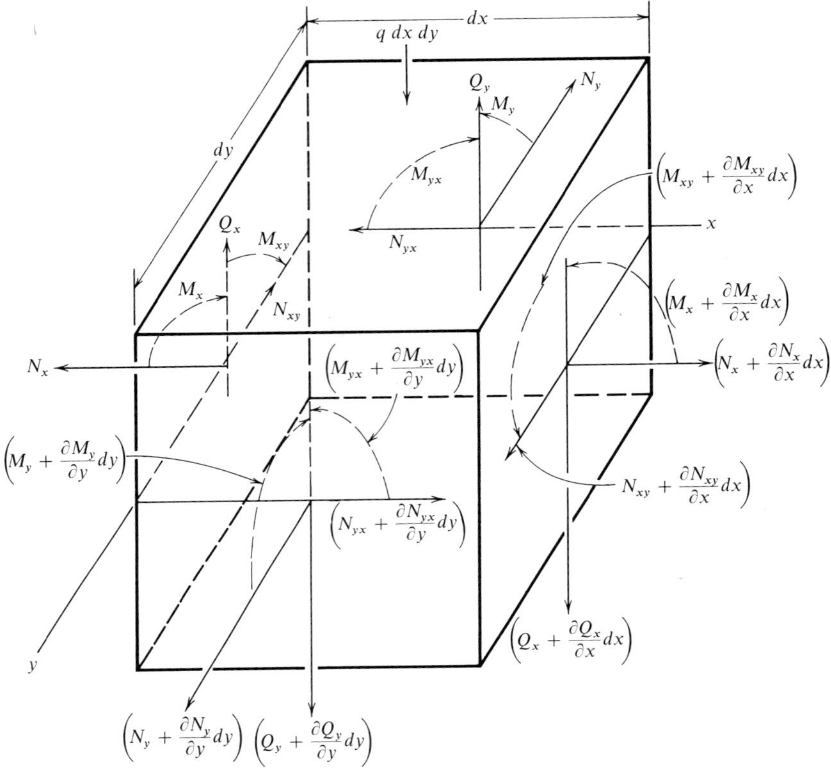

Figure 10.3

The solution of these two equations, together with the compatibility conditions, was discussed in Chapter 9. This chapter is concerned only with plates under pure bending, where the sectional forces N_x, N_y, and N_{xy} are negligible. With $N_x = N_y = N_{xy} = 0$, Equations (10.1.9) give

$$\frac{\partial u_0}{\partial x} = 0 \qquad \frac{\partial v_0}{\partial y} = 0 \qquad \text{and} \qquad \frac{\partial u_0}{\partial y} + \frac{\partial v_0}{\partial x} = 0$$

Using the above relations and Equations (10.1.6) and (10.1.10), we have the following expressions for stresses in terms of bending moments for plates under pure bending:

$$\tau_{xx} = \frac{-Ez}{1-\nu^2}\left(\frac{\partial^2 w}{\partial x^2} + \nu\,\frac{\partial^2 w}{\partial y^2}\right) = \frac{12z}{h^3}\,M_x$$

$$\tau_{yy} = \frac{-Ez}{1-\nu^2}\left(\frac{\partial^2 w}{\partial y^2} + \nu\,\frac{\partial^2 w}{\partial x^2}\right) = \frac{12z}{h^3}\,M_y \qquad (10.1.13)$$

$$\tau_{xy} = -2Gz\,\frac{\partial^2 w}{\partial x\,\partial y} = -\frac{12z}{h^3}\,M_{xy}$$

The conditions of equilibrium normal to the plate are

$$\sum F_z = 0 \qquad \frac{\partial Q_x}{\partial x} + \frac{\partial Q_y}{\partial y} + q = 0 \qquad (10.1.14)$$

$$\sum M_y = 0 \qquad \frac{\partial M_x}{\partial x} + \frac{\partial M_{yx}}{\partial y} - Q_x = 0 \qquad (10.1.15)$$

$$\sum M_x = 0 \qquad -\frac{\partial M_{xy}}{\partial x} + \frac{\partial M_y}{\partial y} - Q_y = 0 \qquad (10.1.16)$$

Eliminating Q_x and Q_y from Equations (10.1.14) to (10.1.16) yields

$$\frac{\partial^2 M_x}{\partial x^2} - 2\,\frac{\partial^2 M_{xy}}{\partial x\,\partial y} + \frac{\partial^2 M_y}{\partial y^2} + q = 0 \qquad (10.1.17)$$

Substituting Equations (10.1.11) into the above, we obtain the *governing equation for the equilibrium of an elastic thin plate*:

$$D\frac{\partial^2}{\partial x^2}\left(\frac{\partial^2 w}{\partial x^2} + \nu\,\frac{\partial^2 w}{\partial y^2}\right) + 2D(1-\nu)\,\frac{\partial^2}{\partial x\,\partial y}\,\frac{\partial^2 w}{\partial x\,\partial y} + D\,\frac{\partial^2}{\partial y^2}\left(\frac{\partial^2 w}{\partial y^2} + \nu\,\frac{\partial^2 w}{\partial x^2}\right) = q$$

$$q = D\nabla^2\nabla^2 w = D\nabla^4 w \qquad (10.1.18)$$

where

$$\nabla^2 = \left(\frac{\partial^2}{\partial x^2} + \frac{\partial^2}{\partial y^2}\right).$$

From Equations (10.1.11), (10.1.15), and (10.1.16), the sectional shearing forces are

$$Q_x = -D\,\frac{\partial}{\partial x}\left(\frac{\partial^2 w}{\partial x^2} + \frac{\partial^2 w}{\partial y^2}\right) = -D\,\frac{\partial}{\partial x}\,\nabla^2 w$$

$$Q_y = -D\,\frac{\partial}{\partial y}\left(\frac{\partial^2 w}{\partial x^2} + \frac{\partial^2 w}{\partial y^2}\right) = -D\,\frac{\partial}{\partial y}\,\nabla^2 w \qquad (10.1.19)$$

The shearing stresses τ_{zx} and τ_{zy} are then found from the condition of equilibrium of a *dx dy dz* elemental volume as shown in Figure 10.4.

Equilibrium along the x direction requires

$$\int_z^{h/2} \frac{\partial \tau_{xx}}{\partial x} dx\, dy\, dz + \int_z^{h/2} \frac{\partial \tau_{xy}}{\partial y} dy\, dx\, dz - \tau_{zx}\, dx\, dy = 0$$

This gives

$$\int_z^{h/2} \frac{\partial \tau_{xx}}{\partial x} dz + \int_z^{h/2} \frac{\partial \tau_{xy}}{\partial y} dz - \tau_{zx} = 0$$

Similarly equilibrium along the y direction gives

$$\int_z^{h/2} \frac{\partial \tau_{yy}}{\partial y} dz + \int_z^{h/2} \frac{\partial \tau_{xy}}{\partial x} dz - \tau_{zy} = 0$$

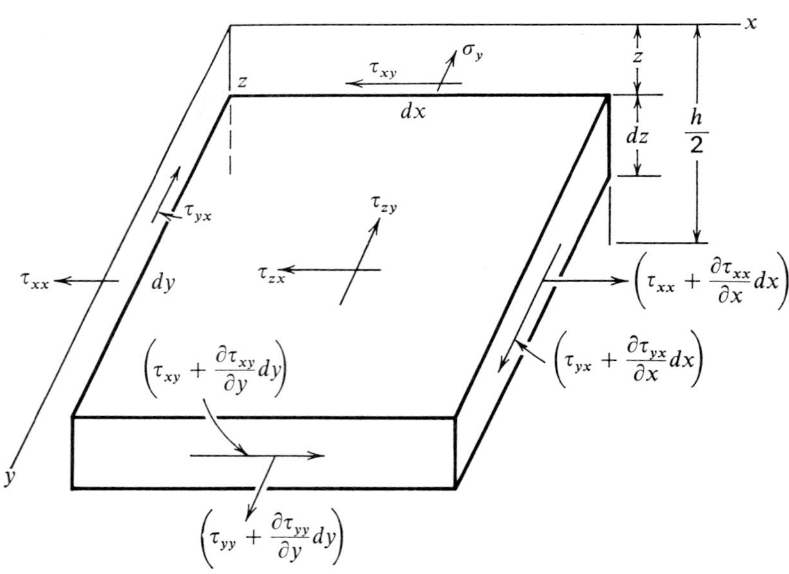

Figure 10.4

Substituting the expressions for τ_{xx} and τ_{xy} from Equation (10.1.13) into the above equation yields

$$\frac{12}{h^3} \frac{\partial M_x}{\partial x} \int_z^{h/2} z\, dz - \frac{12}{h^3} \frac{\partial M_{xy}}{\partial y} \int_z^{h/2} z\, dz - \tau_{zx} = 0$$

Solving for τ_{zx} gives

$$\tau_{zx} = \frac{12}{h^3}\left(\frac{\partial M_x}{\partial x} - \frac{\partial M_{xy}}{\partial y}\right)\int_z^{h/2} z\,dz = \frac{12}{h^3} Q_x \frac{1}{2}\left(\frac{h^2}{4} - z^2\right)$$

$$= \frac{3}{2}\frac{Q_x}{h}\left(1 - 4\frac{z^2}{h^2}\right) \tag{10.1.20}$$

Similarly, we have

$$\tau_{zy} = \frac{3}{2}\frac{Q_y}{h}\left(1 - 4\frac{z^2}{h^2}\right) \tag{10.1.21}$$

In the determination of w, differential Equation (10.1.18) is solved for the given boundary conditions. A few commonly used boundary conditions are shown below.

(1) *Fixed or clamped edge.* Along this edge, both the lateral deflection and the slope of the middle surface normal to this edge are zero. For a fixed edge parallel to the x axis, as shown in Figure 10.5, we have

$$\begin{aligned}(w)_{y=0} &= 0 \\ \left(\frac{\partial w}{\partial y}\right)_{y=0} &= 0\end{aligned} \tag{10.1.22}$$

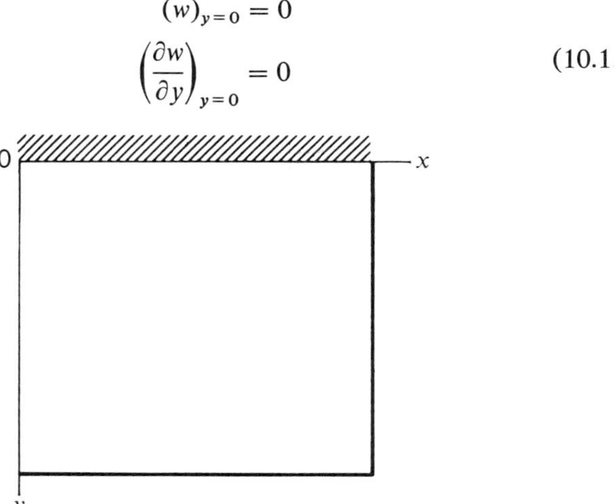

Figure 10.5

(2) *Simply supported edge.* Along this edge, the lateral deflection and the bending moment normal to the edge are zero. For a simply supported edge parallel to the x axis, as shown in Figure 10.6, we have

$$\begin{aligned}(w)_{y=0} &= 0 \\ (M_y)_{y=0} &= -D\left(\frac{\partial^2 w}{\partial y^2} + \nu\frac{\partial^2 w}{\partial x^2}\right)_{y=0} = 0\end{aligned} \tag{10.1.23}$$

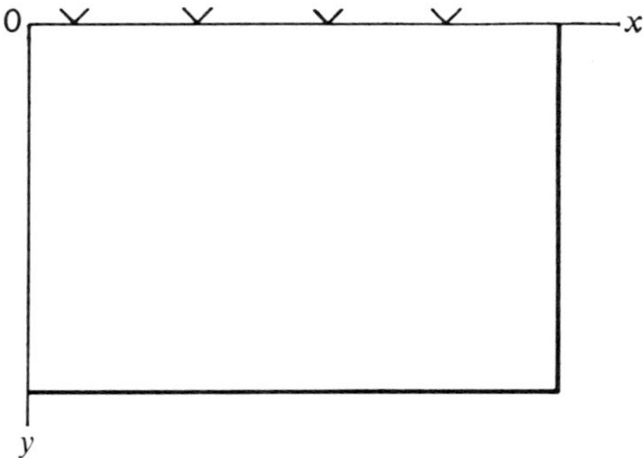

Figure 10.6

(3) *Free edge*. Along such an edge, the bending moment, twisting moment, and vertical shear force are zero. Kirchhoff[3] has shown that these three conditions, obtained from physical reasoning, are too many; that the equations defining a free edge parallel to the x axis at $y = 0$ are

$$(M_y)_{y=0} = -D\left(\frac{\partial^2 w}{\partial y^2} + \nu \frac{\partial^2 w}{\partial x^2}\right)_{y=0} = 0$$

and

$$Q_y - \frac{\partial M_{yx}}{\partial x} = \left(\frac{\partial^3 w}{\partial y^3} + (2 - \nu) \frac{\partial^3 w}{\partial x^2\, \partial y}\right)_{y=0} = 0 \qquad (10.1.24)$$

Simply Supported Rectangular Plates. A rectangular plate with simply supported edges parallel to the coordinate axes is shown in Figure 10.7.

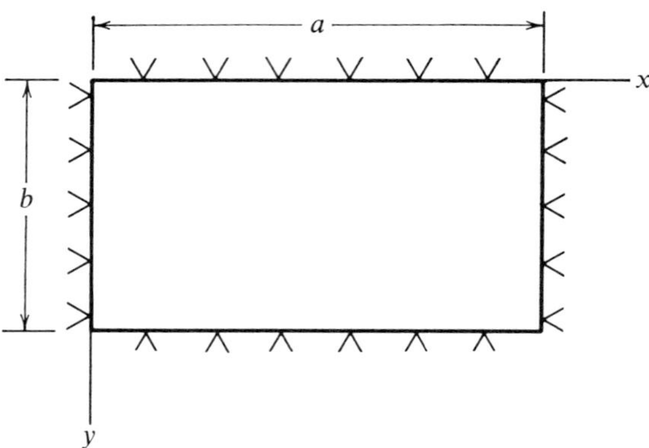

Figure 10.7

An arbitrarily distributed lateral loading is denoted by $q(x, y)$. The differential equation,

$$D\nabla^4 w = q \tag{10.1.18}$$

is solved to satisfy the boundary conditions; at $x = 0$ and $x = a$, we have $w = 0$, $\partial^2 w/\partial x^2 = 0$, and at $y = 0$ and $y = b$, we have $w = 0$, $\partial^2 w/\partial y^2 = 0$. These boundary conditions are satisfied for a deflection w taken in the form of a double sine series,

$$w = \sum_{m=1}^{\infty} \sum_{n=1}^{\infty} b_{mn} \sin \frac{m\pi x}{a} \sin \frac{n\pi y}{b} \tag{10.1.25}$$

where m and n are integers. Substituting Equation (10.1.25) into Equation (10.1.18) gives

$$D \sum_{m=1}^{\infty} \sum_{n=1}^{\infty} b_{mn} \left(\frac{m^4\pi^4}{a^4} + 2\frac{m^2\pi^2}{a^2}\frac{n^2\pi^2}{b^2} + \frac{n^4\pi^4}{b^4} \right) \sin \frac{m\pi x}{a} \sin \frac{n\pi y}{b} = q(x, y) \tag{10.1.26}$$

Multiplying Equation (10.1.26) by $\sin(m'\pi x/a)\sin(n'\pi y/b)\,dx\,dy$ and integrating from $x = 0$ to $x = a$, and from $y = 0$ to $y = b$, we obtain

$$D \sum_{m=1}^{\infty} \sum_{n=1}^{\infty} b_{mn} \left(\frac{m^2\pi^2}{a^2} + \frac{n^2\pi^2}{b^2} \right)^2 \int_0^b \int_0^a \sin \frac{m\pi x}{a} \sin \frac{n\pi y}{b} \sin \frac{m'\pi x}{a} \sin \frac{n'\pi x}{b}\, dx\, dy$$

$$= \int q(x, y) \sin \frac{m'\pi x}{a} \sin \frac{n'\pi x}{b}\, dx\, dy \tag{10.1.27}$$

From the orthogonality property of sine series, we have

$$\begin{aligned} \int_0^a \sin \frac{m\pi x}{a} \sin \frac{m'\pi x}{a}\, dx &= \begin{cases} 0 & \text{if } m \neq m' \\ a/2 & \text{if } m = m' \end{cases} \\ \int_0^b \sin \frac{n\pi x}{b} \sin \frac{n'\pi x}{b}\, dy &= \begin{cases} 0 & \text{if } n \neq n' \\ b/2 & \text{if } n = n' \end{cases} \end{aligned} \tag{10.1.28}$$

and therefore Equation (10.1.27) reduces to

$$Db_{m'n'} \left(\frac{m'^2\pi^2}{a^2} + \frac{n'^2\pi^2}{b^2} \right)^2 \frac{ab}{4} = \int_0^b \int_0^a q(x, y) \sin \frac{m'\pi x}{a} \sin \frac{n'\pi y}{b}\, dx\, dy \tag{10.1.29}$$

Expressing $q(x, y)$ in terms of a Fourier series gives

$$q(x, y) = \sum_{m=1}^{\infty} \sum_{n=1}^{\infty} a_{mn} \sin \frac{m\pi x}{a} \sin \frac{n\pi y}{b}$$

Substitute the Fourier series expansion of q into the right side of Equation (10.1.29) and integrate. Owing to the orthogonality conditions of Equations (10.1.28), we obtain

$$\int_0^b \int_0^a q(x, y) \sin \frac{m'\pi x}{a} \sin \frac{n'\pi y}{b}\, dx\, dy = \frac{ab}{4} a_{m'n'} \qquad (10.1.30)$$

Then Equation (10.1.29) gives

$$Db_{mn}\left(\frac{m^2\pi^2}{a^2} + \frac{n^2\pi^2}{b^2}\right)^2 \frac{ab}{4} = \frac{ab}{4} a_{mn}$$

or, solving for b_{mn}, we have

$$b_{mn} = \frac{a_{mn}}{\pi^4 D(m^2/a^2 + n^2/b^2)^2}$$

Substituting the expression for b_{mn} into Equation (10.1.25) yields

$$w = \frac{1}{\pi^4 D} \sum_{m=1}^{\infty} \sum_{n=1}^{\infty} \frac{a_{mn}}{(m^2/a^2 + n^2/b^2)^2} \sin \frac{m\pi x}{a} \sin \frac{n\pi y}{b} \qquad (10.1.31)$$

where

$$a_{mn} = \frac{4}{ab} \int_0^b \int_0^a q(x, y) \sin \frac{m\pi x}{a} \sin \frac{n\pi y}{b}\, dx\, dy$$

For the particular case of a rectangular plate under a uniformly distributed load q_0, the following results are immediately obtained from Equations (10.1.30) and (10.1.31):

$$a_{mn} = \frac{4q_0}{ab} \int_0^a \int_0^b \sin \frac{m\pi x}{a} \sin \frac{n\pi y}{b}\, dx\, dy$$

$$= \frac{16q_0}{mn\pi^2}$$

when m and n are odd integers; and

$$a_{mn} = 0$$

when m or n is even.

Therefore for a simply supported rectangular plate with sides of lengths a and b, subjected to a uniform load q_0, the deflection is

$$w = \frac{16q_0}{\pi^6 D} \sum_{m=1,3,5,\ldots} \sum_{n=1,3,5,\ldots} \frac{\sin(m\pi x/a) \sin(n\pi y/b)}{mn(m^2/a^2 + n^2/b^2)^2} \qquad (10.1.32)$$

This series converges rapidly; in fact the first few terms generally give satisfactory answers. With w known, M_x, M_y, and M_{xy} may be obtained from Equations (10.1.10), and the stresses may be found from Equations (10.1.13), (10.1.20), and (10.1.21).

For a simply supported rectangular plate, with sides a and b, under a load P applied uniformly over the square shaded area[4] shown in Figure 10.8, we have, from Equation (10.1.30),

$$a_{mn} = \frac{4P}{abc^2} \int_{\xi - c/2}^{\xi + c/2} \int_{\eta - c/2}^{\eta + c/2} \sin \frac{m\pi x}{a} \sin \frac{n\pi y}{b} \, dx \, dy$$

$$= \frac{16P}{\pi^2 mnc^2} \sin \frac{m\pi\xi}{a} \sin \frac{n\pi\eta}{b} \sin \frac{m\pi c}{2a} \sin \frac{n\pi c}{2b} \qquad (10.1.33)$$

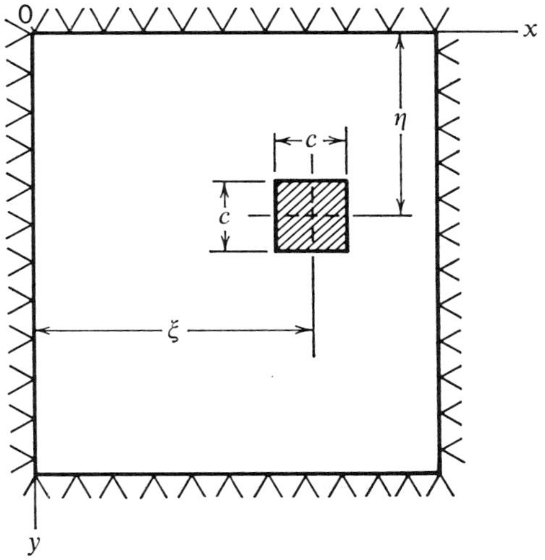

Figure 10.8

For a single concentrated load P acting at the point (ξ, η), the corresponding a_{mn} may be found by letting $c \to 0$ and using L'Hospital's rule, obtaining

$$a_{mn} = \frac{4P}{ab} \sin \frac{m\pi\xi}{a} \sin \frac{n\pi\eta}{b} \qquad (10.1.34)$$

Therefore, for a simply supported rectangular plate, with sides of lengths a and b, subject to a concentrated load P at $x = \xi$ and $y = \eta$, the deflection is

$$w = \frac{4P}{\pi^2 abD} \sum_{m=1}^{\infty} \sum_{n=1}^{\infty} \frac{\sin (m\pi\xi/a) \sin (n\pi\eta/b)}{(m^2/a^2 + n^2/b^2)^2} \sin \frac{m\pi x}{a} \sin \frac{n\pi y}{b} \qquad (10.1.35)$$

The M_x, M_y, and M_{xy} may be obtained from Equations (10.1.11) and the stresses from Equations (10.1.13), (10.1.20), and (10.1.21).

Rectangular Plates with Moments Distributed Along the Edges.[4] Consider a rectangular plate simply supported along the edges $x = 0$ and $x = a$, and $y = \pm b/2$ and subject to distributed edge moments along its edges $y = \pm b/2$ of $\phi_1(x)$ and $\phi_2(x)$, respectively, as shown in Figure 10.9. Since $q = 0$, $\nabla^4 w = 0$, and hence w is a biharmonic function. The boundary conditions are as follows:

(1) Along $x = 0$ and $x = a$, $w = 0$ and $M_x = 0$. These give

$$\frac{\partial w}{\partial y} = 0 \qquad \frac{\partial^2 w}{\partial y^2} = 0$$

$$M_x = 0 \qquad \frac{\partial^2 w}{\partial x^2} = 0$$

(2) Along $y = +b/2$, $w = 0$, and

$$M_y = \phi_1(x) = -D\left(\frac{\partial^2 w}{\partial y^2} + \nu\,\frac{\partial^2 w}{\partial x^2}\right)$$

$$= -D\,\frac{\partial^2 w}{\partial y^2}$$

(3) Along $y = -b/2$, $w = 0$, and

$$M_y = \phi_2(x) = -D\left(\frac{\partial^2 w}{\partial y^2} + \nu\,\frac{\partial^2 w}{\partial x^2}\right) = -D\,\frac{\partial^2 w}{\partial y^2}$$

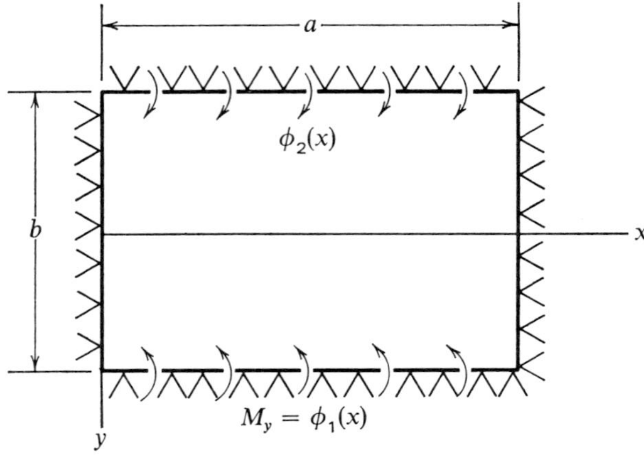

Figure 10.9

For the solution of the biharmonic equation $\nabla^4 w = 0$, assume $w = \sum_{m=1}^{\infty} Y_m \sin m\pi x/a$, where Y_m is a function of y only. This satisfies boundary condition 1. Then $\nabla^4 w = 0$ gives

$$\sum_{m=1}^{\infty}\left(Y_m'''' - 2\,\frac{m^2\pi^2}{a^2}\,Y_m'' + \frac{m^4\pi^4}{a^4}\,Y_m\right)\sin\frac{m\pi x}{a} = 0 \qquad (10.1.36)$$

where each prime denotes a differentiation with respect to y. Multiplying Equation (10.1.36) by $\sin\,(m'\pi x/a)\,dx$ and integrating from $x = 0$ to $x = a$ (recalling the orthogonality relations of the sine series), we obtain

$$Y_m'''' - 2\,\frac{m^2\pi^2}{a^2}\,Y_m'' + \frac{m^4\pi^4}{a^4}\,Y_m = 0 \qquad (10.1.37)$$

The general solution of Equation (10.1.37) is

$$Y_m = A_m \cosh\frac{m\pi y}{a} + B_m \sinh\frac{m\pi y}{a} + C_m\,\frac{m\pi y}{a}\cosh\frac{m\pi y}{a} + D_m\,\frac{m\pi y}{a}\sinh\frac{m\pi y}{a} \qquad (10.1.38)$$

Let

$$\tfrac{1}{2}[\phi_1(x) + \phi_2(x)] = \psi_1(x)$$

$$\tfrac{1}{2}[\phi_1(x) - \phi_2(x)] = \psi_2(x)$$

The edge moments $\phi_1(x)$ and $\phi_2(x)$ may be considered as composed of symmetrical moments $\psi_1(x)$ and antisymmetrical moments $\psi_2(x)$ applied to the two edges. The w resulting from the symmetrical moments $\psi_1(x)$ on the two edges will be symmetrical, hence Y_m is symmetrical,

$$Y_m = A_m \cosh\frac{m\pi y}{a} + D_m\,\frac{m\pi y}{a}\sinh\frac{m\pi y}{a}$$

To satisfy the condition $w = 0$ at $y = \pm b/2$, we have

$$A_m = -\frac{m\pi b}{2a}\,D_m \tanh\frac{m\pi b}{2a}$$

$$w = \sum D_m\left(-\frac{m\pi b}{2a}\tanh\frac{m\pi b}{2a}\cosh\frac{m\pi y}{a} + \frac{m\pi y}{a}\sinh\frac{m\pi y}{a}\right)\sin\frac{m\pi x}{a}$$

$$-D\,\frac{\partial^2 w}{\partial y^2} = D\sum D_m\left[\left(\frac{m\pi b}{2a}\tanh\frac{m\pi b}{2a} - 2\right)\frac{m^2\pi^2}{a^2}\cosh\frac{m\pi y}{a} - \frac{m^3\pi^3 y}{a^3}\sinh\frac{m\pi y}{a}\right]\sin\frac{m\pi x}{a} = \psi_1(x) \qquad (10.1.39)$$

From Equation (10.1.39), the D_m's are determined. The w resulting from the antisymmetrical edge moments $\psi_2(x)$ will be antisymmetrical, and hence

$$w = \sum_{m=1}^{\infty} \left(B_m \sinh \frac{m\pi y}{a} + C_m \frac{m\pi y}{a} \cosh \frac{m\pi y}{a} \right) \sin \frac{m\pi x}{a}$$

To satisfy the condition $w = 0$ at $y = \pm b/2$, we have

$$B_m = -C_m \frac{m\pi b}{2a} \coth \frac{m\pi b}{2a}$$

$$w = \sum C_m \left[-\frac{m\pi b}{2a} \coth \frac{m\pi b}{2a} \sinh \frac{m\pi y}{a} + \frac{m\pi y}{a} \cosh \frac{m\pi y}{a} \right] \sin \frac{m\pi x}{a}$$

$$-D \frac{\partial^2 w}{\partial y^2} = \psi_2(x) = D \sum C_m \left[\left(\frac{m\pi b}{2a} \coth \frac{m\pi b}{2a} - 2 \right) \left(\frac{m\pi}{a} \right)^2 \sinh \frac{m\pi y}{a} - \left(\frac{m\pi}{a} \right)^2 \frac{m\pi y}{a} \cosh \frac{m\pi y}{a} \right] \sin \frac{m\pi x}{a} \tag{10.1.40}$$

The deflection w caused by ϕ_1 and ϕ_2 is obtained by superposition of the deflections due to ψ_1 and ψ_2. The C_m's are determined from Equation (10.1.40).

Since the partial differential equation, $\nabla^4 w = 0$, is linear, the solutions under different loadings may be superposed. In this way, plates with various edge boundary conditions can be analyzed by considering the solutions for simply supported plates only. For example, consider a rectangular plate with two opposite edges simply supported and the other two edges clamped as shown in Figure 10.10. The deflection of this plate under a lateral load q may be obtained by the solution of two simply supported plate problems: one is the simply supported plate with lateral load q only, and the other is the simply supported plate with edge moments M_y applied along the edges, $y = \pm b/2$, which were originally clamped, as shown in Figure 10.11. The solutions of deflections for these two loadings are obtained and superposed; that is, $w = w_1 + w_2$. Since the $y = \pm b/2$ edges are really clamped, the distribution of M_y along the two edges must be such as to produce zero slope $\partial w/\partial y = 0$ at $y = \pm b/2$. The moments and stresses are then readily found.

For a rectangular plate with all edges clamped, the deflection of the plate under lateral loading may be resolved into three parts as shown in Figure 10.12.

From the condition $\partial w/\partial n = 0$, where n is the coordinate normal to the boundary, the distributions of M_x and M_y along the edges are determined.

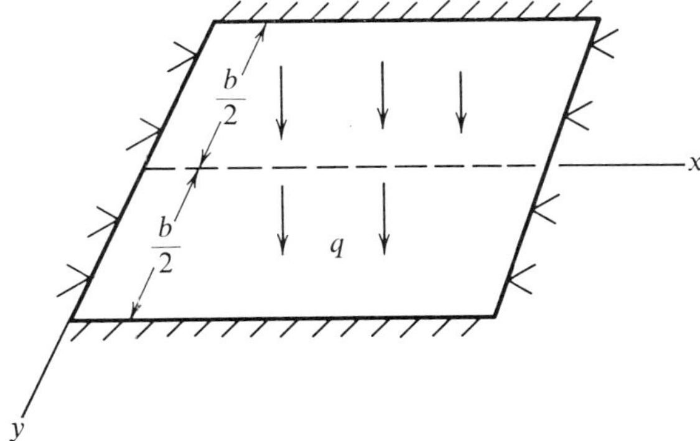

Figure 10.10. A rectangular plate with two parallel edges clamped and the other two simply supported, under lateral loading.

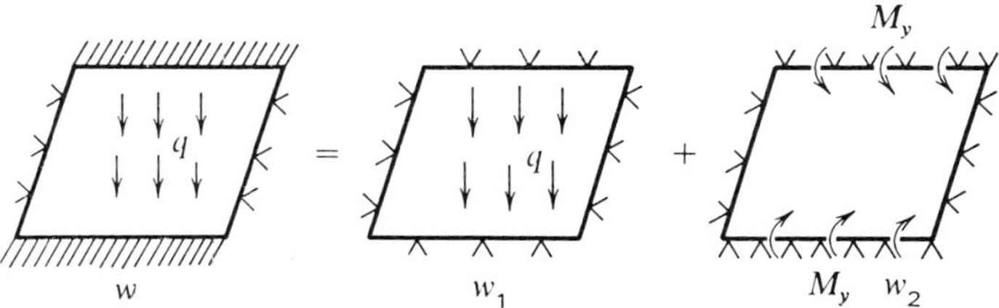

Figure 10.11. Superposition of solutions.

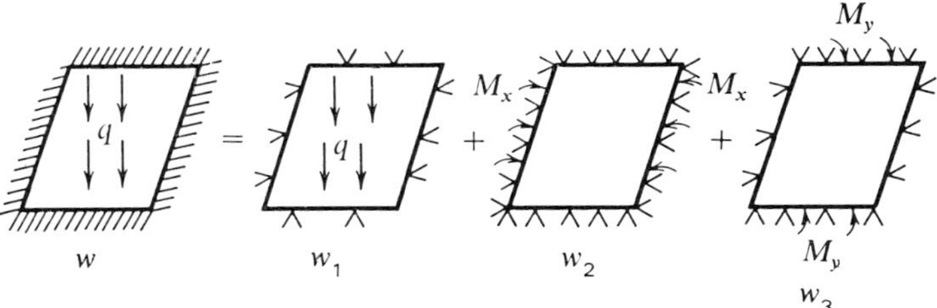

Figure 10.12

The solutions for the deflections are superposed to give the deflection of the actual clamped plate; that is, $w = w_1 + w_2 + w_3$. From w, the moments and stresses in the actual plate are easily calculated.

Circular Plates Under Bending. In the analysis of circular plates, the differential equation $D\nabla^4 w = q$ is expressed in polar coordinates. Referring to Figure 10.13, where x, y are cartesian coordinates and r, θ are polar coordinates, we observe the following relations:

$$r^2 = x^2 + y^2 \qquad \theta = \tan^{-1}\frac{y}{x}$$

$$\frac{\partial w}{\partial x} = \frac{\partial w}{\partial r}\frac{\partial r}{\partial x} + \frac{\partial w}{\partial \theta}\frac{\partial \theta}{\partial x}$$

$$= \left(\cos\theta\frac{\partial}{\partial r} - \frac{1}{r}\sin\theta\frac{\partial}{\partial \theta}\right)w$$

$$\frac{\partial w}{\partial y} = \left(\sin\theta\frac{\partial}{\partial r} + \frac{1}{r}\cos\theta\frac{\partial}{\partial \theta}\right)w$$

$$\frac{\partial^2 w}{\partial x^2} = \left(\cos\theta\frac{\partial}{\partial r} - \frac{1}{r}\sin\theta\frac{\partial}{\partial \theta}\right)\left(\cos\theta\frac{\partial w}{\partial r} - \frac{1}{r}\sin\theta\frac{\partial w}{\partial \theta}\right)$$

$$\frac{\partial^2 w}{\partial y^2} = \left(\sin\theta\frac{\partial}{\partial r} + \frac{1}{r}\cos\theta\frac{\partial}{\partial \theta}\right)\left(\sin\theta\frac{\partial w}{\partial r} + \frac{1}{r}\cos\theta\frac{\partial w}{\partial \theta}\right)$$

$$\frac{\partial^2 w}{\partial x\,\partial y} = \left(\sin\theta\frac{\partial}{\partial r} + \frac{1}{r}\cos\theta\frac{\partial}{\partial \theta}\right)\left(\cos\theta\frac{\partial w}{\partial r} - \frac{1}{r}\sin\theta\frac{\partial w}{\partial \theta}\right)$$

$$\nabla^2 w = \frac{\partial^2 w}{\partial r^2} + \frac{1}{r}\frac{\partial w}{\partial r} + \frac{1}{r^2}\frac{\partial^2 w}{\partial \theta^2}$$

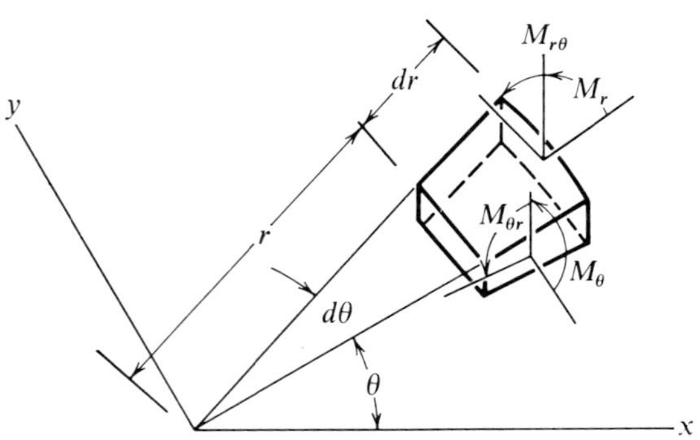

Figure 10.13

Equation (10.1.18), the governing equation of equilibrium, in terms of polar coordinates then becomes

$$\left(\frac{\partial^2}{\partial r^2} + \frac{1}{r}\frac{\partial}{\partial r} + \frac{1}{r^2}\frac{\partial^2}{\partial \theta^2}\right)\left(\frac{\partial^2 w}{\partial r^2} + \frac{1}{r}\frac{\partial w}{\partial r} + \frac{1}{r^2}\frac{\partial^2 w}{\partial \theta^2}\right) = \frac{q}{D} \qquad (10.1.41)$$

The moments are obtained from Equations (10.1.11) as

$$M_r = (M_x)_{\theta=0} = -D\left(\frac{\partial^2 w}{\partial x^2} + \nu\frac{\partial^2 w}{\partial y^2}\right)_{\theta=0}$$

$$= -D\left[\frac{\partial^2 w}{\partial r^2} + \nu\left(\frac{1}{r}\frac{\partial w}{\partial r} + \frac{1}{r^2}\frac{\partial^2 w}{\partial \theta^2}\right)\right] \tag{10.1.42}$$

$$M_\theta = (M_y)_{\theta=0} = -D\left(\frac{\partial^2 w}{\partial y^2} + \nu\frac{\partial^2 w}{\partial x^2}\right)_{\theta=0}$$

$$= -D\left(\frac{1}{r}\frac{\partial w}{\partial r} + \frac{1}{r^2}\frac{\partial^2 w}{\partial \theta^2} + \nu\frac{\partial^2 w}{\partial r^2}\right) \tag{10.1.43}$$

$$M_{r\theta} = -M_{\theta r} = (M_{xy})_{\theta=0} = (1-\nu)D\left(\frac{\partial^2 w}{\partial x\,\partial y}\right)_{\theta=0}$$

$$= (1-\nu)D\left(\frac{1}{r}\frac{\partial^2 w}{\partial r\,\partial \theta} - \frac{1}{r^2}\frac{\partial w}{\partial \theta}\right) \tag{10.1.44}$$

Similarly the shears may be obtained from Equations (10.1.19):

$$Q_r = (Q_x)_{\theta=0} = -D\left[\frac{\partial}{\partial x}\nabla^2 w\right]_{\theta=0} = -D\frac{\partial}{\partial r}\left(\frac{\partial^2 w}{\partial r^2} + \frac{1}{r}\frac{\partial w}{\partial r} + \frac{1}{r^2}\frac{\partial^2 w}{\partial \theta^2}\right)$$

$$Q_\theta = (Q_y)_{\theta=0} = -D\left[\frac{\partial}{\partial y}\nabla^2 w\right]_{\theta=0} = -D\frac{1}{r}\frac{\partial}{\partial \theta}\left(\frac{\partial^2 w}{\partial r^2} + \frac{1}{r}\frac{\partial w}{\partial r} + \frac{1}{r^2}\frac{\partial^2 w}{\partial \theta^2}\right) \tag{10.14.5}$$

The boundary conditions for a simply supported edge are $w = 0$ and $M_r = 0$; for a clamped edge they are $w = 0$ and $\partial w/\partial r = 0$; and for a free edge they are $M_r = 0$ and $Q_r + \partial M_{r\theta}/r\,\partial\theta = 0$.

For a circular plate of uniform thickness with an axisymmetric edge condition under axisymmetric loading, the lateral deflection of the plate will be axisymmetric; that is, w is independent of θ, and Equation (10.1.41) reduces to

$$\left(\frac{d^2}{dr^2} + \frac{1}{r}\frac{d}{dr}\right)\left(\frac{d^2 w}{dr^2} + \frac{1}{r}\frac{dw}{dr}\right) = \frac{q}{D} \tag{10.1.46}$$

Rewriting

$$\left(\frac{d^2}{dr^2} + \frac{1}{r}\frac{d}{dr}\right) \quad \text{as} \quad \frac{1}{r}\frac{d}{dr}\left(r\frac{d}{dr}\right)$$

we may write the above equation as

$$\frac{1}{r}\frac{d}{dr}r\frac{d}{dr}\left[\frac{1}{r}\frac{d}{dr}\left(r\frac{dw}{dr}\right)\right] = \frac{q}{D} \tag{10.1.47}$$

and the shears and moments are

$$Q_r = D\frac{d}{dr}\left(\frac{d\phi}{dr} + \frac{\phi}{r}\right) = D\frac{d}{dr}\left[\frac{1}{r}\frac{d}{dr}(r\phi)\right] \tag{10.1.48}$$

$$M_r = D\left[\frac{d\phi}{dr} + \frac{v}{r}\phi\right] \qquad M_\theta = D\left[\frac{\phi}{r} + v\frac{d\phi}{dr}\right] \tag{10.1.49}$$

where ϕ is $-\dfrac{dw}{dr}$.

10.2 Inelastic Bending of Thin Plates

For plates with inelastic strain, Equation (10.1.1) is changed into

$$\begin{aligned}
e_{xx} &= e'_{xx} + e''_{xx} = \frac{\tau_{xx}}{E} - v\frac{\tau_{yy}}{E} + e_x{}'' \\
e_{yy} &= e'_{yy} + e''_{yy} = \frac{\tau_{yy}}{E} - v\frac{\tau_{xx}}{E} + e_y{}'' \\
e_{xy} &= e'_{xy} + e''_{xy} = \frac{\tau_{xy}}{2G} + e''_{xy}
\end{aligned} \tag{10.2.1}$$

where e''_{ij} denotes the inelastic strain which consists of any combination of thermal, creep, and plastic strains. By expressing stresses in terms of strains, we can write Equations (10.2.1) as

$$\begin{aligned}
\tau_{xx} &= \frac{E}{1-v^2}[(e_{xx} + ve_{yy}) - (e''_{xx} + ve''_{yy})] \\
\tau_{yy} &= \frac{E}{1-v^2}[(e_{yy} + ve_{xx}) - (e''_{yy} + ve''_{xx})] \\
\tau_{xy} &= 2G(e_{xy} - e''_{xy}).
\end{aligned} \tag{10.2.2}$$

Substituting Equations (10.1.5) into Equations (10.2.2) yields

$$\begin{aligned}
\tau_{xx} &= \frac{E}{1-v^2}\left[\frac{\partial u_0}{\partial x} - z\frac{\partial^2 w}{\partial x^2} + v\left(\frac{\partial v_0}{\partial y} - z\frac{\partial^2 w}{\partial y^2}\right) - (e''_{xx} + ve''_{yy})\right] \\
\tau_{yy} &= \frac{E}{1-v^2}\left[\frac{\partial v_0}{\partial y} - z\frac{\partial^2 w}{\partial y^2} + v\left(\frac{\partial u_0}{\partial x} - z\frac{\partial^2 w}{\partial x^2}\right) - (e''_{yy} + ve''_{xx})\right] \\
\tau_{xy} &= G\left(\frac{\partial u_0}{\partial y} + \frac{\partial v_0}{\partial x} - 2z\frac{\partial^2 w}{\partial x\,\partial y} - 2e''_{xy}\right)
\end{aligned} \tag{10.2.3}$$

Substituting Equations (10.2.3) into Equation (10.1.7) and Equation (10.1.8), we obtain the sectional forces

$$N_x = \frac{Eh}{1-\nu^2}\left(\frac{\partial u_0}{\partial x} + \nu\frac{\partial v_0}{\partial y}\right) - \frac{E}{1-\nu^2}\int (e''_{xx} + \nu e''_{yy})\,dz$$

$$= \frac{Eh}{1-\nu^2}\left(\frac{\partial u_0}{\partial x} + \nu\frac{\partial v_0}{\partial y}\right) - N_{x_I}$$

$$N_y = \frac{Eh}{1-\nu^2}\left(\frac{\partial v_0}{\partial y} + \nu\frac{\partial u_0}{\partial x}\right) - N_{y_I}$$

$$N_{xy} = Gh\left(\frac{\partial u_0}{\partial y} + \frac{\partial v_0}{\partial x}\right) - N_{xy_I} \tag{10.2.4}$$

where the inelastic sectional forces are

$$N_{x_I} = \frac{E}{1-\nu^2}\int (e''_{xx} + \nu e''_{yy})\,dz$$

$$N_{y_I} = \frac{E}{1-\nu^2}\int (e''_{yy} + \nu e''_{xx})\,dz \tag{10.2.5}$$

$$N_{xy_I} = 2G\int e''_{xy}\,dz$$

and the sectional moments are

$$M_x = -D\left(\frac{\partial^2 w}{\partial x^2} + \nu\frac{\partial^2 w}{\partial y^2}\right) - M_{x_I}$$

$$M_y = -D\left(\frac{\partial^2 w}{\partial y^2} + \nu\frac{\partial^2 w}{\partial x^2}\right) - M_{y_I} \tag{10.2.6}$$

$$M_{xy} = (1-\nu)D\frac{\partial^2 w}{\partial x\,\partial y} - M_{xy_I}$$

where the inelastic sectional moments are

$$M_{x_I} = \frac{E}{1-\nu^2}\int (e''_{xx} + \nu e''_{yy})z\,dz$$

$$M_{y_I} = \frac{E}{1-\nu^2}\int (e''_{yy} + \nu e''_{xx})z\,dz \tag{10.2.7}$$

$$M_{xy_I} = 2G\int e''_{xy}\,z\,dz$$

Substituting Equation (10.2.7) into Equation (10.1.17) gives the governing *equation for equilibrium of an inelastic thin plate*:

$$D\nabla^4 w = \left(-\frac{\partial^2 M_{x_I}}{\partial x^2} - 2\frac{\partial^2 M_{xy_I}}{\partial x\,\partial y} - \frac{\partial^2 M_{y_I}}{\partial y^2} \right) + q \tag{10.2.8}$$

It is seen that the term in the parentheses has the equivalent effect on w of the lateral load q; therefore this term is called the equivalent load and is denoted by $\bar{q}$. Equations (10.2.8) may then be written as

$$D\nabla^4 w = q + \bar{q} \tag{10.2.9}$$

From this differential equation and the boundary conditions, w is solved.

Let X and Y be the actual body forces along the x and y axes per unit area of the plate. Recall the equilibrium conditions

$$\frac{\partial N_x}{\partial x} + \frac{\partial N_{xy}}{\partial y} + X = 0 \qquad \frac{\partial N_{xy}}{\partial x} + \frac{\partial N_y}{\partial y} + Y = 0 \tag{10.2.10}$$

Substituting Equations (10.2.4) to (10.2.6) into Equation (10.2.10) yields

$$\frac{Eh}{1-\nu^2}\frac{\partial}{\partial x}(e_{xx_0} + \nu e_{yy_0}) - \frac{\partial N_{x_I}}{\partial x} + \frac{Eh}{(1+\nu)}\frac{\partial e_{xy_0}}{\partial y} - \frac{\partial N_{xy_I}}{\partial y} + X = 0$$

$$\frac{Eh}{1+\nu}\frac{\partial e_{xy_0}}{\partial x} - \frac{\partial N_{xy_I}}{\partial x} + \frac{Eh}{1-\nu^2}\frac{\partial}{\partial y}(e_{yy_0} + \nu e_{xx_0}) - \frac{\partial N_{y_I}}{\partial y} + Y = 0$$

where the subscript "0" denotes the value of strain at the middle plane. It is seen that the terms $-\{\partial N_{x_I}/\partial x + \partial N_{xy_I}/\partial y\}$ and $-\{\partial N_{xy_I}/\partial x + \partial N_{y_I}/\partial y\}$ have the same effect on e_{xx_0}, e_{yy_0}, and e_{xy_0} as X and Y, respectively; hence the terms in the parentheses are called equivalent body forces, and are denoted by $\bar{X}$ and $\bar{Y}$. The explicit expressions for $\bar{X}$ and $\bar{Y}$ are

$$\begin{aligned}\bar{X} &= -\frac{E}{1-\nu^2}\frac{\partial}{\partial x}\int (e''_{xx} + \nu e''_{yy})\,dz - 2G\frac{\partial}{\partial y}\int e''_{xy}\,dz \\ \bar{Y} &= -\frac{E}{1-\nu^2}\frac{\partial}{\partial y}\int (e''_{yy} + \nu e''_{xx})\,dz - 2G\frac{\partial}{\partial x}\int e''_{xy}\,dz\end{aligned} \tag{10.2.11}$$

Then

$$\begin{aligned}&\frac{Eh}{1-\nu^2}\frac{\partial}{\partial x}(e_{xx_0} + \nu e_{yy_0}) + \frac{Eh}{1+\nu}\frac{\partial e_{xy_0}}{\partial y} + \bar{X} + X = 0 \\ &\frac{Eh}{1+\nu}\frac{\partial e_{xy_0}}{\partial x} + \frac{Eh}{1-\nu^2}\frac{\partial}{\partial y}(e_{yy_0} + \nu e_{xx_0}) + \bar{Y} + Y = 0\end{aligned} \tag{10.2.12}$$

These partial differential equations together with boundary conditions are used to solve for the strain field e_{xx_0}, e_{yy_0}, and e_{xy_0} in the plate. These are the same as the differential equations for plane stress whose solution is shown in Chapter 9. The stress at any point, from Equation (10.2.3), is

$$\tau_{xx} = \frac{E}{1-\nu^2}\left[(e_{xxo} + \nu e_{yyo}) - z\left(\frac{\partial^2 w}{\partial x^2} + \nu\frac{\partial^2 w}{\partial y^2}\right) - (e''_{xx} + \nu e''_{yy})\right]$$

$$\tau_{yy} = \frac{E}{1-\nu^2}\left[(e_{yyo} + \nu e_{xxo}) - z\left(\frac{\partial^2 w}{\partial y^2} + \nu\frac{\partial^2 w}{\partial x^2}\right) - (e''_{yy} + \nu e''_{xx})\right]$$

$$\tau_{xy} = 2G\left(e_{xyo} - z\frac{\partial^2 w}{\partial x\,\partial y} - e''_{xy}\right) \tag{10.2.13}$$

where w is the solution of the partial differential Equation (10.2.8) which satisfies the given boundary conditions of the problem.

Equations (10.2.8) and (10.2.11) may be derived in an alternative way as follows. Imagine that a plate, before the introduction of inelastic strain, is divided into elemental rectangular areas of Δz thickness. After inelastic strain occurs, these elemental areas are no longer rectangular. Assume that the elemental areas are small enough for the inelastic strains e''_{xx}, e''_{yy}, and e''_{xy} to be considered uniform in each elemental area. To restore each element back to its original shape and size, uniform tractions giving the uniform stresses

$$\tau_{xx} = -\frac{E}{1-\nu^2}(e''_{xx} + \nu e''_{yy}) \qquad \tau_{yy} = -\frac{E}{1-\nu^2}(e''_{yy} + \nu e''_{xx}) \qquad \tau_{xy} = -2Ge''_{xy}$$

are applied to each $\Delta x\,\Delta y$ elemental area. After these tractions are applied, the areas will again match. Imagine that with the applied tractions on they are then welded together perfectly; that is, the original plate is restored. Referring to Figure 10.14, we note that the tractions on two adjacent

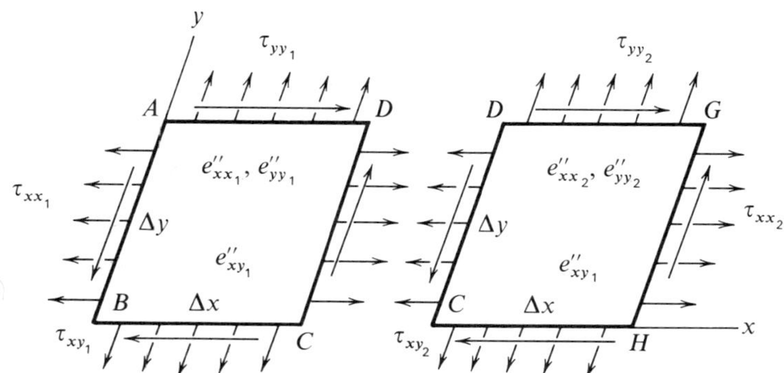

Figure 10.14

areas are not the same, and hence there is an unbalanced force between adjacent elemental areas. Along the boundary DC, there is an unbalanced horizontal force $P_x = -(\tau_{xx_2} - \tau_{xx_1})\,\Delta y\,\Delta z$ and an unbalanced vertical force $P_y = -(\tau_{xy_1} - \tau_{xy_2})\,\Delta y\,\Delta z$. The above expressions for stresses give

$$(\tau_{xx_2} - \tau_{xx_1}) = \frac{-E}{1-\nu^2}\,[(e''_{xx} + \nu e''_{yy})_2 - (e''_{xx} + \nu e''_{yy})_1]$$

$$= \frac{-E}{1-\nu^2}\,\frac{\partial}{\partial x}(e''_{xx} + \nu e''_{yy})\,\Delta x$$

For Δy length, the unbalanced forces on CD are

$$P_x = \frac{E}{1-\nu^2}\left(\frac{\partial e''_{xx}}{\partial x} + \nu\,\frac{\partial e''_{yy}}{\partial x}\right)\Delta x\,\Delta y\,\Delta z$$

$$P_y = 2G\,\frac{\partial e''_{xy}}{\partial x}\,\Delta x\,\Delta y\,\Delta z$$

Similarly, across the horizontal boundary of two adjacent areas, the unbalanced forces on Δx are

$$P_y = \frac{E}{1-\nu^2}\left(\frac{\partial e''_{yy}}{\partial y} + \nu\,\frac{\partial e''_{xx}}{\partial y}\right)\Delta y\,\Delta x\,\Delta z$$

$$P_x = 2G\,\frac{\partial e''_{xy}}{\partial y}\,\Delta y\,\Delta x\,\Delta z$$

For each $\Delta x\,\Delta y$ elemental area, these two sets of forces give the distributed forces per unit volume,

$$F_x = \frac{E}{1-\nu^2}\left(\frac{\partial e''_{xx}}{\partial x} + \nu\,\frac{\partial e''_{yy}}{\partial x}\right) + 2G\,\frac{\partial e''_{xy}}{\partial y}$$
$$F_y = \frac{E}{1-\nu^2}\left(\frac{\partial e''_{yy}}{\partial y} + \nu\,\frac{\partial e''_{xx}}{\partial y}\right) + 2G\,\frac{\partial e''_{xy}}{\partial x} \qquad (10.2.14)$$

Actually there are no such forces, and hence the unbalanced forces must be relaxed by applying equal and opposite forces. These equal and opposite forces are the equivalent body forces $\bar{F}_x$ and $\bar{F}_y$ acting on the plate. Consider a $dx\,dy$ differential element of the plate under the applied distributed lateral load q and the equivalent body forces $\bar{F}_x$ and $\bar{F}_y$ as shown in Figure 10.15. The sectional forces, moments, and shears *due to both the actual loads and the equivalent body forces*[23] are denoted by a bar on top:

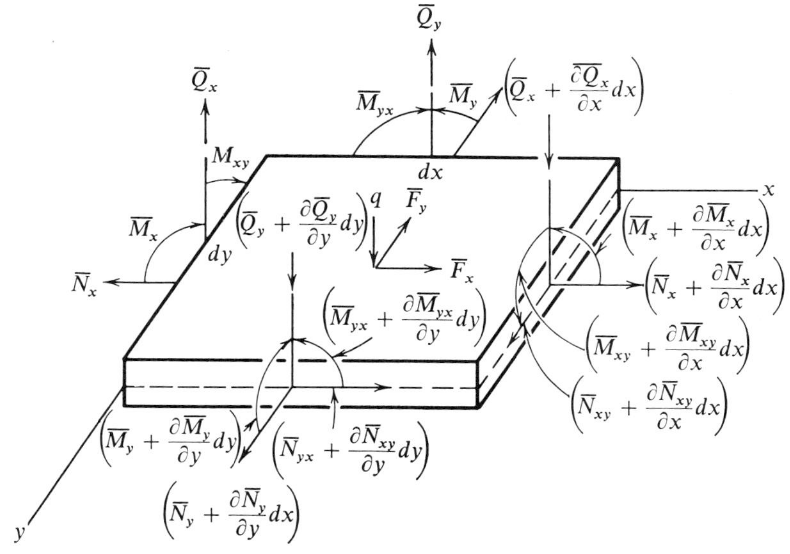

Figure 10.15

$\bar{M}_x$, $\bar{M}_y$, and $\bar{M}_{xy}$. For equilibrium of moments about axes through the centroid and parallel to the x and y axes, we have

$$\frac{\partial \bar{M}_x}{\partial x} + \frac{\partial \bar{M}_{yx}}{\partial y} - \bar{Q}_x + \int \bar{F}_x z \, dz = 0 \tag{10.2.15}$$

$$-\frac{\partial \bar{M}_{xy}}{\partial x} + \frac{\partial \bar{M}_y}{\partial y} - \bar{Q}_y + \int \bar{F}_y z \, dz = 0 \tag{10.2.16}$$

and for equilibrium of vertical forces, we have

$$\frac{\partial \bar{Q}_x}{\partial x} + \frac{\partial \bar{Q}_y}{\partial y} + q = 0 \tag{10.2.17}$$

Differentiating Equation (10.2.15) with respect to x and Equation (10.2.16) with respect to y and substituting the results into Equation (10.2.17), we obtain

$$\frac{\partial^2 \bar{M}_x}{\partial x^2} + \frac{\partial^2 \bar{M}_y}{\partial y^2} - 2\frac{\partial^2 \bar{M}_{xy}}{\partial x \, \partial y} = -q - \frac{\partial}{\partial x}\int \bar{F}_x z \, dz - \frac{\partial}{\partial y}\int \bar{F}_y z \, dz \tag{10.2.18}$$

Expressing sectional moments in terms of lateral deflections w, as given by

Equation (10.1.11), gives the *governing equation of equilibrium for inelastic plates,*

$$D\nabla^4 w = q + \left[\frac{-E}{1-\nu^2}\frac{\partial^2}{\partial x^2}\int(e''_{xx} + \nu e''_{yy})z\,dz - \right.$$

$$\left. - \frac{E}{1-\nu^2}\frac{\partial^2}{\partial y^2}\int(e''_{yy} + \nu e''_{xx})z\,dz - \frac{2E}{1+\nu}\frac{\partial^2}{\partial x\,\partial y}\int e''_{xy}\,z\,dz\right]$$

$$= q + \left[-\frac{\partial^2 M_{x_I}}{\partial x^2} - 2\frac{\partial^2 M_{xy_I}}{\partial x\,\partial y} - \frac{\partial^2 M_{y_I}}{\partial y^2}\right] = q + \bar{q} \qquad (10.2.19)$$

This is identical to Equation (10.2.8).

For equilibrium of forces in the x and y directions, we have

$$\begin{aligned}\frac{\partial \bar{N}_x}{\partial x} + \frac{\partial \bar{N}_{xy}}{\partial y} + \int \bar{F}_x\,dz + X = 0\\ \frac{\partial \bar{N}_{xy}}{\partial x} + \frac{\partial \bar{N}_y}{\partial y} + \int \bar{F}_y\,dz + Y = 0\end{aligned} \qquad (10.2.20)$$

where $\bar{N}_x = N_x + N_{x_I}$; $\bar{N}_y = N_y + N_{y_I}$; $\bar{N}_{xy} = N_{xy} + N_{xy_I}$. It is seen that $\int \bar{F}_x\,dz$ and $\int \bar{F}_y\,dz$ are equivalent to $\bar{X}$ and $\bar{Y}$. From Equations (10.2.16), we have

$$\begin{aligned}\int \bar{F}_x\,dz &= \int\left[-\frac{E}{1-\nu^2}\frac{\partial}{\partial x}(e''_{xx} + \nu e''_{yy}) - 2G\frac{\partial}{\partial y}e''_{xy}\right]dz\\ \int \bar{F}_y\,dz &= \int\left[-\frac{E}{1-\nu^2}\frac{\partial}{\partial y}(e''_{yy} + \nu e''_{xx}) - 2G\frac{\partial}{\partial x}e''_{xy}\right]dz\end{aligned} \qquad (10.2.21)$$

Comparing the above expressions with Equations (10.2.11), we can readily see that

$$\bar{X} = \int \bar{F}_x\,dz \qquad \bar{Y} = \int \bar{F}_y\,dz \qquad (10.2.22)$$

Hence the same results as given by Equations (10.2.8) and (10.2.11) are obtained. The stress in the actual plate may be expressed in terms of $\bar{N}_x$, $\bar{N}_y$, $\bar{N}_{xy}$, $\bar{M}_x$, $\bar{M}_y$, and $\bar{M}_{xy}$ as follows:

$$\begin{aligned}\tau_{xx} &= \frac{\bar{N}_x}{h} + \frac{12\bar{M}_x z}{h^3} - \frac{E}{1-\nu^2}(e''_{xx} + \nu e''_{yy})\\ \tau_{yy} &= \frac{\bar{N}_y}{h} + \frac{12\bar{M}_y z}{h^3} = \frac{E}{1-\nu^2}(e''_{yy} + \nu e''_{xx})\\ \tau_{xy} &= \frac{\bar{N}_{xy}}{h} + \frac{12\bar{M}_{xy} z}{h^3} - 2Ge''_{xy}\end{aligned} \qquad (10.2.23)$$

The previous discussion is valid for arbitrary coordinate systems. For example, referring to Figure 10.13, for polar coordintes r and θ, we have the following expressions for inelastic moments:

$$M_{r_I} = M_{x_I} = \frac{E}{1-\nu^2}\int (e''_{rr} + \nu e''_{\theta\theta}) z\, dz$$

$$M_{\theta_I} = M_{y_I} = \frac{E}{1-\nu^2}\int (e''_{\theta\theta} + \nu e''_{rr}) z\, dz \qquad (10.2.24)$$

$$M_{r\theta_I} = 2G\int e''_{r\theta}\, z\, dz$$

From Equations (10.1.42) through (10.1.44), (10.2.7), and (10.2.10), we obtain

$$M_r = -D\left[\frac{\partial^2 w}{\partial r^2} + \nu\left(\frac{1}{r}\frac{\partial w}{\partial r} + \frac{1}{r^2}\frac{\partial^2 w}{\partial \theta^2}\right)\right] - \frac{E}{1-\nu^2}\int (e_r'' + \nu e_\theta'') z\, dz$$

$$M_\theta = -D\left(\frac{1}{r}\frac{\partial w}{\partial r} + \frac{1}{r^2}\frac{\partial^2 w}{\partial \theta^2} + \nu\frac{\partial^2 w}{\partial r^2}\right) - \frac{E}{1-\nu^2}\int (e''_{rr} + e''_{\theta\theta}) z\, dz$$

$$M_{r\theta} = (1-\nu)D\left(\frac{1}{r}\frac{\partial^2 w}{\partial r\,\partial\theta} - \frac{1}{r^2}\frac{\partial w}{\partial \theta}\right) + 2G\int (e''_{\theta\theta} + \nu e''_{rr}) z\, dz \qquad (10.2.25)$$

Boundary Conditions. The boundary conditions for inelastic plates are the same as for elastic plates. For a fixed or clamped edge, referring to Figure 10.16 we see that the boundary conditions are

$$w = 0 \qquad \text{and} \qquad \frac{\partial w}{\partial n} = 0$$

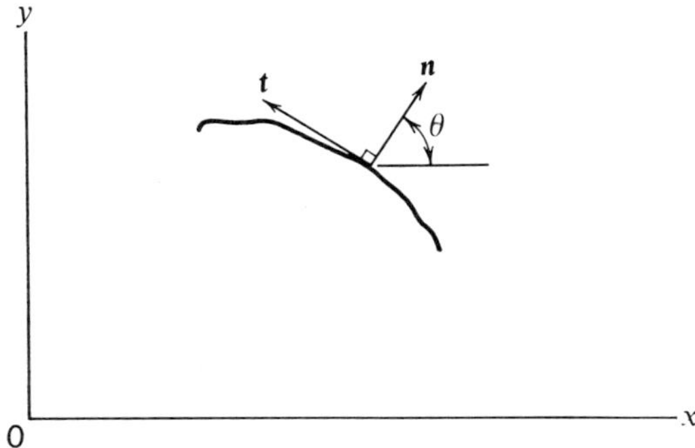

Figure 10.16

where n is the coordinate normal to the boundary. For a rectangular plate with fixed edge along $y = 0$, the above gives

$$(w)_{y=0} = 0 \qquad \left(\frac{\partial w}{\partial y}\right)_{y=0} = 0$$

For a simply supported edge, the boundary conditions are

$$w = 0 \qquad M_n = 0$$

The expression for M_n in terms of w and inelastic moment M_{n_I} is obtained as follows. In terms of τ_{xx}, τ_{yy}, and τ_{yx}, we have τ_{nn} given by

$$\tau_{nn} = \tau_{xx}\cos^2\theta + \tau_{yy}\sin^2\theta + 2\tau_{xy}\sin\theta\cos\theta$$

Since

$$M_n = \int \tau_{nn} z\, dz \qquad \text{and} \qquad M_{xy} = -\int \tau_{xy} z\, dz$$

we obtain

$$M_n = M_x\cos^2\theta + M_y\sin^2\theta - 2M_{xy}\sin\theta\cos\theta \tag{10.2.26}$$

Expressing M_n in terms of deflection w, we see that the boundary condition $M_n = 0$ for a simply supported edge becomes

$$M_n = -D\left(\frac{\partial^2 w}{\partial n^2} + \nu\frac{\partial^2 w}{\partial t^2}\right) - M_{n_I} = 0 \tag{10.2.27}$$

where

$$M_{n_I} = \frac{E}{1-\nu^2}\int (e''_{nn} + \nu e''_{tt}) z\, dz \tag{10.2.28}$$

Along a free edge, the boundary conditions are

$$M_n = 0 \qquad Q_n - \frac{\partial M_{nt}}{\partial t} = 0 \tag{10.2.29}$$

where t is the coordinate along the tangent to the boundary. At any corner in the plate there is a concentrated force R. Referring to Figure 10.17, we note

$$R = [(M_{nt})_2 - (M_{nt})_1]$$

It follows from Equation (10.1.15), after we replace x by n and y by t, that

$$Q_n = \frac{\partial M_n}{\partial n} - \frac{\partial M_{nt}}{\partial t}$$

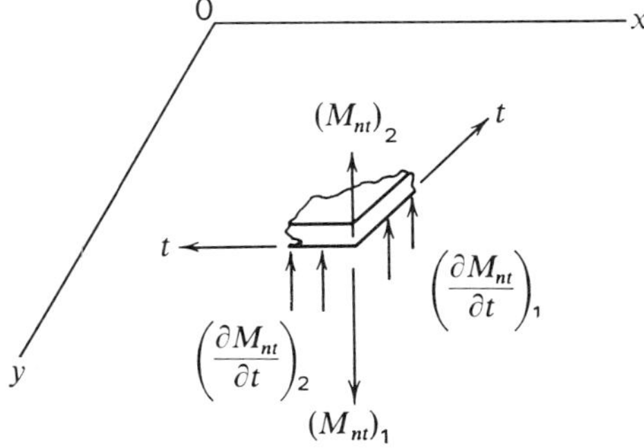

Figure 10.17

Substituting the above expression for Q_n into Equation (10.2.28) gives

$$\frac{\partial M_n}{\partial n} - 2\frac{\partial M_{nt}}{\partial t} = 0 \tag{10.2.30}$$

where

$$M_{nt} = (1-\nu)D\frac{\partial^2 w}{\partial n\, \partial t} + M_{nt_I}$$

and

$$M_{nt_I} = 2G\int e''_{nt}\, z\, dz \tag{10.2.31}$$

Substituting the above equations and Equation (10.2.27) into Equation (10.2.30) gives the second free-edge boundary condition in terms of w and e''_{nn}, e''_{tt}, and e''_{nt}, which may be obtained from e''_{xx}, e''_{yy}, and e''_{xy}.

For plates with $N_x = N_y = N_{xy} = 0$ and, in addition, $N_{x_I} = N_{y_I} = N_{xy_I} = 0$, the deflection due to inelastic strain e''_{xx}, e''_{yy}, and e''_{xy} can be calculated by applying the equivalent load $\bar{q}$ on the plate and the equivalent edge moments M_{n_I}, M_{nt_I} on the boundary. The deflection caused by the actual lateral load q and edge moments may be calculated separately and superposed on that caused by $\bar{q}$, M_{n_I}, and M_{nt_I}. The equivalent lateral load $\bar{q}$ and edge moments M_{n_I} and M_{nt_I} are fictitious and should not be confused with the actual lateral load and edge moments. The deflection of a plate with inelastic strain subject to a given lateral load and edge moments is the same as the deflection of an identical plate with no inelastic strain but with an additional lateral load $\bar{q}$ and edge moments M_{n_I} and M_{nt_I} as given in Equations (10.2.8), (10.2.9), (10.2.28), and (10.2.31). This analogy for plates in bending is tabulated in Table 10.1, with suffixes I and II denoting the actual and fictitious (analogous) plates, respectively.

Table 10.1. ANALOGY BETWEEN LATERAL LOAD AND INELASTIC STRAINS FOR PLATES IN BENDING†

	Plate I (actual)	Plate II (fictitious)
Inelastic strain	$\left.\begin{matrix} e''^{\mathrm{I}}_{xx} \\ e''^{\mathrm{I}}_{yy} \\ e''^{\mathrm{I}}_{xy} \end{matrix}\right\}$ exist	$e''^{\mathrm{II}}_{xx} = 0$ $e''^{\mathrm{II}}_{yy} = 0$ $e''^{\mathrm{II}}_{xy} = 0$
Lateral load	q^{I}	$q^{\mathrm{II}} = q^{\mathrm{I}} + \bar{q}$
Edge moments	M_n^{I} M_{nt}^{I}	$M_n^{\mathrm{II}} = M_n{}^{\mathrm{I}} + M_{n_I}$ $M_{nt}^{\mathrm{I}} = M_{nt}^{\mathrm{I}} + M_{nt_I}$
Stresses	$\tau_{xx}^{\mathrm{I}} = \tau_{xx}^{\mathrm{II}} - \dfrac{E}{1-\nu^2}(e''_{xx} + \nu e''_{yy})$ $\tau_{yy}^{\mathrm{I}} = \tau_{yy}^{\mathrm{II}} - \dfrac{E}{1-\nu^2}(e''_{yy} + \nu e''_{xx})$ $\tau_{xy}^{\mathrm{I}} = \tau_{xy}^{\mathrm{II}} - 2Ge''_{xy}$	$\tau_{xx}^{\mathrm{II}} = \dfrac{M_x^{\mathrm{II}}}{(1/12)h^3} z$ $\tau_{yy}^{\mathrm{II}} = \dfrac{M_y^{\mathrm{II}}}{(1/12)h^3} z$ $\tau_{xy}^{\mathrm{II}} = \dfrac{M_{xy}^{\mathrm{II}}}{(1/12)h^3} z$
Deflection	$w^{\mathrm{I}} = w^{\mathrm{II}}$	
Strain	$e_{ij}^{\mathrm{I}} = e_{ij}^{\mathrm{II}}$	

† This table is for the case $N_x = N_y = N_{xy} = 0$ and $N_{x_I} = N_{y_I} = N_{xy_I} = 0$.

$$\bar{q} = \left[-\frac{E}{1-\nu^2}\frac{\partial^2}{\partial x^2}\int (e''_{xx} + \nu e''_{yy}) z\, dz - \frac{E}{1-\nu^2}\frac{\partial^2}{\partial y^2}\int (e''_{yy} + \nu e''_{xx}) z\, dz \right.$$

$$\left. - \frac{2E}{1+\nu}\frac{\partial^2}{\partial x\, \partial y}\int e''_{xy} z\, dz \right]$$

$$M_{n_I} = \frac{E}{1-\nu^2}\int (e''_{nn} + \nu e''_{tt}) z\, dz \qquad M_{nt_I} = 2G\int e''_{nt} z\, dz$$

10.3 Thermal Bending of Plates

If the inelastic strain is caused by thermal gradient only, i.e., if there is no creep or plastic strain, then the inelastic strains are given by

$$e''_{xx} = e''_{yy} = e''_{zz} = \alpha T \qquad e''_{xy} = e''_{yz} = e''_{xz} = 0 \tag{10.3.1}$$

From Equations (10.2.4) through (10.2.7), (10.2.11), and (10.2.13), we obtain

$$N_{x_I} = N_{y_I} = \frac{E\alpha \int T\,dz}{1-\nu} = \frac{N_T}{1-\nu} \qquad N_{xy_I} = 0 \tag{10.3.2}$$

$$M_{x_I} = M_{y_I} = \frac{E\alpha}{1-\nu}\int Tz\,dz = \frac{M_T}{1-\nu} \qquad M_{xy_I} = 0 \tag{10.3.3}$$

$$\bar{X} = \frac{E\alpha}{1-\nu}\int \frac{\partial T}{\partial x}\,dz \qquad \bar{Y} = \frac{E\alpha}{1-\nu}\int \frac{\partial T}{\partial y}\,dz \tag{10.3.4}$$

Equations (10.3.4) represent the well-known analogy between temperature gradient and body force.[5]

When the actual body forces are zero, $X = Y = 0$, and Equations (10.2.10) are satisfied if we let

$$N_x = \frac{\partial^2 U}{\partial y^2} \qquad N_y = \frac{\partial^2 U}{\partial x^2} \qquad N_{xy} = -\frac{\partial^2 U}{\partial x\,\partial y} \tag{10.3.5}$$

From Equations (10.2.4) through (10.2.6), we find

$$\frac{Eh}{1-\nu^2}\left(\frac{\partial u_0}{\partial x} + \nu\frac{\partial v_0}{\partial y}\right) = \frac{\partial^2 U}{\partial y^2} + \frac{N_T}{1-\nu} \tag{10.3.6}$$

$$\frac{Eh}{1-\nu^2}\left(\frac{\partial v_0}{\partial y} + \nu\frac{\partial u_0}{\partial x}\right) = \frac{\partial^2 U}{\partial x^2} + \frac{N_T}{1-\nu} \tag{10.3.7}$$

$$\frac{Eh}{2(1+\nu)}\left(\frac{\partial u_0}{\partial y} + \frac{\partial v_0}{\partial x}\right) = -\frac{\partial^2 U}{\partial x\,\partial y} \tag{10.3.8}$$

From Equations (10.3.6) and (10.3.7), we obtain

$$\begin{aligned} Eh\frac{\partial u_0}{\partial x} &= \frac{\partial^2 U}{\partial y^2} - \nu\frac{\partial^2 U}{\partial x^2} + N_T \\ Eh\frac{\partial v_0}{\partial y} &= \frac{\partial^2 U}{\partial x^2} - \nu\frac{\partial^2 U}{\partial y^2} + N_T \end{aligned} \tag{10.3.9}$$

The following identify is easily established:

$$\frac{\partial^2}{\partial y^2}\frac{\partial u_0}{\partial x} + \frac{\partial^2}{\partial x^2}\frac{\partial v_0}{\partial y} = \frac{\partial^2}{\partial x\,\partial y}\left(\frac{\partial u_0}{\partial y} + \frac{\partial v_0}{\partial x}\right) \tag{10.3.10}$$

Substituting Equations (10.3.8) and (10.3.9) into Equation (10.3.10) yields

$$-\frac{2(1+\nu)}{Eh}\frac{\partial^2}{\partial x\,\partial y}\frac{\partial^2 U}{\partial x\,\partial y}=\frac{1}{Eh}\frac{\partial^2}{\partial y^2}\left[\frac{\partial^2 U}{\partial y^2}-\nu\frac{\partial^2 U}{\partial x^2}+N_T\right]$$

$$+\frac{1}{Eh}\frac{\partial^2}{\partial x^2}\left[\frac{\partial^2 U}{\partial x^2}-\nu\frac{\partial^2 U}{\partial y^2}+N_T\right]$$

$$\nabla^4 U+\nabla^2 N_T=0 \qquad (10.3.11)$$

where

$$\nabla^2=\left(\frac{\partial^2}{\partial x}+\frac{\partial^2}{\partial y^2}\right)$$

The solution of this differential equation is discussed in Chapter 9. From Equations (10.3.3), we have,

$$M_{x_I}=M_{y_I}=\frac{1}{1-\nu}M_T \qquad M_{xy_I}=0$$

Substituting the above equations into the governing equation of equilibrium, Equation (10.2.8), we have

$$D\nabla^4 w=q-\frac{1}{1-\nu}\nabla^2 M_T \qquad (10.3.12)$$

When the lateral load $q=0$, Equation (10.3.12) becomes

$$D\nabla^4 w=-\frac{1}{1-\nu}\nabla^2 M_T \qquad (10.3.13)$$

For the case in which the temperature varies through the thickness but not along x or y, we have constant M_T. Then

$$\nabla^2 M_T=0$$

and Equation (10.3.13) reduces to $\nabla^4 w=0$. For a rectangular plate with simply supported edges (Figure 10.18), we have, on $x=0$ and $x=a$, $w=0$ and

$$M_x=-D\left(\frac{\partial^2 w}{\partial x^2}+\nu\frac{\partial^2 w}{\partial y^2}+\frac{M_T}{D(1-\nu)}\right)=0$$

On $y=\pm b/2$, $w=0$ and

$$M_y=-D\left(\frac{\partial^2 w}{\partial y^2}+\nu\frac{\partial^2 w}{\partial x^2}+\frac{M_T}{D(1-\nu)}\right)=0$$

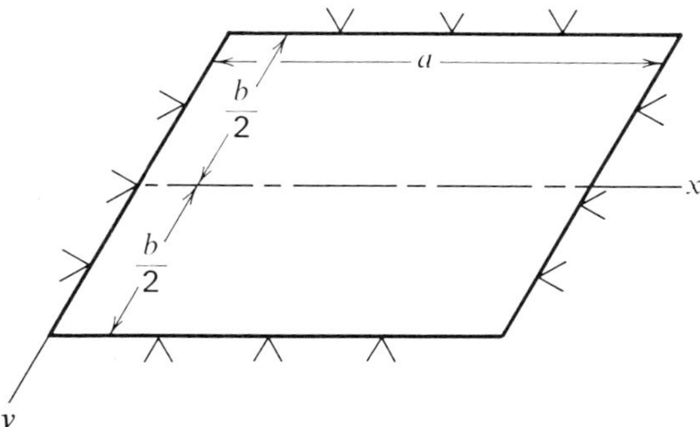

Figure 10.18

The solution of the equation $\nabla^4 w = 0$, together with the above boundary conditions, is[6]

$$w = \frac{6(1+\nu)M_T a^2}{Eh^3}\left[\frac{x}{a}\left(1-\frac{x}{a}\right) - \frac{8}{\pi^3}\sum_{1,3,5,\ldots}^{\infty}\frac{\sin(n\pi x/a)\cosh(n\pi/a)(y-b/2)}{n^3\cosh(n\pi b/2a)}\right] \tag{10.3.14}$$

$$M_x = -4M_T\sum_{1,3,5,\ldots}^{\infty}\frac{\sin(n\pi x/a)\cosh(n\pi/a)(y-b/2)}{n\pi\cosh(n\pi b/2a)} \tag{10.3.15}$$

$$M_y = -4M_T\left[\frac{1}{4} + \sum_{1,3,5,\ldots}^{\infty}\frac{\sin(n\pi x/a)\cosh(n\pi/a)(y-b/2)}{n\pi\cosh(n\pi b/2a)}\right] \tag{10.3.16}$$

Equations (10.3.14) through (10.3.16) give the deflection and moments for a simply supported rectangular plate, whose sides have lengths a and b, with no lateral loading under a temperature distribution which varies only with z.

10.4 Bending of Viscoelastic Plates

From Equations (4.7.10) and (4.7.12), the stress-strain relation for viscoelastic materials whose bulk modulus is time-independent is

$$\sigma(t) = K\theta(t) \tag{10.4.1}$$

$$S_{ij}(t) = 2\mu\left[\varepsilon_{ij}(t) - \int_0^t \phi(t-\tau)\varepsilon_{ij}(t)\,d\tau\right] \tag{10.4.2}$$

where σ is the hydrostatic stress, K the bulk modulus, μ the shear modulus, S_{ij}, ε_{ij} the deviatoric stress and strain components, respectively, θ the dilatation, and ϕ the memory function discussed in Chapters 3 and 4.

Taking the Laplace transform, with respect to time, of Equations (10.4.1) and (10.4.2) gives

$$\bar{\sigma} = K\bar{\theta} \tag{10.4.3}$$

$$\bar{S}_{ij} = 2\mu(1 - \bar{\phi})\bar{\varepsilon}_{ij} \tag{10.4.4}$$

$$\begin{aligned} \bar{\tau}_{ij} &= \bar{S}_{ij} + \delta_{ij}\bar{\sigma} = 2\mu(1 - \bar{\phi})\bar{\varepsilon}_{ij} + \delta_{ij}K\bar{\theta} \\ &= 2\mu(1 - \bar{\phi})\bar{e}_{ij} + \left[K - \frac{2\mu}{3}(1 - \bar{\phi})\right]\delta_{ij}\bar{\theta} \end{aligned} \tag{10.4.5}$$

where use was made of the relations $\sigma_{ij} = S_{ij} + \delta_{ij}\sigma$ and $\varepsilon_{ij} = e_{ij} - \frac{1}{3}\delta_{ij}\theta$, and the bar above a quantity denotes the Laplace transform of the quantity. The elastic stress-strain relation is

$$\tau_{ij} = 2\mu e_{ij} + \delta_{ij}\lambda\theta \tag{10.4.6}$$

Comparing Equations (10.4.5) and (10.4.6), we may write Equation (10.4.5) as

$$\bar{\tau}_{ij} = 2\mu'\bar{e}_{ij} + \delta_{ij}\lambda'\bar{\theta} \tag{10.4.7}$$

where $\mu' = \mu(1 - \bar{\phi})$ and $\lambda' = [K - (2\mu/3)(1 - \bar{\phi})]$. The corresponding Poisson's ratio and Young's modulus are denoted by ν' and E', respectively. The following expressions may be written:

$$K = \frac{E}{3(1 - 2\nu)} = \frac{E'}{3(1 - 2\nu')} \tag{10.4.8}$$

Noting that $\mu = E/2(1 + \nu)$, we obtain

$$\mu' = \frac{E'}{2(1 + \nu')} = \mu(1 - \bar{\phi})$$

Since[7] $E = 9K\mu/(3K + \mu)$ and $\nu = (3K - 2\mu)/2(3K + \mu)$, we obtain

$$E' = \frac{9(1 - \bar{\phi})\mu K}{3K + \mu(1 - \bar{\phi})} \qquad \nu' = \frac{3K - 2\mu(1 - \bar{\phi})}{6K + 2\mu(1 - \bar{\phi})} \tag{10.4.9}$$

Now replace the stress and strain components of Equations (10.1.2) by their transforms $\bar{\tau}_{ij}$ and $\bar{e}_{ij}$, and the elastic constants E and ν by E' and ν'. Then, corresponding to Equation (10.1.2), we have

$$\begin{aligned} \bar{\tau}_{xx} &= \frac{E'}{1 - \nu'^2}(\bar{e}_{xx} + \nu'\bar{e}_{yy}) \\ \bar{\tau}_{yy} &= \frac{E'}{1 - \nu'^2}(\bar{e}_{yy} + \nu'\bar{e}_{xx}) \\ \bar{\tau}_{xy} &= 2\mu'\bar{e}_{xy} \end{aligned} \tag{10.4.10}$$

From this and Equation (10.1.5), the stress in terms of displacements is expressed as

$$\bar{\tau}_{xx} = \frac{E'}{1-\nu'^2}\left[\frac{\partial \bar{u}_0}{\partial x} - z\frac{\partial^2 \bar{w}}{\partial x^2} + \nu'\left(\frac{\partial \bar{v}_0}{\partial y} - z\frac{\partial^2 \bar{w}}{\partial y^2}\right)\right]$$

$$\bar{\tau}_{yy} = \frac{E'}{1-\nu'^2}\left[\frac{\partial \bar{v}_0}{\partial y} - z\frac{\partial^2 \bar{w}}{\partial y^2} + \nu'\left(\frac{\partial \bar{u}_0}{\partial x} - z\frac{\partial^2 \bar{w}}{\partial x^2}\right)\right] \qquad (10.4.11)$$

$$\bar{\tau}_{xy} = \mu'\left[\frac{\partial \bar{u}_0}{\partial y} + \frac{\partial \bar{v}_0}{\partial x} - 2z\frac{\partial^2 \bar{w}}{\partial x\,\partial y}\right]$$

From the moment-stress relations, Equations (10.1.8), we obtain

$$\overline{M}_x = -D'\left(\frac{\partial^2 \bar{w}}{\partial x^2} + \nu'\frac{\partial^2 \bar{w}}{\partial y^2}\right)$$

$$\overline{M}_y = -D'\left(\frac{\partial^2 \bar{w}}{\partial y^2} + \nu'\frac{\partial^2 \bar{w}}{\partial x^2}\right) \qquad (10.4.12)$$

$$\overline{M}_{xy} = D'(1-\nu')\frac{\partial^2 \bar{w}}{\partial x\,\partial y}$$

where

$$D' = \frac{\mu' h^3}{6(1-\nu')} = \frac{h^3}{12}\frac{(1-\bar{\phi})[6K + 2\mu(1-\bar{\phi})]2\mu}{[3K + 4\mu(1-\bar{\phi})]} \qquad (10.4.13)$$

Substituting these expressions into Equation (10.1.17), we obtain the *governing equilibrium equation for bending of viscoelastic plates*,

$$D'\nabla^4 \bar{w} = \bar{q} \qquad (10.4.14)$$

where the bar over a quantity denotes its Laplace transform. For a lateral load which is constant with respect to time, its transform is given by

$$\bar{q} = \int_0^\infty \exp(-st) q\, dt = \frac{q}{s}$$

With given boundary conditions of the plate, Equation (10.4.14) may be solved for $\bar{w}$. Then, by taking the inverse transform of $\bar{w}$, we obtain w vs. time. For cases with the distributed load q independent of time, the deflection mode of the viscoelastic plate is the same, up to a magnitude factor, as that of the corresponding elastic plate.

Example. Consider a simply supported rectangular viscoelastic plate, with sides of lengths "a" and "b" and thickness "h," under a uniformly distributed load "q" per unit area. The material property of the plate is represented by a Maxwell model; that is,

$$[1 - \bar{\phi}(s)] = \frac{s}{s + \alpha} \tag{10.4.15}$$

where α is a constant. It is required to find the deflection of the plate as a function of time.

Substituting Equation (10.4.15) into Equation (10.4.13) yields

$$D' = \frac{h^3}{12} \cdot \frac{4\mu s[3K(s+\alpha) + \mu s]}{[3K(s+\alpha) + 4\mu s](s+\alpha)} \tag{10.4.16}$$

$$\bar{q} = \frac{q}{s} \tag{10.4.17}$$

From Equations (10.4.14), (10.4.16), and (10.4.17), we have

$$4\mu \frac{h^3}{12} \nabla^4 \bar{w} = \frac{q}{s} \frac{[3K(s+\alpha) + 4\mu s](s+\alpha)}{s[3K(s+\alpha) + \mu s]} \tag{10.4.18}$$

Expressing q and w in terms of sine series, as discussed in Section 10.1, gives

$$\bar{q} = \sum_{m=1}^{\infty} \sum_{n=1}^{\infty} \frac{a_{mn}}{s} \sin \frac{m\pi x}{a} \sin \frac{n\pi y}{b} \tag{10.4.19}$$

$$\bar{w} = \sum_{m=1}^{\infty} \sum_{n=1}^{\infty} \bar{b}_{mn} \sin \frac{m\pi x}{a} \sin \frac{n\pi y}{b} \tag{10.4.20}$$

Substituting Equations (10.4.19) and (10.4.20) into (10.4.18) yields

$$\begin{aligned} 4\mu \frac{h^3}{12} \pi^4 \sum_{m=1}^{\infty} \sum_{n=1}^{\infty} \bar{b}_{mn} \left(\frac{m^2}{a^2} + \frac{n^2}{b^2}\right)^2 \sin \frac{m\pi x}{a} \sin \frac{n\pi y}{b} \\ = \sum_{m=1}^{\infty} \sum_{n=1}^{\infty} \frac{a_{mn}}{s} \frac{[3K(s+\alpha) + 4\mu s](s+\alpha)}{s[3K(s+\alpha) + \mu s]} \sin \frac{m\pi x}{a} \sin \frac{n\pi y}{b} \end{aligned} \tag{10.4.21}$$

Multiplying through by $\sin(m'\pi x/a)\sin(n'\pi y/b)$ and integrating from $x = 0$ to $x = a$, and from $y = 0$ to $y = b$, we obtain, as in Section 10.1,

$$\begin{aligned} \bar{b}_{mn} &= \frac{a_{mn}[3K(s+\alpha) + 4\mu s](s+\alpha)}{4\mu(h^3/12)\pi^4(m^2/a^2 + n^2/b^2)^2 s^2[3K(s+\alpha) + \mu s]} \\ &= \frac{a_{mn}}{\pi^4(h^3/12)4\mu(m^2/a^2 + n^2/b^2)^2} \left[\frac{K+\mu}{Ks} + \frac{\alpha}{s^2} - \frac{\mu^2}{K[3K(s+\alpha) + \mu s]}\right] \end{aligned} \tag{10.4.22}$$

whose inverse is

$$b_{mn} = \frac{a_{mn}}{\pi^4(h^3/12)4\mu(m^2/a^2 + n^2/b^2)^2} \times \left[\frac{K+\mu}{K} + \alpha t - \frac{\mu^2}{K(3K+\mu)} \exp\left(-\frac{3K\alpha}{3K\alpha+\mu} t\right)\right] \tag{10.4.23}$$

Hence the deflection as a function of time for this plate is

$$w = \frac{1}{3\pi^4 h^3 \mu}\left[\frac{K+\mu}{K} + \alpha t - \frac{\mu^2}{K(3K+\mu)} \exp\left(-\frac{3K\alpha}{3K\alpha+\mu} t\right)\right] \cdot \sum_{m=1}^{\infty} \sum_{n=1}^{\infty} \frac{a_{mn} \sin(m\pi x/a) \sin(n\pi y/b)}{(m^2/a^2 + n^2/b^2)^2} \tag{10.4.24}$$

Comparing Equations (10.4.24) and (10.1.31), we see that the deflection mode of a viscoelastic plate under a uniformly distributed load is the same as that of its elastic counterpart. The magnitude of deflection of the viscoelastic plate increases with time t.

10.5 Bending of Plates with Arbitrary Creep Characteristics

Since the strain rate of most engineering materials at elevated temperatures is not proportional to stress, its stress-strain-time relation is not viscoelastic. The use of viscoelastic models for these materials will involve certain errors and can only be expected to give approximate solutions to the actual problems. However, in many cases, where exact solutions are difficult to obtain, approximate solutions may suffice for practical purposes. The analysis of bending of viscoelastic plates was given in Section 10.4. In this section, analysis of plate bending with arbitrary creep characteristics is shown.

Kachanov[9] in 1949 and Malinin[10] in 1953 obtained solutions for circular plates of uniform thickness h under axisymmetric loadings. Their analyses considered only steady creep, whose rate was taken as proportional to the nth power of stress. With σ_1, σ_2, and σ_3 denoting the principal stresses; e_1, e_2, and e_3 denoting the principal strains; and $\dot{e}_1''$, $\dot{e}_2''$, and $\dot{e}_3''$ denoting the rates of principal creep strains, the effective stress σ^* and the effective strain rate $\dot{e}^{*''}$ are defined as

$$\sigma^* = \frac{1}{\sqrt{2}}[(\sigma_1 - \sigma_2)^2 + (\sigma_2 - \sigma_3)^2 + (\sigma_3 - \sigma_1)^2]^{1/2}$$

$$\dot{e}^{*''} = \frac{\sqrt{2}}{3}[(\dot{e}_1'' - \dot{e}_2'')^2 + (\dot{e}_2'' - \dot{e}_3'')^2 + (\dot{e}_3'' - \dot{e}_1'')^2]^{1/2}$$

Assuming a power-law relation for steady creep, we have

$$\dot{e}^{*\prime\prime} = B\sigma^{*n}$$

where B and n are material constants. The rates of principal creep strains are (see Section 4.8),

$$\dot{e}_1'' = \frac{\dot{e}^{*\prime\prime}}{\sigma^*}\left(\sigma_1 - \frac{\sigma_2 + \sigma_3}{2}\right)$$

$$\dot{e}_2'' = \frac{\dot{e}^{*\prime\prime}}{\sigma^*}\left(\sigma_2 - \frac{\sigma_1 + \sigma_3}{2}\right)$$

$$\dot{e}_3'' = \frac{\dot{e}^{*\prime\prime}}{\sigma^*}\left(\sigma_3 - \frac{\sigma_1 + \sigma_2}{2}\right)$$

On the basis of these steady creep stress-strain-time relations, solutions of the four cases shown in Figure 10.19 have been obtained by Malinin. The steady-state deflection rates $\dot{w}_0$ at the center of the plate for these four cases are shown. Using variational procedures, Sanders, McComb, and Schlechte[11] in 1957 calculated the deflection of case (a) in Figure 10.19 vs. time, using a similar creep strain-stress-time relation.

Table 10.2. VALUES OF J FOR EQUATIONS (a) THROUGH (d) IN FIGURE 10.19

n	J			
	a	b	c	d
∞	3.65	0.534	0.647	0.429
5	5.51	0.552	0.707	0.429
2.5	8.36	0.574	0.782	0.432
1.667	12.8	0.600	0.877	0.442
1.25	19.5	0.631	0.997	0.549
1.0	30.0	0.667	1.17	0.500

In the above, the strain rate is taken to be a function of stress only. For many actual materials, the strain rate depends also on the amount of creep strain, as discussed in Chapter 3. To include this strain-hardening effect or other material creep properties, the author,[14] in 1960 showed a method for calculating the bending stresses in plates with arbitrary creep characteristics.

Consider the case of a circular plate under axisymmetric loading; the radial and tangential directions are those of the principal stresses and

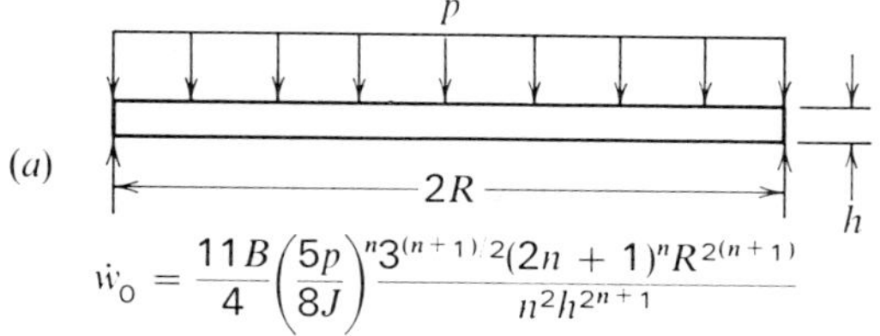

$$\dot{w}_0 = \frac{11B}{4}\left(\frac{5p}{8J}\right)^n \frac{3^{(n+1)/2}(2n+1)^n R^{2(n+1)}}{n^2 h^{2n+1}}$$

(b)

$$\dot{w}_0 = \frac{B}{4}\left(\frac{p}{24J}\right)^n \frac{3^{(n+1)/2}(2n+1)^n R^{2(n+1)}}{n^2 h^{2n+1}}$$

(c)

$$\dot{w}_0 = \frac{7B}{12}\left(\frac{7P}{24\pi J}\right)^n \frac{3^{(n+1)/2}(2n+1)^n R^2}{n^n h^{2n+1}}$$

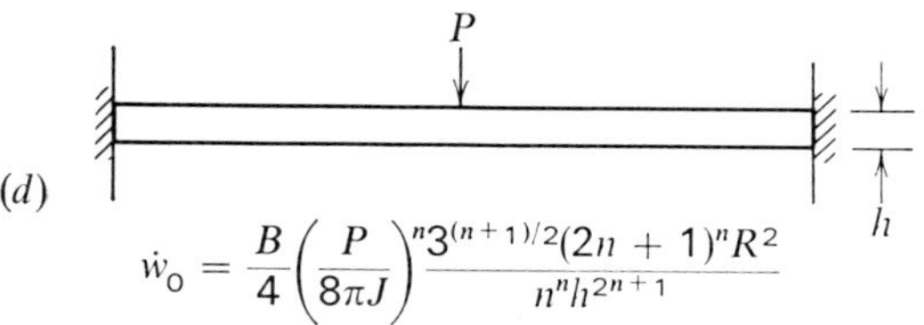

$$\dot{w}_0 = \frac{B}{4}\left(\frac{P}{8\pi J}\right)^n \frac{3^{(n+1)/2}(2n+1)^n R^2}{n^n h^{2n+1}}$$

Figure 10.19. Steady-state creep deflection rate $\dot{w}_0$ at the center of the plate.

strains. The deflection being axisymmetric, $\partial w/\partial\theta = 0$, $e''_{r\theta} = 0$, and Equations (10.2.11) reduce to

$$M_r = -D\left[\frac{d^2w}{dr^2} + \frac{\nu}{r}\frac{dw}{dr}\right] - \frac{E}{1-\nu^2}\int(e''_{rr} + \nu e''_{\theta\theta})z\,dz \qquad (10.5.1)$$

$$M_\theta = -D\left[\frac{1}{r}\frac{dw}{dr} + \nu\frac{d^2w}{dr^2}\right] - \frac{E}{1-\nu^2}\int(e''_{\theta\theta} + \nu e''_{rr})z\,dz \qquad (10.5.2)$$

$$M_{r\theta} = 0$$

Letting

$$\phi = -\frac{dw}{dr}$$

we see that the above equations become

$$M_r = D\left(\frac{d\phi}{dr} + \nu\frac{\phi}{r}\right) - \frac{E}{1-\nu^2}\int (e''_{rr} + \nu e''_{\theta\theta})z\,dz \tag{10.5.3}$$

$$M_\theta = D\left(\frac{\phi}{r} + \nu\frac{d\phi}{dr}\right) - \frac{E}{1-\nu^2}\int (e''_{\theta\theta} + \nu e''_{rr})z\,dz \tag{10.5.4}$$

Consider the condition of equilibrium of an element $r\,d\theta\,dr$ (Figure 10.20). The sum of moments about an axis parallel to the tangent and passing through the centroid of the element is

$$\left(M_r + \frac{dM_r}{dr}dr\right)(r + dr)\,d\theta - M_r r\,d\theta - M_\theta\,dr\,d\theta$$
$$+ Q_r\frac{r}{2}\,d\theta\,dr + \left(Q_r + \frac{\partial Q_r}{\partial r}dr\right)\frac{r}{2}\,d\theta\,dr = 0$$

where Q_r is the vertical shear force per unit length of the circumference $r\,d\theta$. Neglecting higher-order terms and dropping the subscript "r" in Q_r, we see that the previous equilibrium equation reduces to

$$M_r + r\frac{dM_r}{dr} - M_\theta + Qr = 0 \tag{10.5.5}$$

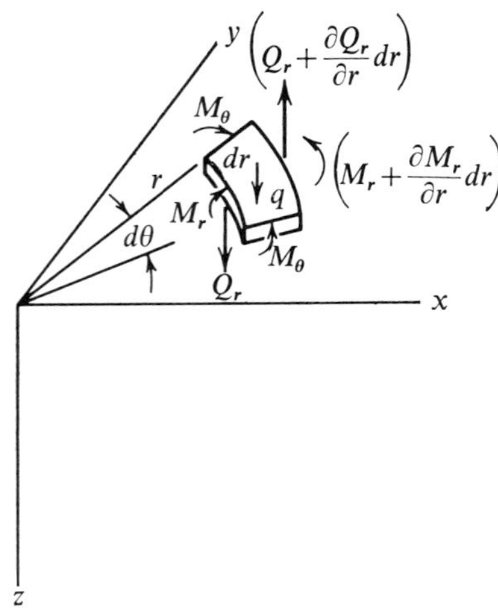

Figure 10.20

Substituting Equations (10.5.3) and (10.5.4) into Equation (10.5.5), we obtain

$$D\left[\frac{d^2\phi}{dr^2} + \frac{1}{r}\frac{d\phi}{dr} - \frac{\phi}{r^2}\right] - \frac{E}{1-\nu^2}\left[\frac{d}{dr}\int(e''_{rr} + \nu e''_{\theta\theta})z\,dz + \frac{1}{r}(1-\nu)\int(e''_{rr} - e''_{\theta\theta})z\,dz\right] + Q = 0 \qquad (10.5.6)$$

Multiplying the above equation by $2\pi r$, we get

$$2\pi rD\left[\frac{d^2\phi}{dr^2} + \frac{1}{r}\frac{d\phi}{dr} - \frac{\phi}{r^2}\right] - \frac{2\pi E}{1-\nu^2}\left[r\frac{d}{dr}\int(e''_{rr} + \nu e''_{\theta\theta})z\,dz + (1-\nu)\int(e''_{rr} - e''_{\theta\theta})z\,dz\right] + 2\pi Qr = 0 \qquad (10.5.7)$$

Differentiating Equation (10.5.7) with respect to r yields

$$2\pi D\frac{d}{dr}\left\{r\left[\frac{d^2\phi}{dr^2} + \frac{1}{r}\frac{d\phi}{dr} - \frac{\phi}{r^2}\right]\right\} - \frac{2\pi E}{1-\nu^2}\frac{d}{dr}\left[r\frac{d}{dr}\int(e''_{rr} + \nu e''_{\theta\theta})z\,dz + (1-\nu)\int(e''_{rr} - e''_{\theta\theta})z\,dz\right] + \frac{d}{dr}(2\pi Qr) = 0 \qquad (10.5.8)$$

Taking $\sum F_z = 0$ for a differential annulus yields

$$2\pi(r + dr)\left(Q + \frac{dQ}{dr}dr\right) = Q2\pi r + 2\pi r\,drq$$

Neglecting higher-order terms, we have

$$\frac{d}{dr}(2\pi Qr)\,dr = 2\pi qr\,dr \qquad (10.5.9)$$

It is seen from the previous equations that

$$-\frac{2\pi E}{1-\nu^2}\frac{d}{dr}\left[\frac{1}{r}\frac{d}{dr}\int(e''_{rr} + \nu e''_{\theta\theta})z\,dz + (1-\nu)\int(e''_{rr} - e''_{\theta\theta})z\,dz\right]$$

is equivalent to $2\pi qr$. Hence the equivalent load $\bar{q}$ in this case is

$$\bar{q} = -\frac{E}{1-\nu^2}\frac{1}{r}\frac{d}{dr}\left[r\frac{d}{dr}\int(e''_{rr} + \nu e''_{\theta\theta})z\,dz + (1-\nu)\int(e''_{rr} - e''_{\theta\theta})z\,dz\right] \qquad (10.5.10)$$

If the plate is simply supported, $M_r = 0$ along its outer edge. Let $\overline{M}_r$ denote the moment caused by both the actual and the equivalent loads. At the boundary,

$$M_r = \overline{M}_r - M_{r_I} = 0$$

from which we obtain

$$\overline{M}_r = M_{r_I} = \frac{E}{1 - \nu^2} \int (e''_{rr} + \nu e''_{\theta\theta}) z \, dz \tag{10.5.11}$$

The deflection surface of the plate is found by determining the deflection of an elastic plate subject to the equivalent distributed load $\bar{q}$ given by Equation (10.5.10), in addition to the applied lateral q and an edge moment M_{r_I} on the boundary.

The stress distribution changes as creep develops, and hence the creep strain rate changes. The stress components at a point in the plate vary with time and are generally represented by smooth curves. These smooth curves may be approximated by small incremental steps as shown in Figure 3.9. Each step consists of a constant stress period Δt followed by an instantaneous increment of stress $\Delta\sigma$. The incremental creep strains $\Delta e''_{xx}$, $\Delta e''_{yy}$, and $\Delta e''_{xy}$ occurring during the Δt period are determined from the multiaxial creep characteristics of the material at the particular temperature, using the stresses—considered as constant—existing at the end of the previous Δt period. The error introduced by considering the stresses constant during Δt vanishes as Δt approaches zero.

Replacing e''_{rr}, $e''_{\theta\theta}$ in Equations (10.5.10) and (10.5.11) by $\Delta e''_{rr}$ and $\Delta e''_{\theta\theta}$, we may obtain the incremental equivalent load and moment, $\Delta\bar{q}$ and $\Delta\overline{M}_r$, for each time interval Δt. From an elastic analysis of the plate, the incremental lateral deflection Δw for each time interval is readily found. With the incremental deflection Δw and the incremental creep strain known, the incremental stress components, and hence the stress components at the beginning of the next time interval, are readily calculated. Repeating the process for successive Δt time intervals gives the deflections at different instants of time after loading. The method is illustrated by the following example.

Example. A simply supported circular plate of 75ST aluminum alloy at 600° F is subjected to a uniformly distributed load of 36 psi. The diameter of the plate is 10 in. and the thickness h is 0.5 in. Young's modulus of elasticity E for this material at 600° F is 5.2×10^6 psi. Poisson's ratio is 0.3 for elastic strains and 0.5 for creep strains. The stress-strain-time relation for constant stress may be represented by[12]

$$e' = e - e'' = \frac{\sigma}{E} \qquad \text{or} \qquad e = \frac{\sigma}{E} + e'' \tag{10.5.12}$$

$$e'' = A \exp(B\sigma) t^K \tag{10.5.13}$$

where e' is the elastic uniaxial strain, e is the total uniaxial strain, e'' the uniaxial creep strain, σ is the uniaxial stress in psi, exp is the base of the natural logarithm, and the values of the constants are $A = 2.64 \times 10^{-7}$, $B = 1.92 \times 10^{-3}$, t is in hr, $K = 0.66$, and h is the thickness of the plate in inches. It is assumed that $\dot{e}'' = f(e'', \sigma)$, as discussed in Section 3.4.

The initial *elastic* deflection for uniform loading q is obtained by solving Equation (10.5.8) with $e'' = e_\theta'' = 0$,

$$2\pi D \frac{d}{dr}\left\{r\left[\frac{d^2\phi}{dr^2} + \frac{1}{r}\frac{d\phi}{dr} - \frac{\phi}{r^2}\right]\right\} + 2\pi q r = 0 \tag{10.5.14}$$

For a simply supported circular plate, the boundary conditions at $r = R$ are

$$(M_r)_{r=R} = D\left[\frac{d\phi}{dr} + \frac{v}{r}\phi\right] = 0 \qquad \text{and} \qquad w = 0 \tag{10.5.15}$$

From Equation (10.5.14), we have

$$2\pi D r\left[\frac{d^2\phi}{dr^2} + \frac{1}{r}\frac{d\phi}{dr} - \frac{\phi}{r^2}\right] = -2\pi\int_0^r qr\,dr = -Q2\pi r \tag{10.5.16}$$

or

$$\frac{d}{dr}\left[\frac{1}{r}\frac{d}{dr}(r\phi)\right] = -\frac{Q}{D} \tag{10.5.17}$$

Using Equation (10.5.9), we may write Equation (10.5.17) as

$$\frac{1}{r}\frac{d}{dr}\left\{r\frac{d}{dr}\left[\frac{1}{r}\frac{d}{dr}\left(r\frac{dw}{dr}\right)\right]\right\} = \frac{q}{D} \tag{10.5.18}$$

For a simply supported circular plate under a uniformly distributed load q, the shear force Q at r is

$$Q = \frac{\pi r^2 q}{2\pi r} = q\,\frac{r}{2}$$

Substituting the last equation into Equation (10.5.17) and integrating with respect to r yields

$$D\frac{1}{r}\frac{d}{dr}(r\phi) = -\frac{qr^2}{4} + C_1$$

Integrating again, we find

$$D\phi = -\frac{qr^3}{16} + C_1\frac{r}{2} + \frac{C_2}{r}$$

For a circular plate with axisymmetric loading $dw/dr = 0$ at $r = 0$, hence $C_2 = 0$, and we obtain

$$D\phi = -\frac{qr^3}{16} + C_1\frac{r}{2} \qquad D\frac{d\phi}{dr} = -\frac{3qr^2}{16} + \frac{C_1}{2}$$

At $r = R$, $M_r = 0$, and, from Equation (10.5.3), we have

$$-\frac{3qR^2}{16} + \frac{C_1}{2} + v\left(-\frac{qR^2}{16} + \frac{C_1}{2}\right) = 0$$

$$(1 + v)\frac{C_1}{2} = \frac{qR^2}{16}(3 + v)$$

$$C_1 = \frac{qR^2}{8}\frac{3 + v}{1 + v}$$

Therefore

$$-D\frac{dw}{dr} = D\phi = -\frac{qr^3}{16} + \frac{qR^2}{16}\frac{3 + v}{1 + v}r$$

which gives, upon integration,

$$Dw = \frac{qr^4}{64} - \frac{qR^2}{16}\frac{3 + v}{1 + v}\frac{r^2}{2} + C_3$$

From the boundary condition $w = 0$ at $r = R$, we find

$$C_3 = -\frac{qR^4}{64} + \frac{qR^4}{32}\frac{(3 + v)}{(1 + v)}$$

Then

$$Dw = \frac{q(r^4 - R^4)}{64} - \frac{qR^2}{32}\frac{3 + v}{1 + v}(r^2 - R^2)$$

or

$$w = \frac{q(R^2 - r^2)}{64D}\left[\frac{5 + v}{1 + v}R^2 - r^2\right] \tag{10.5.19}$$

From Equations (10.5.1) and (10.5.2) with $e'' = e_\theta'' = e''_{r\theta} = 0$ and derivatives of w obtained from the previous equation, we obtain

$$\begin{aligned} M_r &= \frac{q}{16}(3 + v)(R^2 - r^2) \\ M_\theta &= \frac{q}{16}[R^2(3 + v) - r^2(1 + 3v)] \end{aligned} \tag{10.5.20}$$

The initial stresses are then obtained from $\tau_{rr} = M_r z/I$ and $\tau_{\theta\theta} = M_\theta z/I$,

$$\tau_{rr} = \frac{6M_r}{h^2}\frac{z}{h/2} = \frac{3q}{8h^2}\,3.3(R^2 - r^2)\frac{z}{h/2}$$
$$\tau_{\theta\theta} = \frac{6M_\theta}{h^2}\frac{z}{h/2} = \frac{3q}{8h^2}[3.3R^2 - 1.9r^2]\frac{z}{h/2} \qquad (10.5.21)$$

With $R = 5$ in., and $q = 36$ psi, the initial elastic deflection at $t = 0$ is

$$\frac{w}{h} = 0.0118\left[1 - \left(\frac{r}{R}\right)^2\right]\left[4.07 - \left(\frac{r}{R}\right)^2\right] \qquad (10.5.22)$$

The creep strain rates are assumed to be proportional to the corresponding deviatoric stress components as given by Equations (4.5.1),

$$\frac{\dot{\varepsilon}''_{rr}}{S_r} = \frac{\dot{\varepsilon}''_{\theta\theta}}{S_\theta} = \frac{\dot{\varepsilon}''_{r\theta}}{S_{r\theta}} = k \qquad (10.5.23)$$

Johnson's[13] tests have shown that for a number of engineering materials the rate of increase of the second creep strain invariant I_2'' may be expressed as a function of the deviatoric second stress invariant $\bar{J}_2$ and the creep strain invariant I_2'' itself. Using this relation and the uniaxial creep test data, $\dot{e}''_{11} = F(\sigma_{11}, e''_{11})$, in which case $\bar{J}_2 = \sigma_{11}^2/3$ and $I_2'' = \frac{3}{4}e''^2_{11}$, we obtain

$$\frac{2}{\sqrt{3}}\frac{d\sqrt{I_2''}}{dt} = F\left(\sqrt{3\bar{J}_2}, \frac{2}{\sqrt{3}}\sqrt{I_2''}\right) \qquad (10.5.24)$$

where $3\bar{J}_2 = [\sigma_{rr}^2 + \sigma_{\theta\theta}^2 + (\sigma_{rr} - \sigma_{\theta\theta})^2]$ and $I_2'' = e''^2_{rr} + e''_{\theta\theta} + (e''_{rr} + e''_{\theta\theta})^2$ for axisymmetric loading. Generalization of the uniaxial stress-strain-time relation, $e'' = A\exp(B\sigma)t^K$, to the multiaxial state gives $e^{*''} = A\exp(B\sigma^*)t^K$; then

$$\sqrt{\tfrac{4}{3}I_2''} = A\exp\left(B\sqrt{3\bar{J}_2}\right)t^K \qquad (10.5.25)$$

The problem is solved by assuming the stress to remain constant in each incremental time interval Δt; that is, the smooth stress-time curve is approximated by incremental steps as shown in Chapter 3. The first incremental time interval Δt is taken to be $\frac{1}{20}$ hr. During the first time interval, the stresses are assumed to be those at $t = 0$, that is, the stresses from the elastic solution. From the expressions of σ_{rr} and $\sigma_{\theta\theta}$, given by Equation (10.5.21), we obtain S_{rr} and $S_{\theta\theta}$. From Equation (10.5.25), we can calculate I_2'' at the end of the first time interval. From Equation (4.8.7)

$$\frac{e''_{rr}}{S_{rr}} = \frac{e''_{\theta\theta}}{S_{\theta\theta}}$$

and

$$I_2'' = e_{rr}''^2 + e_{\theta\theta}''^2 + e_{rr}'' e_{\theta\theta}''$$

we may solve for e_{rr}'' and $e_{\theta\theta}''$. For the second time interval, the stresses τ_{ij} and the creep strains e_{ij}'' are known at the beginning of the interval. $\bar{J}_2$ and I_2'' may then be calculated. Writing Equation (10.5.24) for a finite increment of time, we have

$$\Delta\sqrt{I_2''} = \sqrt{\tfrac{3}{4}}F(\sqrt{3\bar{J}_2}, \sqrt{\tfrac{4}{3}I_2''})\,\Delta t \tag{10.5.26}$$

For the present case, recalling the expression for I_2'', we have

$$\Delta\sqrt{I_2''} = [(e_{rr}'' + \Delta e_{rr}'')^2 + (e_{\theta\theta}'' + \Delta e_{\theta\theta}'')^2 + (e_{rr}'' + \Delta e_{rr}'')(e_{\theta\theta}'' + \Delta e_{\theta\theta}'')]^{1/2}$$
$$-[e_{rr}''^2 + e_{\theta\theta}''^2 + e_{rr}'' e_{\theta\theta}'']^{1/2} \tag{10.5.27}$$

Multiplying Equation (10.5.23) by Δt, letting $k\,\Delta t = \kappa$, and then substituting into Equation (10.5.27), we obtain

$$\Delta\sqrt{I_2''} = [(e_{rr}'' + \kappa S_{rr})^2 + (e_{\theta\theta}'' + \kappa S_{\theta\theta})^2 + (e_{rr}'' + \kappa S_{rr})(e_{\theta\theta}'' + \kappa S_{\theta\theta})]^{1/2}$$
$$-[e_{rr}''^2 + e_{\theta\theta}''^2 + e_{rr}'' e_{\theta\theta}'']^{1/2} \tag{10.5.28}$$

At the beginning of a Δt time interval, I_2'', S_{rr}, $S_{\theta\theta}$, e_{rr}'', and $e_{\theta\theta}''$ are known. $\Delta\sqrt{I_2''}$ is then obtained from Equation (10.5.26) and inserted into Equation (10.5.28), from which the constant κ is determined. The $\Delta e_{rr}''$ and $\Delta e_{\theta\theta}''$ are thereafter obtained from Equation (10.5.23). From the $\Delta e_{rr}''$ and $\Delta e_{\theta\theta}''$, the incremental equivalent lateral load $\Delta\bar{q}$ is found from Equation (10.5.10).

For the first time interval, the stress, being obtained by an elastic analysis, is proportional to the distance from the middle plane; that is,

$$\sqrt{3\bar{J}_2} \propto z \qquad \sqrt{3\bar{J}_2} = \frac{2}{h}(\sqrt{3\bar{J}_2})_{h/2} \cdot z = Cz$$

where $C = (2/h)(\sqrt{3\bar{J}_2})_{h/2}$. Since we have taken $e'' = A\exp(B\sigma)t^K$, we have, for the first time interval,

$$\frac{(e_{rr}'')_z}{(e_{rr}'')_{h/2}} = \frac{\exp(B\sqrt{3\bar{J}_2})_z}{\exp(B\sqrt{3\bar{J}_2})_{h/2}} = \frac{\exp(C'z)}{\exp[C'(h/2)]} \tag{10.5.29}$$

where $C' = BC$ and exp is the base of the natural logarithm. From Equation (10.5.9), we obtain

$$\int_{-h/2}^{h/2} e''_{rr} z \, dz = 2 \int_{0}^{h/2} e''_{rr} z \, dz$$

$$= 2(e''_{rr})_{h/2} \cdot \left(\frac{h}{2}\right)^2 \frac{1}{(B\sqrt{3\bar{J}_2})_{h/2}} \left[1 - \frac{1}{(B\sqrt{3\bar{J}_2})_{h/2}} \cdot \left\{1 - \frac{1}{\exp\,(B\sqrt{3\bar{J}_2})_{h/2}}\right\}\right] \tag{10.5.30}$$

Similarly, $\int e''_{\theta\theta} z \, dz$ for the first time interval may be found. From these values, $\Delta\bar{q}$ and $(\Delta\bar{M}_r)_{r=R}$ are found. We may consider $\Delta\bar{q} 2\pi r \, \Delta r$ as a load uniformly distributed along a circle of radius r. The solution of this loading is given by Timoshenko and Woinowsky-Krieger.[4]

Dividing the circular plate into two parts at $r = r_1$ [see Figure 10.21(b) and (c)], and taking $\sum F_z = 0$, we find that the shear at $r = r_1$ due to a uniformly distributed load $\Delta\bar{q} 2\pi r_1 \, \Delta r$ at $r = r_1$ is

$$Q_1 = \Delta\bar{q} \, \Delta r$$

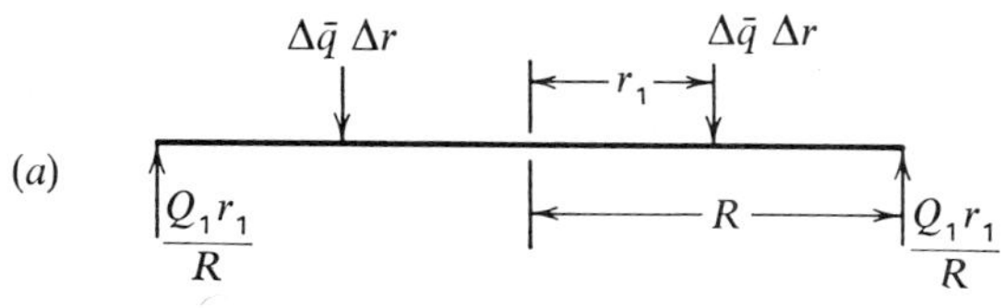

Figure 10.21

From Equation (10.5.18) with $\phi = -(dw/dr)$, the equilibrium equation becomes

$$\frac{1}{r} \frac{d}{dr} \left\{ r \frac{d}{dr} \left[\frac{1}{r} \frac{d}{dr} (r\phi) \right] \right\} = -\frac{q}{D}$$

In Figure 10.21(b) and (c), we have $q = 0$, and the above equation reduces to

$$\frac{d}{dr} \left\{ r \frac{d}{dr} \left[\frac{1}{r} \frac{d}{dr} (r\phi) \right] \right\} = 0$$

From Figure 10.21(c), we have

$$Q = 0 \qquad 0 < r < r_1$$

$$D\frac{d}{dr}\left[\frac{1}{r}\frac{d}{dr}(r\phi)\right] = 0$$

$$M_r = D\left[\frac{d\phi}{dr} + \frac{\nu\phi}{r}\right]$$

Integrating the first of the above equations yields

$$\frac{d}{dr}(r\phi) = C_1 r$$

where C_1 is a constant of integration. Integrating again gives

$$r\phi = C_1\frac{r^2}{2} + C_2 \qquad \phi = \frac{C_1 r}{2} + \frac{C_2}{r}$$

The constant of integration C_2 may be determined from the boundary condition $\phi = 0$ at $r = 0$, which yields $C_2 = 0$; hence $\phi = C_1 r/2$, and we then obtain

$$M_r = D\left[\frac{C_1}{2} + \nu\frac{C_1}{2}\right] = D\frac{C_1}{2}(1 + \nu)$$

From Figure 10.21(b), for $r_1 < r \le R$, the shear force per unit length Q at any radius $r > r_1$ is $Q = Q_1(r_1/r)$. The equilibrium equation is

$$D\frac{d}{dr}\left[\frac{1}{r}\frac{d}{dr}(r\phi)\right] = Q_1\frac{r_1}{r}$$

Integration of this equation gives

$$D\frac{d}{dr}(r\phi) = (Q_1 \ln r + C_3)rr_1$$

Integrating again, we obtain

$$D(r\phi) = Q_1\left(\frac{r^2}{2}\ln r - \frac{r^2}{4}\right) + C_3 r_1\frac{r^2}{2} + C_4$$

$$D\phi = Q_1\frac{r}{2}\left(\ln r - \frac{1}{2}\right) + C_3\frac{rr_1}{2} - \frac{C_4}{r}$$

$$D\frac{d\phi}{dr} = \frac{Q_1}{2}\left(\ln r + \frac{1}{2}\right) + C_3\frac{r_1}{2} - \frac{C_4}{r^2}$$

where C_3 and C_4 are constants of integration. Recalling Equation (10.5.3), we have

$$M_r = D\left[\frac{d\phi}{dr} + \frac{v}{r}\phi\right]$$

$$= Q_1\left(\frac{\ln r}{2} + \frac{1}{4}\right) + C_3\frac{r_1}{2} - \frac{C_4}{r^2} + v\left[Q_1\left(\frac{\ln r}{2} - \frac{1}{4}\right) + \frac{C_3 r_1}{2} + \frac{C_4}{r^2}\right]$$

$$= \frac{Q_1}{2}\left[(1+v)\ln r + \frac{1-v}{2}\right] + C_3\frac{r_1}{2}(1+v) - \frac{C_4}{r^2}(1-v)$$

Since the circular plate is simply supported, $M_r = 0$ at $r = R$, and therefore

$$\frac{Q_1}{2}\left[(1+v)\ln R + \frac{1-v}{2}\right] + \frac{C_3 r_1}{2}(1+v) - \frac{C_4}{R^2}(1-v) = 0 \qquad (10.5.31)$$

Equating M_r for the inner and outer parts at $r = r_1$ gives

$$\frac{Q_1}{2}\left[(1+v)\ln r_1 + \frac{1-v}{2}\right] + \frac{C_3 r_1}{2}(1+v) - \frac{C_4}{{r_1}^2}(1-v) = D\frac{C_1}{2}(1+v) \qquad (10.5.32)$$

For continuity of slope at $r = r_1$, we have

$$\frac{C_1 r_1}{2} = \frac{1}{D}\left[Q_1\frac{r_1}{2}\left(\ln r_1 - \frac{1}{2}\right) + \frac{C_3 {r_1}^2}{2} + \frac{C_4}{r_1}\right] \qquad (10.5.33)$$

where C_1, C_3, and C_4 are determined from Equations (10.5.31) through (10.5.33). Then ϕ is known. Upon integrating ϕ, we obtain

$$w = -\int \phi\, dr + C$$

From the boundary condition $w = 0$ at $r = R$, the constant of integration obtained from the previous equation may be determined. The expression for w when $r > r_1$ is

$$w = \frac{Q_1 r_1}{4D}\left[(R^2 - r^2)\left(1 + \frac{1}{2}\frac{1-v}{1+v}\frac{R^2 - {r_1}^2}{R^2}\right) + ({r_1}^2 + r^2)\ln\frac{r}{R}\right]$$

$$= \frac{Q_1 r_1}{4D}\left[(R^2 - r^2)\left(1 + \frac{0.7}{2.6}\frac{R^2 - {r_1}^2}{R^2}\right) + ({r_1}^2 + r^2)\ln\frac{r}{R}\right] \qquad (10.5.34)$$

where $v = 0.3$ was used. The moments M_r and M_θ for $r > r_1$, from Equations (10.5.3) and (10.5.4), are

$$M_r = -D\left(\frac{d^2w}{dr^2} + \frac{\nu}{r}\frac{dw}{dr}\right)$$

$$= -\frac{Q_1 r_1}{4}\left[0.7 - \frac{0.91}{1.3}\left(1 - \frac{r_1{}^2}{R^2}\right) + 2.6 \ln \frac{r}{R} - 0.7 \frac{r_1{}^2}{r^2}\right]$$

$$M_\theta = -D\left(\frac{1}{r}\frac{dw}{dr} + \nu \frac{d^2w}{dr^2}\right)$$

$$= -\frac{Q_1 r_1}{4}\left[-0.7 - \frac{0.91}{1.3}\left(1 - \frac{r_1{}^2}{R^2}\right) + 0.7 \frac{r_1{}^2}{r^2} + 2.6 \ln \frac{r}{R}\right]$$

Finally, then, the stresses in the outer portion $r > r_1$, due to the equivalent load $\Delta\bar{q}$ at $r = r_1$ on a Δr annulus, may be obtained from Equations (10.5.21) as

$$\tau_{rr} = -\frac{3Q_1 r_1}{2h^2}\left(-0.7 \frac{r_1{}^2}{R^2} + 2.6 \ln \frac{r}{R} - 0.7 \frac{r_1{}^2}{r^2}\right)\frac{z}{h/2}$$

$$\tau_{\theta\theta} = -\frac{3Q_1 r_1}{2h^2}\left[-1.4 + 0.7\left(\frac{r_1{}^2}{R^2} + \frac{r_1{}^2}{r^2}\right) + 2.6 \ln \frac{r}{R}\right]\frac{z}{h/2} \tag{10.5.35}$$

For the inner portion $r < r_1$, the deflection, moments, and stresses are

$$w = \frac{Q_1 r_1}{4D}\left[(r_1{}^2 + r^2)\ln \frac{r_1}{R} + \frac{\left(3.3 - 0.7 \frac{r^2}{R^2}\right)(R^2 - r_1{}^2)}{2.6}\right] \tag{10.3.36}$$

$$M_r = M_\theta = \frac{Q_1 r_1}{4}\left[2.6 \ln \frac{r_1}{R} - 0.7\left(\frac{R^2 - r_1{}^2}{R^2}\right)\right]$$

$$\tau_{rr} = \tau_{\theta\theta} = -\frac{3Q_1 r_1}{2h^2}\left[2.6 \ln \frac{r_1}{R} - 0.7\left(\frac{R^2 - r_1{}^2}{R^2}\right)\right]\frac{z}{h/2} \tag{10.5.37}$$

where $Q_1 = \Delta\bar{q}\,\Delta r$ is applied at $r = r_1$. From Equations (10.5.35) and (10.5.37), the stresses due to equivalent loads $\Delta\bar{q}\,\Delta r$ acting at different radii may be calculated. At the edge $r = R$ the relief in bending moment $\Delta\bar{M}_r$, due to creep, is equal to $[E/(1 - \nu^2)] \int (\Delta e''_{rr} + \nu\, \Delta e''_{\theta\theta})z\, dz$. In order to keep the simply supported edge free of bending moment, a moment equal to ΔM_r must be applied along $r = R$. The equivalent edge moment $\Delta\bar{M}_r$ will result in moments $\Delta M_\theta = \Delta M_r$ and stresses $\Delta\tau_{rr} = \Delta\tau_{\theta\theta} = 6\,\Delta\bar{M}_r z/h^3$.

The stresses at the end of the first time interval are equal to the initial elastic stresses τ_{rr_0} and $\tau_{\theta\theta_0}$ minus the stresses $\Delta\tau''_{rr}$, $\Delta\tau''_{\theta\theta}$ relieved because of creep $\Delta e''_{rr}$ and $\Delta e''_{\theta\theta}$, plus the stresses $\Delta\tau'_{rr}$ and $\Delta\tau'_{\theta\theta}$ caused by the equivalent concentric loads $\Delta\bar{q}$ and edge moment $\Delta\bar{M}_r$. The stresses at the end of the first time interval are

$$\tau_{rr_1} = \tau_{r_0} + \Delta\tau'_{rr} - \Delta\tau''_{rr}$$
$$\tau_{\theta\theta_1} = \tau_{\theta\theta_0} + \Delta\tau'_{\theta\theta} - \Delta\tau''_{\theta\theta} \tag{10.5.38}$$

where

$$\Delta\tau''_{rr} = \frac{E}{1-\nu^2}(\Delta e''_{rr} + \nu\,\Delta e''_{\theta\theta})$$

$$\Delta\tau''_{\theta\theta} = \frac{E}{1-\nu^2}(\Delta e''_{\theta\theta} + \nu\,\Delta e''_{rr})$$

The constant values of stress during the second time interval will be τ_{rr_1} and $\tau_{\theta\theta_1}$. This process is repeated for a second time increment from $\frac{1}{20}$ hr to 1 hr. For the first time interval the stresses τ_{rr} and $\tau_{\theta\theta}$, which were considered as constant during the interval, were obtained from an elastic analysis and hence were proportional to z across each section. However, for the second and subsequent time intervals the variation of stresses τ_{rr} and $\tau_{\theta\theta}$, with respect to z, are not expressible in simple analytical form. We evaluate $\int \Delta e''_{rr}\, z\, dz$ by numerical integration after calculating $\Delta e''_{rr}$ at different values of z. The process is necessarily lengthy, and hence for the present example, to save lengthy numerical calculations, the distribution of incremental creep strain through the thickness is approximated by a power curve in z, as

$$(\Delta e''_{rr})_z = (\Delta e''_{rr})_{h/2}\left(\frac{2z}{h}\right)^{n_1}$$

$$(\Delta e''_{\theta\theta})_z = (\Delta e''_{\theta\theta})_{h/2}\left(\frac{2z}{h}\right)^{n_2}$$

where n_1 and n_2 are constants to be found as follows. At $z = 0$, there is no $\Delta e''_{rr}$ or $\Delta e''_{\theta\theta}$. The values of $\Delta e''_{rr}$ and $\Delta e''_{\theta\theta}$ at $z = h/4$ and $z = h/2$ are calculated by using the exponential creep law. From these two sets of values of $\Delta e''_{rr}$'s and $\Delta e''_{\theta\theta}$'s, we find n_1 and n_2 so that the above equations give the correct values at $h/4$ and $h/2$. Then

$$\int_{-h/2}^{h/2} \Delta e''_{rr}\, z\, dz = \frac{h^2(\Delta e''_{rr})_{h/2}}{2(n_1+2)}$$

In the same way $\int_{-h/2}^{h/2} \Delta e''_{\theta\theta}\, z\, dz$ may be evaluated. The approximate $\Delta e''_{rr}$ and $\Delta e''_{\theta\theta}$ distributions obtained by the above power expressions have been compared to an actual distribution and found to be in close agreement. The deflections and stresses at $t = 0$, $\frac{1}{20}$ hr and 1 hr have been calculated. The deflection curves calculated for these time instants are shown in Figure 10.22. The stresses and the detail calculations are shown in Table 10.3.

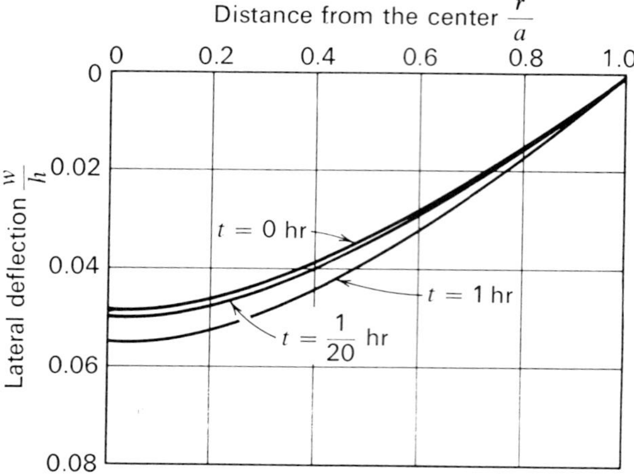

Figure 10.22

The calculated results show that the rate of deflection decreases with time. This is in part due to the fact that the secondary creep rate is less than the primary creep rate and in part to the fact that the stresses in the highly stressed regions are reduced by creep. The accuracy of this method may be improved by decreasing the Δt increments. Gebhart[15] has shown a similar creep analysis for circular plates with clamped edges under a uniform lateral load.

Creep Bending of Rectangular Plates. In the above example of circular plates under axisymmetrical loading, the stress and strain vary only with the radius "r." For rectangular plates, stress and strain vary along two axes, and hence the computations are much lengthier. An example of creep bending of a square plate[16] is presented below.

Example. A 7075-T6 aluminum alloy square plate of dimensions 5 in. × 5 in. × 0.1 in., with four simply supported edges, is loaded with a uniform lateral load of 10 psi at a temperature of 500° F. The uniaxial creep characteristics of this material are assumed to be, as before,

$$e'' = A \exp (B\sigma)t^K$$

The constants A, B, and K at 500° F are as follows:

$$A = 2.0 \times 10^{-6}/\text{hr}$$

$$B = 6.8 \times 10^{-4} \text{ psi}$$

$$K = 0.63$$

$$E = 7 \times 10^6 \text{ psi}$$

The initial elastic lateral deflection w at $t = 0$ is given by Equation (10.1.32). The moments and stresses are obtained by Equations (10.1.10), (10.1.13), and (10.1.21). The creep stress-strain-time relations given by Equations (10.5.25) and (10.5.26) remain the same, except that the values of the constants are changed. In cartesian coordinates, Equation (10.5.23) becomes

$$\frac{\dot{e}''_{xx}}{S_{xx}} = \frac{\dot{e}''_{yy}}{S_{yy}} = \frac{\dot{e}''_{xy}}{S_{xy}} = k \tag{10.5.39}$$

where

$$S_{xx} = \frac{2\tau_{xx} - \tau_{yy}}{3} \qquad S_{yy} = \frac{2\tau_{yy} - \tau_{xx}}{3} \qquad S_{xy} = \tau_{xy}$$

The second deviatoric stress and creep strain invariants in rectangular coordinates are

$$\bar{J}_2 = \tfrac{1}{6}[(\tau_{xx} - \tau_{yy})^2 + \tau_{xx}^2 + \tau_{yy}^2] + \tau_{xy}^2 \tag{10.5.40}$$

$$I_2'' = [e''^2_{xx} + e''^2_{yy} + e''_{xx}e''_{yy} + e''^2_{xy}] \tag{10.5.41}$$

The square plate is divided by grid lines as shown in Figure 10.23.

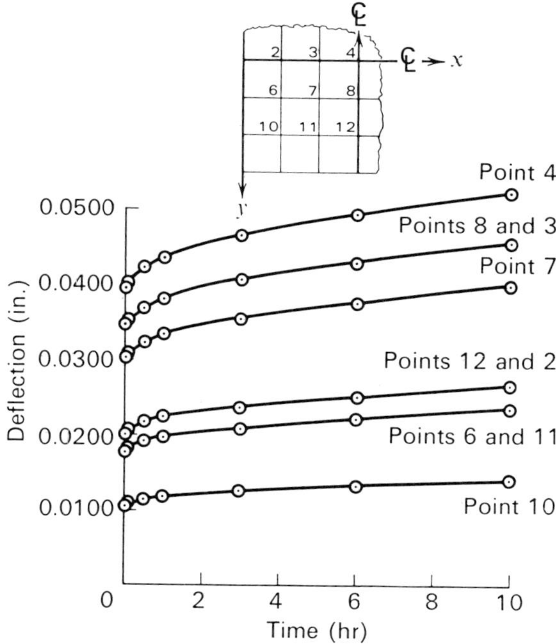

Figure 10.23. Grid point deflections vs. time.

Table 10.3. Creep Stresses and Deflections of a Circular Plate Under Uniform Loading

No.		(1)		(2)		(3)		(4)		(5)		(6)	
1	$\frac{r}{a}$	0		0.2		0.4		0.6		0.8		1.0	
2	$\zeta = \frac{z}{h/2}$	1/2	1	1/2	1	1/2	1	1/2	1	1/2	1	1/2	1
		$t = 0$ to 1/20 hr											
3	w/h	0.0481		0.0457		0.0388		0.0280		0.146		0	
4	σ_r	2230	4460	2140	4280	1915	3830	1425	2850	800	1600	0	0
5	σ_θ	2230	4460	2180	4360	2020	4040	1765	3530	1410	2820	945	1890
6	S_r			700	1400	603	1207	362	724	64	128	−315	−630
7	S_θ			740	1480	708	1417	702	1404	674	1328	630	1260
8	$\sqrt{3\bar{J}_2}$	2230	4460	2160	4320	1970	3940	1620	3240	1225	2450	945	1860
9	$e^B\sqrt{3\bar{J}_2}$	72	5200	63	3980	44	1920	31	500	105	109	6.1	37.6
10	$\sqrt{I_2''} = \sqrt{\frac{3}{4}}\, A \exp(B\sqrt{3\bar{J}_2})\, t^k$			19.6	1260	14	607	9.8	158	3.4	34.5	1.9	12
11	$[S_r^2 + S_\theta^2 + S_r S_\theta]^{1/2}$				2495		2276		1874		1452		1092
12	$\Delta e_r'' \times 10^{-7}$	13.2	950	11.0	710	7.4	323	3.8	61	0.3	3.0	−1.1	−6.8
13	$\Delta e_\theta'' \times 10^{-7}$	13.2	950	11.7	747	8.7	379	7.4	118	3.2	32	2.2	13.7
14	$\Delta\sigma_r''$	9.8	706	8.3	533	5.7	191	3.4	55	0.7	7.3	−2.5	−1.6

15	$\Delta\sigma_\theta''$	9.8	706	8.5	548	6.2	223	4.9	78	1.8	19	1.1	6.7
16	$\int \Delta e_r'' z\,dz \Big/ \left(\frac{h}{2}\right)^2$		196×10^{-7}		151×10^{-7}		74.1×10^{-7}		16.5×10^{-7}		0.50×10^{-7}		-2.78×10^{-7}
17	$\int \Delta e_\theta'' z\,dz \Big/ \left(\frac{h}{2}\right)^2$		196×10^{-7}		158×10^{-7}		87.4×10^{-7}		32×10^{-7}		107×10^{-7}		5.59×10^{-7}
18	$\int \Delta e_r'' + \mu\, \Delta e_\theta'' z\,dz \left(\frac{h}{2}\right)^2$		256×10^{-7}		200×10^{-7}		100×10^{-7}		26.7×10^{-7}		3.7×10^{-7}		$-.28 \times 10^{-7}$
19	$\frac{\partial}{\partial r}(18)$			-280×10^{-7}		-500×10^{-7}		-367×10^{-7}		-115×10^{-7}		$-24 \times 10/^{7}$	
			0		-390×10^{-7}		-434×10^{-7}		-241×10^{-7}		-69.5×10^{-7}		-24×10^{-7}
20	$r(19)$		0		-78×10^{-7}		-173×10^{-7}		-145×10^{-7}		-55.6×10^{-7}		-24×10^{-7}
21	$(1-\mu)\int (\Delta e_r'' - \Delta e_\theta'') z\,dz \Big/ \left(\frac{h}{2}\right)^2$		0		-5.3×10^{-7}		-8.8×10^{-7}		-10.8×10^{-7}		-7.1×10^{-7}		-2.0×10^{-7}
22	(20) + (21)		0		-83×10^{-7}		-182×10^{-7}		-156×10^{-7}		-62×10^{-7}		-26.0×10^{-7}
23	$\frac{2\pi r Q'}{(h/2)^2} \times 10^{-7}$		0		300		652		558		216		78
24	P'			18.9		22.1		−5.8		−21.4		−8.6	
25	$\Delta\sigma_r'$	78	157	58	115	29	58	8	16	0.5	1	−.25	−1.6
26	$\Delta\sigma_\theta'$	78	157	67	134	48	95	30	60	19	39	16	0
27	$\sigma_r + \Delta\sigma_r' - \Delta\sigma_r''$	2298	3911	2190	3892	1936	3697	1431	2811	800	1594	0	0
28	$\sigma_\theta + \Delta\sigma_\theta' - \Delta\sigma_\theta''$	2298	3911	2239	3946	2062	3912	1790	3512	1427	2840	960	1912

Table 10.3 Creep Stresses and Deflections of a Circular Plate Under Uniform Loading (*continued*)

No.		(1)		(2)		(3)		(4)		(5)		(6)	
29	$\Delta w/h$	0.00102		0.00094		0.00073		0.00049		0.00023		0	
30	w/h	0.04912		0.04664		0.03953		0.02849		0.01433		0	
31	$\sqrt{3\bar{J}_2}$	2298	3911	2214	3910	2002	3810	1640	3220	1240	2460	960	1914
32	t_0 in hrs†	0.041	0.245	0.0416	0.164	0.035	0.074	0.077	0.0525	0.0500	0.0704	0.0465	0.0465
					$t = 1/20$ to 1 hr								
33	$t = t_0 + \Delta t$	0.991	1.195	0.992	1.114	0.985	1.024	1.027	1.002	1.00	1.030	0.9965	0.9965
34	$t^{0.66}$	0.994	1.125	0.995	1.074	0.990	1.016	1.018	1.002	1.00	1.013	0.9976	0.9976
35	$\sqrt{I_2'' + \Delta I_2''} \times 10^7$			183	5350	121	4000	63.1	1276	28.5	305	16.6	104
36	S_r			714	1258	604	1161	359	704	58	116		
37	S_θ			763	1343	728	1376	720	1404	685	1362		
38	$\Delta e_r'' \times 10^7$	111	2180	91	2440	57	1790	229	605	2.0	22	−8.5	−52
39	$\Delta e_\theta'' \times 10^7$	111	2180	98	2606	66	2120	445	1208	23	259	17.0	104
40	$\Delta\sigma_r''$	82	1620	69	1842	45	1390	207	553	5.1	56.6	−1.9	−12
41	$\Delta\sigma_\theta''$	82	1620	72	1910	47	1616	294	794	13.5	152	8.3	51
42	$\int \Delta e_r'' z\,dz \Big/ \left(\frac{h}{2}\right)^2$	692×10^{-7}		722×10^{-7}		514×10^{-7}		180×10^{-7}		8.0×10^{-7}		-23×10^{-7}	

43	$\int \Delta e_\theta'' z\, dz \Big/ \left(\frac{h}{2}\right)^2$	692×10^{-7}		771×10^{-7}		605×10^{-7}		357×10^{-7}		95×10^{-7}		45×10^{-7}	
44	$[(42) + \nu\,(43)] \times 10^6$	900		953		696		287		36.5		-10	
45	$\frac{\partial}{\partial r}(44) \times 10^7$		265		-1285		-2045		-1250		-233		
				-510		-1665		-1648		-742		-233	
46	$r(45) \times 10^7$			-510		-666		-990		-593		-233	
47	$(1-\nu)([42) - (43)] \times 10^7$			-34.3		-63.7		-124		-61		-47.6	
48	$[(46) + (47)] \times 10^7$			-136		-729.7		-1114		-654		-281	
49	$\frac{Q' r 2\pi}{(h/2)^2} \times 10^7$			489		2620		4000		2340		1010	
50	$\Delta \frac{Q' r 2\pi}{(h/2)^2} \times 10^7$		489		2131		1380		-1660		-1330		
51	P'		30.6		133		86.4		-104		-83.2		
52	$\Delta\sigma_r'$	348	696	315	629	220	440	88	175	11.0	22.0	-1.9	-12
53	$\Delta\sigma_\theta'$	348	696	330	660	277	554	199	397	133	265	94	187
54	$\sigma_r = (27) - (40) + (50)$	2564	2987	2436	2656	2113	2747	1311	2434	806	1559	0	0
55	$\sigma_\theta = (28) - (41) + (53)$	2564	2987	2497	7696	2292	2850	1295	3115	1546	2953	1046	2048
56	$\Delta w/h$	0.006		0.0056		0.0046		0.0031		0.0015		0	
57	w/h	0.0551		0.0522		0.0441		0.0316		0.0163		0	

† The t_0 was obtained by inserting the value of $\sqrt{I''}$ given by 10 and $\sqrt{3\bar{J}_2}$ given by 31 into Equation 10.5.25 to solve for t. This is t_0.

For the first time interval, stress is proportional to z and, as in Equation (10.5.30), we have

$$\int (e_i'')z\,dz = 2\int_0^{h/2} e_i''z\,dz$$
$$= 2(e_i'')_{h/2}\cdot\left(\frac{h}{2}\right)^2 \frac{1}{(B\sqrt{3\bar{J}_2})_{h/2}}\left[1-\frac{1}{(B\sqrt{3\bar{J}_2})_{h/2}}\cdot\left\{1-\frac{1}{\exp(B\sqrt{3\bar{J}_2})_{h/2}}\right\}\right]$$

where i denotes xx, yy, or xy; that is, e_i'' is either e''_{xx}, e''_{yy}, or e''_{xy}. The integrals for the first time increment are evaluated at different grid points. Along the simply supported edges $x = 0$ and $x = a$, $w = 0$ and $M_x = 0$. Since $w = 0$ along the y axis at $x = 0$ and $x = a$, $\partial w/\partial y$ and $\partial^2 w/\partial y^2$ must also be zero. Since $M_x = D(\partial^2 w/\partial x^2 + v\,\partial^2 w/\partial y^2) = 0$ along $x = 0$ and $x = a$, it then follows that $\partial^2 w/\partial x^2 = 0$ at $x = 0$ and $x = a$. Hence, from Equations (10.1.13), (10.5.39), and (10.2.7), we have $M_{x_I} = M_{y_I} = 0$ along the simply supported edges $x = 0$ and $x = a$. Similarly, M_{y_I} and M_{x_I} are zero along the simply supported edges $y = 0$ and $y = a$. Along the edges, M_{xy} and hence M_{xy_I} may exist.

The equivalent loads $\Delta\bar{q}$ for the first and subsequent time intervals are obtained from Equation (10.5.10). The deflection of a rectangular plate with simply supported edges caused by a single load P acting at (ξ, η), as shown in Figure 10.24, has been given in the double series of Equation (10.1.35). The deflection can also be expressed in single series as[4]

$$w = \frac{Pa^2}{\pi^3 D}\sum_{m=1}^{\infty}\left(1+\beta_m\coth\beta_m - \frac{\beta_m y_1}{b}\coth\frac{\beta_m y_1}{b} - \frac{\beta_m\eta}{b}\coth\frac{\beta_m\eta}{b}\right)$$
$$\times\frac{\sinh\dfrac{\beta_m\eta}{b}\sinh\dfrac{\beta_m y_1}{b}\sin\dfrac{m\pi\xi}{a}\sin\dfrac{m\pi x}{a}}{m^3\sinh\beta_m} \qquad (10.5.42)$$

for $y > \eta$, where

$$\beta_m = \frac{m\pi b}{a} \qquad y_1 = b - y$$

In the region of $y < \eta$, the quantities y_1 and η are replaced by y and $b - \eta$ respectively. These series solutions are lengthy. The fourth-order differential equation $D\,\nabla^4 w = q$ for different locations of a concentrated load P has been numerically solved by the method of finite differences for simply supported rectangular plates. Using finite differences the influence coefficients of the deflection w at points (x, y) caused by a unit load applied at

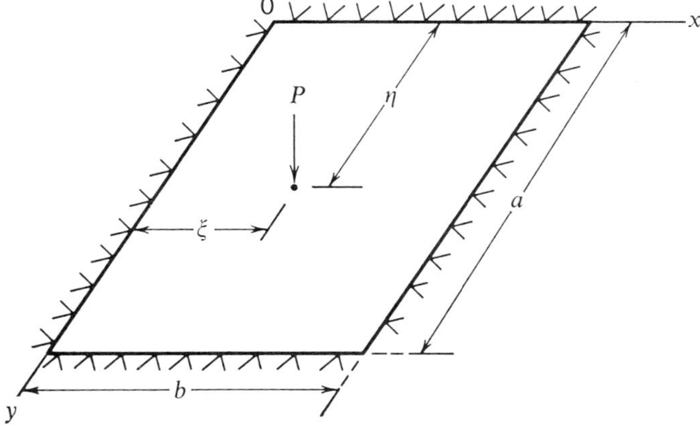

Figure 10.24

(ξ, η) have been given by Tuma, Havner, and French.[17] These tabulated coefficients were used in the numerical calculations. Replacing the distributed load $\Delta\bar{q}$ by concentrated loads $\Delta\bar{P}_{mn}$'s at the (m, n) grid points, the deflections at a point caused by different $\Delta\bar{P}_{mn}$'s are superposed to give the incremental deflection caused by $\Delta\bar{q}$ over the plate. From the increment of deflection, Δw, the incremental moments ΔM_x, ΔM_y, and ΔM_{xy} and then the corresponding incremental stresses denoted by $\Delta\tau'_{xx}$, $\Delta\tau'_{yy}$, and $\Delta\tau'_{xy}$ are obtained from Equation (10.1.13). The stress relaxation due to the incremental creep strains, from Equation (10.2.2), are

$$\begin{aligned}
\Delta\tau''_{xx} &= \frac{E}{1 - \nu^2}(\Delta e''_{xx} + \nu\,\Delta e''_{yy}) \\
\Delta\tau''_{yy} &= \frac{E}{1 - \nu^2}(\Delta e_y'' + \nu\,\Delta e_x'') \qquad (10.5.43) \\
\Delta\tau''_{xy} &= 2G\,\Delta e''_{xy}
\end{aligned}$$

With the initial stresses (at zero time) denoted by τ_{xx_0}, τ_{yy_0}, and τ_{xy_0}, the stresses at the end of the first time interval are

$$\begin{aligned}
\tau_{xx} &= \tau_{xx_0} + \Delta\tau'_{xx} - \Delta\tau''_{xx} \\
\tau_{yy} &= \tau_{yy_0} + \Delta\tau'_{yy} - \Delta\tau''_{yy} \qquad (10.5.44) \\
\tau_{xy} &= \tau_{xy_0} + \Delta\tau'_{xy} - \Delta\tau''_{xy}
\end{aligned}$$

where $\Delta\tau'_{xx}$, $\Delta\tau'_{yy}$, and $\Delta\tau'_{xy}$ are the stresses due to the incremental equivalent load. These stresses are assumed to remain constant during the second time interval. The resulting incremental creep strains for this interval are

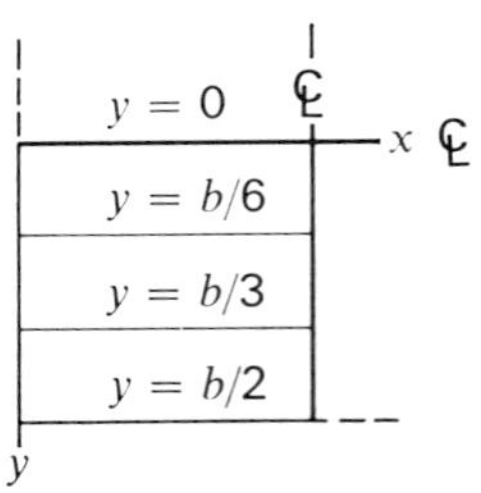

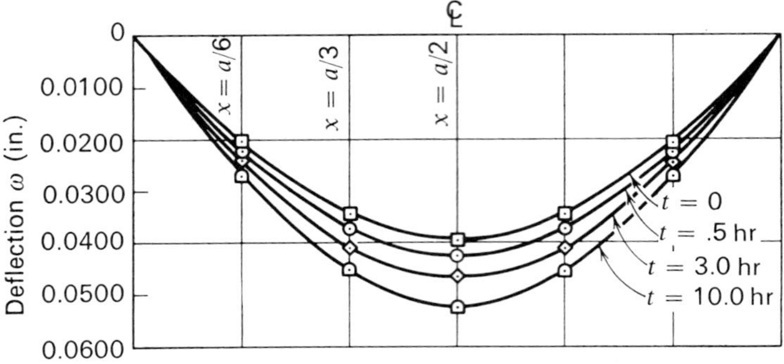

Figure 10.25. Deflection vs. x *at* y = 0 *for various times.*

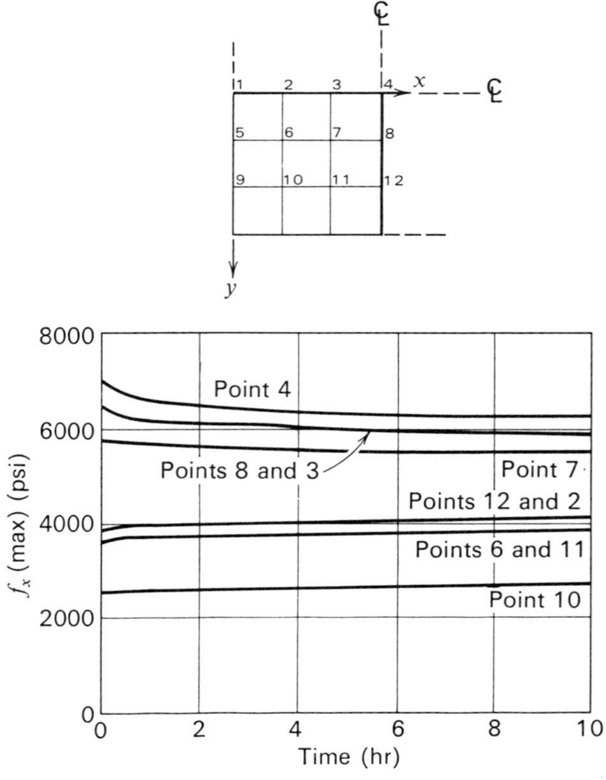

Figure 10.26. Extreme fiber stress vs. time.

calculated by Equations (10.5.25) and (10.5.26), as in the procedure of the previous example, with Equation (10.5.28) replaced by

$$\Delta\sqrt{I_2''} = [(e''_{xx} + \kappa S_{xx})^2 + (e''_{yy} + \kappa S_{yy})^2 + (e''_{xx} + \kappa S_{xx})(e''_{yy} + \kappa S_{yy}) + (e''_{xy} + \kappa S_{xy})^2]^{1/2} - [e''^2_{xx} + e''^2_{yy} + e''_{xx}e''_{yy} + e''^2_{xy}]^{1/2}$$

The $\Delta\bar{q}$'s for the different time intervals may be evaluated by the method used in previous examples. The calculated deflections at different grid points at different instants are shown in Figures 10.23 and 10.25. The variations of the extreme fiber stresses at different grid points are shown in Figure 10.26. It is seen that the stresses in regions which were initially highly stressed are partly relieved, while the stresses in the regions with small initial stress are increased as creep deformation proceeds.

Although in the examples just presented the circular plate and the rectangular plate were both simply supported around their boundaries, the method is easily applied to plates with any combination of boundary edge conditions by the superposition method of Section 10.1.

10.6 Elastoplastic Bending of Plates

Consider a simply supported circular plate of uniform thickness under a uniform lateral load. Because of symmetry, $\tau_{r\theta} = 0$, $e_{r\theta} = 0$, and $\partial w/\partial\theta = 0$.

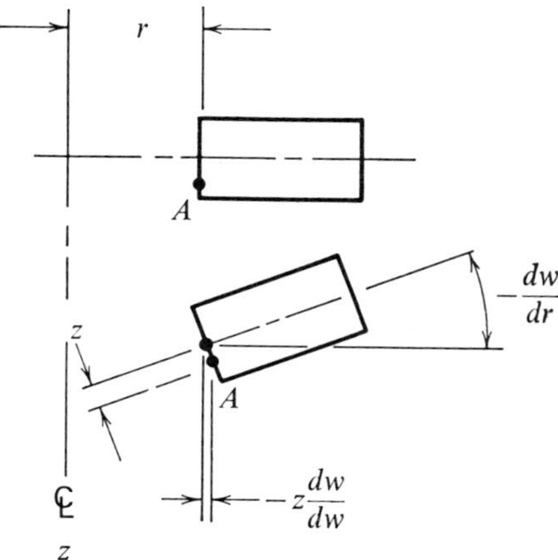

Figure 10.27

The deflection w is assumed to be small enough that the radial displacement of the middle surface is negligible. In such case u_0 is taken to be zero. Referring to Figure 10.27, we observe the following relations:

$$u = -z \frac{dw}{dr} \tag{10.6.1}$$

$$e_{rr} = \frac{du}{dr} = -z \frac{d^2w}{dr^2}$$
$$e_{\theta\theta} = \frac{u}{r} = -\frac{z}{r}\frac{dw}{dr} \tag{10.6.2}$$

If the plate is loaded beyond the elastic limit, plastic strain e_{rr}^p and $e_{\theta\theta}^p$ will exist. The stress-strain relations are

$$\tau_{rr} = \frac{E}{1-\nu^2}\left[e_{rr} + \nu e_{\theta\theta} - (e_{rr}^p + \nu e_{\theta\theta}^p)\right]$$
$$\tau_{\theta\theta} = \frac{E}{1-\nu^2}\left[e_{\theta\theta} + \nu e_{rr} - (e_{\theta\theta}^p + \nu e_{rr}^p)\right] \tag{10.6.3}$$
$$\tau_{r\theta} = 2G(e_{r\theta} - e_{r\theta}^p)$$

where the superscript p on strain denotes plastic strain.

Within the elastic limit, the deflection and stresses are calculated by Equations (10.5.19) and (10.5.21). Plastic strain, once it occurs, will cause an equivalent lateral load $\bar{q}$ as sh own in Equation (10.5.10),

$$\bar{q} = \frac{-E}{1-\nu^2}\frac{1}{r}\frac{d}{dr}\left\{r\frac{d}{dr}\int(e_{rr}^p + \nu e_{\theta\theta}^p)z\,dz + (1-\nu)\int(e_{rr}^p - e_{\theta\theta}^p)z\,dz\right\} \tag{10.6.4}$$

and an equivalent edge moment at the boundary as given by Equation (10.5.11),

$$\overline{M}_r = \frac{E}{1-\nu^2}\int(e_{rr}^p + \nu e_{\theta\theta}^p)z\,dz \tag{10.6.5}$$

Let τ_{rr_0} and $\tau_{\theta\theta_0}$ be the stresses caused by the applied load q and let τ_{rr}^p and $\tau_{\theta\theta}^p$ (corresponding to τ'_{rr} and $\tau'_{\theta\theta}$ in the previous section) be the stresses caused by the equivalent load $\bar{q}$ and edge moment $\overline{M}_r$. Then the stresses at any point are

$$\tau_{rr} = \tau_{rr_0} + \tau_{rr}^p - \frac{E}{1-\nu^2}(e_{rr}^p + \nu e_{\theta\theta}^p)$$
$$\tau_{\theta\theta} = \tau_{\theta\theta_0} + \tau_{\theta\theta}^p - \frac{E}{1-\nu^2}(e_{\theta\theta}^p + \nu e_{rr}^p) \tag{10.6.6}$$

The solution of stresses τ_{rr}^p and $\tau_{\theta\theta}^p$ for a given $\bar{q}$ and $\overline{M}_r$ for an axisymmetrically loaded circular plate may be obtained from Equations (10.5.35) and (10.5.37). For a known distribution of e_{rr}^p and $e_{\theta\theta}^p$, Equations (10.6.6) give the solution of the stresses τ_{rr} and $\tau_{\theta\theta}$ at different points.

The incremental stress-strain relations in the plastic range for plane stress are given by Equation (4.5.9) as

$$\frac{de_{rr}^p}{\frac{2}{3}\tau_{rr} - \frac{1}{3}\tau_{\theta\theta}} = \frac{de_{\theta\theta}^p}{\frac{2}{3}\tau_{\theta\theta} - \frac{1}{3}\tau_{rr}} = \frac{3}{2}\frac{d(e^p)^*}{\sigma^*} \tag{10.6.7}$$

$$de_{rr}^p = \frac{2\tau_{rr} - \tau_{\theta\theta}}{2\sigma^*}\, d(e^p)^* \qquad de_{\theta\theta}^p = \frac{2\tau_{\theta\theta} - \tau_{rr}}{2\sigma^*}\, d(e^p)^* \tag{10.6.8}$$

where

$$(e^p)^* = \sqrt{\tfrac{2}{3}}[(e_{rr}^p)^2 + (e_{\theta\theta}^p)^2 + (e_{rr}^p + e_{\theta\theta}^p)^2]^{1/2} \tag{10.6.9}$$

$$\sigma^* = \sqrt{\tfrac{1}{2}}[\tau_{rr}^2 + \tau_{\theta\theta}^2 + (\tau_{rr} - \tau_{\theta\theta})^2]^{1/2} \tag{10.6.10}$$

The functional relation f between $(e^p)^*$ and σ^* is assumed to be the same as for the tensile test; that is, if $e_{11}^p = f(\tau_{11})$, then $(e^p)^* = f(\sigma^*)$. Hence

$$d(e^p)^* = f'(\sigma^*)\, d\sigma^*$$

and

$$d\sigma^* = \frac{(de^p)^*}{f'(\sigma^*)} = \frac{d\sigma^*}{d(e^p)^*}(de^p)^* \tag{10.6.11}$$

where

$$\frac{d\sigma^*}{(de^p)^*} = \frac{d\tau_{11}}{de_{11}^p}$$

is the slope of the tensile stress vs. plastic strain curve. Equation (10.6.11) gives the incremental plastic stress-strain relation at all points with increasing σ^*. From Equation (10.6.10), we have

$$d\sigma^* = \frac{2\tau_{rr} - \tau_{\theta\theta}}{2\sigma^*}\, d\tau_{rr} + \frac{2\tau_{\theta\theta} - \tau_{rr}}{2\sigma^*}\, d\tau_{\theta\theta} \tag{10.6.12}$$

Writing Equation (10.6.6) in incremental differential form, we obtain

$$\begin{aligned} d\tau_{rr} &= d\tau_{rr_0} + d\tau_{rr}^p - \frac{E}{1 - \nu^2}(de_{rr}^p + \nu\, de_{\theta\theta}^p) \\ d\tau_{\theta\theta} &= d\tau_{\theta\theta_0} + d\tau_{\theta\theta}^p - \frac{E}{1 - \nu^2}(de_{\theta\theta}^p + \nu\, de_{rr}^p) \end{aligned} \tag{10.6.13}$$

With $\tau_{\theta\theta}$, τ_{rr}, and σ^* known at all points, de^p_{rr} and $de^p_{\theta\theta}$ can be expressed in terms of $d(e^p)^*$ in Equation (10.6.8). The $d\tau^p_{rr}$ and $d\tau^p_{\theta\theta}$ depend on the de^p_{rr} and $de^p_{\theta\theta}$ at all points and hence depend on the $d(e^p)^*$ at all points. Then, from Equation (10.6.13), we have $d\tau_{rr}$, $d\tau_{\theta\theta}$, and hence $d\sigma^*$ may be expressed in terms of $d(e^p)^*$ at all points. Therefore, at each point where σ^* is increasing, there is one unknown $d(e^p)^*$ and one incremental stress-strain relation, Equation (10.6.11). Hence the incremental plastic strain $d(e^p)^*$'s may be completely determined by solution of a set of simultaneous equations.

For numerical calculation the following procedure is used. Because of the axisymmetry of a circular plate under a uniform load, deflection, moments, stresses, and strains will be independent of θ. Hence it is only necessary to consider a single rz-section of the plate, all other rz-sections being equivalent. Now divide the rz section of the plate by vertical and horizontal grid lines. The grid point m, n is defined as the intersection of the mth horizontal line with nth vertical line, where the different horizontal lines correspond to different z values and the different vertical lines are associated with different values along the radius. Let $a_{rr_{pqmn}}$ denote the radial stress τ_{rr} at grid location (pq) due to a unit plastic effective strain e^{p*}_{mn} at the grid location (m, n). The grid location (m, n) denotes a circle of radius m located at a distance n from the middle plane of the plate; therefore $a_{rr_{pqmn}}$ is the radial stress at any point on the (pq) circle due to a unit plastic effective strain at every point on the (m, n) circle. Then

$$\begin{aligned} d\tau^p_{rr_{pq}} &= a_{rr_{pqmn}}\, d(e^p)^*_{mn} \\ d\sigma^p_{\theta\theta_{pq}} &= a_{\theta\theta_{pqmn}}\, d(e^p)^*_{mn} \end{aligned} \tag{10.6.14}$$

where $a_{\theta\theta_{pqmn}}$ is the circumferential stress at (pq) due to unit e^{p*}_{mn} at (m, n) and the repetition of the subscript mn denotes summation over all the grid points.

From Equations (10.6.8), (10.6.13), and (10.6.14), we have

$$\begin{aligned} d\tau_{rr_{pq}} &= d\tau_{rro_{pq}} + a_{rr_{pqmn}}\, d(e^p)^*_{mn} \\ &\qquad - \frac{E}{1-\nu^2}\left[\frac{(2\tau_{rr}-\tau_{\theta\theta}) + \nu(2\tau_{\theta\theta}-\tau_{rr})}{2\sigma^*}\, d(e^p)^*\right]_{pq} \\ d\tau_{\theta\theta_{pq}} &= d\tau_{\theta\theta o_{pq}} + a_{\theta\theta_{pqmn}}\, d(e^p)^*_{mn} \\ &\qquad - \frac{E}{1-\nu^2}\left[\frac{(2\tau_{\theta\theta}-\tau_{rr}) + \nu(2\tau_{rr}-\tau_{\theta\theta})}{2\sigma^*}\, d(e^p)^*\right]_{pq} \end{aligned} \tag{10.6.15}$$

Substituting Equation (10.6.15) into Equation (10.6.12) yields $d\sigma^*_{pq}$ expressed as a linear combination of de^{p*}_{mn} terms at all points. This expression may be written as

$$d\sigma^*_{pq} = d\sigma^*_{0_{pq}} + \psi_{pqmn}\, d(e^p)^*_{mn} \qquad (10.6.16)$$

where ψ_{pqmn} varies with the loading state $(\sigma_r, \sigma_\theta, \sigma^*)$. From Equation (10.6.11), we have

$$d\sigma^*_{0_{pq}} + \psi_{pqmn}\, d(e^p)^*_{mn} = \left[\frac{d\sigma^*}{d(e^p)^*}\, d(e^p)^*\right]_{pq} \qquad (10.6.17)$$

For actual calculation, the differential $d(e^p)^*$ is replaced by a finite increment of plastic strain $\Delta(e^p)^*$. Equation (10.6.17) then becomes

$$\Delta\sigma^*_{0_{pq}} + \psi_{pqmn}\, \Delta(e^p)^*_{mn} = \left[\frac{\Delta\sigma^*}{\Delta(e^p)^*}\, \Delta(e^p)^*\right]_{pq}$$

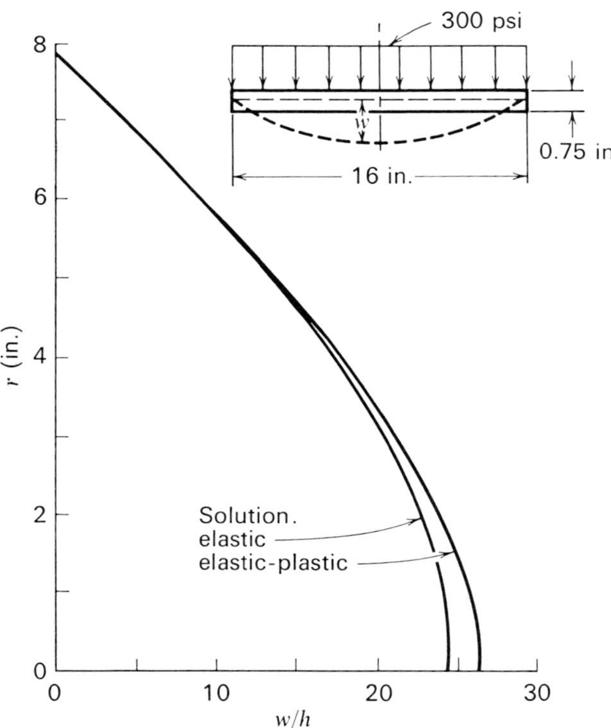

Figure 10.28

This procedure was used by Lackman[18] in his calculation of stresses in a 2024-T4 aluminum simply supported circular plate under uniform lateral load. The diameter of the plate was 16 in., with a thickness of $\frac{3}{4}$ in. After the plate reached the proportional limit, small increments of uniform lateral load were applied. The incremental stresses and deflections were calculated by the above procedure until a uniform lateral load of 300 psi was reached.

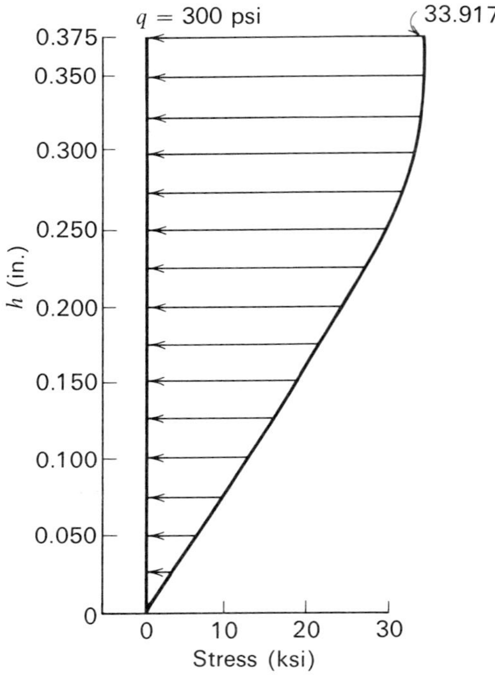

Figure 10.29

The resulting deflections and stresses are shown in Figures 10.28 and 10.29.

This method gives solutions satisfying equilibrium, continuity, and the Reuss-Mises incremental polyaxial stress-strain relation. A similar procedure has been used for the elastoplastic analysis of bending of rectangular plates. However, the computation is much lengthier. The results of elastic-plastic bending of a rectangular plate are shown in Reference 19.

For other works on small deformation of plastic plates, see References 20 through 23. For bending of elastoplastic circular plates with large deflection, Naghdi[24] has presented an iteration method. His calculated deflections agree very well with experimental results.

REFERENCES

1. Kirchhoff, G., "Vorlesungen über Mathematische Physik," *Mechanik*, Vol. 1, 1876.
2. Wang, C. T., *Applied Elasticity*, McGraw-Hill, New York, 1953.
3. Love, A. E. H., *The Mathematical Theory of Elasticity*, 4th Ed., Dover, New York, p. 461, 1944.
4. Timoshenko, S., and S. Woinowsky-Krieger, *Theory of Plates and Shells*, McGraw-Hill, New York, 1959.
5. Duhamel, J. M. C., *Memoires ... par divers Savants*, Vol. 5, p. 440, Paris, 1938.
6. *Thermal Stresses and Deflections in Rectangular Panels*, ASD-TR-61-537, Part I, December 1962.
7. Durelli, A. J., E. A. Phillips, and C. H. Tsao, *Introduction to the Theoretical and Experimental Analyses of Stress and Strain*, McGraw-Hill, New York, 1958.
8. Brull, M. A., *A Structural Theory Incorporating the Effect of Time-Dependent Elasticity*, Ph.D. Dissertation at University of Michigan, 1952, *Proc. First Midwestern Conference on Solid Mechanics*, 1953.
9. Kachanov, L. M., "Certain Problems in the Theory of Creep" (in Russian), *Gos Izdat*, Tekh-Teor. Lit. Leningrad, 1949.
10. Malinin, N. N., "Continuous Creep and Round Symmetrically Loaded Plates" (in Russian), Moskov, *Vyssheye Technicheskoe Uchillishche Tru'da* 26 : 221, 1953.
11. Sanders, J. L., H. G. McComb, Jr., and F. R. Schlechte, *A Variational Theorem for Creep with Applications to Plates and Columns*, NACA TN, 4003, May 1957.
12. Shanley, F. R., *Weight Strength Analysis of Aircraft Structures*, McGraw-Hill, New York, pp. 275–282, 317, 1952.
13. Johnson, A. E., J. Henderson, and V. Mather, "Creep Under Changing Complex Stress Systems," *The Engineer*, **206**, 209, 1958.
14. Lin, T. H., "Bending of a Plate with Nonlinear Strain Hardening Creep," *Proc. International Union of Theoretical and Applied Mechanics, Creep in Structures*, Springer-Verlag, Berlin, Germany, pp. 215–228, 1962.
15. Gebhart, C. E., *Bending of Symmetrically Loaded Circular Plates with Arbitrary Creep Characteristics*, M.S. Thesis, UCLA Engineering Department, 1961.
16. Lin, T. H., and J. K. Ganoung, "Bending of Rectangular Plates with Nonlinear Creep," *International Journal of Mech. Science*, **6**, 337–348, 1964.
17. Tuma, J. J., K. S. Havner, and S. E. French, Jr., *Analysis of Flat Plate by the Algebraic Carryover Method*, Oklahoma State University Engineering Experiment Station, Publications 118 and 119 (1960–1961).
18. Lackman, L. M., "Circular Plates Loaded into the Plastic Region," *J. Eng. Mech. Division*, ASCE, **90**, 4155–4157, 1964.
19. Lin, T. H., and E. Ho, "Elasto-Plastic Bending of a Rectangular Plate," *Eng. Mech. J.*, ASCE. Vol. 94 No. EMI, p. 199, 1968.
20. Ilyushin, A. S., "Some Problems in the Theory of Plastic Deformations," *Prikl. Mat. Mekh.*, **7**, 245–272, 1943.

21. Sokolvsky, W. W., "Elastic-Plastic Bending of Circular and Annular Plates," *Prikl. Mat. Mekh.*, **8**, 141–166, 1944.
22. Trifan, D., *On the Plastic Bending of Circular Plates Under Uniform Transverse Loads*, Tech. Report No. 8, Brown University, Providence, R.I., 1948.
23. Hwang, C., "Plastic Bending of a Work-Hardening Circular Plate with Clamped Edge," *J. Aerospace Sci.*, **27**, 815–820, 1960.
24. Naghdi, P. M., "Bending of Elasto-Plastic Circular Plates with Large Deflection," *J. Appl. Mech.*, **19**, 293–300, 1952.

chapter

11

Inelastic Plates under Combined Lateral and In-Plane Loadings

Introduction. In Chapter 10 inelastic plates under pure bending were considered. This chapter treats inelastic plates under both lateral and in-plane loads. Section 11.1 presents the derivation of the governing equilibrium equations for plates with lateral and in-plane loads. Section 11.2 reviews the elastic analysis of plates with initial curvature under these loads. The analysis of viscoelastic rectangular plates under uniform edge compression is given in Section 11.3. Section 11.4 derives the differential equation of equilibrium in terms of displacements of inelastic circular plates subject to uniform edge compression and uniform lateral load. Sections 11.5 and 11.6 present methods for analyzing circular plates with nonlinear creep characteristics and elastoplastic characteristics, respectively, under lateral and in-plane loads. A method of analysis for rectangular plates with nonlinear creep behavior subject to combined edge compression and lateral load is given in Section 11.7. In the last section the differential equations for large-deflection theory of inelastic plates are derived. Illustrative examples are given.

11.1 Condition of Equilibrium

In Section 10.1, the sectional stresses τ_{xx}, τ_{yy}, and τ_{xy} across the thickness of a plate were represented by sectional moments M_x, M_y, and M_{xy} about axes tangent to the middle surface of the plate and sectional forces N_x, N_y, and N_{xy} acting in the middle surface. When the plate is deflected

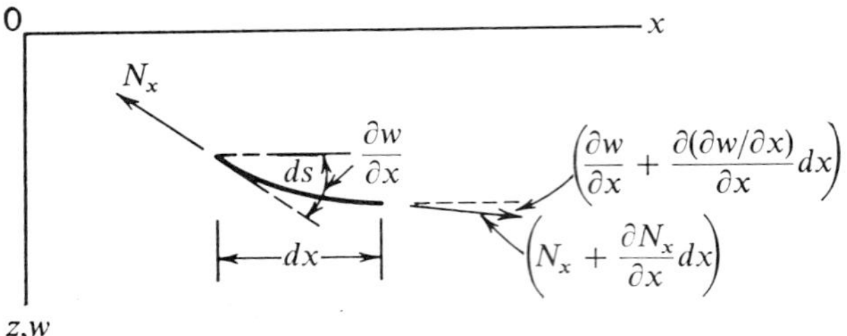

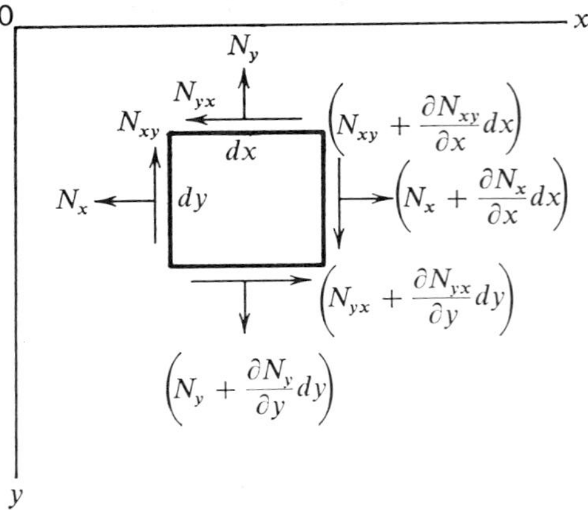

Figure 11.1 A deflected surface with in-plane stresses.

into a curved surface, the sectional forces are no longer horizontal, and have vertical components. These vertical components induce additional lateral load. Consider a differential element of the surface with projection $dx\,dy$ on the xy plane and projection ds on the xz plane, as shown in the lower and upper portions of Figure 11.1. Assuming no body force, we see that equilibrium along the x direction requires

$$\frac{\partial N_x}{\partial x} + \frac{\partial N_{yx}}{\partial y} = 0 \tag{11.1.1}$$

Similarly for equilibrium along the y direction, assuming zero body force, we have

$$\frac{\partial N_y}{\partial y} + \frac{\partial N_{xy}}{\partial x} = 0 \tag{11.1.2}$$

The sectional forces N_x at the left section and $N_x + (\partial N_x/\partial x)\,dx$ at the right (see Figure 11.1) have a net downward z component of

$$-N_x \frac{\partial w}{\partial x}\,dy + \left(N_x + \frac{\partial N_x}{\partial x}\,dx\right)\left(\frac{\partial w}{\partial x} + \frac{\partial^2 w}{\partial x^2}\,dx\right)dy \qquad (11.1.3)$$

Similarly, the z component due to N_y and $N_y + (\partial N_y/\partial y)\,dy$ is

$$-N_y \frac{\partial w}{\partial y}\,dx + \left(N_y + \frac{\partial N_y}{\partial y}\,dy\right)\left(\frac{\partial w}{\partial y} + \frac{\partial^2 w}{\partial y^2}\,dy\right)dx \qquad (11.1.4)$$

Referring to Figure 11.2, we see that the shearing forces N_{xy} on the left and $N_{xy} + (\partial N_{xy}/\partial x)\,dx$ on the right have a net z component of

$$-N_{xy}\,dy\left(\frac{\partial w}{\partial y}\right) + \left(N_{xy} + \frac{\partial N_{xy}}{\partial x}\,dx\right)\left(\frac{\partial w}{\partial y} + \frac{\partial^2 w}{\partial x\,\partial y}\,dx\right)dy \qquad (11.1.5)$$

Similarly, the z component given by N_{yx} and $N_{yx} + (\partial N_{yx}/\partial y)\,dy$ is

$$-N_{yx}\,dx\left(\frac{\partial w}{\partial x}\right) + \left(N_{yx} + \frac{\partial N_{yx}}{\partial y}\,dy\right)\left(\frac{\partial w}{\partial x} + \frac{\partial^2 w}{\partial x\,\partial y}\,dy\right)dx \qquad (11.1.6)$$

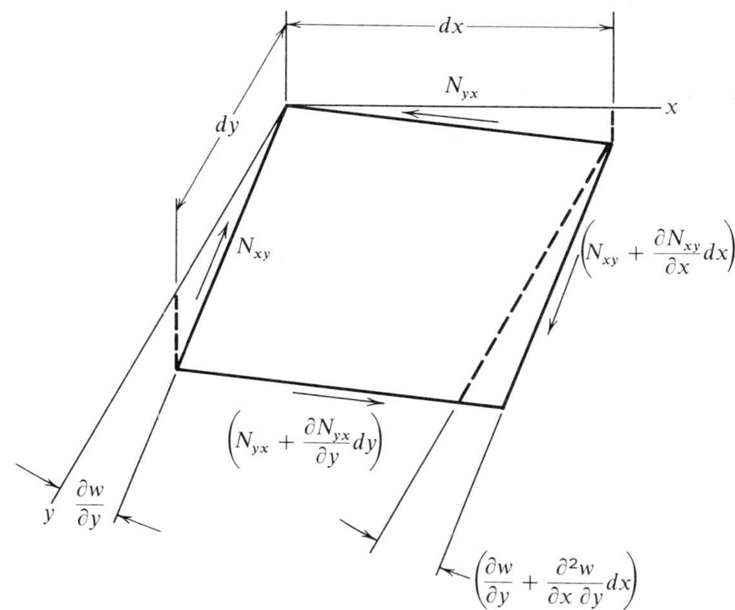

Figure 11.2. Vertical components caused by $\mathbf{N}_{xy}$.

Summing up the z components of N_x, N_y, N_{xy}, and N_{yx} from Equations (11.1.3) through (11.1.6), neglecting terms of high-order differentials, and denoting this sum by $q'\,dx\,dy$, we have

$$\begin{aligned} q' &= N_x \frac{\partial^2 w}{\partial x^2} + \frac{\partial N_x}{\partial x}\frac{\partial w}{\partial x} + N_y \frac{\partial^2 w}{\partial y^2} + \frac{\partial N_y}{\partial y}\frac{\partial w}{\partial y} \\ &\quad + N_{xy}\frac{\partial^2 w}{\partial x\,\partial y} + \frac{\partial N_{xy}}{\partial x}\frac{\partial w}{\partial y} + N_{yx}\frac{\partial^2 w}{\partial y\,\partial x} + \frac{\partial N_{yx}}{\partial y}\frac{\partial w}{\partial x} \\ &= N_x \frac{\partial^2 w}{\partial x^2} + N_y \frac{\partial^2 w}{\partial y^2} + (N_{xy} + N_{yx})\frac{\partial^2 w}{\partial x\,\partial y} \\ &\quad + \frac{\partial w}{\partial x}\left(\frac{\partial N_x}{\partial x} + \frac{\partial N_{yx}}{\partial y}\right) + \frac{\partial w}{\partial y}\left(\frac{\partial N_y}{\partial y} + \frac{\partial N_{xy}}{\partial x}\right) \end{aligned} \tag{11.1.7}$$

The last two terms vanish because of Equations (11.1.1) and (11.1.2). Since $N_{xy} = N_{yx}$, we have

$$q' = N_x \frac{\partial^2 w}{\partial x^2} + N_y \frac{\partial^2 w}{\partial y^2} + 2N_{xy}\frac{\partial^2 w}{\partial x\,\partial y} \tag{11.1.8}$$

This shows that, owing to sectional forces N_x, N_y, and N_{yx}, an additional lateral load q' is induced in the plate. Hence q in Equations (10.1.1) and (10.1.18) becomes $q + q'$ for plates with in-plane, as well as lateral loading q. Then Equation (10.1.17) becomes

$$\frac{\partial^2 M_x}{\partial x^2} + \frac{\partial^2 M_y}{\partial y^2} - 2\frac{\partial^2 M_{xy}}{\partial x\,\partial y} + q + q' = 0 \tag{11.1.9}$$

Equations (11.1.1), (11.1.2), and (11.1.9) are the *governing equilibrium equations* for thin plates with lateral and in-plane loadings.

11.2 Elastic Rectangular Plates with Small Initial Curvature

Substituting the expressions for sectional moments from Equations (10.1.11) into Equations (11.1.9) gives

$$\frac{\partial^4 w}{\partial x^4} + 2\frac{\partial^4 w}{\partial x^2\,\partial y^2} + \frac{\partial^4 w}{\partial y^4} - \frac{1}{D}\left(q + N_x\frac{\partial^2 w}{\partial x^2} + N_y\frac{\partial^2 w}{\partial y^2} + 2N_{xy}\frac{\partial^2 w}{\partial x\,\partial y}\right) = 0$$

The above equation may also be written as

$$D\nabla^4 w = q + q' \tag{11.2.1}$$

For plates with small initial curvature represented by an initial deflection w_0, which is small compared to the plate thickness, additional deflection

w is produced under load. The total deflection of the plate is $w_0 + w$, hence Equation (11.2.1) for a plate with initial deflection $(w_0 + w)$ becomes

$$\frac{\partial^4 w}{\partial x^4} + 2\frac{\partial^4 w}{\partial x^2\,\partial y^2} + \frac{\partial^4 w}{\partial y^4} = \frac{1}{D}\left(q + N_x\frac{\partial^2(w_0 + w)}{\partial x^2} + N_y\frac{\partial^2(w_0 + w)}{\partial y^2} + 2N_{xy}\frac{\partial^2(w_0 + w)}{\partial x\,\partial y}\right) \quad (11.2.2)$$

If a simply supported rectangular plate of length a and width b, with an initial deflection given by

$$w_0 = A_{11}\sin\frac{\pi x}{a}\sin\frac{\pi y}{b} \quad (11.2.3)$$

is subject to the uniformly distributed compressive forces $(-N_x)$ on the edges $x = 0$ and $x = a$ and $(-N_y)$ on the other two edges $y = 0$ and $y = b$, Equation (11.2.2) becomes

$$\frac{\partial^4 w}{\partial x^4} + 2\frac{\partial^4 w}{\partial x^2\,\partial y^2} + \frac{\partial^4 w}{\partial y^4} = \frac{1}{D}\left[A_{11}\left(\frac{N_x}{a^2} + \frac{N_y}{b^2}\right)\pi^2\sin\frac{\pi x}{a}\sin\frac{\pi y}{b} - N_x\frac{\partial^2 w}{\partial x^2} - N_y\frac{\partial^2 w}{\partial y^2}\right] \quad (11.2.4)$$

Let $w = B_{11}\sin(\pi x/a)\sin(\pi y/b)$, and substitute this into Equation (11.2.4), thereby obtaining

$$D\left(\frac{\pi^4}{a^4} + 2\frac{\pi^4}{a^2b^2} + \frac{\pi^4}{b^4}\right)B_{11}\sin\frac{\pi x}{a}\sin\frac{\pi y}{b} = A_{11}\pi^2\left(\frac{N_x}{a^2} + \frac{N_y}{b^2}\right)\sin\frac{\pi x}{a}\sin\frac{\pi y}{b} + \left(N_xB_{11}\frac{\pi^2}{a^2} + N_yB_{11}\frac{\pi^2}{b^2}\right)\sin\frac{\pi x}{a}\sin\frac{\pi y}{b}$$

Hence

$$B_{11} = \frac{A_{11}(N_x/a^2 + N_y/b^2)}{\pi^2D(1/a^2 + 1/b^2)^2 - N_x/a^2 - N_y/b^2} \quad (11.2.5)$$

The total deflection is given by

$$w_0 + w = (A_{11} + B_{11})\sin\frac{\pi x}{a}\sin\frac{\pi y}{b} = \frac{A_{11}\pi^2(1/a^2 + 1/b^2)^2D}{\pi^2D(1/a^2 + 1/b^2)^2 - N_x/a^2 - N_y/b^2}\sin\frac{\pi x}{a}\sin\frac{\pi y}{b} \quad (11.2.6)$$

If the initial deflection is

$$w_0 = A_{mn} \sin \frac{m\pi x}{a} \sin \frac{n\pi y}{b}$$

the differential equation (11.2.4) is satisfied by letting

$$w = B_{mn} \sin \frac{m\pi x}{a} \sin \frac{n\pi y}{b}$$

where

$$B_{mn} = \frac{A_{mn}(N_x m^2/a^2 + N_y n^2/b^2)}{\pi^2 D(m^2/a^2 + n^2/b^2)^2 - N_x m^2/a^2 - N_y n^2/b^2} \tag{11.2.7}$$

Since the differential equation (11.2.4) is linear, its solutions can be superposed. If the initial deflection is expressed as the doubly infinite series

$$w_0 = \sum_{m=1}^{\infty} \sum_{n=1}^{\infty} A_{mn} \sin \frac{m\pi x}{a} \sin \frac{n\pi y}{b} \tag{11.2.8}$$

the additional deflection will be

$$w = \sum_{m=1}^{\infty} \sum_{n=1}^{\infty} B_{mn} \sin \frac{m\pi x}{a} \sin \frac{n\pi y}{b} \tag{11.2.9}$$

where B_{mn} is given by Equation (11.2.7). For the case of uniaxial compression, $N_y = 0$ in Equation (11.2.7) and B_{mn} reduces to

$$B_{mn} = \frac{A_{mn} N_x}{(\pi^2 D/a^2)[m + (n^2/m)(a^2/b^2)]^2 - N_x} \tag{11.2.10}$$

No real plate, of course, is perfectly flat, and hence every plate has an initial curvature w_0. If w_0 vanishes at the edges of a rectangular plate, it may be represented by the double sine series of Equation (11.2.8). Generally, all the coefficients A_{mn} exist. However, for numerical calculation, only a few of the more dominant coefficients are taken to be nonzero. Equation (11.2.10) shows that B_{mn} increases with N_x. In the limit as N_x approaches $(\pi^2 D/a^2)(1 + a^2/b^2)^2$, we see that B_{11} becomes infinitely large. This value of N_x is the critical compressive load of the rectangular plate. It is the same as the critical buckling load of a perfectly flat plate.[1] Hence in the elastic range initial curvature does not change the critical load of a plate. In a similar way, critical values for plates with any combination of N_x and N_y may be determined.

11.3 Viscoelastic Plate with Initial Curvature under Uniform Edge Compression[2]

If the creep characteristic of the material can be represented by viscoelastic models, the stress-strain relations developed for viscoelastic bodies (Section 4.8) may be applied to the analyses. The stress-strain-time relation for viscoelastic materials is given by

$$\bar{S}_{ij} = 2G[1 - \bar{\phi}(s)]\bar{\varepsilon}_{ij} \tag{4.8.12}$$

where S_{ij} and ε_{ij} are the deviatoric stress and strain components, respectively, and the bar above a quantity denotes its Laplace transform. From Equation (10.4.7), we have $\bar{\tau}_{ij} = 2\mu'\bar{\varepsilon}_{ij} + \delta_{ij}\lambda'\bar{\theta}$, where $\lambda' = [K - (2\mu/3)(1 - \bar{\phi})]$ and $\mu' = \mu(1 - \bar{\phi})$. From Equations (10.4.14) and (11.1.10), we have

$$D'\nabla^4\bar{w} = \bar{q} + \bar{q}' \tag{11.3.1}$$

where

$$D' = \frac{h^3}{12}\frac{2\mu(1 - \bar{\phi})[6K + 2\mu(1 - \bar{\phi})]}{[3K + 4\mu(1 - \bar{\phi})]}$$

as given by Equation (10.4.13) and q' is given by Equation (11.1.8). The bulk modulus K is

$$K = \frac{E}{3(1 - 2\nu)} = \frac{2\mu(1 + \nu)}{3(1 - 2\nu)} = \frac{2\mu}{3}f(\nu)$$

where $f(\nu) = (1 + \nu)/(1 - 2\nu)$. Then

$$D' = 2\mu I\,\frac{2f(\nu) + (1 - \bar{\phi})}{f(\nu) + 2(1 - \bar{\phi})} \tag{11.3.2}$$

where $I = h^3/12$. Since no plate is perfectly flat, initial curvature w_0 always exists. The total deflection of the plate at any instant is $w_0 + w$.

Consider a rectangular plate of length a and width b, simply supported along its four edges under a uniform thrust N_x per unit length on the two opposite edges $y = 0$ and $y = b$. The differential equation of equilibrium, Equation (11.3.1), with no lateral load, $q = 0$, gives

$$D'\nabla^4\bar{w} = N_x\left(\bar{w}_{xx} + \frac{1}{s}\,w_{0_{xx}}\right) \tag{11.3.3}$$

Let the initial deflection w_0 be expressed in a double sine series as

$$w_0 = \sum_{m=1}^{\infty}\sum_{n=1}^{\infty} A_{mn}\sin\frac{m\pi x}{a}\sin\frac{n\pi y}{b}$$

Then

$$w_{0_{xx}} = -\sum_{m=1}^{\infty}\sum_{n=1}^{\infty} A_{mn}\frac{m^2\pi^2}{a^2}\sin\frac{m\pi x}{a}\sin\frac{n\pi y}{b}$$

Further, letting $w = \sum_{m=1}^{\infty}\sum_{n=1}^{\infty} B\sin(m\pi x/a)\sin(n\pi y/b)$ and substituting this into Equation (11.3.1), we obtain

$$D'\sum_{m=1}^{\infty}\sum_{n=1}^{\infty}\left\{\left[\left(\frac{m\pi}{a}\right)^4 + 2\left(\frac{m\pi}{a}\right)^2\left(\frac{n\pi}{b}\right)^2 + \left(\frac{n\pi}{b}\right)^4\right] - \frac{N_x}{D'}\left(\frac{m\pi}{a}\right)^2\right\}$$
$$\times \bar{B}_{mn}\sin\frac{m\pi x}{a}\sin\frac{n\pi y}{b} = N_x\sum_{m=1}^{\infty}\sum_{n=1}^{\infty}\frac{A_{mn}}{s}\frac{m^2\pi^2}{a^2}\left(\sin\frac{m\pi x}{a}\sin\frac{n\pi y}{b}\right)$$

(11.3.4)

Multiplying the above equation by $\sin(m'\pi x/a)\sin(n'\pi y/b)\,dx\,dy$ and integrating from $x = 0$ to $x = a$ and $y = 0$ to $y = b$, we obtain, upon making use of the orthogonality property of sine functions,

$$\bar{B}_{m'n'} = \frac{A_{m'n'}(m'\pi/a)^2(1/s)}{[(m'\pi/a)^4 + 2(m'\pi/a)^2(n'\pi/b)^2 + (n'\pi/b)^4](D'/N_x) - (m'\pi/a)^2}$$

(11.3.5)

or

$$\bar{B}_{mn} = \frac{(A_{mn}/s)(m\pi/a)^2 N_x[f(v) + 2(1-\bar{\phi})]}{C_{mn}IG(1-\bar{\phi})[4f(v) + 2(1-\bar{\phi})] - (m\pi/a)^2 N_x[f(v) + 2(1-\bar{\phi})]}$$

(11.3.6)

where

$$C_{mn} = \left[\left(\frac{m\pi}{a}\right)^4 + 2\left(\frac{m\pi}{a}\right)^2\left(\frac{n\pi}{b}\right)^2 + \left(\frac{n\pi}{b}\right)^4\right]$$

The B_{mn} is to be obtained from the inverse transform of $\bar{B}_{mn}$. For simple functions of $1 - \bar{\phi}$, as in the case of the Maxwell model, the inverse transform may be obtained without much difficulty. However, for complicated $\bar{\phi}$ functions, the inverse transform may be difficult to obtain.

Numerical Example. Consider a simply supported viscoelastic plate, 8 in. square and $\frac{3}{16}$ in. thick, subject to a compressive load of 844 lb/in. on the two opposite edges, as shown in Figure 11.3. All four edges are simply supported. The viscoelastic properties of the plate are taken to be represented by a Maxwell model, giving

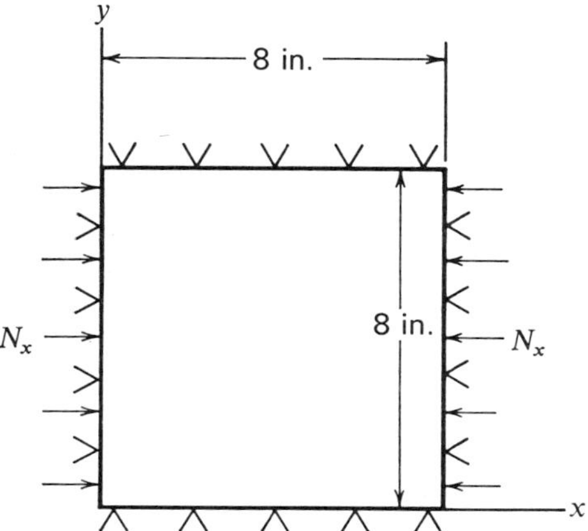

Figure 11.3. Square plate under edge thrust.

$$2G(1 - \bar{\phi}) = \frac{2Gs}{(s + \alpha)} \tag{11.3.7}$$

The values of G and α may be obtained from a uniaxial test of the material. Assume G and α to be 2×10^6 psi and 0.495/hr, respectively. If Poisson's ratio v for elastic deformation is taken to be 0.3, then Equation (11.3.6) gives

$$\bar{B}_{mn} = \frac{A_{mn}/s}{\dfrac{C_{mn} IG \dfrac{s}{(s+\alpha)}\left[4f(v) + \dfrac{2s}{(s+\alpha)}\right]}{\left(\dfrac{m\pi}{a}\right)^2 N_x \left[f(v) + \dfrac{2s}{(s+\alpha)}\right]} - 1} \tag{11.3.8}$$

For this plate,

$$I = \frac{1}{12}\left(\frac{3}{16}\right)^3 \qquad N_x = 844 \text{ lb/in.}$$

$$C_{11} = \frac{\pi^4}{32^2} \qquad C_{11} IG \Big/ \left[\frac{\pi^2}{a^2} N_x\right] = 0.802$$

$$f(v) = (1 + v)/(1 - 2v) = 3.25$$

Substituting the above values of C_{11}, N_x, I, and $f(v)$ into Equation (11.3.8) gives

$$\bar{B}_{11} = \frac{A_{11}/s}{\dfrac{0.802s[13(s+\alpha)+2s]}{(s+\alpha)[3.25(s+\alpha)+2s]} - 1}$$

Upon simplifying and factoring the denominator, this becomes

$$\bar{B}_{11} = \frac{A_{11} \times 0.775(s+\alpha)(s+0.619\alpha)}{s(s+0.849\alpha)(s-0.565\alpha)} \tag{11.3.9}$$

To find the inverse transform of the quotient of two polynomials[3] $p(s)/q(s)$, where $p(s)$ and $q(s)$ are polynomials with no common factors and the degree of $p(s)$ is lower than that of $q(s)$, the following procedure is commonly used. Assume that the factors of $q(s)$ are linear and distinct. Then the mth-degree polynomial $q(s)$ may be written as

$$q(s) = (s-\beta_1)(s-\beta_2)\cdots(s-\beta_m)$$

and the quotient written as

$$\frac{p(s)}{q(s)} = \frac{C_1}{s-\beta_1} + \frac{C_2}{s-\beta_2} + \cdots + \frac{C_m}{s-\beta_m}$$

Multiplying the above equation by $(s-\beta_n)$ and letting s approach β_n, we have

$$\lim_{s\to\beta_n}\left[\frac{s-\beta_n}{q(s)}\,p(s)\right] = C_n$$

Since

$$\lim_{s\to\beta_n}\frac{s-\beta_n}{q(s)} = \lim_{s\to\beta_n}\frac{1}{q'(s)} = \frac{1}{q'(\beta_n)}$$

we have,

$$C_n = \frac{p(\beta_n)}{q'(\beta_n)}$$

Hence the quotient of two polynomials may be written in the easily transform-invertible form,

$$\frac{p(s)}{q(s)} = \sum_{n=1}^{m}\frac{p(\beta_n)}{q'(\beta_n)}\frac{1}{s-\beta_n}$$

The inverse transform of the above is easily seen to be[3]

$$L^{-1}\left[\frac{p(s)}{q(s)}\right] = \sum_{n=1}^{m}\frac{p(\beta_n)}{q'(\beta_n)}\cdot\exp(\beta_n t)$$

Returning to Equation (11.3.9), we have

$$B_{11} = A_{11} \times 0.775 \sum_{1}^{n} \frac{p(\beta_n)}{q'(\beta_n)} \cdot \exp(\beta_n t) \tag{11.3.10}$$

where

$$p(s) = (s + \alpha)(s + 0.619\alpha)$$

$$q(s) = s(s + 0.849\alpha)(s - 0.565\alpha)$$

$$q'(s) = (s + 0.849\alpha)(s - 0.565\alpha) + s(s + 0.849\alpha) + s(s - 0.565\alpha)$$

Hence $\beta_1 = 0$, $\beta_2 = -0.849\alpha$, and $\beta_3 = 0.565\alpha$.

From Equation (11.3.10), for $A_{11} = 0.002$ in., we obtain

$$B_{11} = 0.002 \times 0.775(-1.29 - 0.0289\, e^{-0.849\alpha t} + 2.32\, e^{0.565\alpha t}) \tag{11.3.11}$$

For $m = 1$ and $n = 3$, from Equation (11.3.6), we obtain

$$C_{13} = \frac{\pi^4}{8^4}[1 + 2 \times 1 \times 3^2 + 3^4] = \frac{\pi^4}{8^4} \times 100$$

$$C_{13} IG\left[\frac{\pi^2}{a^2} N_x\right] = 0.802 \times 25$$

From Equation (11.3.8), $\bar{B}_{13}$ is obtained as

$$\bar{B}_{13} = \frac{A_{13}(s + \alpha)(5.25s + 3.25\alpha)/s}{0.802 \times 25s(15s + 13\alpha) - (s + \alpha)(5.25s + 3.25\alpha)}$$

$$= \frac{A_{13}}{56.9} \frac{(s + \alpha)(s + 0.619\alpha)}{s(s + 0.866\alpha)(s - 0.012\alpha)} \tag{11.3.12}$$

For $A_{13} = 0.002$ in., B_{13} is

$$B_{13} = (0.002/56.9)[-59.6 - 0.043\, e^{-0.866\alpha t} + 60.6\, e^{0.012\alpha t}] \tag{11.3.13}$$

In this example the initial curvature was assumed to be symmetrical with respect to the center lines of the square plate, hence only the first two coefficients associated with a symmetric w_0, that is, A_{11} and A_{13}, were taken. The values of B_{11} and B_{13} for values of t from 0 to 8 hr, calculated with an α of 0.495, are given in Table 11.1. The center deflection vs. time curve, due to $A_{11} = A_{13} = 0.002$, is plotted in Figure 11.4.

Table 11.1

(1) t (hr)	(2) B_{11} (in.)	(3) $A_{11}+B_{11}$ (in.)	(4) B_{13} (in.)	(5) $A_{13}+B_{13}$ (in.)	(6) Center Deflection (in.)
0	0.00155	0.00355	0.000035	0.00204	0.00559
2	0.00546	0.00746	0.000071	0.00207	0.00953
4	0.01350	0.01550	0.000106	0.00211	0.01761
6	0.03010	0.03210	0.000142	0.00214	0.03424
8	0.06490	0.06690	0.000178	0.00218	0.06888

Deflection vs. time curves were separately calculated for initial deflections of two different harmonics; one with $m = 1, n = 1$ and the other with $m = 1, n = 3$. As seen in Table 11.1, the initial deflection of the former grows much faster than the latter. A plate with initial deflection composed of different harmonics will have different forms or shapes of deflection surface and hence different stress distributions at different instants.

This analysis also shows that, so long as the stress in the plate is within

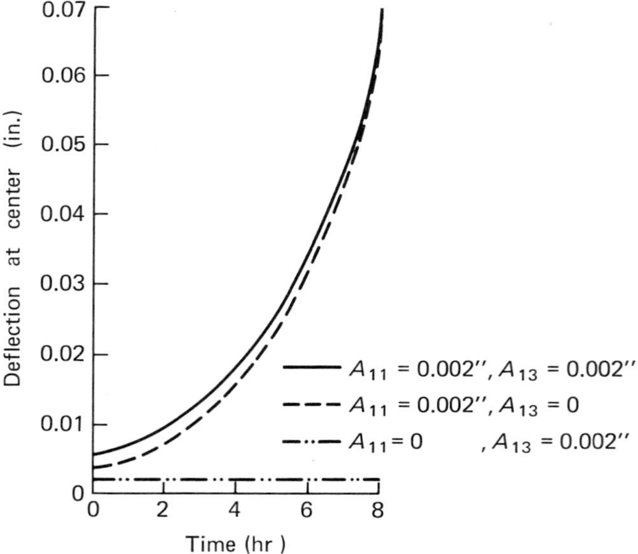

Figure 11.4. Deflection-time curve of a viscoelastic plate 8 in. square, $\frac{3}{4}$ in. thick under edge thrust of 844 lb/in.

the elastic limit of the instantaneous stress-strain curve of the material, infinite rate of deflection would occur only at infinite time. Creep strain in a plate or column is similar to initial curvature in that it does not change the critical compressive strength of the member. For viscoelastic plates with large deflections, see Reference 4.

11.4 Inelastic Circular Plates under Uniform Edge Compression and Uniform Lateral Load

Consider a circular plate of uniform thickness under combined uniform edge compression N_r and uniform lateral load q. Since the plate and loading are axisymmetrical, the displacements, stresses, and strains in the plate are also axisymmetrical. The displacements corresponding to cylindrical coordinates r, θ, and z, are u, v, and w, respectively. The slope ϕ at any point A on the line formed by the intersection of the middle surface with the rz plane is $-dw/dr$. The curvature for small deflections[5] (Figure 11.5) is

$$-\frac{d^2w}{dr^2} = \frac{d\phi}{dr}$$

where ϕ is the angle between the normal to the deflected surface at A and the axis of symmetry OC. This is one principal curvature[5] at A. The second principal curvature[5] lies in the section through the normal AC and perpendicular to the rz plane. Since the normals, like AC, for all points in a circle with radius r from the axis of symmetry form a conical surface with C as its apex, AC is the radius of the second principal curvature, and is equal to

$$\frac{1}{AC} = -\frac{1}{r}\frac{dw}{dr} = \frac{\phi}{r}$$

With u denoting the displacement along the radial direction, we have, from Kirchhoff's assumption (Section 10.1), the following relations:

$$u = u_0 - z\frac{dw}{dr}$$

$$e_{rr} = \frac{du_0}{dr} - z\frac{d^2w}{dr^2}$$

$$e_{\theta\theta} = \frac{u}{r} = \frac{u_0}{r} - \frac{z}{r}\frac{dw}{dr}$$

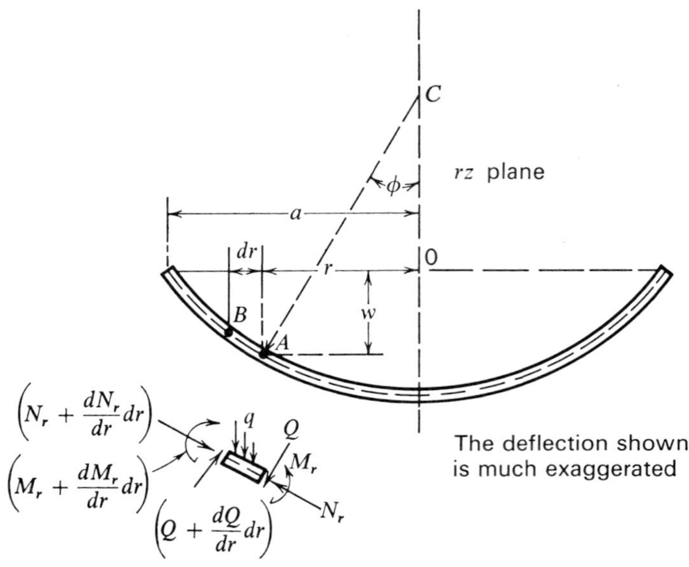

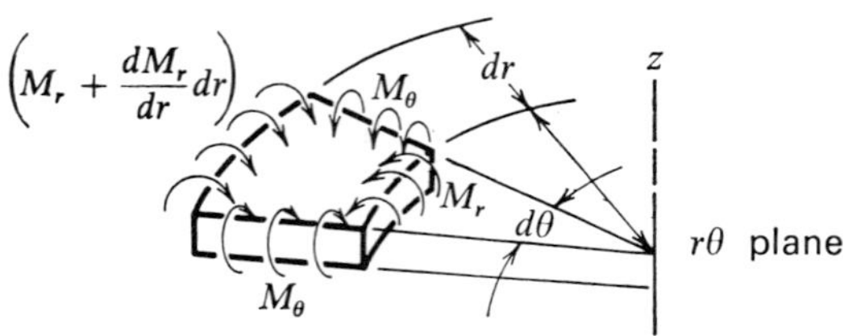

Figure 11.5. A plate element r dr dθ *and loads.*

$$e_{r\theta} = 0 \tag{11.4.1}$$

$$\tau_{rr} = \frac{E}{1-\nu^2}\left[\frac{du_0}{dr} + \nu\frac{u_0}{r} - z\left(\frac{d^2w}{dr^2} + \frac{\nu}{r}\frac{dw}{dr}\right) - e''_{rr} - \nu e''_{\theta\theta}\right]$$

$$\tau_{\theta\theta} = \frac{E}{1-\nu^2}\left[\nu\frac{du_0}{dr} + \frac{u_0}{r} - z\left(\nu\frac{d^2w}{dr^2} + \frac{1}{r}\frac{dw}{dr}\right) - e''_{\theta\theta} - \nu e''_{rr}\right]$$

$$\tau_{r\theta} = 0 \tag{11.4.2}$$

The sectional forces, in cylindrical coordinates, are given by

$$\begin{aligned} N_r &= \frac{Eh}{1-\nu^2}\left(\frac{du_0}{dr} + \nu\frac{u_0}{r}\right) - N_{r_I} \\ N_\theta &= \frac{Eh}{1-\nu^2}\left(\nu\frac{du_0}{dr} + \frac{u_0}{r}\right) - N_{\theta_I} \end{aligned} \tag{11.4.3}$$

where the inelastic forces are

$$\begin{aligned} N_{r_I} &= \frac{E}{1-\nu^2}\int_{-h/2}^{h/2} (e''_{rr} + \nu e''_{\theta\theta})\, dz \\ N_{\theta_I} &= \frac{E}{1-\nu^2}\int_{-h/2}^{h/2} (e''_{\theta\theta} + \nu e''_{rr})\, dz \end{aligned} \tag{11.4.4}$$

The sectional moments in cylindrical coordinates are the same as given in Equations (10.5.3) and (10.5.4). They may be written as

$$\begin{aligned} M_r &= -D\left(\frac{d^2w}{dr^2} + \frac{\nu}{r}\frac{dw}{dr}\right) - M_{r_I} \\ M_\theta &= -D\left(\nu\frac{d^2w}{dr^2} + \frac{1}{r}\frac{dw}{dr}\right) - M_{\theta_I} \end{aligned} \tag{11.4.5}$$

where the inelastic moments are

$$\begin{aligned} M_{r_I} &= \frac{E}{1-\nu^2}\int_{-h/2}^{h/2} (e''_{rr} + \nu e''_{\theta\theta}) z\, dz \\ M_{\theta_I} &= \frac{E}{1-\nu^2}\int_{-h/2}^{h/2} (e''_{\theta\theta} + \nu e''_{rr}) z\, dz \end{aligned} \tag{11.4.6}$$

Equilibrium Conditions. Consider an element $r\, dr\, d\theta$ of the circular plate; the condition of equilibrium in the plane of the plate gives

$$\left(N_r + \frac{dN_r}{dr}\,dr\right)(r + dr)\,d\theta - N_r r\,d\theta - 2N_\theta\,dr\,\frac{d\theta}{2} + Rr\,dr\,d\theta = 0$$

$$\frac{dN_r}{dr} + \frac{N_r - N_\theta}{r} + R = 0 \tag{11.4.7}$$

where R is the body force per unit area in the radial direction.

Referring to Figure 11.5 and equating to zero, the sum of the vertical forces acting on a center portion of the plate with radius r, we have

$$2\pi r(Q\cos\phi - N_r\sin\phi) = \pi r^2 q$$

For small ϕ we have $\cos\phi \cong 1$, and $\sin\phi \cong \phi$, and hence

$$Q = \frac{rq}{2} + N_r\phi \tag{11.4.8}$$

For small deflections, the condition of equilibrium of moment on the element $r\,dr\,d\theta$ yields

$$\left(M_r + \frac{dM_r}{dr}\,dr\right)(r + dr)\,d\theta - M_r r\,d\theta - M_\theta\,dr\,d\theta + Qr\,dr\,d\theta = 0$$

$$\frac{dM_r}{dr} + \frac{M_r - M_\theta}{r} + Q = 0 \tag{11.4.9}$$

where Q is the vertical shear force per unit length.

Substituting Equations (11.4.3) and (11.4.5), into Equations (11.4.7) and (11.4.9), we obtain

$$\frac{Eh}{1-\nu^2}\left(\frac{d^2u_0}{dr^2} + \frac{1}{r}\frac{du_0}{dr} - \frac{u_0}{r^2}\right) + f(r) + R = 0 \tag{11.4.10}$$

$$D\left(\frac{d^2\phi}{dr^2} + \frac{1}{r}\frac{d\phi}{dr} - \frac{\phi}{r^2}\right) + g(r) + Q = 0 \tag{11.4.11}$$

where

$$f(r) = -\frac{E}{1-\nu^2}\left[\frac{d}{dr}\int(e''_{rr} + \nu e''_{\theta\theta})\,dz + \frac{1-\nu}{r}\int(e''_{rr} - e''_{\theta\theta})\,dz\right] \tag{11.4.12}$$

$$g(r) = -\frac{E}{1-\nu^2}\left[\frac{d}{dr}\int(e''_{rr} + \nu e''_{\theta\theta})z\,dz + \frac{1-\nu}{r}\int(e''_r - e''_\theta)z\,dz\right] \tag{11.4.13}$$

Substituting Q, from Equation (11.4.8), into Equation (11.4.11) we obtain

$$D\left(\frac{d^2\phi}{dr^2} + \frac{1}{r}\frac{d\phi}{dr} - \frac{\phi}{r^2}\right) + g(r) + N_r\phi = -\frac{qr}{2} \tag{11.4.14}$$

Equilibrium Equations (11.4.10) and (11.4.14) must be satisfied throughout the plate.

On the boundary $r = a$ we have $N_r = N_{r_a}$ and $M_r = M_{r_a}$, Then, from Equations (11.4.3) and (11.4.5), we have

$$\frac{Eh}{1-\nu^2}\left(\frac{du_0}{dr} + \nu\,\frac{u_0}{r}\right) = N_{r_a} + N_{r_I}$$
$$-D\left(\frac{d^2w}{dr^2} + \frac{\nu}{r}\frac{dw}{dr}\right) = M_{r_a} + M_{r_I} \tag{11.4.15}$$

or, using $\phi = -dw/dr$, we see that this last expression becomes

$$D\left(\frac{d\phi}{dr} + \frac{\nu\phi}{r}\right) = M_{r_a} + M_{r_I} \tag{11.4.16}$$

For a simply supported edge, $M_{r_a} = 0$, and Equation (11.4.16) reduces to

$$D\left(\frac{d\phi}{dr} + \nu\,\frac{\phi}{r}\right) = M_{r_I} \tag{11.4.17}$$

The N_{r_I} and M_{r_I} are the equivalent edge sectional forces and moments due to inelastic strain (see Section 2.2). The solution of an inelastic plate under edge compression and lateral load must satisfy the differential equilibrium Equations (11.4.7) and (11.4.9), as well as the boundary conditions (11.4.15).

Circular Elastic Plates. For circular elastic plates, the above equations are much simplified, since in this case Equations (11.4.12) and (11.4.13) give

$$f(r) = g(r) = 0$$

If, in addition, there are no body forces, $R = 0$, and equilibrium Equation (11.4.10) reduces to

$$\frac{d^2u_0}{dr^2} + \frac{1}{r}\frac{du_0}{dr} - \frac{u_0}{r^2} = 0 \tag{11.4.18}$$

This is satisfied by letting

$$\frac{du_0}{dr} = \frac{u_0}{r} = \text{const.}$$

This, according to Equation (11.4.3), corresponds to constant N_r. The remaining equilibrium Equation (11.4.14) becomes

$$D\left(\frac{d^2\phi}{dr^2} + \frac{1}{r}\frac{d\phi}{dr} - \frac{\phi}{r^2}\right) + N_r\phi = -\frac{qr}{2} \tag{11.4.19}$$

The particular solution of Equation (11.4.19) is $\phi = -qr/2N_r$, which is easily verified by substitution. To find the general solution,[1] let $N_r/D = \alpha^2$ and equate the left side of the differential equation to zero, thereby obtaining

$$r^2\frac{d^2\phi}{dr^2} + r\frac{d\phi}{dr} + (\alpha^2 r^2 - 1)\phi = 0$$

Replacing r by v/α in the above equation, we obtain the well-known Bessel Equation,

$$v^2\frac{d^2\phi}{dv^2} + v\frac{d\phi}{dv} + (v^2 - 1)\phi = 0 \tag{11.4.20}$$

The solution of Bessel's differential equation is

$$\phi = AJ_1(v) + BY_1(v)$$

where A and B are arbitrary constants, and $J_1(v)$ and $Y_1(v)$ are Bessel functions of first order of the first and second kinds, respectively. The complete solution of Equation (11.4.16) is then

$$\phi = AJ_1(\alpha r) + BY_1(\alpha r) - \frac{q}{2N_r}r$$

The $Y_1(\alpha r)$ becomes infinite as αr approaches zero. Since the loading is axisymmetric, $\phi = 0$ at $r = 0$, and hence B must vanish. Therefore

$$\phi = AJ_1(\alpha r) - \frac{q}{2N_r}r \tag{11.4.21}$$

For a clamped edge circular plate, $\phi = 0$ at $r = a$, this gives $A = qa/2N_rJ_1(\alpha a)$. Finally, then, the slope of a circular plate under uniform lateral load and edge compression N_r is

$$\phi = \frac{qa}{2N_r}\left[\frac{J_1(\alpha r)}{J_1(\alpha a)} - \frac{r}{a}\right] \tag{11.4.22}$$

This gives the elastic solution of a circular plate with clamped edge, subject to uniform lateral load and edge compression.

11.5 Circular Plates with Arbitrary Creep Characteristics

To calculate the stress and creep strain of a circular plate under uniform lateral load q and edge compression N_r, we proceed as follows. The stress vs. time curves, for each point in the plate, are approximated by the incremental step curves shown in Figure 3.9. Each step consists of a constant stress period Δt followed by an instantaneous increment of stress $\Delta\sigma$. The incremental creep strains $\Delta e''_{rr}$, $\Delta e''_{\theta\theta}$, and $\Delta e''_{r\theta}$ in the period Δt are determined from the multiaxial creep characteristics of the material at the particular temperature with the particular history of loading. This procedure is the same as given in Section 10.5.

Equations (10.5.26) and (10.5.28) enable us to calculate the increment of creep strain after each time interval. These creep strain increments, added to the existing creep strains, give the total creep strain at the end of each time increment. By treating the total creep strains as equivalent loads, edge forces, and moments, as given in Equations (10.5.10), (10.5.11), (11.4.4), and (11.4.6), we may determine the deflection curve and strain distribution of the plate at various times.

This method gives a solution satisfying the equilibrium conditions, compatibility equations, and the given creep characteristics of the material at the given temperature.

Numerical Example. A 7075-T6 aluminum alloy circular plate is simply supported at a temperature of 600° F under a uniformly distributed lateral load of 1.0 psi and a middle plane compression of 1500 lb/in. The plate has a thickness of 0.5 in. and a diameter of 10 in. The uniaxial stress-strain-time relations for constant stress are given by

$$e = \frac{\sigma}{E} + A \sinh (B\sigma) t^k \tag{11.5.1}$$

$$e'' = A \sinh (B\sigma) t^k \tag{11.5.2}$$

where t is time in hours, σ is stress in psi, and the values of the constants are

$$A = 5.28 \times 10^{-7} \qquad B = 1.92 \times 10^{-3} \qquad k = 0.66$$

$$E = 5.2 \times 10^{6} \quad \text{psi} \qquad \nu = 0.33$$

Expressing Equation (11.5.2) in terms of stress and strain invariants, we have

$$\sqrt{\tfrac{4}{3} I_2''} = A \sinh \left(B\sqrt{3\bar{J}_2}\right) t^k \tag{11.5.3}$$

Because of the symmetry of the plate and loading, deflections, stresses and strains are independent of θ. Hence it is only necessary to consider a typical rz plane. On a typical rz plane we have the following grid system. From $r = 0$ to $r = 5$, vertical lines are drawn at 0.5-in. intervals, while the thickness is divided by nine horizontal lines at equal intervals of 0.005 in. We use $n = 1$ to denote $r = 0$, that is, the center of the circular plate, and $n = 11$ is at r = 5 in., that is, at the outer edge.

By the method of finite differences we have, at the nth station,

$$\left(\frac{du_0}{dr}\right)_n = \frac{u_{0_{n+1}} - u_{0_{n-1}}}{2\Delta r}$$
$$\left(\frac{d^2u_0}{dr^2}\right)_n = \frac{u_{0_{n+1}} - 2u_{0_n} + u_{0_{n-1}}}{(\Delta r)^2} \tag{11.5.4}$$

Substituting Equation (11.5.4) into Equation (11.4.10) and substituting 0.5 in. for Δr, we get

$$\left(4 + \frac{1}{r_n}\right)u_{0_{n+1}} - \left(8 + \frac{1}{r_n^{\,2}}\right)u_{0_n} + \left(4 - \frac{1}{r_n}\right)u_{0_{n-1}} = \frac{1}{h}(\alpha_n + \beta_{n+1} - \beta_{n-1}) \tag{11.5.5}$$

where

$$\alpha = \frac{1-\nu}{r}\int (e''_{rr} - e''_{\theta\theta})\,dz$$
$$\beta = \int (e''_{rr} + \nu e''_{\theta\theta})\,dz \tag{11.5.6}$$

Similarly, Equation (11.4.11) gives

$$\left(4 + \frac{1}{r_n}\right)\phi_{n+1} - \left(8 + \frac{1}{r_n^{\,2}} - \frac{N_{r_n}}{D}\right)\phi_n + \left(4 - \frac{1}{r_n}\right)\phi_{n-1}$$
$$= -\frac{q}{2D}r_n + \frac{E}{D(1-\nu^2)}(\gamma_n + \eta_{n+1} - \eta_{n-1}) \tag{11.5.7}$$

where

$$\gamma = \frac{1-\nu}{r}\int (e''_{rr} - e''_{\theta\theta})z\,dz$$
$$\eta = \int (e''_{rr} + \nu e''_{\theta\theta})z\,dz \tag{11.5.8}$$

The boundary conditions are

$$u_{0_1} = 0$$
$$N_{r_{11}} = 1500$$
$$\phi_1 = 0$$
$$M_{r_{11}} = 0 \tag{11.5.9}$$

Expressing Equations (11.4.3) and (11.4.5) in finite difference form, we have, at the boundary,

$$\left(2+\frac{\nu}{r_{11}}\right)u_{0_{11}}-2u_{0_{10}}=\frac{1-\nu^2}{Eh}N_{r_{11}}+\frac{1}{h}\beta_{11}$$
$$D\left(2+\frac{\nu}{r_{11}}\right)\phi_{11}-2D\phi_{10}=\frac{E}{1-\nu^2}\eta_{11} \tag{11.5.10}$$

At time $t=0$, all terms involving creep strains vanish. Equations (11.5.5) and (11.5.10) are used to solve for u_0 at different stations. The N_r's and N_θ's are calculated by Equation (11.4.3). Then Equations (11.5.7) and (11.5.10) are solved for the ϕ's. The deflections are calculated by numerical integration. The stresses and strains are then obtained from Equations (11.4.1) and (11.4.2).

For the first time increment, Equation (11.5.3) gives

$$\Delta\sqrt{\tfrac{4}{3}I_2''}=A\sinh\left(B\sqrt{3\bar{J}_2}\right)\Delta(t^k) \tag{11.5.11}$$

For the first time increment the creep strains e_{rr}'' and $e_{\theta\theta}''$ are zero; hence, from Equation (10.5.28), the initial κ is

$$\kappa=\frac{\Delta\sqrt{I_2''}}{(S_r{}^2+S_\theta{}^2+S_rS_\theta)^{1/2}} \tag{11.5.12}$$

At the beginning of any subsequent time increment with creep strain invariant I_2'' and second deviatoric stress invariant $\bar{J}_2$, the equivalent time corresponding to this I_2'' under the constant $\bar{J}_2$ is obtained from Equation (11.5.3) as

$$\bar{t}=\left[\frac{\sqrt{\tfrac{4}{3}I_2''}}{A\sinh\left(B\sqrt{3\bar{J}_2}\right)}\right]^{1/k}$$

Substituting this expression for t into Equation (11.5.11), we obtain

$$\Delta\sqrt{\tfrac{4}{3}I_2''}=A\sinh\left(B\sqrt{3\bar{J}_2}\right)k\bar{t}^{k-1}\,\Delta t$$
$$=\frac{k\left[A\sinh\left(B\sqrt{3\bar{J}_2}\right)\right]^{1/k}}{\left[\sqrt{\tfrac{4}{3}I_2''}\right]^{(1/k)-1}}\cdot\Delta t \tag{11.5.13}$$

With $\Delta\sqrt{I_2''}$ known, κ is calculated from Equation (10.5.28). Once κ is determined, the creep strain increments may be obtained from Equations (10.5.23). These creep strain increments, added to the existing ones, give the total creep strain components at the end of each time increment. By taking total creep strains into consideration, u_0 and ϕ may be obtained at the end of each time interval. The above procedure was repeated a sufficient number of times to cover the total time to be investigated. The results were computed with the aid of an IBM computer.

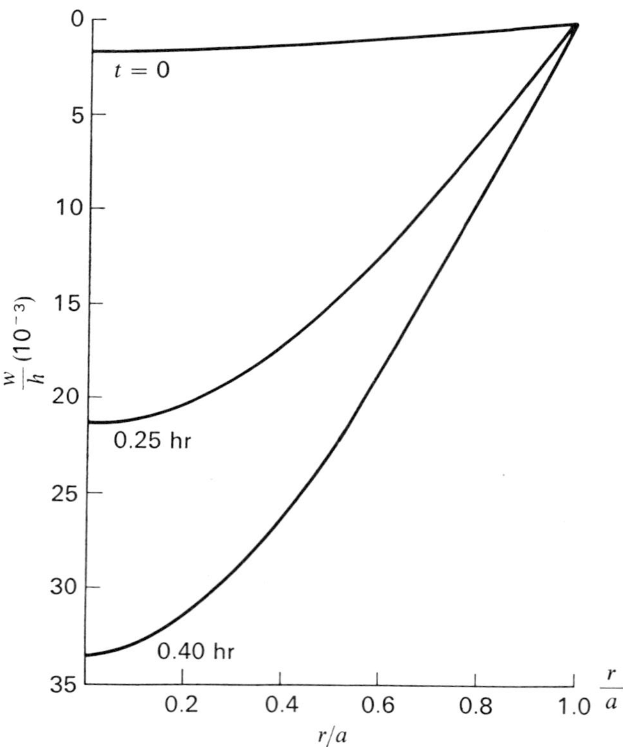

Figure 11.6. Deflection curve of the circular plate at different time instants.

The deflection curves of this circular plate at different instants are shown in Figure 11.6. The radial tangential sectional forces N_r and N_θ vs. r for two instants of time are shown in Figure 11.7. It is noted that N_r at the center decreases with time as creep increases.

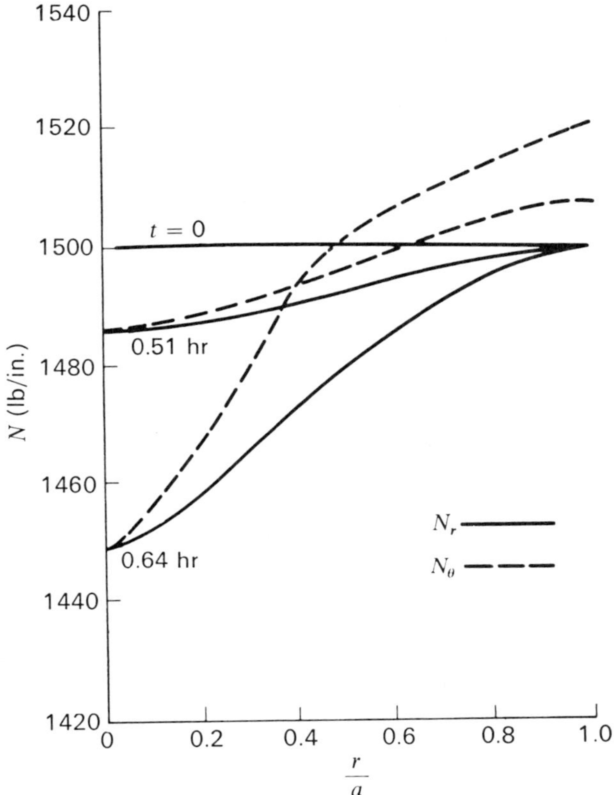

Figure 11.7. Middle surface compressions at different time instants.

11.6 Elastoplastic Circular Plates under Uniform Lateral Load and Edge Compression

When a circular plate is axisymmetrically loaded beyond its elastic limit, plastic strain occurs, and Equation (11.4.2) becomes

$$\tau_{rr} = \frac{E}{1-\nu^2}\left[\frac{du_0}{dr} + \nu\frac{u_0}{r} + z\left(\frac{d\phi}{dr} + \nu\frac{\phi}{r}\right) - e_{rr}^p - \nu e_{\theta\theta}^p\right]$$

$$\tau_{\theta\theta} = \frac{E}{1-\nu^2}\left[\nu\frac{du_0}{dr} + \frac{u_0}{r} + z\left(\nu\frac{d\phi}{dr} + \frac{\phi}{r}\right) - e_{\theta\theta}^p - \nu e_{rr}^p\right] \qquad (11.6.1)$$

$$\tau_{r\theta} = 0$$

where $\phi = -(dw/dr)$ and the superscript "p" denotes plastic strain. The incremental stresses are given as

$$\Delta\tau_{rr} = \frac{E}{1-v^2}\left[\frac{d\,\Delta u_0}{dr} + v\frac{\Delta u_0}{r} + z\left(\frac{d\,\Delta\phi}{dr} + \frac{v}{r}\Delta\phi\right) - \Delta e_{rr}^p - v\,\Delta e_{\theta\theta}^p\right]$$

$$\Delta\tau_{\theta\theta} = \frac{E}{1-v^2}\left[v\frac{d\,\Delta u_0}{dr} + \frac{\Delta u_0}{r} + z\left(v\frac{d\,\Delta\phi}{dr} + \frac{\Delta\phi}{r}\right) - \Delta e_{\theta\theta}^p - v\,\Delta e_{rr}^p\right]$$

(11.6.2)

Writing Equations (11.4.10) and (11.4.11) in incremental form for an incremental loading, we obtain

$$\frac{Eh}{1-v^2}\left(\frac{d^2\,\Delta u_0}{dr^2} + \frac{1}{r}\frac{d\,\Delta u_0}{dr} - \frac{\Delta u_0}{r^2}\right) + \Delta f(r) + \Delta R = 0 \qquad (11.6.3)$$

$$D\left(\frac{d^2\,\Delta\phi}{dr^2} + \frac{1}{r}\frac{d\,\Delta\phi}{dr} - \frac{\Delta\phi}{r^2}\right) + \Delta g(r) + \Delta Q = 0 \qquad (11.6.4)$$

where

$$\Delta f(r) = \frac{-E}{1-v^2}\left[\frac{d}{dr}\int(\Delta e_{rr}^p + \Delta e_{\theta\theta}^p)\,dz + \frac{1-v}{r}\int(\Delta e_{rr}^p - \Delta e_{\theta\theta}^p)\,dz\right] \qquad (11.6.5)$$

$$\Delta g(r) = -\frac{E}{1-v^2}\left[\frac{d}{dr}\int(\Delta e_{rr}^p + v\,\Delta e_{\theta\theta}^p)z\,dz + \frac{1-v}{r}\int(\Delta e_{rr}^p - \Delta e_{\theta\theta}^p)z\,dz\right]$$

(11.6.6)

and, from Equation (11.4.8), we have

$$\Delta Q = \Delta(N_r\phi) + \frac{r}{2}\Delta q = N_r\,\Delta\phi + \phi\,\Delta N_r + \frac{r}{2}\Delta q \qquad (11.6.7)$$

From Equation (10.6.12), the incremental effective stress is

$$\Delta\sigma^* = \frac{2\tau_{rr} - \tau_{\theta\theta}}{2\sigma^*}\Delta\tau_{rr} + \frac{2\tau_{\theta\theta} - \tau_{rr}}{2\sigma^*}\Delta\tau_{\theta\theta} = \frac{d\sigma^*}{d(e^p)^*}\Delta(e^p)^*$$

$$= \frac{E}{1-v^2}\left\{\frac{2\tau_{rr} - \tau_{\theta\theta}}{2\sigma^*}\left[\frac{d\,\Delta u_0}{dr} + v\frac{\Delta u_0}{r} + z\left(\frac{d\,\Delta\phi}{dr} + \frac{v}{r}\Delta\phi\right) - \Delta e_{rr}^p - v\,\Delta e_{\theta\theta}^p\right]\right.$$

$$\left. + \frac{2\tau_{\theta\theta} - \tau_{rr}}{2\sigma^*}\left[v\frac{d\,\Delta u_0}{dr} + \frac{\Delta u_0}{r} + z\left(v\frac{d\,\Delta\phi}{dr} + \frac{\Delta\phi}{r}\right) - \Delta e_{\theta\theta}^p - v\,\Delta e_{rr}^p\right]\right\}$$

(11.6.8)

When the plate is loaded up to the elastic limit, the u_0's, ϕ's, N_r's, $\tau_{\theta\theta}$'s, and τ_{rr}'s throughout the plate are obtained from an elastic analysis. Now consider an increment of load. The points having the highest σ^* are likely to have $(e^p)^*$. For each point n at which $\sigma_n{}^*$ exceeds the elastic limit and $d\sigma_n{}^* > 0$, we will have a $\Delta(e^p)_n{}^*$. Equation (11.6.8) gives a relation between the $\Delta(e^p)_n{}^*$'s, the $\Delta(u_0)_n$'s, and the $\Delta\phi_n$'s. Considering the $\Delta(e^p)^*$'s, Δu_0's, and $\Delta\phi$'s as unknowns, we see that there are as many equations as unknowns. These equations at each point are Equations (11.6.3) and boundary conditions for Δu_0's, Equations (11.6.4) and boundary conditions for $\Delta\phi$'s, and Equations (11.6.8) for $\Delta(e^p)^*$'s.

The $\Delta f(r)$ and $\Delta g(r)$ are linear functions of the $\Delta(e^p)^*$'s. Expressing ΔQ as shown by Equation (11.6.7), $\Delta\tau_{rr}$'s and $\Delta\tau_{\theta\theta}$'s in terms of $\Delta\sigma^*$'s, by Equation (10.6.12), Δe^p_{rr}'s and $\Delta e^p_{\theta\theta}$'s in terms of $\Delta(e^p)^*$'s by Equation (10.6.8), we see that all the equations are linear in terms of Δu_0's, $\Delta\phi$'s, and $\Delta(e^p)^*$'s. From this set of linear equations, Δu_0's, $\Delta\phi$'s, and $\Delta(e^p)^*$'s are completely determined. Finally, the incremental stresses may be computed from Equation (11.6.2).

This procedure has been applied to the analysis of a simply supported elastoplastic circular plate, loaded with uniform lateral pressure q and edge compression N_r. The plate whose diameter and thickness are 10 in. and $\frac{1}{2}$ in., respectively, is of 2024-T3 aluminum alloy. The tensile stress-strain curve is assumed to be represented by[6]

$$e = \frac{\sigma}{E}, \qquad \text{when } \sigma < \sigma_y$$

$$e = \frac{\sigma}{E} + \left(\frac{\sigma}{B}\right)^n - \left(\frac{\sigma_y}{B}\right)^n, \qquad \text{when } \sigma \geq \sigma_y \tag{11.6.9}$$

where $n = 10$, $B = 72{,}300$ psi, $E = 10 \times 10^6$ psi, e is the tensile strain, σ is tensile stress in psi, and σ_y is the tensile yield stress. This gives a uniaxial plastic strain-stress relation for monotonic loading as

$$e^p = \left(\frac{\sigma}{B}\right)^n - \left(\frac{\sigma_z}{B}\right)^n \qquad \text{when } \sigma > \sigma_y \tag{11.6.10}$$

No plastic strain occurs when the stress is less than tensile yield stress. The tensile yield stress σ_y is taken to be 45,000 psi. Equation (11.6.10) gives an incremental stress-strain relation for multiaxial stresses as

$$\frac{\Delta(e^p)^*}{\Delta\sigma^*} = \frac{n}{B}\left(\frac{\sigma^*}{B}\right)^{n-1} \qquad \sigma^* > Y \quad \text{and} \quad \Delta\sigma^* \geq 0 \tag{11.6.12}$$

and

$$\frac{\Delta(e^p)^*}{\Delta\sigma^*} = 0 \qquad \sigma^* < Y \quad \text{or} \quad \Delta\sigma^* < 0 \tag{11.6.13}$$

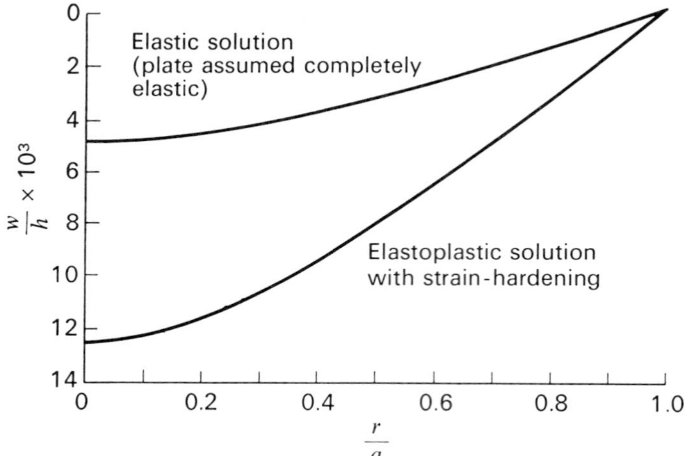

Figure 11.8. Deflection curves at $N_r = 19095$.

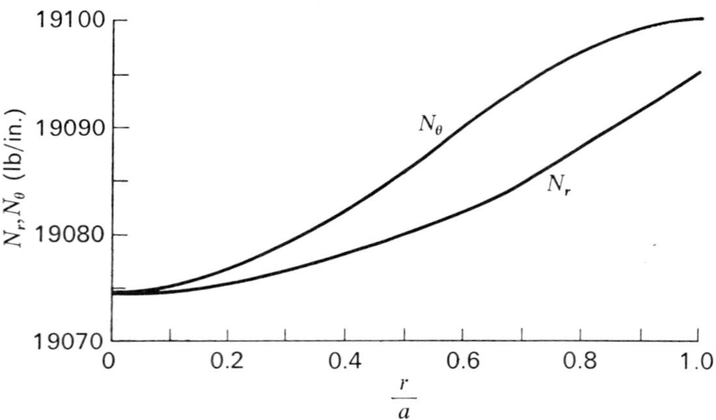

Figure 11.9. Variation of in -plane compression.

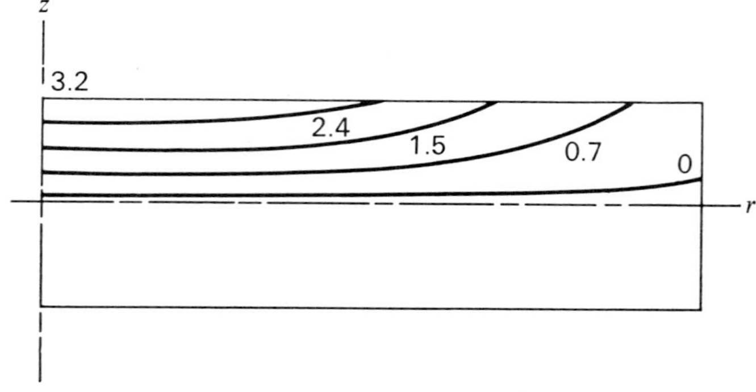

Figure 11.10. Contour lines of plastic strain $e''_{rr} \times 10^5$ ($N_r = 19095$).

Using Equations (11.6.12) and (11.6.13) to obtain the ratio $\Delta\sigma^*/\Delta(e^p)^*$ in Equation (11.6.8), we solve the Δu_0's, and $\Delta(e_p{}^*)$'s, and the incremental stresses and deflections are then obtained. The computed deflections are shown in Figure 11.8. The variation of N_r and N_θ along the radius is shown in Figure 11.9, and the distribution of plastic strain e''_{rr} is shown in Figure 11.10.

Section 11.4 shows the closed-form solution of stresses in elastic circular plates under uniform lateral load and edge compression. The solutions for identical plates with creep or plastic strain require finite difference numerical methods as explained in the preceding and present sections.

11.7 Rectangular Plates with Nonlinear Creep Subject to Uniform Edge Compression and Lateral Load

We consider here a rectangular plate, with nonlinear creep behavior, under a uniform edge compression N_x and a lateral load distribution q. Since creep strain is zero at $t = 0$, that is, at the time of load application, the initial deflection and stresses are obtained from elastic analysis. The initial deflection, w, may be obtained from solution of Equation (11.2.1) and the boundary conditions of the plate. From the initial deflection w, moments M_x, M_y, and M_{xy} are obtained from Equation (10.1.11). Referring to Equation (10.1.13), we see that the stresses are

$$\begin{aligned} \tau_{xx} &= \frac{12z}{h^3} M_x + \frac{N_x}{h} \\ \tau_{yy} &= \frac{12z}{h^3} M_y \\ \tau_{xy} &= \frac{12z}{h^3} M_{xy} \end{aligned} \tag{11.7.1}$$

where N_x is the applied uniform edge compression. From these initial stress components and the creep characteristics of the material, the creep strains produced in the first time increment (assuming that the stresses are constant during this time increment) are readily found, as illustrated in Section 10.5. With the creep strains at the end of the first time increment known, the equivalent forces $\bar{X}$ and $\bar{Y}$, given by Equations (10.2.11), and the equivalent sectional moments M_{x_I}, M_{y_I} and M_{xy_I}, given by Equations (10.2.7), are determined. With $\bar{X}$ and $\bar{Y}$ known, the distribution of sectional forces N_x, N_y, and N_{xy} in the plane of the plate may be obtained by elastic plane stress analysis as discussed in Section 9.7. With N_x, N_y, and N_{xy} distributions over the plate known, the deflection w may be determined by using Equation (11.2.1), together with the applicable boundary conditions. With w and the equivalent sectional moments known, the distribution of section moments M_x, M_y, and M_{xy} are given from Equation (10.2.6). Finally, knowing the sectional forces and moments, and the creep strain distribution e''_{ij}, we can obtain the stresses τ_{ij} at the end of the first time interval from Equation (10.2.23). Assuming these stresses to remain constant during the second time increment, we again obtain the incremental creep strain distribution from the creep characteristics of the material. Adding these incremental creep strains to the previous creep strains yields the total creep strain at the end of the second time increment. From these creep strains the equivalent body forces, $\overline{X}$ and $\overline{Y}$, and sectional moments M_{x_I}, M_{y_I}, and M_{xy_I} for the first plus second time increment are obtained. From these values, the stress distribution is obtained as before. This procedure is repeated for successive increments of time until the desired life of the plate is obtained. The procedure is necessarily lengthy. However, with the aid of modern computers, the buckling of rectangular plates with nonlinear creep may be analyzed.

Elastoplastic analysis of circular plates subject to uniform edge compression and uniform distributed lateral load has been shown in Section 11.6. Similar analyses for rectangular plates under the same loading would be much more difficult. The case with zero lateral load and uniform edge compression applied along only two opposite edges has been studied by various distinguished investigators.[7–21] Through these studies, the understanding of this problem has been greatly enhanced. Formulas, in reasonably good agreement with experiments, for predicting the plastic buckling strength of plates have been proposed.[21] However, the mechanism of plastic buckling of plates is not resolved.

11.8 Inelastic Plates with Large Deflection

Referring to Figure 11.11, consider a ds element of a section of the plate in the xz plane. After deflection occurs, the length of $A'B'$ is

$$ds^2 = \left(1 + \frac{\partial u_0}{\partial x}\right)^2 dx^2 + \left(\frac{\partial w}{\partial x}\right)^2 dx^2$$

$$ds = \left[1 + 2\frac{\partial u_0}{\partial x} + \left(\frac{\partial u_0}{\partial x}\right)^2 + \left(\frac{\partial w}{\partial x}\right)^2\right]^{1/2} dx$$

By the binomial theorem, we obtain

$$ds \cong dx + \frac{\partial u_0}{\partial x} dx + \frac{1}{2}\left(\frac{\partial w}{\partial x}\right)^2 dx$$

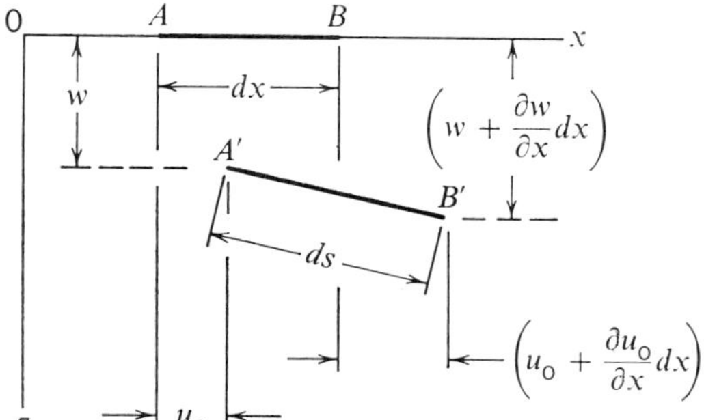

Figure 11.11

where higher-order terms of $(\partial u_0/\partial x)$ and $(\partial w/\partial x)^2$ have been neglected. Using $e_{xx_0} = (ds - dx)/dx$ yields

$$e_{xx_0} = \frac{\partial u_0}{\partial x} + \frac{1}{2}\left(\frac{\partial w}{\partial x}\right)^2 \tag{11.8.1}$$

Similarly,

$$e_{yy_0} = \frac{\partial v_0}{\partial y} + \frac{1}{2}\left(\frac{\partial w}{\partial y}\right)^2 \tag{11.8.2}$$

Referring to Figure 11.12, we see that the element OA moves to O_1A_1 and OB to O_1B_1. The direction cosines of O_1A_1 are

$$\frac{1}{[1 + (\partial w/\partial x)^2]^{1/2}}, \qquad 0, \qquad \frac{\partial w/\partial x}{[1 + (\partial w/\partial x)^2]^{1/2}}$$

and the direction cosines of O_1B_1 are

$$0, \qquad \frac{1}{[1 + (\partial w/\partial y)^2]^{1/2}}, \qquad \frac{\partial w/\partial y}{[1 + (\partial w/\partial y)^2]^{1/2}}$$

hence

$$\cos\theta = \frac{(\partial w/\partial x)(\partial w/\partial y)}{[1 + (\partial w/\partial y)^2]^{1/2}[1 + (\partial w/\partial x)^2]^{1/2}} \cong \frac{\partial w}{\partial x}\frac{\partial w}{\partial y}$$

$$\cos\theta = \sin\left(\frac{\pi}{2} - \theta\right) = \sin 2e_{xy_0} \cong 2e_{xy_0}$$

Thus the contribution to $2e_{xy_0}$ due to w is

$$2e_{xy_0} = \frac{\partial w}{\partial x}\frac{\partial w}{\partial y}$$

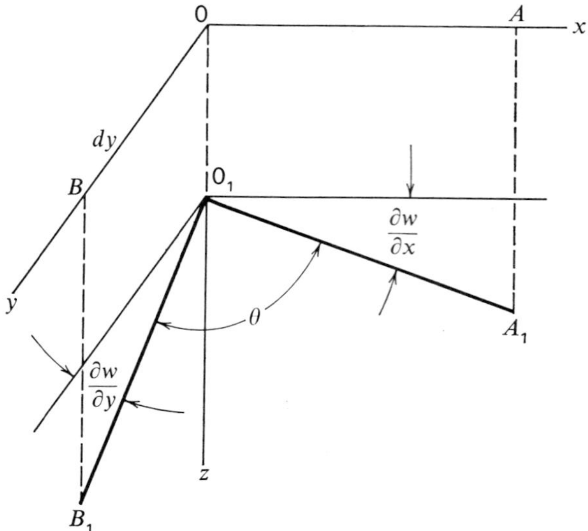

Figure 11.12

With the presence of u_0 and v_0, we see that $2e_{xy_0}$ is

$$2e_{xy_0} = \frac{\partial u_0}{\partial y} + \frac{\partial v_0}{\partial x} + \frac{\partial w}{\partial x}\frac{\partial w}{\partial y} \tag{11.8.3}$$

From Kirchhoff's assumptions, Equations (10.1.5) may be written as

$$\begin{aligned} e_{xx} &= e_{xx_0} - z\frac{\partial^2 w}{\partial x^2} \\ e_{yy} &= e_{yy_0} - z\frac{\partial^2 w}{\partial y^2} \\ e_{xy} &= e_{xy_0} - z\frac{\partial^2 w}{\partial x\,\partial y} \end{aligned} \tag{11.8.4}$$

and Equations (10.2.3) become

$$\begin{aligned} \tau_{xx} &= \frac{E}{1-\nu^2}\left[e_{xx_0} + \nu e_{yy_0} - z\frac{\partial^2 w}{\partial x^2} - \nu z\frac{\partial^2 w}{\partial y^2} - (e''_{xx} + \nu e''_{yy})\right] \\ \tau_{yy} &= \frac{E}{1-\nu^2}\left[e_{yy_0} + \nu e_{xx_0} - z\frac{\partial^2 w}{\partial y^2} - \nu z\frac{\partial^2 w}{\partial x^2} - (e''_{yy} + \nu e''_{xx})\right] \\ \tau_{xy} &= 2G\left(e_{xy_0} - z\frac{\partial^2 w}{\partial x\,\partial y} - e''_{xy}\right) \end{aligned} \tag{11.8.5}$$

From Equations (10.2.4) through (10.2.6), we have the following expressions for the sectional forces:

$$
\begin{aligned}
N_x &= \frac{Eh}{1-\nu^2}(e_{xx_0} + \nu e_{yy_0}) - N_{x_I} \\
N_y &= \frac{Eh}{1-\nu^2}(e_{yy_0} + \nu e_{xx_0}) - N_{y_I} \qquad (11.8.6) \\
N_{xy} &= Gh(2e_{xy_0}) - N_{xy_I}
\end{aligned}
$$

The expressions for sectional moments M_x, M_y, and M_{xy} remain the same as in (10.2.7). Equations (11.8.6), when solved for the strains, give

$$
\begin{aligned}
e_{xx_0} &= \frac{1}{Eh}[N_x + N_{x_I} - \nu(N_y + N_{y_I})] = \frac{\partial u_0}{\partial x} + \frac{1}{2}\left(\frac{\partial w}{\partial x}\right)^2 \\
e_{yy_0} &= \frac{1}{Eh}[N_y + N_{y_I} - \nu(N_x + N_{x_I})] = \frac{\partial v_0}{\partial y} + \frac{1}{2}\left(\frac{\partial w}{\partial y}\right)^2 \qquad (11.8.7) \\
e_{xy_0} &= \frac{1}{2Gh}[N_{xy} + N_{xy_I}] = \frac{1}{2}\left(\frac{\partial u_0}{\partial y} + \frac{\partial v_0}{\partial x}\right) + \frac{1}{2}\frac{\partial w}{\partial x}\frac{\partial w}{\partial y}
\end{aligned}
$$

Differentiating the first of Equations (11.8.7) twice with respect to y, the second twice with respect to x, and the third with respect to x and then y, we obtain

$$
\frac{\partial^2 e_{xx_0}}{\partial y^2} + \frac{\partial^2 e_{yy_0}}{\partial x^2} - 2\frac{\partial^2 e_{xy_0}}{\partial x\,\partial y} = \left(\frac{\partial^2 w}{\partial x\,\partial y}\right)^2 - \frac{\partial^2 w}{\partial x^2}\frac{\partial^2 w}{\partial y^2} \qquad (11.8.8)
$$

Substituting Equation (11.8.7) into Equation (11.8.8), we see that the left-hand side becomes

$$
\begin{aligned}
\frac{1}{Eh}\Bigg\{&\frac{\partial^2}{\partial y^2}(N_x + N_{x_I}) + \frac{\partial^2}{\partial x^2}(N_y + N_{y_I}) - 2\frac{\partial^2}{\partial x\,\partial y}(N_{xy} + N_{xy_I}) \\
&- \nu\frac{\partial}{\partial y}\left(\frac{\partial N_y}{\partial y} + \frac{\partial N_{xy}}{\partial x}\right) - \nu\frac{\partial}{\partial x}\left(\frac{\partial N_x}{\partial x} + \frac{\partial N_{xy}}{\partial y}\right) \\
&- \nu\frac{\partial}{\partial y}\left(\frac{\partial N_{y_I}}{\partial y} + \frac{\partial N_{xy_I}}{\partial x}\right) - \nu\frac{\partial}{\partial x}\left(\frac{\partial N_{x_I}}{\partial x} + \frac{\partial N_{xy_I}}{\partial y}\right)\Bigg\}
\end{aligned}
$$

Because of the equilibrium condition, Equations (11.1.1) and (11.1.2), the fourth and fifth terms vanish. Letting

$$
N_x = h\frac{\partial^2 U}{\partial y^2} \qquad N_y = h\frac{\partial^2 U}{\partial x^2} \qquad \text{and} \qquad N_{xy} = -h\frac{\partial^2 U}{\partial x\,\partial y}
$$

we see that Equations (11.1.1) and (11.1.2) are satisfied, and Equation (11.8.8) then becomes

$$\nabla^2\nabla^2 U + \frac{1}{h}\left[\frac{\partial^2 N_{x_I}}{\partial y^2} + \frac{\partial^2 N_{y_I}}{\partial x^2} - 2\frac{\partial^2 N_{xy_I}}{\partial x\,\partial y}\right] - \frac{\nu}{h}\left[\frac{\partial^2 N_{x_I}}{\partial x^2} + \frac{\partial^2 N_{y_I}}{\partial y^2} + 2\frac{\partial^2 N_{xy_I}}{\partial x\,\partial y}\right]$$

$$= E\left[\left(\frac{\partial^2 w}{\partial x\,\partial y}\right)^2 - \frac{\partial^2 w}{\partial x^2}\frac{\partial^2 w}{\partial y^2}\right] \tag{11.8.9}$$

Expressing the sectional forces N_x, N_y, and N_{xy} in terms of Airy's stress function U, and substituting Equations (11.1.8) and (10.2.9) into Equation (11.1.9), we obtain

$$D\nabla^2\nabla^2 w = \left(q + \bar{q} + h\frac{\partial^2 U}{\partial y^2}\frac{\partial^2 w}{\partial x^2} + h\frac{\partial^2 U}{\partial x^2}\frac{\partial^2 w}{\partial y^2} - 2h\frac{\partial^2 U}{\partial x\,\partial y}\frac{\partial^2 w}{\partial x\,\partial y}\right) \tag{11.8.10}$$

where

$$\bar{q} = \left(-\frac{\partial^2 M_{x_I}}{\partial x^2} - 2\frac{\partial^2 M_{xy_I}}{\partial x\,\partial y} - \frac{\partial^2 M_{y_I}}{\partial y^2}\right)$$

and, as before, the inelastic moments are

$$M_{x_I} = \frac{E}{1-\nu^2}\int (e''_{xx} + \nu e''_{yy})z\,dz$$

$$M_{y_I} = \frac{E}{1-\nu^2}\int (e''_{yy} + \nu e''_{xx})z\,dz$$

$$M_{xy_I} = 2G\int e''_{xy}\,z\,dz$$

Equations (11.8.9) and (11.8.10) are the two differential equations for plates with large deflection and inelastic strains e''_{ij}, which may be any combination of thermal, creep, and plastic strains. For the case with thermal strain only,

$$e''_{ij} = \delta_{ij}\alpha T$$

and Equations (10.3.2) and (10.3.3) become

$$N_{x_I} = N_{y_I} = \frac{N_T}{1-\nu}$$

$$M_{x_I} = M_{y_I} = \frac{M_T}{1-\nu}$$

Equations (11.8.9) and (11.8.10) reduce to

$$\nabla^2\nabla^2 U + \frac{1}{h}\nabla^2 N_T = E\left[\left(\frac{\partial^2 w}{\partial x\,\partial y}\right)^2 - \frac{\partial^2 w}{\partial x^2}\frac{\partial^2 w}{\partial y^2}\right] \tag{11.8.11}$$

$$\nabla^2\nabla^2 w + \frac{\nabla^2 M_T}{D(1-\nu)} = \frac{q}{D} + \frac{h}{D}\left[\frac{\partial^2 U}{\partial y^2}\frac{\partial^2 w}{\partial x^2} + \frac{\partial^2 U}{\partial x^2}\frac{\partial^2 w}{\partial y^2} - 2\frac{\partial^2 U}{\partial x\,\partial y}\frac{\partial^2 w}{\partial x\,\partial y}\right] \tag{11.8.12}$$

These equations were obtained by Nowacki.[22] For cases with no thermal, creep or plastic strain, $e''_{ij} = 0$ and hence

$$N_{x_I} = N_{y_I} = N_{xy_I} = 0$$

$$M_{x_I} = M_{y_I} = M_{xy_I} = 0$$

Equations (11.8.9) and (11.8.10) then reduce to

$$\nabla^2\nabla^2 U = E\left[\left(\frac{\partial^2 w}{\partial x\,\partial y}\right)^2 - \frac{\partial^2 w}{\partial x^2}\frac{\partial^2 w}{\partial y^2}\right] \tag{11.8.13}$$

$$\nabla^2\nabla^2 w = \frac{q}{D} + \frac{h}{D}\left[\frac{\partial^2 U}{\partial y^2}\frac{\partial^2 w}{\partial x^2} + \frac{\partial^2 U}{\partial x^2}\frac{\partial^2 w}{\partial y^2} - 2\frac{\partial^2 U}{\partial x\,\partial y}\frac{\partial^2 w}{\partial x\,\partial y}\right] \tag{11.8.14}$$

which are the Von Karmon equations for large deflection of plates subject to lateral load and plane stresses N_x, N_y, and N_{xy}, as given in many textbooks.[17,22].

REFERENCES

1. Timoshenko, S., and J. Gere, *Theory of Elastic Stability*, McGraw-Hill, New York, pp. 332–345, 389–390, 1961.
2. Lin, T. H., "Creep Deflection of Visco-elastic Plate Under Uniform Edge Compressions," *J. Aerospace Sci.*, **23**, 883–887, 1956.
3. Churchill, R. V., *Modern Operational Mathematics in Engineering*, McGraw-Hill, New York, pp. 44–45, 1944.
4. McComb, H. G., Jr., "Analysis of the Creep Behavior of a Square Plate Loaded in Edge Compression," NACA Tech. Note 4398, September 1958.
5. Timoshenko, S., *Theory of Plates and Shells*, McGraw-Hill, New York, pp. 55–56, 1940.
6. Shanley, F. R., *Strength of Materials*, McGraw-Hill, New York, p. 159, 1957.
7. Handleman, E. H., and W. Prager, "Plastic Buckling of a Rectangular Plate Under Edge Thrusts," NACA Tech. Report 946, 1949.

8. Ilyushin, A. A., "Stability of Plates and Shells Stressed Beyond the Proportional Limit," *Prikl. Mat. i. Mekh.*, **8**, p. 337, 1944, NACA Tech. Memo No. 1116, October 1947.
9. Bijlaard, P. P., "Theory of Plastic Buckling of Plates and Applications to Simply Supported Plates Subject to Bending or Eccentric Compression in Shear Plane," *J. Appl. Mech.*, **23**, 27–34, 1956.
10. Bijlaard, P. P., C. F. Kollbrunner, and F. Strussi, "Theorie und Versuche über das plastische Ausbinlen von Richteckplatten unter Blechmässig verteiler Langsdruck," International Association for Bridge and Structural Engineering, Third Congress, Liege, Belgium, September 1948. Preliminary publication 1948, pp. 119–128.
11. Schuette, E. H., and J. C. McDonald, "Prediction and Reduction to Minimum Properties of Plate Compressive Curves," *J. Aerospace Sci.*, **15**, 23–27, 1948.
12. Bijlaard, P. P., "Theory and Tests on Plastic Stability of Plates and Tests on Plastic Stability of Plates and Shells," *J. Aerospace Sci.*, **16**, 529–541, 1949.
13. Pride, R. A., and G. J. Heimerl, "Plastic Buckling of Simply Supported Compressed Plates," NACA Tech. Note 1817, April 1949.
14. Pride, R. A., "Plastic Buckling of Simply Supported Plates in Compressions," *J. Aerospace Sci.*, **19**, 69–70, 1952.
15. Teodosiadis, R., H. L. Langhaar, and J. O. Smith, "Inelastic Buckling of Flat Plates," *Proc. 1st Midwestern Conf. Solid Mech.*, 1953.
16. Stowell, E. Z., "A Unified Theory of Plastic Buckling of Columns and Plates," NACA Tech. Report No. 898, 1948.
17. Pearson, C. E., "Bifurcation Criterion and Plastic Buckling of Plates and Columns," *J. Aerospace Sci.*, **17**, 425–455, 1950.
18. Stowell, E. Z., "Compressive Strength of Flanges," NACA Tech. Report No. 1029, 1951.
19. Gerard, G., "Secant Modulus Method for Determining Plate Instability Above the Proportional Limit," *J. Aerospace Sci.*, **13**, 38–44, 1946.
20. Peters, Roger W., "Buckling of Long Square Tubes in Combined Compression and Torsion and Comparison with Flat-Plate Buckling Theories," NACA Tech. Note 3184, 1954.
21. Gerard, G., and H. Becker, "Buckling of Flat Plates," NACA Tech. Note 3781, July 1957.
22. Nowacki, W., *Thermoelasticity*, Pergamon, London, England, pp. 488–490, 1962.
23. Switzky, H., M. J. Forray, and M. Newman, "Thermo-Structural Analysis Manual," Tech. Report No. WADD-TR-60-517, Vol. 1, pp. 6–10, August 1962.

chapter

12

Inelastic Bending of Shells

Introduction. This chapter presents a method of analysis for inelastic shells of revolution. The differential equations of equilibrium (in terms of displacements) are derived in general for inelastic shells of revolution under axisymmetric loadings, and for inelastic circular cylindrical shells subject to arbitrary loading conditions. The inelastic analysis of thin circular cylindrical shells under axisymmetric loading and zero longitudinal sectional force is discussed in detail and then illustrated by an example. The method given here may be extended to shells of other shapes under various loadings.

12.1 Equilibrium Conditions of Circular Cylindrical Shells

A circular cylindrical shell is a body bounded by two coaxial circular cylindrical surfaces, where the distance between the surfaces, called the thickness of the shell, is small compared with the other dimensions and radii of curvature of the surfaces. The surface that lies at equal distance from the two boundary circular surfaces is called the middle surface of the shell. Consider the differential shell element $OABC$ shown in Figure 12.1. On the middle surface, $OABC$, the length of element OC or AB is dx, the radius is "a," and the arc length of AO or BC is $a\,d\phi$. The corresponding arc at z distance from the middle surface is $(a + z)\,d\phi$. Consider the forces acting on a ϕz-plane section. The force resultants of the stresses acting on any such section, per unit length of arc BC of the middle surface, are

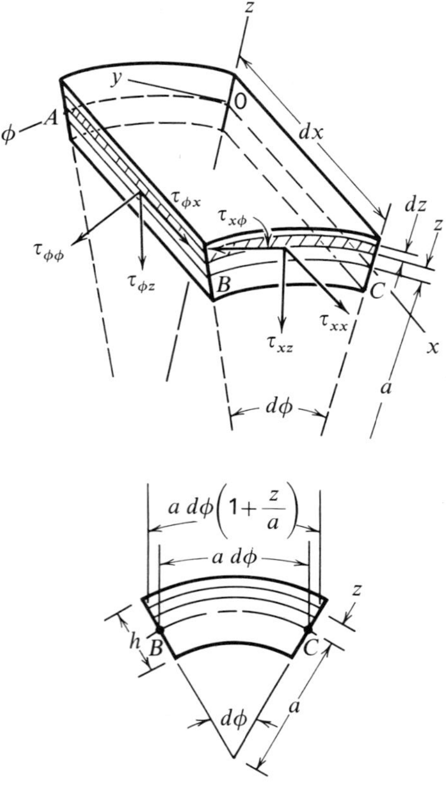

Figure 12.1. Element of a circular cylindrical shell.

$$
\begin{aligned}
N_x &= \int_{-h/2}^{h/2} \tau_{xx}\left(1 + \frac{z}{a}\right) dz \\
N_{x\phi} &= \int_{-h/2}^{h/2} \tau_{x\phi}\left(1 + \frac{z}{a}\right) dz \\
Q_x &= \int_{-h/2}^{h/2} \tau_{xz}\left(1 + \frac{z}{a}\right) dz
\end{aligned}
\tag{12.1.1}
$$

The force resultants of stresses acting on any xz-plane section, per unit length along the x axis, are

$$
\begin{aligned}
N_\phi &= \int_{-h/2}^{h/2} \tau_{\phi\phi}\, dz \qquad N_{\phi x} = \int_{-h/2}^{h/2} \tau_{\phi x}\, dz \\
Q_\phi &= \int_{-h/2}^{h/2} \tau_{\phi z}\, dz
\end{aligned}
\tag{12.1.2}
$$

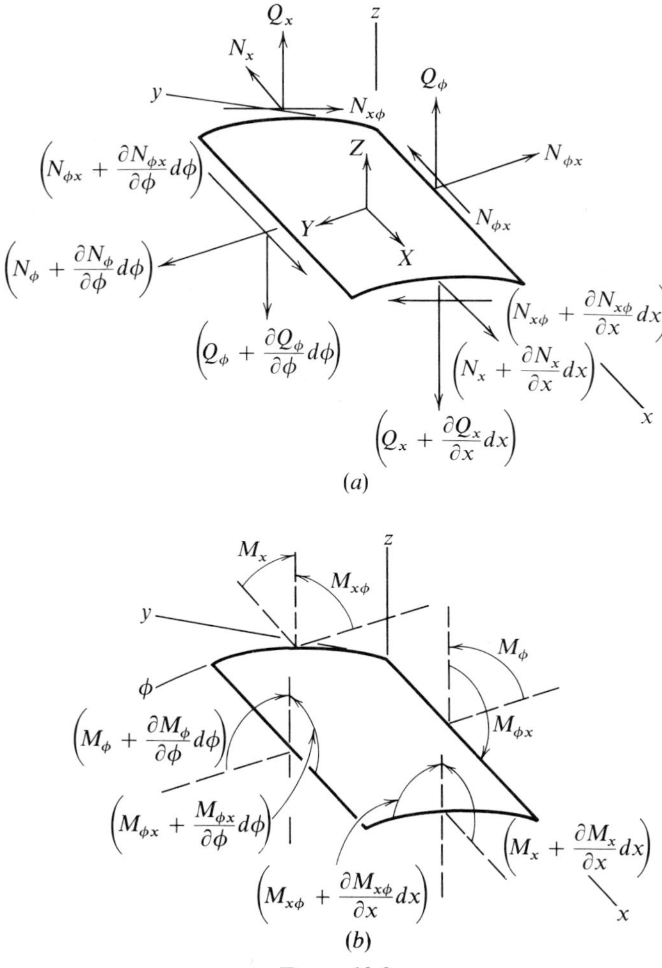

Figure 12.2

The above sectional forces are shown in Figure 12.2(a). The sectional moments per unit length of section, shown in Figure 12.2(b), may be similarly written as

$$
\begin{aligned}
M_x &= -\int_{-h/2}^{h/2} \tau_{xx}\left(1 + \frac{z}{a}\right) z \, dz \\
M_{x\phi} &= -\int_{-h/2}^{h/2} \tau_{x\phi}\left(1 + \frac{z}{a}\right) z \, dz
\end{aligned}
\tag{12.1.3}
$$

$$M_\phi = -\int_{-h/2}^{h/2} \tau_{\phi\phi} z\, dz$$
$$M_{\phi x} = -\int_{-h/2}^{h/2} \tau_{\phi x} z\, dz \qquad (12.1.4)$$

The signs of these sectional forces and moments are shown in Figure 12.2(a) and 12.2(b). The applied forces per unit area of the middle surface along the axial, circumferential, and radial directions are denoted by X, Y, and Z, respectively.

Referring to Figure 12.2(a), we see that equilibrium of forces in the x, ϕ, and z directions requires that

$$\frac{\partial N_x}{\partial x} a\, d\phi\, dx + \frac{\partial N_{\phi x}}{\partial \phi} d\phi\, dx + Xa\, d\phi\, dx = 0$$

or

$$\frac{\partial N_x}{\partial x} + \frac{\partial N_{\phi x}}{a\, \partial \phi} + X = 0 \qquad (12.1.5)$$

$$\frac{\partial N_\phi}{\partial \phi} + a \frac{\partial N_{x\phi}}{\partial x} - Q_\phi + Ya = 0 \qquad (12.1.6)$$

$$\frac{\partial Q_\phi}{\partial \phi} + a \frac{\partial Q_x}{\partial x} + N_\phi - Za = 0 \qquad (12.1.7)$$

Referring to Figure 12.2(b), we see that equilibrium of moments about axes through the center of the element and parallel to x, y, and z give, respectively,

$$\frac{\partial M_\phi}{\partial \phi} + a \frac{\partial M_{x\phi}}{\partial x} - aQ_\phi = 0 \qquad (12.1.8)$$

$$a \frac{\partial M_x}{\partial x} + \frac{\partial M_{\phi x}}{\partial \phi} - aQ_x = 0 \qquad (12.1.9)$$

$$aN_{x\phi} - aN_{\phi x} + M_{\phi x} = 0 \qquad (12.1.10)$$

Eliminating Q_x and Q_ϕ in the above equations, we obtain[2]

$$\frac{\partial N_x}{\partial x} + \frac{\partial N_{\phi x}}{a\, \partial \phi} + X = 0 \qquad (12.1.5)$$

$$\frac{\partial N_\phi}{a\, \partial \phi} + \frac{\partial N_{x\phi}}{\partial x} - \frac{\partial M_\phi}{a^2\, \partial \phi} - \frac{\partial M_{x\phi}}{a\, \partial x} + Y = 0 \qquad (12.1.11)$$

$$\frac{\partial^2 M_\phi}{a^2\, \partial \phi^2} + \frac{\partial^2 M_{x\phi}}{a\, \partial x\, \partial \phi} + \frac{\partial^2 M_{\phi x}}{a\, \partial x\, \partial \phi} + \frac{\partial^2 M_x}{\partial x^2} + \frac{N_\phi}{a} - Z = 0 \qquad (12.1.12)$$

$$a(N_{x\phi} - N_{\phi x}) + M_{\phi x} = 0 \qquad (12.1.13)$$

Equations (12.1.5) and (12.1.11) through (12.1.13) are the four *governing equations of equilibrium for circular cylindrical shells.*

12.2 Sectional Forces and Moments in Terms of Displacement for Inelastic Circular Cylindrical Shells

The displacements along the x, ϕ, and z axes (Figure 12.1) are denoted by u, v, and w respectively. Let the subscript "0" refer to the middle surface. The thickness of the shell is taken to be small as compared to the radii of curvature of the middle surface. It is assumed[2,3] in the following analysis that (1) normals to the middle surface before deformation remain straight and normal to the deformed middle surface; and (2) the length of elements normal to the middle surface remain unchanged. The first derivatives of the displacement are taken to be negligible compared to unity. These assumptions are similar to those used for thin-plate theory, and lead to the following displacement relations:

$$w = w_0 \tag{12.2.1}$$

$$u = u_0 - z\frac{\partial w}{\partial x} \tag{12.2.2}$$

$$v = v_0\frac{a+z}{a} - \frac{z}{a}\frac{\partial w}{\partial \phi} \tag{12.2.3}$$

The strain-displacement relations are

$$e_{xx} = \frac{\partial u}{\partial x} = \frac{\partial u_0}{\partial x} - z\frac{\partial^2 w}{\partial x^2} \tag{12.2.4}$$

$$e_{\phi\phi} = \frac{1}{(a+z)}\frac{\partial v}{\partial \phi} + \frac{w}{a+z} = \frac{\partial v_0}{a\,\partial \phi} - \frac{z}{a(a+z)}\frac{\partial^2 w}{\partial \phi^2} + \frac{w}{a+z} \tag{12.2.5}$$

$$\begin{aligned} 2e_{x\phi} &= \frac{\partial v}{\partial x} + \frac{1}{a+z}\frac{\partial u}{\partial \phi} \\ &= \frac{1}{a+z}\frac{\partial u_0}{\partial \phi} - \frac{z}{a+z}\frac{\partial^2 w}{\partial x\,\partial \phi} + \frac{a+z}{a}\frac{\partial v_0}{\partial x} - z\frac{\partial^2 w}{a\,\partial x\,\partial \phi} \\ &= \frac{1}{a+z}\frac{\partial u_0}{\partial \phi} + \frac{a+z}{a}\frac{\partial v_0}{\partial x} - \frac{\partial^2 w}{\partial x\,\partial \phi}\left(\frac{z}{a} + \frac{z}{a+z}\right) \end{aligned} \tag{12.2.6}$$

Hereafter the subscript "0" on the displacement will be deleted. Then u, v, and w will refer to middle-surface displacements. The elastic stress-strain relation is

$$e'_{xx} = \frac{\tau_{xx}}{E} - \nu\frac{\tau_{\phi\phi}}{E} \qquad e'_{\phi\phi} = \frac{\tau_{\phi\phi}}{E} - \nu\frac{\tau_{xx}}{E} \qquad e'_{x\phi} = \frac{1+\nu}{E}\tau_{x\phi} \tag{12.2.7}$$

Upon solving these expressions for the stresses, we obtain

$$\tau_{xx} = \frac{E}{1-\nu^2}(e'_{xx} + \nu e'_{\phi\phi}) \qquad \tau_{\phi\phi} = \frac{E}{1-\nu^2}(e'_{\phi\phi} + \nu e'_{xx})$$
$$\tau_{x\phi} = 2Ge'_{x\phi} = \frac{E}{1+\nu} e'_{x\phi} \tag{12.2.8}$$

Taking the elastic strain as the difference between the total and the inelastic strains, that is, $e'_{ij} = e_{ij} - e''_{ij}$, we see that Equations (12.2.8) become

$$\tau_{xx} = \frac{E}{1-\nu^2}[e_{xx} + \nu e_{\phi\phi} - e''_{xx} - \nu e''_{\phi\phi}]$$
$$\tau_{\phi\phi} = \frac{E}{1-\nu^2}[e_{\phi\phi} + \nu e_{xx} - e''_{\phi\phi} - \nu e''_{xx}] \tag{12.2.9}$$
$$\tau_{x\phi} = \frac{E}{1+\nu}(e_{x\phi} - e''_{x\phi})$$

Substituting Equations (12.2.4) through (12.2.6) into Equation (12.2.9) and then substituting in the expressions given by Equation (12.1.1) through (12.1.4) yields

$$N_\phi = \int_{-h/2}^{h/2} \tau_{\phi\phi}\, dz = \frac{E}{1-\nu^2}\int [e_{\phi\phi} + \nu e_{xx} - e''_{\phi\phi} - \nu e''_{xx}]\, dz$$
$$= \frac{E}{a(1-\nu^2)}\left[\left(\frac{\partial v}{\partial \phi} + \nu a\frac{\partial u}{\partial x}\right)h - \frac{\partial^2 w}{\partial \phi^2}\left(h - a\ln\frac{2a+h}{2a-h}\right)\right.$$
$$\left. + aw\ln\frac{2a+h}{2a-h}\right] - \frac{E}{1-\nu^2}\int (e''_{\phi\phi} + \nu e''_{xx})\, dz \tag{12.2.10}$$

where ln denotes the natural logarithm. Denoting the last term of Equation (12.2.10) by N_{ϕ_I} and expanding the logarithms in powers of h/a, we have, upon neglecting terms with higher-order powers,

$$N_\phi + N_{\phi_I} = \frac{Eh}{a(1-\nu^2)}\left[\frac{\partial v}{\partial \phi} + w + \nu a\frac{\partial u}{\partial x} + \left(\frac{\partial^2 w}{\partial \phi^2} + w\right)\frac{h^2}{12a^2}\right] \tag{12.2.11}$$

In a similar manner,[4] the other sectional force expressions are found to be

$$N_x + N_{x_I} = \frac{Eh}{a(1-\nu^2)}\left[a\frac{\partial u}{\partial x} + \nu\frac{\partial v}{\partial \phi} + \nu w\right] - \frac{Eh^3}{12(1-\nu^2)a}\frac{\partial^2 w}{\partial x^2} \tag{12.2.12}$$

$$N_{\phi x} + N_{\phi x_I} = \frac{Eh}{2a(1+\nu)}\left(\frac{\partial u}{\partial \phi} + a\frac{\partial v}{\partial x}\right) + \frac{Eh^3}{24(1+\nu)a^2}\left(\frac{\partial u}{a\,\partial \phi} + \frac{\partial^2 w}{\partial x\,\partial \phi}\right) \tag{12.2.13}$$

$$N_{x\phi} + N_{x\phi_I} = \frac{Eh}{2a(1+\nu)}\left(\frac{\partial u}{\partial \phi} + a\frac{\partial v}{\partial x}\right) + \frac{Eh^3}{24(1+\nu)a^2}\left(\frac{\partial v}{\partial x} - \frac{\partial^2 w}{\partial x\,\partial \phi}\right) \tag{12.2.14}$$

where

$$\begin{aligned} N_{\phi_I} &= \frac{E}{1-\nu^2}\int (e''_{\phi\phi} + \nu e''_{xx})\,dz \\ N_{x_I} &= \frac{E}{1-\nu^2}\int (e''_{xx} + \nu e''_{\phi\phi})\left(1+\frac{z}{a}\right)dz \\ N_{\phi x_I} &= \frac{E}{1+\nu}\int e''_{\phi x}\,dz \\ N_{x\phi_I} &= \frac{E}{1+\nu}\int e''_{x\phi}\left(1+\frac{z}{a}\right)dz \end{aligned} \tag{12.2.15}$$

The expressions given by Equation (12.2.15) represent the effect of inelastic strains and are called "inelastic sectional forces." By a similar procedure, the following moment-displacement equations are easily obtained:

$$\begin{aligned} M_\phi + M_{\phi_I} &= \frac{Eh^3}{12(1-\nu^2)a^2}\left(w + \frac{\partial^2 w}{\partial \phi^2} + \nu a^2\frac{\partial^2 w}{\partial x^2}\right) \\ &= D\left(\frac{\partial^2 w}{a^2\,\partial\phi^2} + \frac{w}{a^2} + \nu\frac{\partial^2 w}{\partial x^2}\right) \end{aligned} \tag{12.2.16}$$

$$M_x + M_{x_I} = D\left(\frac{\partial^2 w}{\partial x^2} + \frac{\nu}{a^2}\frac{\partial^2 w}{\partial \phi^2} - \frac{1}{a}\frac{\partial u}{\partial x} - \frac{\nu}{a^2}\frac{\partial v}{\partial \phi}\right) \tag{12.2.17}$$

$$M_{\phi x} + M_{\phi x_I} = \frac{D(1-\nu)}{a^2}\left(a\frac{\partial^2 w}{\partial x\,\partial\phi} + \frac{1}{2}\frac{\partial u}{\partial \phi} - \frac{a}{2}\frac{\partial v}{\partial x}\right) \tag{12.2.18}$$

$$M_{x\phi} + M_{x\phi_I} = \frac{D(1-\nu)}{a^2}\left(a\frac{\partial^2 w}{\partial x\,\partial\phi} - a\frac{\partial v}{\partial x}\right) \tag{12.2.19}$$

where $D = Eh^3/12(1-\nu^2)$ is called the flexural rigidity of a shell, and

$$\begin{aligned} M_{\phi_I} &= -\frac{E}{1-\nu^2}\int (e''_{\phi\phi} + \nu e''_{xx})z\,dz \\ M_{x_I} &= -\frac{E}{1-\nu^2}\int (e''_{xx} + \nu e''_{\phi\phi})\left(1+\frac{z}{a}\right)z\,dz \\ M_{\phi x_I} &= -\frac{E}{1+\nu}\int e''_{\phi x}\,z\,dz \\ M_{x\phi_I} &= -\frac{E}{1+\nu}\int e''_{x\phi}\left(1+\frac{z}{a}\right)z\,dz \end{aligned} \tag{12.2.20}$$

are the so-called "inelastic sectional moments" due to inelastic strains. Substituting Equations (12.2.11) through (12.2.19) into the four differential equations of equilibrium, (12.1.5) and (12.1.11) through (12.1.13), we obtain three equations of equilibrium and an identity. Noting $e_{\phi x} = e_{x\phi}$, we see that the identity results from Equation (12.1.13). The resulting three differential equations of equilibrium in terms of the unknown displacements u, v, and w are

$$a^2 \frac{\partial^2 u}{\partial x^2} + \frac{1-\nu}{2} \frac{\partial^2 u}{\partial \phi^2} + \frac{1+\nu}{2} a \frac{\partial^2 v}{\partial x \, \partial \phi} + \nu a \frac{\partial w}{\partial x} + D\left[\frac{1-\nu}{2} \frac{\partial^2 u}{\partial \phi^2} - a^3 \frac{\partial^3 w}{\partial x^3} + \frac{1-\nu}{2} a \frac{\partial^3 w}{\partial x \, \partial \phi^2}\right] + \frac{Xa^2(1-\nu^2)}{Eh} - \frac{a^2(1-\nu^2)}{Eh}\left(\frac{\partial N_{xI}}{\partial x} + \frac{\partial N_{\phi x I}}{a \, \partial \phi}\right) = 0 \qquad (12.2.21)$$

$$\frac{1+\nu}{2} a \frac{\partial^2 u}{\partial x \, \partial \phi} + \frac{\partial^2 v}{\partial \phi^2} + \frac{1-\nu}{2} a^2 \frac{\partial^2 v}{\partial x^2} + \frac{\partial w}{\partial \phi} + D\left[\frac{3}{2}(1-\nu) a^2 \frac{\partial^2 v}{\partial x^2} - \frac{3-\nu}{2} a^2 \frac{\partial^3 w}{\partial x^2 \, \partial \phi}\right] + \frac{Ya^2(1-\nu^2)}{Eh} - \frac{(1-\nu^2)}{Eh}\left(\frac{a \, \partial N_{\phi I}}{\partial \phi} + \frac{a^2 \, \partial N_{x\phi I}}{\partial x} - \frac{\partial M_{\phi I}}{\partial \phi} - \frac{a \, \partial M_{x\phi I}}{\partial x}\right) = 0 \qquad (12.2.22)$$

$$\frac{a \, \partial u}{\partial x} + \frac{\partial v}{\partial \phi} + w + D\left[\frac{1-\nu}{2} a \frac{\partial^3 u}{\partial x \, \partial \phi^2} - \frac{a^3 \, \partial^3 u}{\partial x^3} - \frac{(3-\nu)}{2} \frac{a^2 \, \partial^3 v}{\partial x^2 \, \partial \phi} + a^4 \frac{\partial^4 w}{\partial x^4} + 2a^2 \frac{\partial^4 w}{\partial x^2 \, \partial \phi^2} + \frac{\partial^4 w}{\partial \phi^4} + 2 \frac{\partial^2 w}{\partial \phi^2} + w\right] - \frac{Za^2(1-\nu^2)}{Eh} - \frac{(1-\nu^2)}{Eh}\left[\frac{\partial^2 M_{\phi I}}{\partial \phi^2} + \frac{a \, \partial^2 M_{x\phi I}}{\partial x \, \partial \phi} + \frac{a \, \partial^2 M_{\phi x I}}{\partial x \, \partial \phi} + \frac{a^2 \, \partial^2 M_{xI}}{\partial x^2} + aN_{\phi I}\right] = 0 \qquad (12.2.23)$$

These three equations are the governing differential *equations of equilibrium for the bending of inelastic circular cylindrical shells.* It is seen that the bracketed terms containing N_I's and M_I's in Equations (12.2.21) through (12.2.23) are similar to the distributed forces X, Y, and Z; therefore these bracketed terms represent the so-called "*equivalent*" *forces.*

In addition to the above equilibrium equations, a set of boundary-conditions must be satisfied by u, v, and w. As in inelastic plate theory, equivalent sectional forces and moments due to inelastic strains will result at the boundaries. These will lead to a set of "modified" boundary conditions.

A set of functions u, v, and w which satisfy both the equations of equilibrium and the boundary conditions comprise the solution to an inelastic circular cylindrical shell boundary value problem. Hence an inelastic shell may be treated as an identical elastic shell with "modified" boundary conditions and the additional equivalent forces.

Circular Cylindrical Shell Under Axisymmetrical Loading. Consider the case of symmetrical loading. Then

$$N_{\phi x} = 0 \qquad M_{x\phi} = 0 \qquad N_{x\phi} = 0 \qquad M_{\phi x} = 0$$

For a circular cylindrical shell under axisymmetric loading, all displacements, forces, moments, stresses, and strains are independent of ϕ. Then equilibrium Equations (12.1.5) and (12.1.11) through (12.1.13) reduce to

$$\frac{dN_x}{dx} + X = 0 \tag{12.2.24}$$

$$a\,\frac{d^2 M_x}{dx^2} + N_\phi - aZ = 0 \tag{12.2.25}$$

For the case $X = 0$ and $N_x = 0$, the above set of equilibrium equations further reduce to the single equilibrium Equation (12.2.25). Recall Equaton (12.2.12),

$$N_x = \frac{Eh}{a(1-\nu^2)}\left(a\,\frac{du}{dx} + \nu w\right) - \frac{D}{a}\frac{d^2 w}{dx^2} - N_{x_I} = 0$$

or

$$\frac{Eh}{1-\nu^2}\frac{du}{dx} = N_{x_I} - \frac{Eh\nu}{a(1-\nu^2)}\,w + \frac{D}{a}\frac{d^2 w}{dx^2} \tag{12.2.26}$$

In the case of axisymmetric loading. Equations (12.2.11) and (12.2.17), upon use of Equation (12.2.26), reduce to

$$\begin{aligned} N_\phi &= \frac{Eh}{a(1-\nu^2)}\left(w + \nu a\,\frac{du}{dx} + w\,\frac{h^2}{12a^2}\right) - N_{\phi_I} \\ &= \frac{Eh}{a^2(1-\nu^2)}\left(1 + \frac{h^2}{12a^2}\right) w + \nu\left(N_{x_I} - \frac{Eh\nu}{a(1-\nu^2)}\,w + \frac{D}{a}\frac{d^2 w}{dx^2}\right) - N_{\phi_I} \end{aligned} \tag{12.2.27}$$

$$M_x = -M_{x_I} + D\left(\frac{d^2w}{dx^2} - \frac{1}{a}\frac{du}{dx}\right)$$

$$= -M_{x_I} + D\frac{d^2w}{dx^2} - \frac{h^2}{12a}\left(N_{x_I} - \frac{Eh\nu}{a(1-\nu^2)}w + \frac{D}{a}\frac{d^2w}{dx^2}\right) \qquad (12.2.28)$$

Since there can be no ϕ dependence for this axisymmetric problem, N_{x_I}, N_{ϕ_I}, and M_{x_I} are functions of x only. Upon substituting these known inelastic forces and moments into Equations (12.2.27) and (12.2.28), which in turn are substituted into Equation (12.2.25), a fourth-order differential equation of equilibrium in w is obtained. We then seek the solution w of this equation which satisfies the modified boundary conditions.

In general a solution of the above equilibrium equation is difficult to obtain. It then becomes necessary to make some approximations which will lead to a set of equilibrium equations more amenable to solution. For very thin shells it is possible to develop an approximate theory which yields equations in simplified form, as shown in the next section.

12.3 Approximation Theory of Inelastic Thin Circular Cylindrical Shells

If the shell is thin enough that we may neglect z/a and h/a as compared with unity, Equations (12.2.4) to (12.2.6) then reduce to

$$\begin{aligned} e_{xx} &= \frac{\partial u}{\partial x} - z\frac{\partial^2 w}{\partial x^2} \\ e_{\phi\phi} &= \frac{\partial v}{a\,\partial\phi} + \frac{w}{a} - \frac{z}{a^2}\frac{\partial^2 w}{\partial\phi^2} \\ 2e_{x\phi} &= \frac{1}{a}\frac{\partial u}{\partial\phi} + \frac{\partial v}{\partial x} - 2\frac{z}{a}\frac{\partial^2 w}{\partial x\,\partial\phi} \end{aligned} \qquad (12.3.1)$$

Equations (12.2.10) through (12.2.13) and Equations (12.2.15) through (12.2.18), under the same assumptions, reduce to

$$\begin{aligned} N_\phi &= \frac{Eh}{(1-\nu^2)}\left(\frac{\partial v}{a\,\partial\phi} + \frac{w}{a} + \nu\frac{\partial u}{\partial x}\right) - N_{\phi_I} \\ N_x &= \frac{Eh}{1-\nu^2}\left(\frac{\partial u}{\partial x} + \nu\frac{\partial v}{a\,\partial\phi} + \nu\frac{w}{a}\right) - N_{x_I} \\ N_{\phi x} = N_{x\phi} &= \frac{Eh}{1-\nu^2}\frac{(1-\nu)}{2a}\left(\frac{\partial u}{\partial\phi} + a\frac{\partial v}{\partial x}\right) - N_{x\phi_I} \end{aligned} \qquad (12.3.2)$$

$$M_\phi = D\left(\frac{\partial^2 w}{a^2\,\partial\phi^2} + \nu\frac{\partial^2 w}{\partial x^2}\right) - M_{\phi_I}$$

$$M_x = D\left(\frac{\partial^2 w}{\partial x^2} + \nu\frac{\partial^2 w}{a^2\,\partial\phi^2}\right) - M_{x_I} \qquad (12.3.3)$$

$$M_{\phi x} = M_{x\phi} = \frac{D(1-\nu)}{a}\frac{\partial^2 w}{\partial x\,\partial\phi} - M_{\phi x_I}$$

From Equations (12.2.9) and (12.3.1), we have

$$\tau_{xx} = \frac{E}{1-\nu^2}\left[\frac{\partial u}{\partial x} + \nu\left(\frac{\partial v}{a\,\partial\phi} + \frac{w}{a}\right) - z\left(\frac{\partial^2 w}{\partial x^2} + \nu\frac{\partial^2 w}{a^2\,\partial\phi^2}\right) - (e''_{xx} + \nu e''_{\phi\phi})\right]$$

$$\tau_{\phi\phi} = \frac{E}{1-\nu^2}\left[\frac{\partial v}{a\,\partial\phi} + \frac{w}{a} + \nu\frac{\partial u}{\partial x} - z\left(\frac{\partial^2 w}{a^2\,\partial\phi^2} + \nu\frac{\partial^2 w}{\partial x^2}\right) - (e''_{\phi\phi} + \nu e''_{xx})\right]$$

$$\tau_{x\phi} = \frac{E}{1+\nu}\left[\frac{\partial u}{2a\,\partial\phi} + \frac{\partial v}{2\,\partial x} - \frac{z}{a}\frac{\partial^2 w}{\partial x\,\partial\phi} - e''_{x\phi}\right] \qquad (12.3.4)$$

From Equations (12.3.2) through (12.3.4), we obtain

$$\tau_{xx} = \frac{N_x + N_{x_I}}{h} - \frac{12z}{h^3}(M_x + M_{x_I}) - \frac{E}{1-\nu^2}(e''_{xx} + \nu e''_{\phi\phi})$$

$$\tau_{\phi\phi} = \frac{N_\phi + N_{\phi_I}}{h} - \frac{12z}{h^3}(M_\phi + M_{\phi_I}) - \frac{E}{1-\nu^2}(e''_{\phi\phi} + \nu e''_{xx})$$

$$\tau_{x\phi} = \frac{N_{x\phi} + N_{x\phi_I}}{h} - 12\frac{z}{h^3}(M_{\phi x} + M_{\phi x_I}) - \frac{E}{1+\nu}e''_{x\phi} \qquad (12.3.5)$$

These equations are the same as Equation (10.2.24). Substitution of Equations (12.3.2) and (12.3.3), for sectional forces and moments into Equations (12.1.5) and (12.1.11) through (12.1.13) yields the following equilibrium equations:

$$a^2\frac{\partial^2 u}{\partial x^2} + \frac{1-\nu}{2}\frac{\partial^2 u}{\partial\phi^2} + \frac{1+\nu}{2}a\frac{\partial^2 v}{\partial x\,\partial\phi} + \nu a\frac{\partial w}{\partial x}$$
$$+ \frac{Xa^2(1-\nu^2)}{Eh} - \frac{a^2(1-\nu^2)}{Eh}\left(\frac{\partial N_{x_I}}{\partial x} + \frac{\partial N_{\phi x_I}}{a\partial\phi}\right) = 0 \qquad (12.3.6)$$

$$\frac{1+\nu}{2}a\frac{\partial^2 u}{\partial x\,\partial\phi} + \frac{\partial^2 v}{\partial\phi^2} + \frac{1-\nu}{2}a^2\frac{\partial^2 v}{\partial x^2} + \frac{\partial w}{\partial\phi}$$
$$- \frac{h^2}{12a^2}\left(a^2\frac{\partial^3 w}{\partial x^2\,\partial\phi} + \frac{\partial^3 w}{\partial\phi^3}\right) + \frac{Ya^2(1-\nu^2)}{Eh}$$
$$- \frac{(1-\nu^2)}{Eh}\left(a\frac{\partial N_{\phi_I}}{\partial\phi} + \frac{a^2\,\partial N_{x\phi_I}}{\partial x} - \frac{\partial M_{\phi_I}}{\partial\phi} - \frac{a\,\partial M_{x\phi_I}}{\partial x}\right) = 0 \qquad (12.3.7)$$

$$\nu a \frac{\partial u}{\partial x} + \frac{\partial v}{\partial \phi} + w + \frac{h^2}{12a^2}\left(a^4 \frac{\partial^4 w}{\partial x^4} + 2a^2 \frac{\partial^4 w}{\partial x^2\, \partial \phi^2} + \frac{\partial^4 w}{\partial \phi^4}\right) - \frac{Za^2(1-\nu^2)}{Eh}$$
$$- \frac{(1-\nu^2)}{Eh}\left[\frac{\partial^2 M_{\phi_I}}{\partial \phi^2} + \frac{a\, \partial^2 M_{x\phi_I}}{\partial x\, \partial \phi} + \frac{a\, \partial^2 M_{\phi x_I}}{\partial x\, \partial \phi} + a^2 \frac{\partial^2 M_{x_I}}{\partial x^2} + aN_{\phi_I}\right] = 0 \tag{12.3.8}$$

These are similar to the displacement equations given by Poritsky[5] in his calculation of the effect of creep on stresses in cylindrical shells.

Axisymmetric Loading. In cases with axisymmetric loading, all forces, displacements, stresses, strains, and moments are independent of ϕ, and hence the displacements v, as well as $N_{x\phi}$, $N_{\phi x}$, $M_{x\phi}$, are equal to zero. Equations (12.3.6) to (12.3.8) then reduce to

$$\frac{d^2 u}{dx^2} + \frac{\nu}{a}\frac{dw}{dx} + \frac{(1-\nu^2)X}{Eh} - \frac{(1-\nu^2)}{Eh}\frac{dN_{x_I}}{dx} = 0 \tag{12.3.9}$$

$$\frac{\nu}{a}\frac{du}{dx} + \frac{w}{a^2} + \frac{h^2}{12}\frac{d^4 w}{dx^4} - \frac{(1-\nu^2)}{Eh} Z - \frac{1-\nu^2}{Eh}\left(\frac{d^2 M_{x_I}}{dx^2} + \frac{N_{\phi_I}}{a}\right) = 0 \tag{12.3.10}$$

For the particular case of zero longitudinal sectional force, N_x and $X = 0$. Then Equation (12.3.2) gives

$$N_x = \frac{Eh}{1-\nu^2}\left(\frac{du}{dx} + \nu \frac{w}{a}\right) - N_{x_I} = 0$$

and

$$N_\phi = \frac{Eh}{1-\nu^2}\left(\frac{w}{a} + \nu \frac{du}{dx}\right) - N_{\phi_I}$$

Multiplying the first of the above equations through by $-\nu$ and then adding it to the second equation gives

$$N_\phi = Eh \frac{w}{a} + \nu N_{x_I} - N_{\phi_I} \tag{12.3.11}$$

From Equation (12.3.3), we obtain

$$M_x = D \frac{d^2 w}{dx^2} - M_{x_I}.$$

Substituting this and Equation (12.3.11) into Equation (12.2.25), we obtain

$$D \frac{d^4 w}{dx^4} + Eh \frac{w}{a^2} - \left(\frac{d^2 M_{x_I}}{dx^2} - \frac{\nu}{a} N_{x_I} + \frac{N_{\phi_I}}{a}\right) - Z = 0 \tag{12.3.12}$$

The parenthesized term has the same effect on the radial deflection as Z and hence is called the *equivalent radial loading*. This gives the differential equation of thin, inelastic, circular cylindrical shells under axisymmetric loading and zero longitudinal sectional force. The application of this equation is illustrated in the following example.

Example. A thin circular cylindrical shell of thickness h, radius "a," and length $2l$ with its two ends $x = \pm l$ free, is subject to a temperature distribution T, represented by

$$T = a_2\left(\frac{x}{l}\right)^2 + a_4\left(\frac{x}{l}\right)^4 + \frac{z}{h}\left[b_2\left(\frac{x}{l}\right)^2 + b_4\left(\frac{x}{l}\right)^4\right] \tag{12.3.13}$$

Let α be the coefficient of thermal expansion of the cylinder material. It is required to find the thermal radial deflection and stresses in this shell due to the given temperature distribution.

The thermal expansion is isotropic and the inelastic thermal strains are

$$e''_{xx} = e''_{\phi\phi} = \alpha T \qquad e''_{x\phi} = 0$$

Since there is no radial loading, $Z = 0$. Moreover, z/a, being small compared to unity, is neglected. Then, from Equations (12.2.15) and (12.2.20), we have

$$N_{x_I} = N_{\phi_I} = \frac{E(1+\nu)}{1-\nu^2}\int_{-h/2}^{h/2} \alpha T\, dz = \frac{Eh\alpha}{1-\nu}\left[a_2\left(\frac{x}{l}\right)^2 + a_4\left(\frac{x}{l}\right)^4\right] \tag{12.3.14}$$

$$M_{x_I} = M_{\phi_I} = -\frac{E(1+\nu)}{1-\nu^2}\int_{-h/2}^{h/2} \alpha T z\, dz = -\frac{E\alpha}{1-\nu}\left[b_2\left(\frac{x}{l}\right)^2 + b_4\left(\frac{x}{l}\right)^4\right]\frac{h^2}{12} \tag{12.3.15}$$

Substitution of Equations (12.3.14) and (12.3.15) into Equation (12.3.12) yields

$$\begin{aligned} D\frac{d^4 w}{dx^4} + \frac{Eh}{a^2}w &= \frac{d^2 M_{x_I}}{dx^2} - \frac{\nu}{a}N_{x_I} + \frac{1}{a}N_{\phi_I} \\ &= -\frac{E\alpha}{6(1-\nu)}\frac{h^2}{l^2}\left[b_2 + 6b_4\left(\frac{x}{l}\right)^2\right] + \frac{Eh\alpha}{a}\left[a_2\left(\frac{x}{l}\right)^2 + a_4\left(\frac{x}{l}\right)^4\right] \end{aligned} \tag{12.3.16}$$

The general solution of this differential equation is

$$\begin{aligned} w = {} & K_1 \sinh \beta x \sin \beta x + K_2 \cosh \beta x \cos \beta x \\ & + K_3 \sinh \beta x \cos \beta x + K_4 \cosh \beta x \sin \beta x + F(x) \end{aligned} \tag{12.3.17}$$

where

$$\beta = \left(\frac{Eh}{4a^2D}\right)^{1/4} = \left[\frac{3(1-\nu^2)}{a^2h^2}\right]^{1/4} \tag{12.3.18}$$

The particular solution of Equation (12.3.16), is $F(x)$, and K_1 through K_4 are the constants of integration. Here the particular solution $F(x)$ is found to be

$$F(x) = -\frac{\alpha h a^2}{6(1-\nu)l^2}\left[b_2 + 6b_4\left(\frac{x}{l}\right)^2\right] + \alpha a\left[a_2\left(\frac{x}{l}\right)^2 + a_4\left(\frac{x}{l}\right)^4\right] - \frac{24\,D\alpha a}{l^4}a_4 \tag{12.3.19}$$

The solution of Equation (12.3.16) given by Equations (12.3.17) through (12.3.19) can easily be checked by direct substitution. Since the temperature T is symmetrical with respect to the $x = 0$ plane, the deflection must also be symmetrical about this plane. This condition of symmetry requires that K_3 and K_4 be zero. Hence the deflection w is

$$w = K_1 \sinh\beta x \sin\beta x + K_2 \cosh\beta x \cos\beta x + F(x) \tag{12.3.20}$$

Since the two ends $x = \pm l$ are free, the end conditions at $x = \pm l$ are

$$M_x = D\frac{d^2w}{dx^2} - M_{xI} = 0 \tag{12.3.21}$$

$$Q_x = \frac{dM_x}{dx} = D\frac{d^3w}{dx^3} - \frac{dM_{xI}}{dx} = 0 \tag{12.3.22}$$

From Equation (12.3.20), we obtain

$$\frac{d^2w}{dx^2} = 2\beta^2[K_1\cosh\beta x\cos\beta x - K_2\sinh\beta x\sin\beta x] + F''(x) \tag{12.3.23}$$

$$\begin{aligned}\frac{d^3w}{dx^3} = 2\beta^3[&K_1\sinh\beta x\cos\beta x - K_1\cosh\beta x\sin\beta x\\ &- K_2\cosh\beta x\sin\beta x - K_2\sinh\beta x\cos\beta x] + F'''(x)\end{aligned} \tag{12.3.24}$$

where $F''(x)$ and $F'''(x)$ denote the second and third derivatives of F with respect to x, respectively. That is,

$$F''(x) = -\frac{\alpha h a^2}{6(1-\nu)l^2}\frac{12b_4}{l^2} + \frac{\alpha a}{l^2}\left[2a_2 + 12a_4\left(\frac{x}{l}\right)^2\right] \tag{12.3.25}$$

$$F'''(x) = \frac{\alpha a}{l^3}24a_4\left(\frac{x}{l}\right) \tag{12.3.26}$$

Substituting the expressions for d^2w/dx^2, d^3w/dx^3, M_{x_I}, and dM_{x_I}/dx, given by Equations (12.3.16), (12.3.23), and (12.3.24) into the boundary conditions, Equations (12.3.21) and (12.3.22), gives two equations for K_1 and K_2. With K_1 and K_2 known, w is easily obtained from Equation (12.3.20), N_ϕ is obtained from Equation (12.3.11), M_ϕ and M_x are obtained from Equation (12.3.3), and finally then the stresses are found from Equations (12.3.5).

12.4 Inelastic Shells of Revolution under Axisymmetrical Loading

A differential element of a shell of revolution loaded symmetrically with respect to its axis is shown in Figure 12.3. Consider an element $ABCD$ cut by two adjacent meridian planes and two sections perpendicular to the meridians. Because of the axisymmetry of the shell and loading,

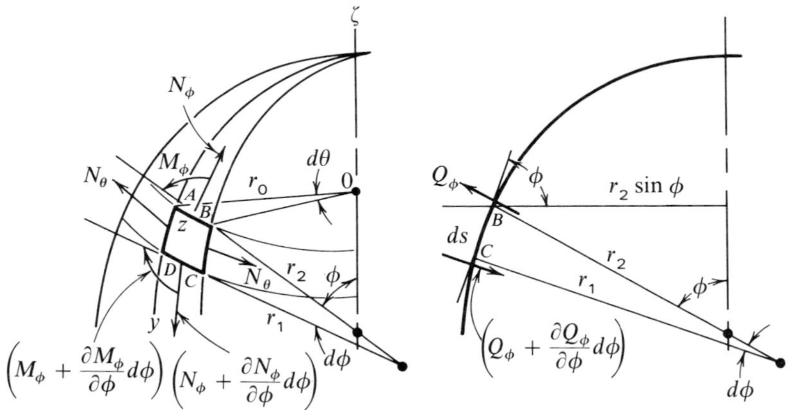

Figure 12.3. A portion of a shell of revolution under axisymmetrical loading.

there is no shear stress on the meridian planes AD and BC. Planes AD and BC have the length $r_1\, d\phi$. The resultant of the stresses acting on AD or BC may be represented by $N_\theta r_1\, d\phi$ and $M_\theta r_1\, d\phi$, where N_θ and M_θ are independent of θ. On AB and DC, the direct stresses give a resultant force $N_\phi r_2 \sin\phi\, d\theta$ and a resultant moment $M_\phi r_2 \sin\phi\, d\theta$. On these same sides the shearing stresses give a resultant shear force $Q_\phi r_2 \sin\phi\, d\theta$. The external load on this element may be resolved into two components, $Yr_1r_2 \sin\phi\, d\phi\, d\theta$ (tangent to the meridian) and $Zr_1r_2 \sin\phi\, d\phi\, d\theta$ (normal to the shell). The N_θ and N_ϕ are assumed to be much less than their critical values for buckling, so the change of geometry is neglected in deriving the equations of equilibrium.

Equilibrium of the element requires that forces tangent to the meridian coordinate and forces normal to the element be zero. These conditions give the force *equilibrium equations*,

$$\frac{d}{d\phi}(N_\phi r_0) - N_\theta r_1 \cos\phi - r_0 Q_\phi + r_0 r_1 Y = 0 \tag{12.4.1}$$

$$N_\phi r_0 + N_\theta r_1 \sin\phi + \frac{d(Q_\phi r_0)}{d\phi} + Z r_1 r_0 = 0 \tag{12.4.2}$$

Equating the sum of moments of all the forces acting on the element about an axis parallel to AB and CD to zero, we obtain†

$$\left(M_\phi + \frac{dM_\phi}{d\phi}\,d\phi\right)\left(r_0 + \frac{dr_0}{d\phi}\,d\phi\right)d\theta - M_\phi r_0\,d\theta$$

$$- M_\theta r_1\,d\phi(\cos\phi\,d\theta) - Q_\phi r_1 r_2 \sin\phi\,d\phi\,d\theta = 0$$

After simplifying, we obtain

$$\frac{d}{d\phi}(M_\phi r_0) - M_\theta r_1 \cos\phi - Q_\phi r_1 r_0 = 0 \tag{12.4.3}$$

Let v be the displacement along the tangent to the meridian and w be the displacement along the direction normal to and toward the interior of the shell. The shell is thin and it is here assumed that the term z/a is negligible as compared to unity in terms such as those given in Equations (12.1.1), (12.1.3), (12.2.1) through (12.2.6); the strain components are[4]

$$\begin{aligned} e_{\phi\phi} &= \frac{1}{r_1}\frac{dv}{d\phi} - \frac{w}{r_1} \\ e_{\theta\theta} &= \frac{v}{r_2}\cot\phi - \frac{w}{r_2} \end{aligned} \tag{12.4.4}$$

Following the procedure given in Section 12.2,

$$N_\phi + N_{\phi_I} = \frac{Eh}{1-\nu^2}\left[\frac{1}{r_1}\left(\frac{dv}{d\phi} - w\right) + \frac{\nu}{r_2}(v\cot\phi - w)\right] \tag{12.4.5}$$

$$N_\theta + N_{\theta_I} = \frac{Eh}{1-\nu^2}\left[\frac{1}{r_2}(v\cot\phi - w) + \frac{\nu}{r_1}\left(\frac{dv}{d\phi} - w\right)\right] \tag{12.4.6}$$

† The angle between AD and BC is $\cos\phi\,d\theta$.

where the inelastic sectional forces are

$$N_{\phi_I} = \frac{E}{1-\nu^2}\int_{-h/2}^{h/2} (e''_{\phi\phi} + \nu e''_{\theta\theta})\, dz$$
$$N_{\theta_I} = \frac{E}{1-\nu^2}\int_{-h/2}^{h/2} (e''_{\theta\theta} + \nu e''_{\phi\phi})\, dz \qquad (12.4.7)$$

It has been shown[4] that the rotation of side AB is

$$\frac{v}{r_1} + \frac{dw}{r_1\, d\phi}$$

and that of CD is

$$\frac{v}{r_1} + \frac{dw}{r_1\, d\phi} + \frac{d}{d\phi}\left(\frac{v}{r_1} + \frac{dw}{r_1\, d\phi}\right) d\phi$$

Hence the change of curvature of the meridian is

$$\chi_\phi = \frac{1}{r_1}\frac{d}{d\phi}\left(\frac{v}{r_1} + \frac{dw}{r_1\, d\phi}\right) \qquad (12.4.8)$$

Both the left and right sides of an element rotate in their meridian plane by the amount $v/r_1 + dw/r_1\, d\phi$. A unit vector $\vec{y}$ along the y-axis may be resolved into a radial component $\cos\phi$ along OA and a component $-\sin\phi$ along the axis of the shell 0ζ. A unit vector $\vec{t}$ tangent to AB at B has a component $-\sin(d\theta) \cong -d\theta$ along OA but no component along 0ζ. The cosine of the angle between $\vec{y}$ and $\vec{t}$ is $-\cos\phi\, d\theta$. Referring to Figure 12.3, we note that the rotation of the left side AD has no component along the y axis, and the rotation on the right side along the y axis is

$$-\left(\frac{v}{r_1} + \frac{dw}{r_1\, d\phi}\right)\cos\phi\, d\theta$$

which gives the change of curvature

$$\chi_\theta = \left(\frac{v}{r_1} + \frac{dw}{r_1\, d\phi}\right)\frac{\cos\phi}{r_0} \quad \text{or} \quad \left(\frac{v}{r_1} + \frac{dw}{r_1\, d\phi}\right)\frac{\cot\phi}{r_2} \qquad (12.4.9)$$

Using Equations (12.4.8) and (12.4.9), we see that the sectional moments are

$$M_\phi + M_{\phi_I} = -D\left[\frac{1}{r_1}\frac{d}{d\phi}\left(\frac{v}{r_1} + \frac{dw}{r_1\, d\phi}\right) + \frac{\nu}{r_2}\left(\frac{v}{r_1} + \frac{dw}{r_1\, d\phi}\right)\cot\phi\right] \qquad (12.4.10)$$

$$M_\theta + M_{\theta_I} = -D\left[\left(\frac{v}{r_1} + \frac{dw}{r_1\, d\phi}\right)\frac{\cot\phi}{r_2} + \frac{\nu}{r_1}\frac{d}{d\phi}\left(\frac{v}{r_1} + \frac{dw}{r_1\, d\phi}\right)\right] \qquad (12.4.11)$$

where

$$M_{\phi_I} = \frac{E}{1-\nu^2} \int_{-h/2}^{h/2} (e''_{\phi\phi} + \nu e''_{\theta\theta}) z \, dz$$
$$M_{\theta_I} = \frac{E}{1-\nu^2} \int_{-h/2}^{h/2} (e''_{\theta\theta} + \nu e''_{\phi\phi}) z \, dz \tag{12.4.12}$$

are the inelastic sectional moments. Eliminating $Q_\phi r_0$ from Equations (12.4.1) through (12.4.3), we obtain the two equilibrium equations

$$\frac{d}{d\phi}(N_\phi r_0) - N_\theta r_1 \cos\phi - \frac{1}{r_1}\frac{d}{d\phi}(M_\phi r_0) + M_\theta \cos\phi + r_0 r_1 Y = 0 \tag{12.4.13}$$

$$N_\phi r_0 + N_\theta r_1 \sin\phi + \frac{d}{d\phi}\left[\frac{1}{r_1}\frac{d}{d\phi}(M_\phi r_0) - M_\theta \cos\phi\right] + Z r_1 r_0 = 0 \tag{12.4.14}$$

Substituting Equations (12.4.5), (12.4.6), (12.4.10), and (12.4.11) into the above two equations gives two equilibrium equations in v, w, N_{ϕ_I}, N_{θ_I}, M_{ϕ_I}, and M_{θ_I}. With the distribution of inelastic strains $e''_{\phi\phi}$ and $e''_{\theta\theta}$ known, the distribution of N_{ϕ_I}, N_{θ_I}, M_{ϕ_I}, and M_{θ_I} may readily be determined. These inelastic sectional forces and moments N_{ϕ_I}, N_{θ_I}, M_{ϕ_I}, and M_{θ_I} give terms similar to the external loads Y and Z exactly as in the inelastic plate theory of Chapters 10 and 11. Inserting these N_{ϕ_I}, N_{θ_I}, M_{ϕ_I}, and M_{θ_I} into the above two differential equations of equilibrium yields two equations in the two unknown v and w. We then seek the solution to these two equations which satisfies the boundary conditions of the problem. As in the previous analyses of inelastic plates, the boundary conditions must include equivalent boundary forces due to inelastic strains. Hence the analyses of inelastic bending of shells of revolution under axisymmetric loading reduces to the analyses of the elastic bending of the identical shell with the additional set of equivalent forces and "modified" boundary conditions. This method may be extended to shells of other shapes under arbitrary loadings.[6,7] For large inelastic deformation of axisymmetric shells, the reader is referred to the numerical iteration method given by Stricklin, Hsu and Pian.[8]

REFERENCES†

1. Love, A. E. H., *Mathematical Theory of Elasticity*, Dover, New York, p. 528, 1944.
2. Flugge, W., *Stresses in Shells*, Springer-Verlag, Berlin, Chap. 5, 1960.
3. Novozhilov, V. V., *The Theory of Thin Shells*, Noordhoff, Groningen, The Netherlands, 1959.
4. Timoshenko, S., and S. Woinowsky-Krieger, *Theory of Plates and Shells*, McGraw-Hill, New York, pp. 430–432, 446, 508–513, 535, 1959.
5. Poritsky, H., *Effect of Creep on Stresses in Cylindrical Shells*, Proc. IUTAM Colloquium Creep in Structures, Springer-Verlag, Berlin, p. 229–248, 1962.
6. Stern, G. S., *Creep Bending of Rotational Shell Structures Subject to Nonsymmetric Loads*, Ph.D. Dissertation, University of California, Los Angeles, 1964.
7. Roberts, S. B., *The Analysis of a Shallow Spherical Shell with an Eccentric Circular Hole in the Presence of Nonlinear Creep*, Ph.D. Dissertation, University of California, Los Angeles, 1965.
8. Stricklin, J. A., P. T. Hsu and T. H. H. Pian, *Large Elastic, Plastic and Creep Deflections of Curved Beams and Axisymmetric Shells*, Jour. Am. Inst. Aeronautics and Astronautics Vol. 2, No. 9, pp. 1613–1620, 1964.

† For the analysis of inelastic shells of more complicated geometry, see References 6 and 7.

Index